Water Relations
of Plants and Soils

Water Relations
of Plants and Soils

Paul J. Kramer
8 May 1904–24 May 1995
Department of Botany
Duke University
Durham, North Carolina

John S. Boyer
College of Marine Studies and College of Agriculture
University of Delaware, Sharp Campus
Lewes, Delaware

Academic Press
San Diego   New York   Boston
London   Sydney   Tokyo   Toronto

This book is printed on acid-free paper. ∞

Copyright © 1995, 1983 by ACADEMIC PRESS, INC.

All Rights Reserved.
No part of this publication may be reproduced or transmitted in any form or by any means, electronic or mechanical, including photocopy, recording, or any information storage and retrieval system, without permission in writing from the publisher.

Academic Press, Inc.
A Division of Harcourt Brace & Company
525 B Street, Suite 1900, San Diego, California 92101-4495

United Kingdom Edition published by
Academic Press Limited
24-28 Oval Road, London NW1 7DX

Library of Congress Cataloging-in-Publication Data

Kramer, Paul Jackson, date.
    Water relations of plants and soils / by Paul J. Kramer, John S.
Boyer.
        p.   cm.
    Includes bibliographical references and index.
    ISBN 0-12-425060-2
    1. Plant-water relationships.   2. Plants, Effect of soil moisture
on.   I. Boyer, John S. (John Srickland), date.   II. Title.
QK870.K72   1995
581.1--dc20   94-48901
                                          CIP

PRINTED IN THE UNITED STATES OF AMERICA
95 96 97 98 99 00 BC 9 8 7 6 5 4 3 2 1

Contents

7 TRANSPIRATION AND THE ASCENT OF SAP

8 STOMATA AND GAS EXCHANGE

Preface

Everyone who grows plants, whether a single geranium in a flower pot or hundreds of acres of corn or cotton, is aware of the importance of water for successful growth. Water supply not only affects the yield of gardens and field crops, but also controls the distribution of plants over the earth's surface, ranging from deserts and grasslands to rain forests, depending on the amount and seasonal distribution of precipitation. However, few people understand fully why water is so important for plant growth. This book attempts to explain its importance by showing how water affects the physiological processes that control the quantity and quality of growth. It is a useful introduction for students, teachers, and investigators in both basic and applied plant science, including botanists, crop scientists, foresters, horticulturists, soil scientists, and even gardeners and farmers who desire a better understanding of how their plants grow. An attempt has been made to present the information in terms intelligible to readers with various backgrounds. If the treatment of some topics seems inadequate to specialists in certain fields, they are reminded that the book was not written for specialists, but as an introduction to the broad field of plant water relations. As an aid in this respect, a laboratory manual is available with detailed instructions for some of the more complex methods (J. S. Boyer *in* "Measuring the Water Status of Plants and Soils," Academic Press, San Diego, 1995).

We begin with a brief review of the research on plant water relations from Aristotle to the 20th century, including the development of such basic concepts as plant water balance, the soil–plant–atmosphere continuum, and the Klebs concept showing that both genetic potentialities and environmental factors modify growth through their effects on physiological processes and conditions. Some current questions, such as the role of roots as sensors of water stress and the increasing importance of investigations at the cellular and molecular level, are mentioned briefly in preparation for later discussion. Succeeding chapters are devoted to the unique properties of water and to cell water relations, providing an opportunity to define some of the terminology and units used in later chapters. Cell water relations are discussed in detail because they are basic to later discussions of plant water relations. Soil is discussed as a reservoir for water and a medium for root growth, root structure and growth are discussed with respect to the absorption of water and minerals, and the transport of water to the transpiring shoots is discussed in detail. Considerable attention is given to transpiration and the role of stomata in controlling it because transpiration often dominates plant water relations. Finally, we discuss the effects of water deficits on various physiological processes that control growth and yield of plants.

There is considerable cross referencing among chapters but there is also some repetition of material in various chapters. This is intended to make each chapter a fairly complete unit that can be read without excessive referral to other chapters, facilitating use of the book as a reference.

The need for a book summarizing modern views on plant water relations has been increased by the large volume of publications and the changes in viewpoint that have occurred in recent years. A number of books on plant water relations have appeared, but most of them are collections of papers on special topics. This book attempts to present the entire field of water relations in an organized manner, using current concepts and a consistent, simple terminology. Emphasis is placed on the interdependence of various processes. For example, the rate of water absorption is closely linked to the rate of transpiration by the sap stream in the vascular system, and it also is affected by resistance to water flow into roots and by soil factors affecting the availability of water. The rate of transpiration depends primarily on the energy supply, the stomatal opening, and the leaf water supply. Proper functioning of the physiological processes involved in growth requires a favorable water balance, which is controlled by relative rates of water absorption and water loss by transpiration. These complex interrelationships are discussed and described in modern terminology.

The large volume of publications in recent years makes it impossible to cite all of the relevant literature and many good papers have been omitted. Nevertheless, the bibliography is extensive enough to serve as an introduction to the literature in most areas of plant water relations research.

The present lively research activity is producing significant changes in explanations of various phenomena, and some long held views may have to be modified or abandoned. Examples of new trends are the recent emphasis on roots rather than leaves as primary sensors of water stress, the role of cell wall metabolism versus turgor in cell expansion, questions concerning the validity of water potential as a measure of plant water status, and the importance of osmotic adjustment. Even the cohesion theory of the ascent of sap is being questioned. Another new concept is the ecological importance of "hydraulic lift," or the supplying of water to shallow rooted plants by deeper rooted plants. Research at the molecular level is providing a better understanding of the reasons for such phenomena as differences in drought tolerance among species and cultivars. It also suggests the possibility of increasing plant drought tolerance by the genetic engineering of crop plants to minimize the effects of water stress.

Differences in opinion among various investigators are discussed, and in some instances the authors have indicated their preference, but it is pointed out that in many instances more research is needed before conclusions can be reached. We hope the uncertainty about some phenomena will challenge investigators to develop better explanations. Readers are reminded that so-called scientific facts often are merely the most logical explanations that can be developed from the available information. As additional research provides more information, it frequently becomes necessary to revise generally accepted explanations, and some that seem logical today may become untenable later.

This book owes more to interactions with other scientists than can be easily identified. We are indebted to our many graduate students and postdoctoral research associates and to our colleagues for their valuable suggestions. We especially acknowledge E. L. Fiscus, M. R. Kaufmann, J. S. MacFall, and C. D. Raper, Jr., for their useful comments on several chapters. We also acknowledge the assistance of Peggy Conlon in Dr. Boyer's office in typing several chapters and in preparing the bibliography, Dr. An-Ching Tang for preparing the artwork in several chapters, and the secretaries in the Duke University Department of Botany office who patiently typed and revised several chapters so many times.

Paul J. Kramer
John S. Boyer

Historical Review

1

INTRODUCTION

This chapter presents a brief historical review of progress in the field of plant water relations because the authors feel that it is impossible to fully understand the present without some knowledge of the past. As the Danish philosopher Kierkegaarde wrote, "Life can only be understood backward, but it can only be lived forward," and this also is true of science. The present generation needs to be reminded that some generally accepted concepts have their origin in ideas of 17th or 18th century writers and although others were suggested many decades ago, they were neglected until recently.

As might be expected, the importance of water to plant growth was recognized by prehistoric farmers because irrigation systems already existed in Egypt, Babylonia (modern Iraq), and China at the beginning of recorded history, and the first European explorers found extensive irrigation systems in both North and South America. However, irrigation was not used extensively in agriculture in the United States until after the middle of the 19th century and little research on plant water relations occurred until the 20th century.

Early Research

Although plant water relations appear to have been the first area of plant physiology to be studied, progress was slow from Aristotle who died in 322 B.C. to the middle of the 19th century. According to Aristotle, plants absorbed their food ready for use from the soil, and plant nutrition was controlled by a soul or vital principle that allowed plants to absorb only those substances useful in growth. This idea only began to be questioned in the 17th century by Jung, van Helmont, Mariotte, and others, and it persisted into the 19th century. The early contributions to an understanding of plant water relations were discussed by Sachs (1875, Eng. trans. 1890) and Green (1914). Many early contributions are overlooked. For example, Leonardo da Vinci (1452–1519) observed that the thickness of a tree trunk equals the combined thickness of the branches above it (Zimmermann, 1983, p. 66) and this is supported by later observations (see Chapter 7).

Perhaps the earliest attempt to explain absorption of water by plants was made by the Italian physician and herbalist Andrea Cesalpino in his "De Plantis" published in 1583. After considering and rejecting magnetism and suction, he concluded that roots absorb liquid in the same manner as a piece of linen or a sponge. This must have been one of the earliest attempts to explain a plant process in terms of a physical process. Another physician, J. B. van Helmont (1577–1644), attacked the Aristotelian doctrine of plant nutrition by growing a willow tree for 5 years in a covered pot to which nothing was added but water. As the willow increased in weight by 164 lb. and the soil weight decreased only 2 oz. he concluded that plants use water as food, but he did not try to explain how the water was absorbed or assimilated. An understanding of plant mineral nutrition did not develop until the 19th and early 20th century.

Harvey's publication in 1628 on the circulation of blood in animals increased interest in the possible circulation of sap in plants, and in about 1670 the Royal Society of London sponsored experiments by Ray and Willoughby to learn if sap circulates in plants as it does in animals. Their results were inconclusive, but they did demonstrate that water could move either way in stems. Probably the most important contributions of the 17th century were the anatomical observations of Nehemiah Grew (The Anatomy of Plants, 1682) and Marcello Malpighi (Anatome Plantarum, 1675). Both were interested in physiological processes, and Grew in particular speculated about water absorption and the ascent of sap. Both seemed to think that the sap underwent some change in the leaves that increased is nutritive value, after which it moved to fruits or down to roots, and Malpighi is said to have observed stomata. Although tempting, it is dangerous to reinterpret those old writings in terms of modern plant physiology because of the difficulty in understanding exactly what the writers meant.

The Work of Stephen Hales

The first truly quantitative experiments on plant water relations apparently were those of Stephen Hales, published in his "Vegetable Staticks" in 1727. Hales was a clergyman and humanitarian as well as a physiologist and is known to animal physiologists for his pioneering measurements of blood pressure, published in "Haemostaticks" in 1733. He placed great emphasis on the importance of plant sap, probably influenced by his work on blood pressure in animals, and made numerous observations on the absorption and translocation of water. Hales wrote, "the most likely way therefore, to get any insight into those parts of the creation, which come within our observation, must in all reason be to number, weigh and measure." This he did to an extent never previously attempted on plants.

Hales measured transpiration of plants of a number of species and calculated root and leaf surfaces from measurements of representative samples, an early use of a sampling technique. He observed that the transpiration rate varied with the species, temperature, time of day, and brightness of the sun, and that although sap exuded from cut grape stems and some trees with considerable force in the spring, exudation ceased later in the season after leaves expanded and water was lost by transpiration, or perspiration as it was then termed. He also observed that water was absorbed through cut stems of transpiring plants. Apparently Hales understood that there is a relationship between absorption and transpiration and regarded absorption as a physical process. There even is some indication that he regarded roots as differentially permeable. His experiments led him to discard the view that sap circulates in plants much as in animals. He seems to have regarded the chief function of leaves as the attraction of sap to the growing regions, although he suggested that light striking leaves might contribute in some way to the nutrition of plants. His observations were published in "Vegetable Staticks" in 1727.

THE CENTURY AFTER HALES

Nearly a century passed before contributions as important as those of Hales were made to the understanding of plant water relations. Early in the 19th century de Saussure (1804) found that the absorption of minerals was not proportional to the absorption of water and that different solutes were not absorbed in the same proportions in which they occurred in the soil, suggesting that roots differ in permeability to various solutes. An example of the unsatisfactory state of physiological knowledge was the wide acceptance of the spongiole theory of A. P. de Candolle (1832). He thought root caps were contractile-absorbing organs and this view was widely accepted for several decades, although an examination of the root tips should have indicated its improbability. Rational explorations of many physical phenomena based on physical and chemical principles were hindered in the early part of the 19th century by the belief that plant and animal processes were controlled by a mysterious "vital force," an idea that apparently originated with Aristotle (Sachs, 1875, pp. 504–509). Perhaps the most important contribution to plant water relations in the first half of the 19th century was development of an osmotic theory by Dutrochet (1837). This was used to explain a variety of phenomena, including the escape of spores from sporangia and the uptake of water by plants exhibiting root pressure. He also attributed the general condition that we term "turgor" to osmosis. Unfortunately, Dutrochet did not understand the role of differentially permeable membranes in osmosis, but this was developed later in the century by Traube (1867), De Vries (1877), and Pfeffer (1877). The concept of osmosis was philosophically important because important processes previously poorly explained as manifestations of a "vital principle" could now be explained in terms of sound physical processes.

Toward the end of the 18th and the first half of the 19th century advances in chemistry facilitated research on photosynthesis and the mineral nutrition of plants. Identification of oxygen and carbon dioxide and development of methods for measuring gases led to an understanding of gas exchange in photosynthesis by de Saussure and others. The research of Liebig (1841) and later investigators resulted in abandonment of the Aristotelian theory of plant nutrition and development of an understanding of nutrition that forms the basis for modern agricultural practices. However, that lies outside the scope of this book.

THE SECOND HALF OF THE 19th CENTURY

Progress was more rapid after the middle of the 19th century because of advances in chemistry and because more and better trained investigators were involved. Fick (1855) made the first mathematical treatment of diffusion, resulting in Fick's law which still is in use (see Nobel, 1991, pp. 10–12). Graham (1862) differentiated between colloidal and noncolloidal material (crystalloids) and Traube (1867) demonstrated that cell membranes are colloidal. Traube (1867) described how to form artificial membranes that could completely distinguish between small solutes and water, and he proposed a sieve theory of differential permeability and improved on Dutrochet's theory of osmosis. Pfeffer (1877) was the first to accurately measure osmotic pressure, using the membranes of Traube (1867). He recognized that osmotic pressure is determined by the solute concentration in an osmometer (Pfeffer, 1877) but considered cells to have a variable osmotic pressure, sometimes called the turgor pressure (Pfeffer, 1900). He correctly considered the pressure inside cells to arise from the resistance of the cell walls to stretching, which he called turgor. Gibbs (1875–1876) already recognized that, at equilibrium, osmotic pressure balanced the difference in osmotic potential across a membrane, and Pfeffer (1900) occasionally used the potential concept. The idea that protoplasmic membranes can bring about active uptake of solutes against a concentration gradient seems to have originated later with Reid (1890). De Vries (1877) first explained the conditions necessary for development of cell turgor and established a relationship between molecular weight and osmotic pressure that was used by the Swedish chemist Arrhenius to support his ionization theory, which with modifications is in use today.

Modern plant physiology is often said to have begun with the work of Sachs (1882a,b) and his contemporaries in the middle of the 19th century, and this is particularly true of water relations. Sachs (1882a,b) made important investigations of the effects of soil moisture, soil aeration, and soil temperature on water absorption and root growth, and showed that water absorption of warm season plants such as cucurbits is reduced more by cooling the soil than absorption by cool season plants. He also determined that clay soil contains more available water than sandy soil. Sachs demonstrated that water absorption occurs chiefly in the root hair zone instead of through the root tips or spongioles, as proposed by de Candolle. He also believed that transpiration and water absorption are closely related, a view fully developed late in the 19th century by Askenasy (1895) and Dixon and Joly (1895). Sachs mistakenly thought that sap rises in the walls of xylem elements rather than in the lumina because the lumina are often blocked by gas bubbles, but this error was soon corrected by Elfving (1882), Errera (1886), Vesque (1884), and Strasburger (1891). This is discussed further in Chapter 7 in connection with cavitation.

EARLY PLANT PHYSIOLOGY IN THE UNITED STATES

The study of plant physiology in the American colonies and in the early days of the United States lagged far behind England and Europe, partly because the abundance of new plants collected as new regions of the country were explored emphasized plant taxonomy. Among the botanists of prerevolutionary days were Cadwallader Colden of New York who introduced the Linnaean system of classification and John Bartram of Philadelphia whose garden still exists. The latter collected from New York to Florida and sent many specimens to European botanists, including Linnaeus who regarded him as the leading botanist in North America. European visitors such as the Englishman Nuttall, and the Frenchman Michaux who collected in the Southeast, added to the expanding knowledge of American plants. This phase of American botany culminated in the career of Asa Gray who by the middle of the 19th century was recognized in Europe as the dean of American botanists.

Perhaps the earliest research on plant water relations in America was done by Samuel Williams, a native of Vermont, who described it in his book, "The Natural and Civil History of Vermont," published in 1794, and in an enlarged edition in 1809. He wrote that in 1789 he measured the water loss from a detached maple twig, counted the leaves on a maple tree and the number of trees per acre, and estimated the water loss for 12 hr at 3875 gallons per acre, quite reasonable for the methods available. Miller's data (1938, p. 412) indicate an average water loss from an acre of Kansan corn of 3240 gallons per day. Williams also discussed the volume of sap obtained by tapping maple trees, determined the age of trees by counting tree rings, and considered the probable importance of trees in purifying the polluted air of cities. This indicates that air pollution is an old problem, especially in cities where every household burned coal. William's approach to plant water relations suggests that he may have been familiar with the work of Hale, published in 1727.

The botanical textbooks of the 19th century were mostly taxonomic manuals with short discussions of structure. However, Gray's widely used "Lessons in Botany and Vegetable Physiology" with a Preface dated 1857, revised in 1868, contains some physiology. For example, in Chapter XXV he states that most water absorption occurs through the newest root surfaces and that expansion of the root surface is necessary to supply water to the expanding leaf surface. In Chapter XXVI he states, "For in a leafy plant or tree the sap is not forced up from below, but is drawn up from above," a remarkably modern-sounding statement for the time. In Chapter XXV he had stated that most of the water escapes through the stomata. It is evident that he was familiar with the views of European physiologists of his time and incorporated them in his book.

As the supply of new plants dwindled, interest shifted from taxonomy to other areas of botany. In the third quarter of the 19th century W. S. Clark (1874, 1875) made important observations on sap flow and root growth. For example, he found that a squash plant growing in a greenhouse bench produced nearly 16 miles of roots (25 km) and his observations on sap flow from wounds in trees are acceptable today. There seems to have been little further notable work on plant water relations in the United States in the 19th century, although it is possible some interesting work is hidden in unpublished reports and has been overlooked.

THE 20th CENTURY

Burton E. Livingston started his productive career in teaching and research early in the 20th century. His first book, "The Role of Diffusion and Osmosis in Plants" (Livingston, 1903), was quite influential and he later developed apparatus to measure environmental factors, including porous porcelain atmometers to measure evaporation, soil point cones to measure the "water supplying power" of the soil, and lithium chloride clips to measure transpiration. The senior author's first research on plant water relations developed from E. N. Transeau's questions concerning a paper by Livingston (1927). In collaboration with Forrest Shreve (1921), Livingston wrote an important book on the role of climatic factors in plant distribution (Livingston and Shreve, 1921). Most important, he inspired and trained many young people to go into physiological and ecological research, as did the ecologist F. E. Clements. Osterhout also did important research on cell permeability during the early part of this century. Shull (1916) measured the force with which water is held by the soil by comparing the amount of water absorbed by *Xanthium* seeds from soil and from solutions of various osmotic pressures. He contributed to the development of plant physiology by founding the journal *Plant Physiology.*

Considerable progress was made in Europe in the early part of the 20th century, beginning in 1900 with Brown and Escombe's study of movement by diffusion through small pores such as stomata as a purely physical process. Unfortunately, Brown and Escombe's experiments were conducted in quiet air where the boundary layer resistance is as great as stomatal resistance, and their results indicated that large changes in stomatal aperture have little effect on transpiration. This led to a long and unproductive argument concerning the importance of stomatal control of transpiration. However, later experiments by Stålfelt (1932) and Bange (1953), summarized in Slatyer (1967, pp. 260–269), indicate that stomatal closure has a large effect on transpiration in moving air where the boundary layer resistance is low. Stomatal behavior is discussed in Chapter 8. Lundegårdh's "Environment and Plant Development" (1931) emphasized that the environment acts on plants through its effects on their physiological process.

From Osmosis to Water Potential

One of the important developments of the middle of the 20th century in the field of plant water relations was the shift in emphasis from the osmotic pressure to the water potential of plant tissue as an indicator of plant water status. During the early part of the century thousands of measurements were made of osmotic pressure as an indicator of plant water stress (Korstian, 1924, for example). Some of these data were summarized by Harris (1934). However, it began to be realized that water movement in plants cannot be explained in terms of osmotic pressure gradients, but rather in terms of what is now termed water potential (see Chapter 2). Also there was some uncertainty concerning the reliability of measurements of osmotic pressure. Initially, there was confusion concerning the best terminology and nearly as many terms were used as there were investigators. These included Saugkraft or suction force (Ursprung and Blum, 1916), water-absorbing power (Thoday, 1918), Hydratur (Walter, 1931, 1965), net osmotic pressure (Shull, 1930), diffusion pressure deficit (Meyer, 1938, 1945), and finally water potential (Owen, 1952; Slatyer and Taylor, 1960; Tang and Wang, 1941).

Acceptance of the water potential concept was slow because of the confusion regarding terminology, the lack of convenient methods for measuring it, and the inadequate training of plant physiologists in physical chemistry. As a result, plant water status seldom was measured during the second quarter of the 20th century. Development of thermocouple psychrometers (Monteith and Owen, 1958; Richards and Ogata, 1958; Spanner, 1951) and pressure equilibration by Scholander and his colleagues (1964, 1965) made measurement of water potential relatively easy, and they are the measurements used most often today to characterize plant water status. Thus although osmotic pressure measurements dominated the first half of the 20th century, the second half was dominated by measurements of water potential. However, interest in osmotic pressure was revived by concern with osmotic adjustment in water-stressed plants (Morgan, 1984; Turner and Jones in Turner and Kramer, 1980). Water potential is now under attack because it is not always well correlated with physiological processes (Sinclair and Ludlow, 1985), but it is defended by Kramer (1988) and Boyer (1989). This is discussed in Chapter 2.

The Permanent Wilting Percentage

In 1911 Briggs and Shantz published their concept of the permanent wilting percentage, which was refined by Richards and Wadleigh (1952), Veihmeyer and Hendrickson (1928, 1950), and others and became widely used. Slatyer (1957) pointed out that instead of being a soil constant, the permanent wilting percentage really depends on the water potential at which leaves lose their turgor, which depends on their osmotic properties and the meteorological conditions affecting transpiration, and on soil conditions that affect water absorption. Although there is no sharply defined lower limit for availability of water to plants (Gardner and Nieman, 1964; Hagan *et al.*, 1959), the permanent wilting percentage is a useful concept. It is discussed in more detail in Chapter 4.

The Absorption of Water

Although water absorption has been discussed since the time of Aristotle, textbook discussions in the first half of the 20th century were disappointingly vague concerning the mechanism. For example, Miller (1938) stated that imbibition, osmosis, and passive absorption apparently were involved, but he doubted if these forces provided an adequate explanation. This is surprising because Renner (1912, 1915) had already distinguished between the "active" absorption mechanism operating in slowly transpiring plants and the "passive" mechanism operating in rapidly transpiring plants. Renner's views were reinvestigated and expanded by Kramer (1932, 1937, 1939, 1940b) and it is now generally accepted that although roots of slowly transpiring plants act as osmometers, water is "pulled" in passively through the roots of rapidly transpiring plants. It also became possible to distinguish between effects of environmental stresses such as low temperature and deficient aeration on root resistance (conductance or permeability) to water flow and on the driving force causing water movement. Water absorption is discussed in more detail in Chapter 6.

SOME GENERAL CONCEPTS

By the middle of the 20th century much valuable information was available, but it was poorly organized. However, several general concepts became available which have been useful in organizing information and research on plant water relations.

Plant Water Balance

One useful idea is the concept of plant water balance, proposed by Montfort (1922) and Maximov (1929). This treats plant water status as dependent on the relative rates of water absorption and water loss in transpiration. Although these two processes are coupled by the cohesive columns of xylem sap, they do not always operate synchronously because of the capacitance factor provided by water stored in the parenchyma cells of leaves, stems, and roots. Thus water absorption often lags behind transpiration in the morning (Fig. 6.1), resulting in loss of turgor and temporary wilting by midday, even in plants growing in moist soil. This led to the view that plant water status cannot be predicted reliably from measurements of soil moisture, but must be measured directly on the plant (Kramer, 1963).

Soil–Plant–Atmosphere Continuum and Ohm's Law Analogy

Another useful, unifying concept is that of the soil–plant–atmosphere continuum (SPAC) which seems to have been proposed by Huber (1924), developed by Gradmann (1928) and Honert (1948), and brought into general use by Slatyer and Taylor (1960), Rawlins (1963), and others. This concept emphasizes the interrelationships of soil, plant, and atmospheric factors in determining plant water status. As the equation for water flow through the SPAC resembles that for flow of electricity in a conducting system it is often termed the Ohm's Law analogy. Cowan (1965) used this concept to produce a model showing resistances and capacitances in the SPAC. Although very convenient for teaching and analyzing the way in which various factors affect water flow through the soil–plant–atmosphere continuum, it is an oversimplification because it assumes steady-state flow and constant resistance, conditions that seldom exist. Some of the difficulties with the water potential terminology were discussed by Passioura (1982), and Fiscus and Kaufmann (in Stewart and Nielson, 1990, p. 228) argue that the Ohm's Law analogy has no advantage over a fluid transport law and should be abandoned. However, we think it is too useful to be abandoned. Johnson *et al.* (1991) discuss the integration of new concepts into a satisfactory model of water flow.

Kleb's Concept

A third useful concept, shown in Fig. 1.1, indicates how hereditary potentialities and environmental factors such as water stress act through physiological processes to determine the quantity and quality of plant growth. This concept is based on ideas of a German physiologist, Klebs (1910, 1913). Kleb's concept

HEREDITARY POTENTIALS	ENVIRONMENTAL FACTORS
Depth and extent of root systems	SOIL. Texture, structure, depth, chemical composition and pH, aeration, temperature, waterholding capacity, and water conductivity
Size, shape and total area of leaves, and ratio of internal to external surface	
Number, location, and behavior of stomata	ATMOSPHERIC. Amount and seasonal distribution of precipitation
	Ratio of precipitation to evaporation
	CO_2 concentration
	Radiant energy, wind, vapor pressure, and other factors affecting evaporation and transpiration
	BIOTIC. Competing plants, diseases, insects

PLANT PROCESSES AND CONDITIONS

Water absorption

Ascent of sap

Transpiration

Internal water balance as reflected in
water potential, turgidity, stomatal opening,
and cell enlargement

Effects on kinds and amounts of growth regulators,
photosynthesis, respiration, carbohydrate and
nitrogen metabolism, and other metabolic processes

QUANTITY AND QUALITY OF GROWTH

Size of cells, organs, and plants

Dry weight, succulence, kinds and amounts of
various compounds produced and accumulated

Root-shoot ratio

Vegetative versus reproductive growth

Figure 1.1   A diagram showing how the quantity and quality of plant growth are controlled by hereditary potentialities and environmental factors operating through the physiological processes and conditions of plants, with special emphasis on effects of water.

emphasizes the fact that environmental factors such as water stress operate through physiological processes to affect the quantity and quality of plant growth, within the limits permitted by heredity. It is useful in organizing ideas and emphasizes the fact that environmental factors such as drought or irrigation, cultural practices, and plant breeding affect crop yield by modifying the efficiency of essential physiological processes.

THE SITUATION TODAY

There has been great progress in our understanding of plant water relations during the 20th century and in the development of new instrumentation that facilitates research (Hashimoto *et al.*, 1990). As a result of the increase in research and publication, one of the most troublesome problems today is that of keeping up with the current literature. While computerized bibliographies are helpful, they cannot distinguish between important and trivial papers. Furthermore, what seems trivial to one reader may be important to another or when read in a different context.

Changing Viewpoints

The accumulation of new information often requires the revision of existing explanations. Many so-called scientific explanations are "true," only in the sense that they represent the most logical conclusions that can be drawn from our present information, and the acquisition of additional information may require their revision. For example, it has been assumed for many years that shoot and leaf turgor control cell expansion, stomatal conductance, and photosynthesis. However, research by Shackel *et al.* (1987), Zhu and Boyer (1992), and others indicates that cell enlargement depends on cell wall metabolism as well as turgor. Several experiments discussed by Schulze (1986a), Turner (1986, pp. 13–15), Davies and Zhang (1991), and Davies *et al.* (1994) suggest that stomatal conductance and photosynthesis are better correlated with soil and root water status than with leaf water status. Kramer (1988) expressed reservations concerning the broad application of the conclusions from split-root pot experiments to plants in nature. In any event there is increasing interest in the importance of biochemical signals such as cytokinin and abscisic acid (ABA) transmitted from roots to shoots that affect physiological processes in the shoots. This is discussed in more detail in Chapters 6, 9, and 11.

Also, after several decades of emphasis on the plant water potential as a good indicator of plant water status, some investigators are claiming that physiological processes are better related to relative water content than to water potential (Sinclair and Ludlow, 1985). As pointed out earlier, there has been considerable interest in osmotic adjustment as a factor in drought tolerance (Morgan, 1984) but this idea also is now being questioned (Munns, 1988). Considerable uncertainty concerning the relative importance of stomatal and nonstomatal inhibition of photosynthesis in water-stressed plants also exists (Farquhar and Sharkey, 1982; also see Chapter 10). Even the cohesion theory of the ascent of sap is under attack by some investigators (see Balling and Zimmermann, 1990; Zimmerman *et al.*, 1993). This is discussed in Chapter 7.

Bound water was once regarded as an important factor in drought and cold tolerance (Kramer, 1955), but it has been neglected in recent years. However, development of new methods for measuring it, such as the water sorption isotherm (Vertucci and Leopold, 1987a,b), nuclear magnetic resonance (NMR) spectroscopy (Burke *et al.*, 1974; Kramer *et al.* in Hashimoto *et al.*, 1990; Roberts, 1984), and NMR imaging (G. A. Johnson *et al.*, 1987; Veres *et al.*, 1991), are likely to result in a renewed study of bound water as a factor in the tolerance of dehydration and freezing.

Increasing Emphasis at the Molecular Level

There is likely to be more research on the effects of water deficits on physiological processes at the molecular level. For example, water deficits seem to increase some enzymatically mediated processes but to decrease others (Hsiao and Bradford in Taylor *et al.*, 1983). Chapter 9 discusses enzyme responses and suggests a hypothesis to account for them (see Regulator Hypothesis of Enzyme Control). Root/shoot communication is being increasingly studied at the molecular level, and Chapter 9 discusses an example for the enzyme nitrate reductase where control of the shoot enzyme is determined by the delivery of nitrate from the roots. Ho and Mishkind (1991) reported that a water deficit caused appearance of some new mRNAs and disappearance of others in water-deficient tomato leaves, and cited other research showing the appearance of new proteins in these plants. Molecular work also opens up the prospect of increasing the tolerance to unfavorable conditions by altering the genetic potential of plants. Tarczynski *et al.* (1993) increased the salt tolerance of tobacco plants by introducing a bacterial gene that increased their capacity to synthesize mannitol, and Potrykus (1991) surveyed methods used to transfer genes and some results obtained. The direction of future research is indicated by some papers in Close and Bray (1993). These include discussion of the role of special proteins such as dehydrin and osmotin and accumulation of proline, glycine betaine, and quaternary ammonium and tertiary sulfonium compounds in water-deficient and salinized plants that are also discussed in Chapters 3 and 12. Other examples of current trends are the study by Thomas and Bohnert (1993) of the induction by salt stress of plant growth regulators, the effects of phosphorus deficiency on root metabolism (Johnson *et al.*, 1994), and the study of the varying role of hormones on root development of water-deficient *Arabidopsis* (Vartanian *et al.*, 1994). Chrispeels and Maurel (1994) found that special proteins form water channels in membranes (see Chapter 3). These studies show that molecular genetics is being increasingly used in physiological research, and the May, 1994 issue of *Plant, Cell and Environment* contains 16 papers on the use of mutants and transgenic plants in physiological studies.

The current interest in research at the molecular level is very important, but it should not obscure the fact that there is still need for research on water relations at the whole plant level. The success of crop plants depends primarily on the avoidance of stress, which depends on the success of leaves in controlling water loss and the effectiveness of roots in absorbing water in competition with other plants. It is true that tolerance of dehydration depends on characteristics at the cellular and molecular level, such as osmotic adjustment, the water-binding capacity of tissues, and the manner in which water stress affects enzyme-mediated processes, and it is likely that research at that level will make important contributions in this area. However, the successful growth of economic plants depends more on the avoidance than on the tolerance of severe water stress, and severely water-stressed crops seldom are profitable. Avoidance of severe water stress requires coordination at the whole plant level between control of water loss from transpiring shoots and water absorption through root systems. Thus from a practical standpoint, research at the whole plant level remains very important. Furthermore, a good understanding of whole plant structure and physiology is essential for the effective application of molecular biology to the solution of problems in the field and forest. This is discussed in Chapter 12.

SUMMARY

For nearly 2000 years after the observation of the early Greek philosophers, Aristotle and Theophrastus, there were no important contributions to plant physiology. In the 17th century Grew and Malpighi published on plant anatomy and speculated about plant processes, but the first quantitative studies of plant water relations were not published until 1727, by Hales. Advances in chemistry toward the end of the 18th and early 19th century permitted a better understanding of photosynthesis and mineral nutrition.

Modern plant physiology began to develop after the middle of the 19th century with the work of Sachs, Strasburger, and other German plant scientists. Early plant research in North America was concerned chiefly with the classification of plants brought home by explorers. However, Gray's textbooks, published after the middle of the 19th century, contained some physiology, apparently derived largely from European publications. After the beginning of the 20th century there was a worldwide increase in research on plant water relations, including work on soil moisture, root systems, and measurement of plant water status. In the United States the early research occurred chiefly in agricultural experiment stations, at a few universities, and at the Desert Laboratory at Tucson, Arizona.

Toward the middle of the 20th century emphasis began to shift from osmotic pressure to water potential as an indicator of plant water status. This was facilitated by the introduction of pressure equilibration and thermocouple psychometers to measure water potential. At present there is expanding research at the cellular and molecular level which will increase our understanding of how water stress injures plants and how to increase tolerance of water stress. Several concepts were developed which are useful in organizing our information concerning plant water relations. One is the concept of the plant water balance as controlled by the relative rates of water absorption and loss, another is the treatment of water as constituting a continuum in the soil–plant–atmosphere system. Another more general concept states that heredity and environment act through physiological processes to determine the quantity and quality of growth. These concepts have assisted in organizing information more effectively and in planning research.

In conclusion, we are in a period of change when some long established concepts are being questioned and new explanations are being proposed. Furthermore, research emphasis is shifting from the whole plant to the cellular and molecular level. The authors have attempted to evaluate some conflicting explanations and to support those that seem most logical. However, as more information becomes available it may become necessary to modify or even abandon some of our present beliefs and explanations.

SUPPLEMENTARY READING

Clark, W. S. (1875). Observations upon the phenomena of plant life. *Mass. State Board Agric. Ann. Report* **22**, 204–312.

Close, T. J., and Bray, E. A., eds. (1993). "Plant Responses to Cellular Dehydration During Environmental Stress," Am. Soc. Plant Physiologists Ser., Vol. 10, Rockville, MD.

Crafts, A. S., Currier, H. B., and Stocking, C. R. (1949). "Water in the Physiology of Plants." Chronica Botanica Co., Waltham, MA.

de Candolle, A. P. (1832). "Physiologie Végétale." Bechet Jeune, Paris.

de Saussure, N. T. (1804). "Researches Chimiques sur la Végétation." Madame Huzard, Paris.

Dutrochet, H. H. (1837). "Memoires Pour Servir a l'Historie Anatomique et Physiologique des Vegetaux et des Animaux." J. B. Baillière et Fils, Paris.

Evelyn, J. (1670). "Sylva." J. Martyn and J. Allestry, London.

Gray, A. (1868). "Lessons in Botany and Vegetable Physiology." Irison, Blakeman, and Taylor, New York.

Green, J. R. (1914). "A History of Botany in the United Kingdom from the Earliest Times to the End of the 19th Century." J. M. Dent & Sons, Ltd., London.

Haberlandt, G. (1914). "Physiological Plant Anatomy," English translation by M. Drummond. Macmillan & Co. Ltd, London.

Hales, S. (1727). "Vegetable Staticks." A. and J. Innys and T. Woodward, London. Reprinted by Scientific Book Guild, London.

Klebs, G. (1910). Alterations in the development and forms of plants as a result of environment. *Proc. Roy. Soc. London* **82B**, 547–558.

Klebs, G. (1913). Über das verhaltnis der Aussenwelt zur Entwicklung der Pflanzen. *Sitzber, Heidelberg Akad. Wiss. Abt* **B5**, 1–47.

Liebig, J. von (1843). "Chemistry in Its Application to Agriculture and Physiology," 3rd ed. English translation, Peterson, Philadelphia.

Lundegårdh, H. (1931). "Environment and Plant Development." English translation, Edward Arnold & Co., London, 1961.

Miller, E. C. (1938). "Plant Physiology," 2nd Ed. McGraw-Hill, New York.

Pfeffer, W. (1877). "Osmotic Investigations: Studies on Cell Mechanics." English translation by
   G. R. Kepner and E. J. Stadelmann, Von Nostrand Reinhold, New York, 1985.
Pfeffer, W. (1900). "The Physiology of Plants," 2nd Ed. English translation by Ewart, Oxford University Press.
Reed, H. S. (1942). "A Short History of the Plant Sciences." Chronica Botanica Co. Waltham, MA.
Somero, G. N., Osmond, C. B., and Bolis, C. L., eds. (1992). "Water and Life." Springer-Verlag,
   Berlin.
Strasburger, E. (1891). "Über den Bau und Verrichtungen des Leitungsbahnen in den Pflanzen." G.
   Fischer, Jena.
von Sachs, J. (1875). "History of Botany." English translation by Garnsey, Oxford University Press,
   1890.

Functions and Properties of Water

INTRODUCTION

Water is one of the most common and most important substances on the earth's surface. It is essential for the existence of life, and the kinds and amounts of vegetation occurring on various parts of the earth's surface depend more on the quantity of water available than on any other single environmental factor. The importance of water was recognized by early civilizations, and it occupied a prominent place in ancient cosmologies and mythologies. The early Greek philosopher Thales asserted that water was the origin of all things, and it was one of the four basic elements (earth, air, fire, water) recognized by later Greek philosophers such as Aristotle. It was also one of the five elemental principles (water, earth, fire, wood, metal) of early Chinese philosophers. Today it is realized that the availability of water not only limits the growth of plants but can also limit the growth of cities and industries. This chapter discusses the ecological and physiological importance of water, its unique properties, and the properties of aqueous solutions.

Ecological Importance of Water

The distribution of plants over the earth's surface is controlled chiefly by water and temperature (see Chapters 9 and 12), and where temperature permits plants to grow, chiefly by the quantity and distribution of precipitation. Where rainfall is abundant and well distributed we find the lush vegetation of tropical rain forests, the Olympic Peninsula, and the cove forests of the southern Appalachians. Strong seasonal variations as in Mediterranean climates with hot dry summers result in the shrubby vegetation of the Mediterranean and southern California coasts and parts of the west coast of southern South America. Severe summer droughts result in forests being replaced by grasslands as in the steppes of Eurasia, the prairies of the United States and the pampas of Argentina, and finally in the driest areas by deserts. In contrast, where poor drainage results in more or less permanently saturated soil the vegetation characteristic of swamps and bogs occurs.

Even in humid climates, most of the year-to-year variation in the diameter growth of trees can be related to the variation in rainfall (Zahner in Kozlowski, 1968). From this relationship arose the study of dendrochronology (Fritts, 1976), dealing with the use of variation in width of tree rings to determine rainfall conditions in the past and the age of ancient buildings (Giddings, 1962). For example, the reduction in ring width of timbers in the cliff dwellings of the southwestern United States indicates that a severe drought occurred in that region in the 13th century which was responsible for abandonment of many settlements. Variations in width of tree rings in 1600-year-old bald cypress trees growing in the Coastal Plain of North Carolina indicate that wet and dry periods with an average duration of about 30 years have occurred in that region during the past 1600 years (Stahle *et al.*, 1988). The narrower rings in trees growing on fault lines, caused by root disturbance during earthquakes, were used by Jacoby *et al.* (1988) to date the occurrence of minor earthquakes in California.

The effects of temperature on vegetation are partly exerted through water relations because an amount of rainfall sufficient to maintain forests in a cool climate where the rate of evapotranspiration is low can only maintain grasslands in a warmer climate where the rates of evaporation and transpiration are much higher. As a result, the distribution of forests in the eastern and central United States is much better correlated with the ratio of rainfall to evaporation than with rainfall alone (Transeau, 1905). Currie and Paquin (1987) reported that three-fourths of the variation in numbers of tree species in North America, Great Britain, and Ireland could be explained by differences in annual evapotranspiration, and Rosenzweig (1968) concluded that differences in net annual aboveground productivity are well correlated with variations in evapotranspiration as calculated by the method of Thornthwaite and Mather (1957). These relationships exist because evapotranspiration depends on the combined effects of temperature, irradiation, and precipitation.

Physiological Importance of Water

The ecological importance of water is the result of its physiological importance. The only way in which an environmental factor such as water can affect

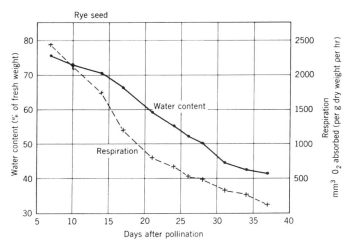

Figure 2.1 Decrease in water content and in rate of respiration during maturation of rye seed. From Kramer (1983), after Shirk (1942).

plant growth is by influencing physiological processes and conditions, as shown in Fig. 1.1.

Almost every plant process is affected directly or indirectly by the water supply. Many of these effects will be discussed later, but it can be emphasized here that within limits metabolic activity of cells and plants is closely related to their water content. For example, the respiration of young, maturing seeds is quite high, but it decreases steadily during maturation as water content decreases (see Fig. 2.1). The respiration rate of air-dry seeds is very low and increases slowly with increasing water content up to a critical point, at which there is a rapid increase in respiration with a further increase in water content (Fig. 2.2). The growth of plants is controlled by rates of cell division and enlargement and by the supply of organic and inorganic compounds required for the synthesis of new protoplasm and cell walls. Cell enlargement is particularly dependent on at least a minimum degree of cell turgor, and stem and leaf elongations are quickly checked or stopped by water deficits, as shown in Fig. 2.3. A decrease in water content inhibits photosynthesis (Fig. 2.3) and usually reduces the rate of respiration and other enzyme-mediated processes.

In summary, decreasing water content is accompanied by loss of turgor and wilting, cessation of cell enlargement, closure of stomata, reduction in photosynthesis, and interference with many other basic metabolic processes. Eventually, continued dehydration causes disorganization of the protoplasm and death of most organisms. The effects of water deficits on physiological processes are

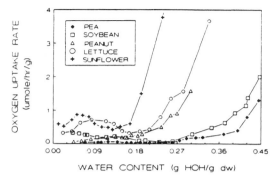

Figure 2.2 Relationship between water content and oxygen uptake for seeds of five species. At some critical water content the rate of oxygen uptake increases rapidly. Water probably is bound too firmly at lower contents to be available for physiological processes. From Vertucci and Roos (1990).

discussed in more detail in Chapters 9–11. So important are the effects of water on physiological processes that McIntyre (1987) suggested that water should be regarded as a major factor in the regulation of plant growth, with some effects of hormones being produced through the control of water status.

FUNCTIONS OF WATER IN PLANTS

The importance of water can be summarized by listing its most important functions under four general headings.

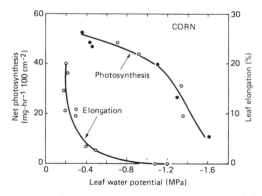

Figure 2.3 Relationship among leaf water potential, leaf elongation, and photosynthesis of corn. Note that leaf elongation almost ceases before there is much reduction in photosynthesis. From Kramer (1983), after Boyer (1970).

Constituent

Water is as important quantitatively as it is qualitatively, constituting 80–90% of the fresh weight of most herbaceous plant parts and over 50% of the fresh weight of woody plants. Some data on water content of various plant structures are shown in Table 2.1. Water is as important a part of the protoplasm as the protein and lipid molecules which constitute the protoplasmic framework, and the reduction of water content below some critical level is accompanied by changes in structure and ultimately in death. A few plants and plant organs can be dehydrated to the air-dry condition, or even to the oven-dry condition in the case of some kinds of seeds and spores, without loss of viability, but a marked decrease in physiological activity always accompanies a decrease in tissue water content (see Chapters 9 and 10). The relationship be-

Table 2.1   Water Content of Various Plant Tissues Expressed as Percentages of Fresh Weight[a]

	Plant parts	Water content (%)	Reference
Roots	Barley, apical portion	93.0	Kramer and Wiebe (1952)
	Pinus taeda, apical portion	90.2	Hodgson (1953)
	P. taeda, mycorrhizal roots	74.8	Hodgson (1953)
	Carrot, edible portion	88.2	Chatfield and Adams (1940)
	Sunflower, average of entire root system	71.0	Wilson *et al.* (1953)
Stems	Asparagus stem tips	88.3	Daughters and Glenn (1946)
	Sunflower, average of entire stems on 7-week-old plant	87.5	Wilson *et al.* (1953)
	Pinus banksiana	48.0–61.0	Raber (1937)
	Pinus echinata, phloem	66.0	Huckenpahler (1936)
	P. echinata, wood	50.0–60.0	Huckenpahler (1936)
	P. taeda, twigs	55.0–57.0	McDermott (1941)
Leaves	Lettuce, inner leaves	94.8	Chatfield and Adams (1940)
	Sunflower, average of all leaves on 7-week-old plant	81.0	Wilson *et al.* (1953)
	Cabbage, mature	86.0	Miller (1938)
	Corn, mature	77.0	Miller (1938)
Fruits	Tomato	94.1	Chatfield and Adams (1940)
	Watermelon	92.1	Chatfield and Adams (1940)
	Strawberry	89.1	Daughters and Glenn (1946)
	Apple	84.0	Daughters and Glenn (1946)
Seeds	Sweet corn, edible	84.8	Daughters and Glenn (1946)
	Field corn, dry	11.0	Chatfield and Adams (1940)
	Barley, hull-less	10.2	Chatfield and Adams (1940)
	Peanut, raw	5.1	Chatfield and Adams (1940)

[a]From Kramer (1983).

tween water content and protein structure has been discussed by Tanford (1963, 1980), Kuntz and Kauzmann (1974), Edsall and McKenzie (1978), and others. There is further discussion of the role of water in growth in Chapters 11 and 12.

Solvent

A second essential function of water in plants is as the solvent in which gases, minerals, and other solutes enter plant cells and move from cell to cell and organ to organ. The relatively high permeability of most cell walls and protoplasmic membranes to water results in a continuous liquid phase, extending throughout the plant, in which translocation of solutes occurs.

Reactant

Water is a reactant or substrate in many important processes, including photosynthesis and hydrolytic processes such as the amylase-mediated hydrolysis of starch to sugar in germinating seeds. It is just as essential in this role as carbon dioxide in photosynthesis or nitrate in nitrogen metabolism. There also is increasing interest in water as a ligand in chemical reactions (Rand, 1992).

Maintenance of Turgidity

Another role of water is in the maintenance of the turgor which is essential for cell enlargement and growth and for maintaining the form of herbaceous plants. Turgor is also important in the opening of stomata and the movements of leaves, flower petals, and various specialized plant structures. Inadequate water to maintain turgor results in an immediate reduction of vegetative growth, as shown in Fig. 2.3.

PROPERTIES OF WATER

The importance of water in living organisms results from its unique physical and chemical properties. These unusual properties were recognized in the 19th century (see Edsall and McKenzie, 1978, for references), and their importance was discussed early in the 20th century by Henderson (1913), Bayliss (1924), and Gortner (1938). Even today there is some uncertainty about the structure of water and some of its properties, as will be seen later. However, there is no doubt that water has the largest collection of anomalous properties of any common substance.

Unique Physical Properties

A substance with the molecular weight of water should exist as a gas at room temperature and have a melting point of below $-100°C$. Instead, water is a liquid at room temperature and its melting point is 0°C. It has the highest specific heat of any known substance except liquid ammonia, which is about 13% higher. The high specific heat of water tends to stabilize temperatures and is reflected in the relatively uniform temperature of islands and land near large bodies of water. This is important with respect to agriculture and natural vegetation. The standard unit for measuring heat, the calorie (cal), is 4.18 joules (J) and is based on the specific heat of water or the amount of energy required to warm 1 gram (g) of water 1°, from 14.5° to 15.5°C. The heat of vaporization is the highest known, 540 cal/g at 100°C, and the heat of fusion, 80 cal/g, is also unusually high. Because of the high heat of vaporization, evaporation of water has a pronounced cooling effect and condensation has a warming effect. Water is also an extremely good conductor of heat compared with other liquids and nonmetallic solids, although it is poor compared with metals. Water is transparent to visible radiation (390–760 nm). It allows light to penetrate bodies of water and makes it possible for algae to carry on photosynthesis and grow to considerable depths. It is nearly opaque to longer wavelengths in the infrared range so that water filters are fairly good heat absorbers (see Fig. 2.4).

Water has a much higher surface tension than most other liquids because of the high internal cohesive forces between molecules. This provides the tensile strength required by the cohesion theory of the ascent of sap. Water also has a high density and is remarkable in having its maximum density at 4°C instead of at the freezing point. Even more remarkable is the fact that water expands on freezing, so that ice has a volume about 9% greater than the liquid water from which it was formed (Fig. 2.5). This explains why ice floats and pipes and radiators burst when the water in them freezes. Incidentally, if ice sank, bodies of

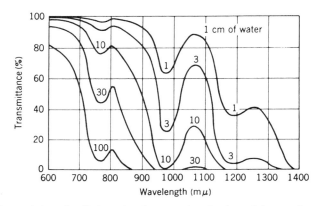

Figure 2.4 Transmission of radiation of various wavelengths through layers of water of different thicknesses. Radiation between 390 and 760 mμ is visible to the human eye. The numbers on the curves refer to the thickness of the layers in centimeters. Transmission is much greater at short than at long wavelengths. From Kramer (1983), after Hollaender (1956, p. 195).

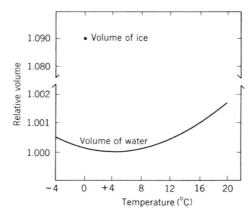

Figure 2.5   Change in volume of water with change in temperature. The minimum volume is at 4°C, and below that temperature there is a slight increase in volume as more molecules are incorporated into the lattice structure. The volume increases suddenly when water freezes because all molecules are incorporated into a widely spaced lattice. Above 4°C there is an increase in volume caused by increasing thermal agitation of the molecules. From Kramer (1983).

water in the cooler parts of the world would all be filled permanently with ice, with disastrous effects on the climate and on aquatic organisms.

Water is very slightly ionized; only one molecule in 55.5×10^7 is dissociated. It also has a high dielectric constant (ability to neutralize attraction between electrical charges) which contributes to its behavior as an almost universal solvent. It is a good solvent for electrolytes because the attraction of ions to the partially positive and negative charges on water molecules results in each ion being surrounded by a shell of water molecules which keeps ions of opposite charge separated (Fig. 2.6). It is a good solvent for many nonelectrolytes be-

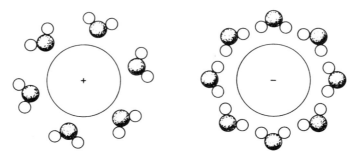

Figure 2.6   Diagram showing approximate arrangement of water molecules in shells oriented around ions. These shells tend to separate ions of opposite charge and enable them to exist in solution. They also disrupt the normal structure of water and slightly increase the volume. From Kramer (1983), after Buswell and Rodebush (1956).

cause it can form hydrogen bonds with N in amino groups and O in carbonyl groups. It tends to be adsorbed, or bound strongly, to the surfaces of clay micelles, cellulose, protein molecules, and many other substances. This characteristic is of great importance in soil and plant water relations.

Explanation of Unique Properties

It was realized early in this century that the unusual combination of properties found in water could not exist in a system consisting of individual H_2O molecules. At one time, it was proposed that water vapor is monomeric H_2O, that ice is a trimer $(H_2O)_3$ consisting of three associated molecules, and that liquid water is a mixture of a dimer $(H_2O)_2$ and a trimer. Now the unusual properties are explained by assuming that water molecules are associated in a more or less ordered structure by hydrogen bonding. Ice is characterized by an open crystalline lattice (Fig. 2.7). Liquid water has increasing disorder, and in the vapor phase the individual molecules are not associated at all. The properties and structure of water have been treated in many articles and books, including Kavanau (1964), Eisenberg and Kauzmann (1969), and a multivolume com-

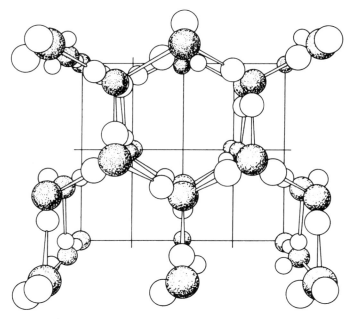

Figure 2.7   Diagram showing approximately how water molecules are bound together in a lattice structure in ice by hydrogen bonds. The dark spheres are oxygen atoms, and the light spheres are hydrogen atoms. From Kramer (1983), after Buswell and Rodebush (1956).

pendium edited by Franks (1975). More recent views are presented by Edsall and McKenzie (1978), Stillinger (1980), and Benson and Siebert (1992).

To explain the unusual properties of water requires a brief review of the kinds of electrostatic forces that operate among atoms and molecules. These include the strong ionic or electrovalent bonds and covalent bonds, weaker attractive forces known as van der Waals or London forces, and hydrogen bonds. Ionic bonds result from electrostatic attraction between oppositely charged partners, as between sodium and chlorine atoms in sodium chloride (NaCl). Such compounds usually ionize readily. Covalent bonds are formed by sharing electrons, as between oxygen and hydrogen atoms in water and carbon and hydrogen atoms in organic compounds. Such compounds do not ionize readily. Covalent bonds are strong, about 110 kcal/mol for the O-H bond in water, but they may be broken during chemical reactions.

If ionic or covalent bonds were the only types of bonding, there would be no liquids or solids because these do not allow individual molecules to interact with each other. However, there are intermolecular binding forces called van der Waals or London forces and hydrogen bonds that operate between adjacent molecules and affect the behavior of gases and liquids. Some molecules are polar or electrically asymmetric because they have partially positive and negative areas caused by an unequal sharing of electrons between atoms. These charged areas attract one molecule to another. Water shows this dipole effect rather strongly because the two hydrogen atoms form a bond angle of 104.5° with oxygen to give a V-shaped molecule resulting in the hydrogen bonding discussed later. Substances such as carbon tetrachloride and methane do not show permanent dipole effects because their molecules have no asymmetric distribution of electrons and consequently no charged areas.

Even electrically neutral molecules show anomalous properties, and in 1873 van der Waals suggested that the nonideal behavior of gases is caused by weak attractive forces operating between such molecules. In about 1930, London developed an explanation for these attractions based on the assumption that even those molecules that on the average are electrically symmetrical or neutral develop momentary or instantaneous dipoles from the motion of their electrons. These dipoles induce temporary dipoles in neighboring molecules, causing the instantaneous or momentary attraction between them known as van der Waals or London forces. This attraction is weak, about 1 kcal/mol, and effective only if molecules are very close together. In general, the physical properties of liquids such as the boiling point, heat of vaporization, and surface tension depend on the strength of intermolecular bonding. For example, gases condense into liquids when cooled enough so that the van der Waals and other attractions between molecules exceed the dispersive effect of their kinetic energy. For small molecules the size of water, temperatures usually must be very low to form liquids.

The peculiar physical properties of water result from additional intermolecular forces much stronger than van der Waals forces. These strong attractive forces are hydrogen bonds that result from the weak electrostatic attraction of the partially positively charged hydrogen atoms of one water molecule to the partially negatively charged oxygen atoms of adjacent molecules. They operate over considerable distances and have a binding force of about 1.3–4.5 kcal/mol in water. The forces produced by the asymmetric distribution of charges on water molecules bind them in the symmetrical crystalline lattice structure of ice, shown diagrammatically in Fig. 2.7. The water molecules in ice are arranged in a lattice with unusually wide spacing, resulting in a density lower than that of liquid water.

As ice melts, 13–15% of the bonds break, and about 8% of the molecules escape from the lattice. This results in a partial collapse of the lattice into a more disorderly but also more compact structure and an increase to maximum density at 4°C. As the temperature rises above 4°C, further increases in breakage and deformation of hydrogen bonds result in an increase in volume (see Fig. 2.5). There has been, and still is, some uncertainty about the structure of liquid water, i.e., the manner in which the molecules are oriented in relation to one another (Amato, 1992; Benson and Siebert, 1992). Incidentally, the concept of structure refers only to average positions of molecules because they are continually in motion and exchanging bonds. At one time it was believed that liquid water consisted of "flickering clusters" or "icebergs" of structured water molecules surrounded by unstructured molecules (Némethy and Scheraga, 1962). However, Stillinger (1980) stated that recent studies tend to rule out the iceberg concept. He regards liquid water as a three-dimensional network of hydrogen-bonded molecules showing a tendency toward tetrahedral geometry but containing many strained or broken bonds. Benson and Siebert (1992) propose that liquid water consists of hydrogen-bonded cyclic tetramers and octamers. Because hydrogen bonds have a half-life of only about 2×10^{-10} sec, these structures are difficult to detect.

Only part of the structure is destroyed by heating, and about 70% of the hydrogen bonds found in ice remain intact in liquid water near 100°C. The high boiling point of water results from the large amount of energy required to break the remaining hydrogen bonds and vaporize liquid water. The structure is somewhat modified by the pH, because it affects the distance between the hydrogen and the oxygen atoms, and by ions, because of their attraction for water molecules.

Ions also form dipole bonds with water molecules. The result is that the ions become surrounded by firmly bound shells of water molecules (Fig. 2.6). In fact, Bernal (1965) described ions, protein molecules, and cell surfaces as being coated with "ice," i.e., with layers of structured water molecules. It is now considered unlikely that water can form a uniform layer over the surface of protein

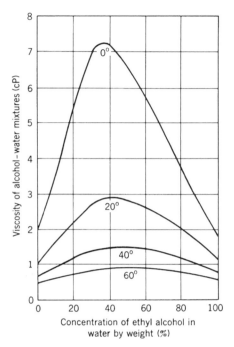

Figure 2.8 The effect of various concentrations of ethanol on the viscosity of water at four temperatures. Mixtures of water with polar organic liquids often show large increases in viscosity at low temperatures because they have a more tightly packed structure than the component liquids alone. From Kramer (1983), after Bingham and Jackson (1918).

molecules, but it probably can form such layers on the surfaces of cellulose and other substances having a more uniform distribution of bonding sites. Solutions of alcohols, amides, or other polar liquids in water result in a more strongly structured system than occurs in the separate substances. This is seen in the high viscosity of such solutions. For example, the viscosity of an ethyl alcohol–water mixture at 0°C is four times that of water or alcohol alone. However, this structure is easily broken by high temperatures (see Fig. 2.8). The addition of nonpolar substances such as benzene or other hydrocarbons to water breaks bonds and produces "holes" or disorganized areas in the structure which are surrounded by areas with a tighter structure. The water bound to large molecules such as proteins has an important effect on their structure; Tanford (1963, 1980) cites evidence suggesting that the relative stability of the structure of viruses, DNA, and globular proteins is affected by the water associated with them.

It was mentioned earlier that the changes in volume of water during freezing and thawing are caused by changes in the proportion of water molecules bound in an organized lattice by hydrogen bonds. The high boiling point results principally from the large amount of energy required to break hydrogen bonds, nearly two of which must be broken for each molecule evaporated. Methane (CH_4) has almost the same molecular weight as water, but it boils at $-161°C$ because no hydrogen bonding occurs and only a small amount of energy is required to break the weak van der Waals forces holding the molecules together in the liquid.

The unusually high viscosity and surface tension of water also result from the fact that hydrogen bonds between water molecules resist rearrangement. Water wets and adheres to glass, clay micelles, cellulose, and other substances having exposed oxygen atoms at the surface with which hydrogen bonds can be formed. It does not wet paraffin and other hydrocarbons because it cannot form hydrogen bonds with the C atoms in them, but it wets cotton because it forms numerous hydrogen bonds with the O atoms of the cellulose molecules.

According to Jenniskens and Blake (1994), the ice formed on the interstellar dust of molecular clouds and comets is formed from less structured or amorphous water. They observed three forms of amorphous water ice during warming from 15 to 188 K and stated that these changes in structure explain some anomalous properties of astrophysical ice.

Bound Water

In the period from 1920 to 1950 considerable attention was given to what was termed "bound water" as a factor in cold and drought tolerance. Gortner (1938) discussed the origin of the term and methods of measuring it and Kramer (1955) reviewed the literature. The concept of bound water originated from observations that a variable fraction of water in both living and nonliving systems has a lower vapor pressure, remains unfrozen at $-20°$ or $-25°C$, does not function as a solvent, and seems to be less available physiologically than bulk water. For example, the rate of respiration of seeds is low at a low water content, but suddenly increases rapidly above a certain water content, when the amount of free water increases, as shown in Fig. 2.2.

Attempts to explain cold or drought tolerance in terms of differences in bound water content proved disappointing (Levitt, 1980, pp. 179–182; Burke *et al.*, 1976) and bound water was neglected for several decades. However, improvements in the technology used to measure it has resulted in renewed interest in its possible importance. Burke *et al.* (1974) used nuclear magnetic resonance spectroscopy to observe increase in water binding in cold acclimating dogwood stems and NMR imaging is now being used. G. A. Johnson *et al.* (1987) were able to follow simultaneous changes in total and bound water in stems of transpiring geranium plants and Veres *et al.* (1991) followed dehydration and rehydration in fern stems. Leopold and his colleagues (Vertucci and Leopold, 1987a,b) studied bound water in seeds and other tissues using sorption isotherms. They found that tissues easily injured by dehydration had a lower water-binding capacity than more tolerant tissues and conclude that binding of water has a protective effect. Rascio *et al.* (1992) found that the more drought tolerant of two varieties of wheat also bound more water. Parsegian and his co-workers (Colombo *et al.*, 1992) found that some proteins such as hemoglobin show a change in the amount of water bound when they bind substrate, and propose that this change is an important part of catalysis. It seems probable that there will be more research by use of the new technology on the possible effects of bound water.

Isotopes of Water

The three isotopes of hydrogen having atomic weights of 1, 2, and 3 make it possible to differentiate tracer water from ordinary water. In the 1930s, heavy water [water containing deuterium (hydrogen of atomic weight 2)] became available and was used widely in biochemical studies. It was also used extensively in studies of permeability of animal and plant membranes (e.g., Ordin and Kramer, 1956; Ussing, 1953). However, deuterium was largely supplanted as a tracer by water containing tritium (hydrogen of atomic weight 3), e.g., in the experiments of Raney and Vaadia (1965a,b) and Couchat and Lasceve (1980). Tritium is radioactive and therefore more convenient as a label, being easier to detect than deuterium, which requires use of a mass spectrometer.

A stable isotope of oxygen with an atomic weight of 18 makes it possible to study the role of oxygen in water. An example is the series of experiments with $H_2^{18}O$ which demonstrates that the oxygen released during photosynthesis comes from water instead of from carbon dioxide (Ruben *et al.*, 1941).

The ratio of deuterium to hydrogen in water has been used in various ways ranging from reconstruction of past climates to study of the fractionation of water in leaves and cells. As $H_2^{16}O$ evaporates more rapidly than $H_2^{18}O$, leaf water has higher ratios of ^{18}O than root or stem water, although the enrichment is not strictly proportional to the rate of transpiration (Farris and Strain, 1978; Flanagan *et al.*, 1991). Yakir *et al.* (1990) and Yakir (1992) concluded from studying the effects of high and low rates of transpiration on the isotopic ratio of water in leaves that water must exist in three compartments: a fraction in the veins, a symplastic pool, and a pool from which evaporation occurs. Pedersen (1993) used tritiated water to demonstrate root pressure and guttation in submerged flowering plants. The use of stable isotopes of water for research on plants was reviewed by White (1989) and in a book edited by Ehleringer *et al.* (1993).

Unorthodox Views Concerning Water

It is generally believed that although most of the water in cells possesses the structure and properties of bulk water, a small amount is adsorbed on the surfaces of membranes and macromolecules. This is the ice of Bernal (1965) and the bound water of Gortner (1938) and others. However, according to Ling (1969), Cope (1967), and a few other physiologists, a significant amount of the water in living cells has a structure different from bulk water. This vicinal or associated water is said by them to affect the accumulation of ions and eliminate the classical role of cell membranes and their associated ion pumps (see Hazelwood and others in Drost-Hansen and Clegg, 1979). This concept has not been widely accepted (Kolata, 1979), and it seems especially doubtful that it could apply to plant cells with their large volume of vacuolar water.

It also was suggested by Drost-Hansen (1965) and others that there are anomalies in the physical properties of water at about 15°, 30°, and 45°C. For example, it is claimed that there are peaks in the disjoining pressure and viscosity of water adsorbed on surfaces, caused by phase transitions in vicinal water, i.e., water adsorbed on surfaces of macromolecules in cells (Etzler and Drost-Hansen in Drost-Hansen and Clegg, 1979). It was also claimed by these writers and by Nishiyama (1975) and Peschel (in Lange *et al.,* 1976) that there are peaks in seed germination, growth of microorganisms, and other biological processes which are related to these anomalies. These claims have been received skeptically (Eisenberg and Kauzmann, 1969), and Falk and Kell (1966) concluded that the reported discontinuities in physical properties are no greater than the errors of measurement. It seems likely that discontinuities in biological processes are related more to phase transitions in membranes than to phase transitions or discontinuities in the properties of water.

Another anomaly is the polywater reported by Russian investigators in the 1960s. This was believed to be a polymeric form of water with anomalous properties, but it later turned out to be water containing a high concentration of solute (Davis *et al.,* 1971). The story of polywater was told in detail by Franks (1981). It has also been claimed by Russian investigators that water from freshly melted snow stimulates certain biological processes. Other Russian investigators claimed that water boiled to remove all dissolved gas and then quickly cooled not only has greater density, viscosity, and surface tension but also stimulates plant and animal growth; concrete prepared with it is stronger than that prepared with ordinary water (Maugh, 1978). These claims have not been verified elsewhere and should be treated with caution.

PROPERTIES OF AQUEOUS SOLUTIONS

In plant physiology we seldom deal with pure water because the water in plants and in their root environment contains a wide range of solutes. Therefore, it is necessary to understand how the properties of water in solution differ from those of pure water. Only a brief discussion is possible here, and readers are referred to physical chemistry texts for full development of these ideas.

The characteristics of water in solution can be shown concisely by tabulation

Table 2.2   Colligative Properties of a Molal Solution of a Nonelectrolyte
Compared with Water

	Pure water	Molal solution
Vapor pressure	0.61 kPa at 0°C	Decreased according to Raoult's law
	101.3 kPa at 100°C	
Boiling point	100°C	100.518°C
Freezing point	0°C	−1.86°C
Osmotic pressure	0	2.27 MPa at 0°C
Chemical potential	Reference	2.27 MPa below reference at 0°C

of its colligative properties, i.e., the properties associated with the concentration of solutes dissolved in it. These are shown in Table 2.2 and include the effects on vapor pressure, boiling and freezing points, osmotic pressure, and water potential. They occur because the addition of solute dilutes or lowers the concentration of the water.

Pressure Units

Many of the colligative properties of water are described in terms of pressure, which is the force applied to a unit area of the enclosing surface, and the primary units are dynes per square centimeter or newtons per square meter. Pressure units also are equivalent to energy per unit volume such as ergs per cubic centimeter or joules per cubic meter because a unit of work or energy is a dyne-cm (erg) or newton-meter (joule). Vapor pressure often is expressed in millimeters (mm) of mercury or millibars (mbar) and atmospheric pressure in bars (1 bar = 0.987 atm). However, there is a strong tendency toward the use of SI (Système International) units (Incoll *et al.*, 1977), and for the most part these units are used in this book. The primary pressure unit is then the pascal (Pa = 1 newton per square meter), and 1 bar = 10^5 Pa, 100 kPa, or 0.1 MPa (megapascal). In general, megapascals will be used in place of bars and kPa in place of millibars. One millibar is equal to 0.1 kPa, and standard atmospheric pressure is 101.3 kPa (760 mmHg, or 1013 mbar).

Vapor Pressure

The vapor pressure of water in solution is decreased essentially because the water is diluted by the addition of solutes. This is shown by Raoult's law, which states that the vapor pressure of solvent vapor in equilibrium with a dilute solution is proportional to the mole fraction N_w of solvent in the solution of non-dissociating solute

$$e = e_o\, N_w = e_o\, \frac{n_w}{n_w + n_s}, \tag{2.1}$$

where e is the vapor pressure of the solution, e_o is the vapor pressure of pure solvent, n_w is the number of moles of solvent, n_s is the number of moles of solute, and $N_w = n_w/(n_w + n_s)$. This is strictly applicable only to dilute molal solutions, i.e., those prepared with a mole or less of solute per 1000 g of water, and the relation holds because the surface of the solution is occupied by fewer water molecules due to the presence of solute at the surface.

Boiling and Freezing Points

The effects of solutes on the boiling and freezing points are exerted through their effects on the vapor pressure of water (see Fig. 2.9). The addition of solute lowers its freezing point because it dilutes the water and lowers its vapor pressure, thereby decreasing the temperature at which the vapor, liquid, and solid phases are in equilibrium. It can be calculated that the vapor pressure at freezing of a molal solution of a nonelectrolyte in water is decreased from 0.610 to 0.599 kPa and that a reduction in temperature of 1.86° below zero is required to bring about freezing.

Water boils when its vapor pressure is raised to that of the atmosphere. When the vapor pressure has been lowered by the addition of solute, the water in a

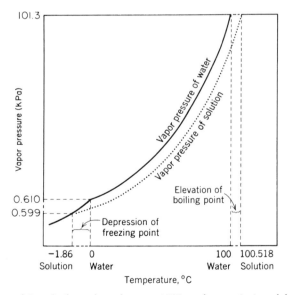

Figure 2.9   Effects of 1 mol of nonelectrolyte per 1000 g of water (a 1 molal solution) on the freezing and boiling points and vapor pressure of the solution. Note that this diagram is not drawn to scale. From Kramer (1983).

solution must be heated to a higher temperature than pure water to produce the required increase in vapor pressure. At high altitudes water boils below 100°C, sometimes causing practical problems in cooking food.

Osmotic Pressure or Osmotic Potential

Raoult's law shows that the vapor pressure of water in a solution is lowered in proportion to the extent to which the mole fraction of water in the solution is decreased by adding solute. Therefore, if water is separated from a solution by a membrane permeable to water but impermeable to the solute, water will move across the membrane along a gradient of decreasing water mole fraction into the solution. The pressure, which must be applied to the solution to prevent movement in a system such as that shown in Fig. 2.10, is termed the osmotic pressure. It is often denoted by the symbol π.

The osmotic pressure is present only when a balancing pressure is applied to the solution as in Fig. 2.10. The fundamental property causing the movement is the osmotic potential resulting from the dilution of water by the addition of solute (Gibbs, 1931) and it is always present for a solution. When in the apparatus in Fig. 2.10, the osmotic potential is opposed by the pressure, and the osmotic potential Ψ_s, is equal to the negative of the osmotic pressure at balance, that is, $\Psi_s = -\pi$. Outside of the apparatus, there is no osmotic pressure and the solution is affected only by Ψ_s. The basis for the use of the term potential is discussed in the following section (Chemical Potential of Water).

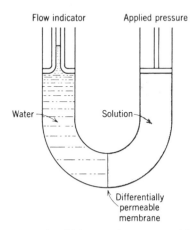

Figure 2.10   Diagram of osmometer in which a membrane permeable to water but impermeable to solute separates pure water from a solution. The osmotic pressure of the solution is equal to the pressure that must be applied to prevent movement of water into it. Water movement is observed by a change in the level of water in the capillary tube on the left. From Kramer (1983).

Van't Hoff developed an equation relating osmotic pressure to solute concentration in the solution (see Appendix 2.1). Mathematically expressed,

$$\pi V = n_s RT, \tag{2.2}$$

where π is the osmotic pressure in megapascals, V is the volume of solvent in cubic meters, n_s is the moles of solute, R is the gas constant (8.32×10^{-6} MPa $\cdot$ m^3 $\cdot$ mol^{-1} $\cdot$ K^{-1}), and T is the temperature in degrees Kelvin (K). For 1 mole of solute in 1 liter of solvent at 273 K (0°C), this equation gives a value for π of 22.7×10^5 Pa or 2.27 MPa (22.4 atm or 22.7 bars).

Direct measurements have shown that this relationship is approximately correct for dilute solutions of nondissociating substances. However, there are large deviations from the theoretical value for electrolytes that ionize in solution and release more particles than nondissociating substances. Thus, the osmotic pressure of a molal solution of NaCl is approximately 4.32 MPa (43.2 bars) instead of the theoretical 2.27 MPa (22.7 bars) of a nonelectrolyte. Assuming complete dissociation of the NaCl, the osmotic pressure should be 4.54 MPa and the discrepancy can probably be attributed chiefly to van der Waals forces operating between ions. Some nondissociating molecules become hydrated or bind water molecules. This binding of water reduces the effective concentration of water and increases the observed osmotic pressure. An example is sucrose solution in which each sucrose molecule apparently binds six molecules of water, and the osmotic pressure of a molal solution of sucrose is approximately 2.51 MPa (25.1 bars) instead of the expected 2.27 MPa (22.7 bars).

The relationships among concentration, vapor pressure, freezing point, and osmotic pressure make it possible to calculate the osmotic pressure of a solution from the freezing point depression. Since the theoretical depression of the freezing point of an ideal molal solution is 1.86°C and the osmotic pressure is 2.27 MPa at 0°C, the osmotic pressure of a solution can be calculated from the depression of the freezing point (T) by the following equation:

$$\pi = \frac{\Delta T}{1.86} \times 2.27 \text{ MPa}, \tag{2.3}$$

where ΔT is the observed depression. The derivation of this relationship can be found in Crafts *et al.* (1949) or in a physical chemistry text. The freezing point method is widely used, and if suitable corrections are made for supercooling, it gives accurate results. The problems encountered in using it on plant sap are discussed in Chapter 7 of Crafts *et al.* (1949) and by Barrs (in Kozlowski, 1968).

The osmotic pressure can also be calculated from the vapor pressure or, more readily, from the relative vapor pressure or relative humidity, $e/e_o \times 100$, according to Raoult's law [see later and Eq. (2.1) and Appendix 2.1]. In recent years, thermocouple psychrometers have been used extensively to measure e/e_o.

Their operation is discussed in Chapter 3 in connection with the measurement of the water status of plants.

Chemical Potential of Water

The properties of solutions are described by the thermodynamic concept of the potential first put forth by J. Willard Gibbs (1875–1876). The chemical potential μ of a substance is a measure of the capacity of 1 mol of that substance to do work (see Appendix 2.2). It is equal to the partial molal Gibbs free energy (Nobel, 1991, pp. 76–77; Slatyer, 1967; Spanner, 1964). The mole basis of the potential indicates that the free energy is expressed for Avogadro's number of molecules or, in other words, on a per molecule basis. The concept of energy or work that can be done by a molecule has many applications, and the energy can be affected by many factors, including the concentration of water molecules, that is, the mole fraction of the water. The degree to which the presence of solute reduces the chemical potential of the water in the solution below that of pure free water is shown by

$$(\mu_w - \mu_o) = RT\ln N_w, \tag{2.4}$$

where μ_w is the chemical potential of water in the solution, μ_o is the chemical potential of pure water at the same temperature, R and T have the usual meaning, and N_w is the mole fraction of water. For use with ionic solutions, the mole fraction is replaced by the activity of water, a_w, and for general use, where water may not be in a simple solution, by the relative vapor pressure, e/e_o [see Eq. (2.1)]. Equation (2.4) is then written

$$(\mu_w - \mu_o) = RT\ln\frac{e}{e_o}. \tag{2.5}$$

When the vapor pressure of the water in the system under consideration is the same as that of pure free water, $\ln(e/e_o)$ is zero, and the potential difference $(\mu_w - \mu_o)$ is also zero. When the vapor pressure of the system is less than that of pure water, $\ln(e/e_o)$ is a negative number; hence, the potential of the system is less than that of pure free water and the potential difference is expressed as a negative number.

The expression of chemical potential in energy units such as joules per mole is inconvenient in discussions of cell water relations. It is more convenient to use units of pressure. Dividing both sides of Eq. (2.5) by the partial molal volume of water $\overline{V}_w$ (m³/mol) gives energy in units of joules/m³ which is equivalent to pressure in newtons per square meter or megapascals in SI units:

$$\Psi_w = \frac{\mu_w - \mu_o}{\overline{V}_w} = \frac{RT}{\overline{V}_w}\ln\frac{e}{e_o}. \tag{2.6}$$

The resulting term is called the water potential Ψ_w. For the conversion to pressure units, $\overline{V}_w$ is a constant in dilute solutions ($18 \times 10^{-6}\,\text{m}^3/\text{mol}$) and Ψ_w is proportional to the chemical potential (see Appendix 2.2).

The water potential in any system is decreased by those factors that reduce the relative vapor pressure, including:

1. Addition of solutes which dilute the water and decrease its activity by hydration of the solute molecules or ions.
2. Addition of porous solids which displace water and have surface forces and microcapillary forces. These forces together are called matric force and are found in soils, cell walls, protoplasm, and other substances that adsorb or bind water (for a fuller description, see Appendix 2.3).
3. Application of negative pressure or tensions such as those in the xylem of transpiring plants.

The water potential in any system is increased by those factors that increase the relative vapor pressure, including:

1. Dilution or removal of solutes.
2. Hydration of matrices.
3. Application of pressure above atmospheric, such as that exerted by the elastic cell wall on the cell contents (turgor).

Temperature affects the water potential according to the term T in Eq. (2.6), although there also is a slight effect of T on $\overline{V}_w$ because of changes in water density. Decreasing T makes Ψ_w less negative whereas increasing T causes the opposite effect. Because T has units of K, the effect is not large over the biological range of temperatures.

The water status of plants usually is described in terms of water potential because it is physically defined and allows experiments to be easily repeated. Also, water moves through the soil–plant system because of the water potential or forces included in the water potential. Sinclair and Ludlow (1985) argued that relative water content should be used instead because it is better correlated with physiological processes. As pointed out in later chapters, decreases in cell water content can concentrate cell constituents and probably cause changes in enzyme action. However, the water content often fails as an indicator of plant water status, as seen for example in Fig. 10.2 which shows the difference in photosynthesis between *Fucus*, a marine plant, and sunflower, a land plant, as relative water content or water potential varies. Because water content is based on a biological and thus variable reference, *Fucus* and sunflower appear to have maximum photosynthesis at a similar water status (Fig. 10.2A). In reality, the seawater that fully hydrates *Fucus* is markedly inhibitory to sunflower (Fig. 10.2B). The relative water content does not adequately detect this difference and, while water content is important, the water potential is preferred.

Throughout the discussion, we have explained the colligative properties of solutions using the example of the lowering of the concentration of the solvent, water, by the addition of solute. However, Hammel (1976) and Hammel and Scholander (1976) argued that the addition of a solute lowers the chemical potential of the solvent by creating a negative pressure or tension on the solvent molecules. Andrews (1976) discussed their arguments in detail and concluded that there is no mechanism by which solvent and solute molecules can sustain different pressures. The writers agree that the classical solvent dilution theory adequately explains the behavior of solutions.

SUMMARY

Water plays essential roles in plants as a constituent, a solvent, a reactant in various chemical processes, and in the maintenance of turgidity. The physiological importance of water is reflected in its ecological importance; plant distribution over the earth's surface is controlled by the availability of water wherever temperature permits growth. Its importance is a result of its numerous unique properties, most of which arise from the fact that water molecules are organized into a definite structure held together by hydrogen bonds. Furthermore, the water bound to proteins, cell walls, and other hydrophilic surfaces has important effects on their physiological activity.

Water in plants and soils contains solutes that modify its colligative properties by diluting it. As a result of this dilution, the chemical potential, vapor pressure, osmotic potential, and freezing point are lowered in proportion to the concentration of solute present. The best measure of the energy status of water in plants and soil is the water potential (Ψ_w), which is the amount by which its chemical potential is reduced below that of pure water. Some writers argue that relative water content is a better indication of plant water status than water potential with respect to physiological processes, but the water potential has the advantage of being physically defined and the force that causes water movement. Thus, the water potential seems preferable.

SUPPLEMENTARY READING

Alberty, R. A., and Daniels, F. (1979). "Physical Chemistry," 5th Ed. Wiley, New York.

Andrews, F. C. (1976). Colligative properties of simple solutions. *Science* **194**, 567–571.

Edsall, J. T., and McKenzie, H. A. (1978). Water and proteins. I. The significance and structure of water: Its interaction with electrolytes and non-electrolytes. *Adv. Biophys.* **10**, 137–207.

Eisenberg, D., and Kauzmann, W. (1969). "The Structure and Properties of Water." Oxford University Press, London/New York.

Franks, F., ed. (1982). "Water: A Comprehensive Treatise," Vol. 7. Plenum Press, New York.

Gibbs, J. W. (1931). "The Collected Works of J. Willard Gibbs," Vol. 1. Longmans, Green and Co., New York.

Hammel, H. T. (1976). Colligative properties of a solution. *Science* **192**, 748–756.

Somero, G. N., Osmond, C. B., and Bolis, C. T., eds. (1992). "Water and Life." Springer-Verlag, Berlin.

Stillinger, F. H. (1980). Water revisited. *Science* 209, 451–457.

Tanford, C. (1980). "The Hydrophobic Effect." Wiley, New York.

Tinoco, I., Sauer, K., and Wang, J. C. (1978). "Physical Chemistry: Principles and Applications: Biological Sciences." Prentice-Hall, Englewood, NJ.

Appendix 2.1: The van't Hoff Relation

The van't Hoff equation is a simplified version of the equation relating the solute concentration to the chemical potential according to the mole fraction of the water

$$(\mu_w - \mu_o) = RT \ln \frac{n_w}{n_s + n_w}, \qquad (2.7)$$

where R is the gas constant (8.32×10^{-6} MPa $\cdot$ m$^3 \cdot$ mol$^{-1} \cdot$ K^{-1}), T is the Kelvin temperature (K), n_s is the number of moles of solute, n_w is the number of moles of water, and the mole fraction of water is $n_w/(n_s + n_w)$.

For dilute solutions, the mole fraction of water can be approximated by

$$\frac{n_w}{n_s + n_w} = 1 - \frac{n_s}{n_s + n_w} \approx 1 - \frac{n_s}{n_w}, \qquad (2.8)$$

and the natural logarithm can be further approximated with the Taylor expansion for a dilute solution

$$\ln \left[1 - \frac{n_s}{n_w} \right] \approx - \frac{n_s}{n_w} \qquad (2.9)$$

so that

$$(\mu_w - \mu_o) \approx -RT \frac{n_s}{n_w}. \qquad (2.10)$$

Dividing both sides of Eq. 2.10 by the partial molal volume of water, $\bar{V}_w = V/n_w$, gives the osmotic potential

$$\Psi_s = \frac{\mu_w - \mu_o}{\bar{V}_w} \approx -RTn_s/V, \qquad (2.11)$$

and since the concentration of solute $C = n_s/V$, Eq. 2.11 becomes

$$\Psi_s \approx -RTC, \qquad (2.12)$$

and because $\Psi_s = -\pi$,

$$\pi \approx RTC, \qquad (2.13)$$

which was first derived by van't Hoff. This equation only works for dilute solutions of nondissociating solutes but it shows the practical result that Ψ_s is proportional to the concentration of solute under these conditions. It also shows that at lower temperatures the osmotic potential becomes less negative and at absolute zero the osmotic potential becomes zero.

Appendix 2.2: The Chemical Potential

When we consider molecules of any kind, we know that all of them contain energy in their atoms and chemical bonds that can be exchanged with the surroundings by their motions, chemical reactions, and radiational exchanges (here we assume that the isotopic composition remains stable). The energy exchanges always result in a rearrangement of chemical or atomic structure that in itself requires energy. Thus, a fraction of the energy goes to the rearrangements and a fraction goes to the surroundings; the latter fraction can be made to do work. The rearrangement energy is termed the entropy and the energy available for work is the free energy.

It readily can be seen that the amount of work is determined by the number of molecules exchanging energy. Doubling the number of molecules doubles the work, all other factors remaining constant. Often, however, it is more desirable to know the work per molecule or per mole of molecules than the total work. Gibbs (1931) recognized this and defined the term "potential" and symbol μ as the way to describe the work that a mole of molecules can do.

The work is not known in absolute terms because the total amount of energy in molecules is not known. Therefore, the work is determined by comparing the chemical potential of the system with a reference potential. For liquid water, the reference has been chosen to be pure unrestrained water at atmospheric pressure, a defined gravitational position, and the same temperature as the system being compared. If we define the chemical potential of the system to be measured as μ_w and the chemical potential of the reference as μ_o, $(\mu_w - \mu_o)$ is the comparison we wish to make. When the system is not pure water, the μ_w is lower than μ_o, and $(\mu_w - \mu_o)$ is negative. When the system is pure water, $(\mu_w - \mu_o)$ is zero.

The $(\mu_w - \mu_o)$ is the energy state of the molecules. It does not matter how the molecules get to that state, the energy is the same whenever $(\mu_w - \mu_o)$ is at the same level. The energy represents the maximum work that can be done if the molecules are part of an ideal machine. Pure water moving through a selective membrane into a solution on the other side is a machine allowing work to be done. If the membrane allows water to pass but not solute (the membrane reflects solute), more water will move to the solution side than to the other side because the free energy of the pure water is higher than in the solution. The work is determined by the potential difference on the two sides of the membrane and the net volume of water moved. The work can be measured by opposing this movement with a potential that counters the movement, such as with pressure.

If the membrane is not reflective for solute, the volume of water moving into the solution is essentially the same as the volume of water and solute moving in the opposite direction. No work is done because there is no net volume change. Nevertheless, at the beginning, the $(\mu_w - \mu_o)$ is the same as when the reflective membrane was present. Thus, the ability to do work is identical but the work actually done depends on the characteristics of the machine.

This example illustrates that the $(\mu_w - \mu_o)$ is an intrinsic property of the molecules. The membrane simply determines the work extracted from the molecules. The reflectiveness of the membrane is the determining factor and is usually described by the reflection coefficient, which is 1 for a perfectly reflective membrane but 0 for a nonreflective one. The osmotic effectiveness of a solution is determined by the membrane reflectiveness for the solute even though large concentration differences exist on the two sides of the membrane. In other words, the $(\mu_w - \mu_o)$ will bring about osmosis only if the membrane is reflective for the solute *even though $(\mu_w - \mu_o)$ can be measured in a vapor pressure osmometer or similar apparatus.*

Because the chemical potential is normally given in energy units per mole but can be converted to pressure units according to

$$\Psi_w = \frac{(\mu_w - \mu_o)}{\bar{V}_w}, \tag{2.14}$$

the units of the water potential Ψ_w are energy per volume = force per unit area = pressure. The pressure is usually expressed in megapascals where 1 megapascal = 10^6 pascals = 10^6 newtons $\cdot$ m^{-2} = 10^6 joule $\cdot$ m^{-3} = 10^7 ergs $\cdot$ cm^{-3} = 10^7 dynes $\cdot$ cm^{-2} = 10 bars = 9.87 atmospheres or 145 pounds per square inch.

This works well when $\bar{V}_w$ is a constant. The $\bar{V}_w$ is the volume of a mole of liquid water mixed with other molecules in the system and is nearly a constant 18 cm$^3 \cdot$ mol^{-1} over most of the temperatures and water contents of cells and soils. Therefore, the Ψ_w is simply $(\mu_w - \mu_o)$ divided by a constant. In concentrated solutions, dry soils, dry seeds, and other systems of low water content, this simplification may not hold because interactions between water and the other molecules can be so extensive that 1 mol of water no longer occupies 18 cm^3. In this case, the proportionality breaks down, and $(\mu_w - \mu_o)$ should be used without converting to pressure units. Pressure units become meaningless anyway when systems are dry enough not to have a continuous liquid phase, usually whenever Ψ_w is below about -10 MPa.

Appendix 2.3: Matric Potentials

Matric potentials occur because the surface of a liquid has properties that differ from those in the interior. At the surface, the molecules interact with other liquid molecules only on the liquid side of the surface. In bulk liquid, they interact in all directions. At the surface, the molecules tend to be pulled into the bulk by their attraction to other molecules, and this forms a surface tension. In the bulk liquid, there is no surface tension. A porous medium extends the surface as water enters the pores, and thus increases the amount of water having surface properties. Often the surface of the medium is wettable mostly because of hydrogen bonding between water and O and OH groups on the surfaces. Also, surface charges attract the water dipole and ions in the water. The total effect is to constrain water and solute next to the surfaces. The wettability and solute constraint attracts water from the air or any aqueous liquid and, because of bonding between water molecules, liquid water tends to be dragged into the pores, filling the matrix. The meniscus in the pore exhibits surface tension that keeps the pore filled. Electrically constrained ions next to the pore surfaces also move water into the matrix with osmotic-like force. As a consequence, the water content of the matrix can become very large, as seen in many gels. Pressures are generated next to the surfaces and the whole matrix can swell.

For most plant cells, the walls are the major site of the matric potential (Boyer, 1967b). The surfaces are highly wettable, and water fills the pores. Because of their small diameter, tensions (negative pressures) of as much as -75 MPa can be present without draining them. Note that the tension on the pore water can vary between zero and the tensions that just cause the pores to drain. Thus, as plants generate tensions, the wall pores remain water filled as long as the tensions do not exceed about -75 MPa.

When the matrix dehydrates enough to drain the pores, the remaining water lines the surfaces and is strongly constrained. It is capable of attracting liquid water with great force. As a consequence, dry seeds can absorb water against large external forces, even though the cell membranes are nonfunctional.

Cell Water
Relations

INTRODUCTION

Cells are the basic structural units of organisms, and plant organization varies from single cells to aggregations of cells to complex multicellular structures. With increasing complexity there are increasingly sophisticated systems for absorbing water, moving it large distances, and conserving it but fundamentally the cell remains the central unit that controls the plant response to water. The driving forces for water movement are generated in the cells, and growth and metabolism occur in the aqueous medium provided by the cells. The cell properties can change and result in acclimation to the water environment. As a consequence, many features of complex multicellular plants can be understood only from a knowledge of the cell properties. This chapter is concerned with those properties and how they are measured. Later chapters will consider the whole organism more fully and will use the principles described here for the cells.

STRUCTURE

The plant cell consists of a multicompartmented cytoplasm bounded on the outside by a membrane and cell wall (Fig. 3.1A). There usually is a dilute solution on the outside, but in some instances there may be a moderately concentrated solution as in seawater or around embryonic cells. On the inside, there always is a concentrated solution that contains metabolites, inorganic salts, and macromolecules in varying concentrations, depending on the location.

The membrane bounding the outside of the cytoplasm is the plasmalemma which is highly permeable to water but only slowly and selectively permeable to solutes. The cell wall outside the plasmalemma is porous and permits water and solutes of low molecular weight to move rapidly to and from the plasmalemma. Inside the cytoplasm, there are compartments or organelles such as the vacuole, mitochondria, nucleus, and plastids, each bounded by a membrane similar to the plasmalemma and capable of exchanging water and solute with the surrounding cytosol. Each organelle contains its own unique composition of solute. The plasmalemma is thus the primary barrier controlling the molecular traffic into and out of the cell, but the cell wall and internal membranes also play a role.

The high concentration of solute inside the cell dilutes the internal water compared to that outside and water enters in response, causing the cell to swell. The plasmalemma has insufficient strength to resist the swelling but it is supported by the structurally tough and often rigid wall, which resists enlargement. As a result, the swelling causes the wall to stretch and become turgid, and turgor pressure develops inside the cell. Without the wall, the plasmalemma would rupture but with the support of the wall, the plasmalemma is pressed tightly against the wall microstructure (Figs. 3.2A and 3.2B). The wall sometimes can stretch by a considerable amount, and the membrane inside must be capable of stretching as well. Stretching of the wall will be treated in detail later but it is worth pointing out that the resistance to stretch gives structural rigidity that contributes to the form and strength of tissues. Much of the form of leaves and stems of herbaceous plants results from the turgor pressure developing in their cells.

Cells lose water when solute concentrations are high outside or when evaporation occurs, and they shrink as the volume of water decreases inside (Fig. 3.1B). The membranes cannot resist the shrinkage, and the organelles become distorted when dehydration is severe (Fig. 3.1B). The cell wall often develops folds as the cell shrinks (Fig. 3.1B), and the folding deforms the adjacent plasmalemma. In some cells, the walls are stiffened by the deposition of layers of rigid wall material and folding does not occur. In such cells, the wall resists shrinkage and the cell contents may come under tension (Boyer, 1995).

The primary walls develop in young cells and are composite porous structures consisting of cellulose microfibrils embedded in a matrix of related oligosaccharides and some structural proteins (Fig. 3.2). The microfibrils contain clusters of crystalline cellulose totaling about 10 nm in diameter. They provide much of the tensile strength of the wall. The matrix binds the microfibrils and holds them in an organized fashion. The orientation of the microfibrils may control cell growth by restricting enlargement to particular directions. The matrix probably affects the rate at which enlargement occurs (see Chapter 11). As

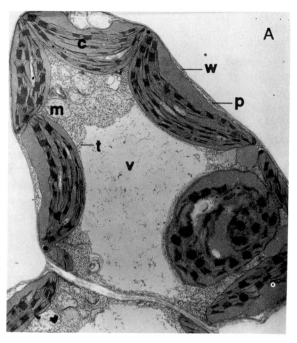

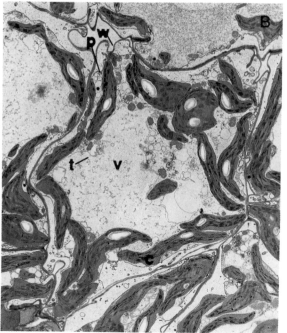

the cell ages, growth stops and additional layers of wall are often deposited on the inside of the primary wall. These secondary wall layers may contain lignins, suberins, and other compounds that give the wall special characteristics of rigidity, imperviousness to water, and so on. The secondary walls account for most of the properties of different woods, tree bark, nutshells, and other specialized plant parts.

There are two kinds of pores in the wall. A few large pores, the plasmodesmata, are filled with protoplasm and lined with the plasmalemma (Fig. 3.2A). These pores connect the protoplasts of adjacent cells and probably transmit water and solutes directly between the protoplasts. The second kind of pore is much smaller and more numerous (Fig. 3.2B) and is not filled with protoplasm but instead is filled with the solution contacting the cell exterior. This type of pore is distributed throughout the wall and has a diameter variously estimated to be 4.0 to 6.5 nm (Baron-Epel et $al.$, 1988; Carpita, 1982; Carpita et $al.$, 1979; Tepfer and Taylor, 1981). It freely transmits water with its diameter of only about 0.4 nm, sugars and amino acids with their diameters of 1 to 1.5 nm, and smaller proteins, but the passage of molecules with weights larger than about 60,000 D (diameters larger than 8.5 nm) is generally blocked. The plasmalemma crosses the ends of these pores and is unsupported there, but it is 4.5 to 25 nm thick and thus can support itself over the small diameter of these pores (Fig. 3.2B).

Because small solutes can pass readily through the small wall pores, the solutes can move to the surface of the plasmalemma where they are selectively transported into the cell. The uptake often requires metabolic energy and, once inside, additional metabolic activity may modify them or they may be further transported into the vacuole or other organelles of the cell. Macromolecules generally do not account for much of the internal solute because they are present in comparatively small concentrations. For example, proteins typically exist in micromolar concentrations whereas the small metabolites and inorganic ions have concentrations totaling 0.5 to 1 molal in the cytoplasm. Nevertheless, many of the macromolecules are enzymes or nucleic acids that regulate the metabolites and the properties of the membranes, as well as the nature of the cell

Figure 3.1 Structure of a typical plant cell. (A) Mesophyll cell of a sunflower leaf having a high water potential (-0.44 MPa) and relative water content (99%). Cell wall (w), chloroplast (c), plasmalemma (p), mitochondrion (m), vacuole (v), and vacuolar membrane (tonoplast, t). Magnification, 6300×. (B) Same as in (A) but having a low water potential (-2.11 MPa) and relative water content (35%). Note shrunken vacuole, folded cell wall, and contorted chloroplasts in this cell. In some cells, there was evidence of plasmalemma and/or tonoplast breakage, and loss of cell contents. Magnification, 3800×. In order to preserve cell structure in these micrographs, the osmotic potential of the fixative was adjusted to equal the water potential of the cells (see Appendix 3.1). Adapted from Fellows and Boyer (1978).

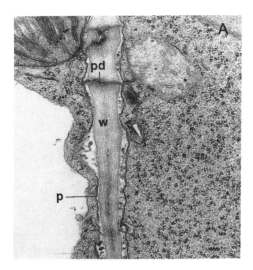

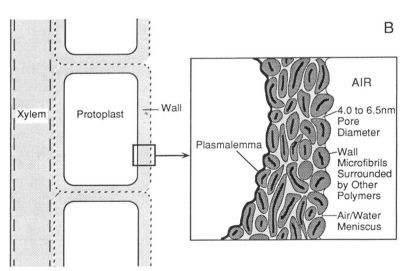

Figure 3.2   (A) Enlarged view of the cell wall (w) and plasmalemma (p) of a mesophyll cell in a sunflower leaf (37,400×). Note the plasmodesma (pd) extending through the wall to form a symplast between the adjacent protoplasts (R. J. Fellows and J. S. Boyer, unpublished). The microfibrillar structure of the wall is also apparent. (B) Diagrammatic representation of the apoplast (shaded). The microfibrillar structure of the wall is shown in the enlarged inset together with the air/water menisci between the microfibrils and matrix polymers. Not shown are cross links between the polymers. The plasmalemma is pressed against the wall substructure by the pressure inside and tension outside. The tension in the wall passes into the xylem through the 4 to 6.5 nm pores distributed throughout the wall.

walls. As a consequence, the water relations of the cell are set in motion by the macromolecules but water is affected most immediately by the small solutes and membranes.

OSMOSIS

Osmosis is the net flow of water across a differentially permeable membrane separating two solutions of differing solute concentration (also see Chapter 2). This situation occurs commonly in plant cells because of the differences in solute concentrations across the plasmalemma. The solute difference inevitably causes a corresponding but opposite difference in water concentration. Since water can cross the membrane but the solute cannot, more water molecules move toward the side with the lower water concentration than in the opposite direction. Without a compensating flow of solute, this net flow causes water to be transferred toward the side with lower water concentration and enlarges the volume on that side.

The solute concentration inside plant cells is typically 0.5 to 1 molal greater than outside, causing a strong tendency for water to enter. The resulting increase in volume of the inner solution is opposed by the resistance of the wall to stretching. Turgor pressure develops inside and can increase until it completely opposes the osmotic force causing water to enter (see Chapter 2). For a concentration of 1 molal inside the cell and 0 molal outside, the pressure calculated from the van't Hoff relation is 2.27 MPa at 273 K and 2.47 MPa at 298 K (see Chapter 2). Thus, the pressure inside equals the osmotic pressure and in this instance is about 10 times the pressure in an automobile tire!

This example is essentially that of an ideal osmometer when pure water is on one side of the membrane and a solution on the other (see Fig. 2.11). Note that the pressure is the same as is developed by 1 mol of an ideal gas (2.27 MPa at 273 K and 2.47 MPa at room temperature of 298 K). Thus, the osmotic pressure is numerically equal to the pressure calculated for an ideal gas but the mechanism is entirely different. Mainly the analogy with the gas gives us a convenient way to remember how the osmotic pressure is related to solute concentration.

Although the pressure can be large inside cells, in most circumstances it does not achieve the theoretical osmotic pressure of the cell solution for several reasons. First, the water outside normally is not pure but contains solute that reduces the internal pressure needed for balance. These concentrations in multicellular plants are in the range of 10 to 20 millimolal with few exceptions (Boyer, 1967a; Jachetta *et al.*, 1986; Klepper and Kaufmann, 1966; Nonami and Boyer, 1987; Scholander *et al.*, 1964, 1965, 1966). Second, tensions often are present in the solution outside because of the porous structure of the wall (see Appendix 2.3). These can be considerable and further diminish the pressures for balance inside. Finally, in growing cells the wall enlarges and it appears that this can prevent the internal pressure from developing fully (see Chapter 11).

Together these effects cause the turgor pressure to vary in cells, sometimes rapidly and in large amounts, although the osmotic potential of the cell solution is relatively stable. Some confusion exists on this point because some authors use the term osmotic pressure to mean the osmotic potential of the solution (Nobel, 1974, 1983, 1991; Slatyer, 1967; Steudle, 1989). The osmotic potential is a solution property regardless of whether membranes or pressures are present but osmotic pressure depends on the presence of differentially permeable membranes and is a pressure. It is readily apparent that the osmotic pressure is more closely related to the turgor pressure than to the osmotic potential and it seems most appropriate to use the term potential to refer to the osmotic property of solutions, as Gibbs (1931) originally did (see also Chapter 2).

Typically, osmotic potentials are uniform throughout the cell. The internal compartments are bounded by membranes of negligible strength, and an increase in solute concentration in them is immediately followed by water entry from the surrounding cytosol. The compartment swells until it re-equilibrates with the cytosol. An example is the large central vacuole. In young cells, this organelle has negligible volume and most of the cell compartment is filled with cytosol containing other organelles. As the cell grows, the vacuole enlarges as it accumulates salts and some metabolites that act as reserves. A few enzymes and secondary products of metabolism also may be found in it. In response to the accumulating solute, water enters and keeps the vacuole in osmotic balance with the peripheral cytoplasm. The vacuole eventually becomes so large that it is the dominant organelle in the protoplast (Fig. 3.1A).

Osmotic balance among cells probably is enhanced by the plasmodesmata, and cells in tissues tend to behave osmotically as though there is one highly interconnected protoplasm (Fig. 3.2A). The plasmodesmatal pores are large enough in diameter to allow small solutes and even some macromolecules to pass with water so that concentration differences generally remain moderate between the cells. The plasmalemma lining the pore is continuous with the plasmalemma of the adjacent cells. Thus, it is possible for most cells in a uniform tissue to be surrounded by one continuous plasmalemma and to act as a unit, the symplast. The cell wall surrounding the symplast is termed the apoplast (Fig. 3.2B). The xylem also is part of the apoplast. For reviews of plasmodesmata and symplast function, see Lucas *et al.* (1993), Olesen and Robards (1990), and Robards and Lucas (1990).

Osmotic balance becomes more difficult when plants are subjected to dehydration or high salinity. Because water is lost but not solute, the concentration of many cell constituents increases. Cell structures are distorted (Fig. 3.1B) and the plasmalemma and vacuolar membrane (tonoplast) can break or become leaky. Fellows and Boyer (1978) observed breakage of these membranes and

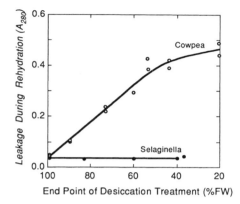

Figure 3.3 Leakage of proteins from leaf cells that had been desiccated to varying degrees and rehydrated. Proteins were detected in the rehydrating solution by measuring the absorbancy of the solution at 280 μm (A_{280}) after 20 min. Desiccation-sensitive cowpea showed a large leakage but desiccation-tolerant *Selaginella* did not. Adapted from Leopold *et al.* (1981).

loss of the cell contents. To make these measurements, special precautions were essential to preserve cell structure in the electron microscope; interested readers can find them detailed in Appendix 3.1. Leopold *et al.* (1981) showed that a species such as cowpea, which is unable to tolerate desiccation, loses cell contents to the external medium (Fig. 3.3) but a desiccation-tolerant species did not show this loss (Fig. 3.3, *Selaginella*). This suggests that desiccation tolerance may be determined at least in part by membrane properties that decrease leakage or disruption. Crowe *et al.* (1984, 1986, 1987, 1988), Crowe and Crowe (1986), Caffrey *et al.* (1988), Koster and Leopold (1988), and Madin and Crowe (1975) propose that membranes are protected by high concentrations of certain sugars such as sucrose and trehalose whose hydrogen bonding with the membrane is sterically similar to that of water. Accordingly, the bonding holds membrane constituents in an ordered fashion resembling that in water, protecting the membrane. Williams and Leopold (1989) found that certain sugars enter a glassy, candy-like state at low water contents and suggest that this could further protect the molecular structure of desiccated membranes.

WATER STATUS

It is apparent that osmosis is the central process that moves water into and through plants and that the plasmalemma is the key to the process. Indeed, if the plasmalemma is disrupted by external factors (e.g., freezing and thawing or chemical agents), water transport is abruptly diminished and the plant rapidly desiccates to the air-dry state despite the presence of concentrated solutions in the cells. Osmosis brings about water absorption that normally maintains cell water content but the osmotic conditions vary in and around cells and it is desirable to have some way to measure their water status. As pointed out in Chapter 2, the water status is most usefully characterized in terms of the chemical potential as defined by J. Willard Gibbs (1875–1876, 1931) who applied it to membrane systems and porous media. His concepts provided much of the foundation for physical chemistry and solution thermodynamics and thus to cells. Slatyer and Taylor (1960) proposed practical expressions for the chemical potential of water in plants and soils, which gave considerable impetus to adoption of the Gibbs concepts.

The main advantage is that the water status is based on a physically defined reference rather than a biological one. This avoids some of the variation inherent in biological systems and allows the water status to be reproduced at any time or place, a great advantage for experimentation. In addition, described in this way the water status indicates the force that moves water from place to place. This permits water movement to be predicted and resistances to movement to be measured. When expressed in pressure units, the potential is termed the water potential (see Chapter 2).

The water potential is determined by several components important for cells and their surroundings. The components originate from the effects of solute, pressure, solids (especially porous solids), and gravity on the cell water potential. We will follow the practice of Gibbs (1931) and consider solutes to be all dissolved molecules whether they are aggregated or not as long as they do not precipitate, pressures to be from external forces, porous solids to cause surface effects that differ from those in the bulk medium, and gravity to be important in vertical water columns. Accordingly, the components are expressed as

$$\Psi_w = \Psi_s + \Psi_p + \Psi_m + \Psi_g, \tag{3.1}$$

where the subscripts s, p, m, and g represent the effects of solute, pressure, porous matrices, and gravity, respectively. Each potential refers to the same point in the solution, and each component is additive algebraically according to whether it increases (positive) or decreases (negative) the Ψ_w at that point compared to the reference potential. The reference potential is pure, free water at atmospheric pressure and a defined gravitational position, at the same temperature as the system of interest.

The components affect Ψ_w in specific ways. Solute lowers the chemical potential of water by diluting the water and decreasing the number of water molecules able to move compared to the reference, pure water. In a similar way, matrices that are wettable have surface attractions that decrease the number of water molecules able to move (see Appendix 2.3). External pressure above atmospheric increases the ability of water to move but below atmospheric decreases ⌒ravity similarly increases or decreases the ability of water to move depending on whether local pressure is increased or decreased by the weight of water. Pressures are high at the bottom of the ocean and tensions can develop in siphons for this reason.

In dealing with cells, gravitational potentials often can be ignored because they become significant only at heights greater than 1 m in vertical water columns, as in trees. In this case, Eq. (3.1) becomes

$$\Psi_w = \Psi_s + \Psi_p + \Psi_m. \tag{3.2}$$

The presence of the interior and exterior of the plasmalemma in single cells and the symplast and apoplast in tissues means that Eq. (3.2) cannot be applied to cells without some consideration of structure. At its simplest, the cell consists of two compartments: the protoplast or symplast inside and the external solution or apoplast outside (Fig. 3.2). Equation (3.2) is then applied to each compartment. The protoplast contains a solution under pressure (turgor) applied by the walls. The protoplast water potential is then

$$\Psi_{w(p)} = \Psi_{s(p)} + \Psi_{p(p)}, \tag{3.3}$$

where the subscript (p) denotes the protoplast compartment. We can ignore Ψ_m because the water content generally is high and there are no air–water interfaces (Fig. 3.2).

The apoplast contains a solution in the porous cell wall subjected to local pressures generated by surface effects of the wall matrix (Fig. 3.2 and also see Appendix 2.3). The apoplast water potential is

$$\Psi_{w(a)} = \Psi_{s(a)} + \Psi_{m(a)}, \tag{3.4}$$

where the subscript (a) denotes the apoplast compartment. We can ignore Ψ_p because the external pressure is atmospheric. Figures 3.4A and 3.4B are diagrams of the potentials showing that there is a concentrated solution ($\Psi_{s(p)}$) and a turgor ($\Psi_{p(p)}$) in the protoplast but a dilute solution ($\Psi_{s(a)}$) and a matric potential ($\Psi_{m(a)}$) in the apoplast (Boyer, 1967b). The water potential is the algebraic sum of the component potentials with due regard for positive or negative quantities indicating whether the component increases or decreases the potential. In the example of Fig. 3.4A, the cell having a $\Psi_{s(p)}$ of -0.9 MPa and a $\Psi_{p(p)}$ of 0.3 MPa would have a $\Psi_{w(p)}$ of -0.6 MPa ($= (-0.9) + (+0.3)$).

In a unicellular marine alga, water surrounds the cell and saturates the porous cell wall. In this situation, the matric component can be ignored and the water potential in the apoplast is simply

$$\Psi_{w(a)} = \Psi_{s(a)}. \tag{3.5}$$

Water moves readily into and out of cells (see later) according to the water potential differences between the protoplast and apoplast compartments. The water potentials need not differ much across membranes to create large flows

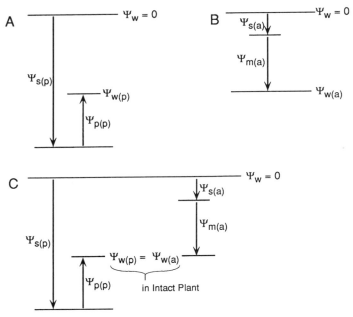

Figure 3.4   Water potentials in plant cells with component potentials shown by arrows (decreasing potentials are downward pointing, increasing potentials are upward pointing). The water potential of zero is shown by upper horizontal bar. (A) Protoplast (symplast) water potential consisting of the osmotic potential ($\Psi_{s(p)}$) and the turgor pressure ($\Psi_{p(p)}$), (B) cell wall (apoplast) water potential consisting of the osmotic potential ($\Psi_{s(a)}$) and matric potential ($\Psi_{m(a)}$), (C) equilibrium between the protoplast and apoplast water potentials. Note that the difference in osmotic potential is large across the plasmalemma ($\Psi_{s(p)} - \Psi_{s(a)}$). Also, the matric potential consists mostly of tension (negative pressure) in the pores of the apoplast. Therefore, the pressure difference across the plasmalemma also is large ($\Psi_{p(p)} - \Psi_{m(a)}$). At equilibrium, the difference in osmotic potential ($\Psi_{s(p)} - \Psi_{s(a)}$) equals the difference in pressure ($\Psi_{p(p)} - \Psi_{m(a)}$).

(see later). For flows commonly present, water potential differences across membranes are so small that a near equilibrium (local equilibrium) exists between the protoplast and its cell wall (Molz and Ferrier, 1982; Molz and Ikenberry, 1974). As a consequence, it is assumed that

$$\Psi_{w(a)} = \Psi_{w(p)} \qquad (3.6)$$

in many situations, and substituting Eqs. (3.3) and (3.4) in Eq. (3.6) gives

$$\Psi_{s(a)} + \Psi_{m(a)} = \Psi_{s(p)} + \Psi_{p(p)}, \qquad (3.7)$$

which shows that the components of the water potential in the protoplasts are balanced by the components in the apoplast at equilibrium. This result, shown in Fig. 3.4C, indicates that there is a large difference in the solute concentration across the membrane with the inside being much more concentrated. Also, the turgor in the cells is positive ($\Psi_{p(p)}$) but the water in the apoplast is under tension ($\Psi_{m(a)}$) in a multicellular plant. This causes a large pressure difference across the plasmalemma. Were it not for the restraining effect of the wall, the plasmalemma would burst.

Measuring Water Status

These water potentials can be measured with a thermocouple psychrometer that detects the vapor pressure of water because there is a relationship between vapor pressure and potential (Chapter 2). A sample of cells of unknown potential is sealed into a chamber containing a droplet of solution of known vapor pressure (Fig. 3.5A). The apparatus is surrounded by a heat sink and insulation in order to keep temperatures uniform. If evaporation occurs from the water in the solution, it is detected as a cooling of the solution by using a thermocouple. The solution can be replaced by another until one is found whose vapor pressure is the same as that of the water in the cells. In this case, the droplet is neither cooled by evaporation nor warmed by condensation and equilibrium exists. The solution is isopiestic (equal in vapor pressure) with the sample and it has the same water potential (Boyer and Knipling, 1965). Since the water potential of the solution is known, the water potential of the tissue is then known.

The psychrometer measures the water potential in the cell walls because the water surface of the sample is located there and the vapor pressure develops there. The water potential of the walls is the same as the protoplasts (Fig. 3.4) and thus the potential applies to the entire cell. Figure 3.5B shows the water potential of some cells and tissues measured with this technique. The water potential of pollen is always lower than in the stigmas (silks) or leaves of the same maize plants, and it decreases through the day. The mature pollen is not attached to the vascular supply of the plant and readily dehydrates. The leaves and silks are supplied with water and do not dehydrate as much.

It is also possible to measure the osmotic potential in the apoplast by applying pressure to the cells to force water from the protoplasts into the apoplast. With tissues, a pressure chamber (Scholander et al., 1965) can apply the pressure as shown in Fig. 3.6A, displacing the original wall solution into the xylem from which it exudes onto the cut surface of the xylem. The exudate is collected and its osmotic potential is measured in the thermocouple psychrometer to obtain $\Psi_{s(a)}$ of Eq. (3.4) (Fig. 3.6B). The pressure P_{gas} necessary to displace the water gives $\Psi_{m(a)}$ because it opposes the tensions pulling water into the wall pores. Thus, $-P_{gas} = \Psi_{m(a)}$ (Fig. 3.6B). The water potential measured in the wall with the psychrometer ($\Psi_{m(a)}$) can then be checked by these two additional potentials ($\Psi_{s(a)} + \Psi_{m(a)}$) according to Eq. (3.4) (Boyer, 1967a).

For the protoplast compartment, the osmotic potential can be measured by

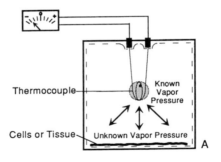

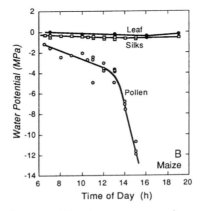

Figure 3.5   Thermocouple psychrometer (A) and measurements of water potentials in cells and tissues of maize made with a psychrometer (B). The droplet of known vapor pressure on the thermocouple can exchange water with the unknown sample on the bottom of the chamber and thus cool or warm the thermocouple. The solution neither cooling nor warming the thermocouple has a vapor pressure (water potential) equal to that of the sample. Since the solution water potential is known and is the same as that of the sample, the sample water potential is then known. The measurement is in the apoplast in equilibrium with the protoplasts. Typical measurements in maize (B) show that the water potential decreased only slightly during the day in the plant, but decreased markedly in pollen grains collected at various times from the same plant. The leaves and stigmas (silks) were connected through the xylem to the water supply in the soil but the mature pollen was not. Adapted from Westgate and Boyer (1986a).

applying pressure to cells that have been frozen and thawed to break the plasmalemma (Ehlig, 1962). The pressure removes the cell solution and the osmotic potential is measured with a psychrometer to give $\Psi_{s(p)}$ of Eq. (3.3). Since $\Psi_{w(p)}$ is known from the just-mentioned measurements in the apoplast, the turgor pressure $\Psi_{p(p)}$ can be calculated from Eq. (3.3). If necessary, the osmotic potential of the solution can be corrected for the effect of mixing with solution in the apoplast by noting the volumes of the wall and protoplast solutions and assuming complete mixing (Boyer, 1995; Boyer and Potter, 1973).

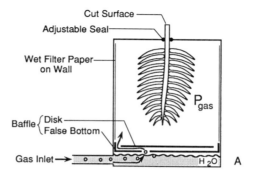

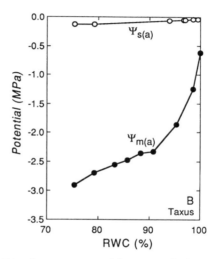

Figure 3.6 Pressure chamber (A) and measurements of the pressure in the apoplast of plant tissues using the pressure chamber (B). The incoming gas is humidified by bubbling through water, and the external pressure increases until it forces the xylem solution onto the cut surface. The pressure is adjusted to maintain the solution at the surface with no flow into or out of the tissue. This balancing pressure (P_{gas}) measures the internal pressure (tension $\Psi_{m(a)}$) on the apoplast solution according to $-P_{gas} = \Psi_{m(a)}$. In (B), the tension becomes more negative as the relative water content (RWC) decreases in the tissue (*Taxus* branch), indicating that a greater pull is being exerted by the leaves on the water in the xylem. Also shown is the osmotic potential of the apoplast solution ($\Psi_{s(a)}$) measured on xylem solution from the same *Taxus* branch. Note that $\Psi_{s(a)}$ is a small component at all RWC. The water potential of the apoplast solution is ($\Psi_{m(a)} + \Psi_{s(a)}$). Adapted from Boyer (1967b).

The $\Psi_{p(p)}$ can be checked by measuring it directly with a pressure probe (Fig. 3.7A, Hüsken *et al.*, 1978). The probe has a microcapillary whose tip can be inserted into a cell. Using a metal rod controlled by a micrometer screw, the pressure on oil in the microcapillary can be changed until the cell solution is

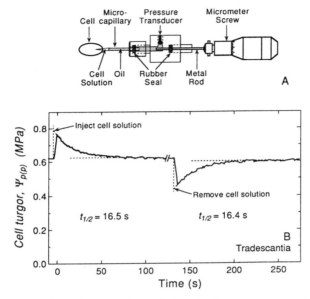

Figure 3.7 Pressure probe (A) for measuring and changing the turgor pressure inside plant cells (B). The probe is mostly filled with silicone oil (shaded), and a meniscus is visible between the cell solution and the oil in the tip of the microcapillary. When there is liquid continuity between the cell and the microcapillary, the pressure in the cell extends into the microcapillary and is sensed by the pressure transducer. The accurate measurement of cell turgor requires the meniscus to be returned to the position prior to entering the cell. Turning the micrometer screw forces the metal rod into the oil and moves the meniscus by changing the internal volume. The volume change causes the pressure to change as solution is injected into or removed from the cell (B) in a *Tradescantia* leaf. The volume of solution removed from or injected into the cell is determined from the distance the meniscus moves and the diameter of the microcapillary. Adapted from Tyerman and Steudle (1982).

returned to the original position close to the cell. The pressure inside the probe is then the same as the turgor in the cell and is measured with a pressure transducer (Fig. 3.7A).

With these methods, all of the potentials in Eqs. (3.6) and (3.7) can be measured. The methods give similar results when they are compared (Boyer, 1967a; Murphy and Smith, 1994; Nonami and Boyer, 1987, 1989, 1993; Nonami et al., 1987) and can be used with a wide range of tissues. The psychrometer also can measure the water potential of soil. Boyer (1995) gives a detailed description of these methods.

In the plant cell, the protoplast and apoplast measurements are straightforward but require us to distinguish between pressures of different origins. Some authors (e.g., Nobel, 1974, 1983; Passioura, 1980b; Steudle, 1989) combine pressures such as those arising from turgor or matric potentials regardless of origin. However, matric effects are not totally explained by pressures (see Appendix 2.3). Dehydrated matrices may contain so little liquid that local pressures on the liquid molecules are meaningless. In plants, these conditions occur in desiccated seeds, dry pollen, and various tissues of desiccation-tolerant plants. They also occur in porous media such as soil, wood, or paper. Therefore, it is important to distinguish between matric potentials and external pressures and this practice is followed in this book.

Negative potentials are common in nature because water often contains solutes or is held in a matrix. To move into a cell, the potential inside the cell must be even more negative. Depending on the system, the driving force may be some component of Ψ_w or the total Ψ_w. In a cell containing viable membranes, the force usually is the difference in water potential across the plasmalemma, but not all systems contain differentially permeable membranes that can harness the osmotic potential. In soil, water moves mostly because of matric force, and solutes have little effect. Similarly in the xylem, membranes are absent at maturity and water moves because of pressure differences developed by the surrounding cells. Thus, although water always moves toward the more negative potential, the critical potential depends on the physical system. Consideration of the system often can indicate what component potentials are important.

MECHANISM OF OSMOSIS

One of the most interesting aspects of osmosis is that solutes and pressures cause equivalent flows through plant membranes. It is not intuitive why this should occur but the effect can be plainly seen with a pressure probe for single cells. Figure 3.7B shows that a pressure probe can first inject a cell solution, then remove it. When the cell solution is injected, the turgor increases above that for balancing the osmotic potential and water is driven out of the cell by the extra pressure. When the pressure is reduced and the cell solution is removed, the turgor falls below the balance point and water enters because of the excess osmotic potential. The rate (half-time $t_{1/2}$) is the same for the outward pressure-driven and inward solute-driven flows although they are opposite in direction (Fig. 3.7B).

This behavior was addressed by Ray (1960) who proposed that biological membranes contain pores inside of which pressures exist that drive water through the membranes. He reasoned that experiments had shown that osmosis could occur faster than water could diffuse across the membrane and thus water-filled pores must exist in the membrane. He also recognized that if the membrane excludes solute from the pores there must be pressures in the pores. These were simplifications because membranes transport solute at low rates, often by active processes. However, once inside the cell, the solutes do not readily leak out and he reasoned that the slow rates and lack of leakage indicated that the solutes likely were in different channels and could be ignored. His

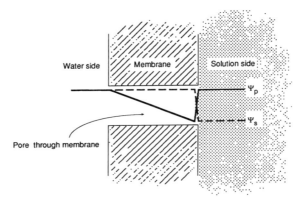

Figure 3.8   Osmotic flow through plant membranes according to Ray (1960). The osmotic potential (Ψ_s) undergoes an abrupt decrease at the pore entrance on the solution side of the membrane because no solute enters the pores. The pressure (Ψ_p) is kept at atmospheric on both sides. Because there is only water in the pore, Ray (1960) proposed that a pressure gradient exists inside the pore when osmotic flow occurs. The pressure decreases toward the solution side and the flow is driven by this pressure gradient. Adapted from Ray (1960).

concept is illustrated in Fig. 3.8 where a membrane separates a concentrated solution from pure water. The solution ends abruptly at the solution face of the membrane because solute cannot enter the pore. Water extends into the membrane pore. There is a jump downward in osmotic potential at the solution side of the membrane. A compensating pressure jump exists inside the pore to match the jump in osmotic potential at the solution side (Fig. 3.8). Because the external pressures are the same on both sides of the membrane, flow is driven by the pressure gradient in the pore.

This elegant logic received experimental support from Robbins and Mauro (1960) who used artificial membranes to measure osmotically driven flow through artificial membranes with a range of water conductances. Water was labeled with deuterium and supplied to one side to allow water diffusion to be measured. At conductances in the range for plant cells, diffusion was only a minor component of the total flow, and bulk flow predominated. This indicated that the membrane contained water-filled pores.

The presence of pressures in the pores was demonstrated by Mauro (1965) by enclosing the water on the water side of the membrane in a rigid compartment. As water moved through the membrane to the outer solution, the pressure decreased in the compartment. The pressure dropped until it prevented water from entering the membrane pores. Mauro (1965) reasoned that, in this equilibrium state, the pressure would be the same everywhere in the water. Since the water extended into the membrane pores from the water side, the pressure must also be the same inside the pores. Mauro (1965) found that large tensions developed inside the rigid container and thus in the pores (Fig. 3.9A).

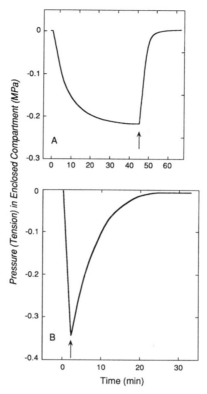

Figure 3.9 Demonstration of pressure gradient in membrane pores. The system is the same as in Fig. 3.8 except that the pressure is measured in a rigid compartment enclosing the water on the water side of the membrane (left side of Fig. 3.8). (A) As water flows into solution on the other side of the membrane, pressure in the rigid compartment falls until flow stops. The negative pressure (tension) in the compartment is the same everywhere including the membrane pores and becomes equal to the osmotic potential (-0.21 MPa) on the solution side (right side of Fig. 3.8). The solution was replaced with water at the arrow. (B) Large tensions can form rapidly in the membrane pores when a concentrated solution is present on the other side of the membrane and flow is occurring. The solution is removed at the arrow. In these graphs, zero pressure is atmospheric. After Mauro (1965).

The existence of negative pressures in the pores of membranes indicates that tensions arise in the plasmalemma and can be transmitted to various places in plants (e.g., the xylem and apoplast) much as they were transmitted to the rigid container of Mauro (1965). The pores must be very small in diameter so water is retained even under large tensions. Large tensions and rapid water movement were seen by Mauro (1965) as shown in Fig. 3.9B. In this way, the osmotic force is developed by the solute at the inner face of the plasmalemma where the pores contact the cell solution, and the force is transmitted nearly instantaneously

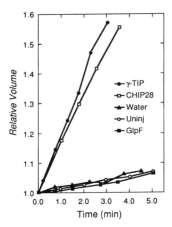

Figure 3.10 Volumes during osmotic swelling of frog (*Xenopus*) oocytes that had overproduced water channel proteins for the tonoplast membrane for 72 hr (γ-TIP) or that had the normal complement of water channel proteins (uninjected or water injected). The γ-TIP increased water transport. Also shown is the effect of a plasmalemma water channel protein from humans (CHIP28) which also increased water transport and a glycerol transport protein from bacteria called glycerol facilitator (GlpF) which did not transport water. The cells were injected individually with messenger RNA for one of the proteins and the mRNA was translated for 72 hr during which the protein was accumulated in the plasmalemma. The cells then were transferred to a dilute medium and osmotic swelling occurred as shown. Faster swelling indicates a more conductive plasmalemma. Adapted from Maurel *et al.* (1993).

as a tension through the membranes to the cell wall pores and apoplast and throughout the plant. On land, the tension can extend out of the plant and into the soil.

There is increasing evidence that special proteins form the water transport pores in plant and animal membranes. Maurel *et al.* (1993) injected messenger RNA (mRNA) for one of the plant membrane proteins (γ-TIP, tonoplast intrinsic protein) into *Xenopus* (frog) oocytes. After enough time for the oocyte to make protein and for the protein to incorporate into the plasmalemma, the conductivity of the membranes increased markedly for water (Fig. 3.10). When mRNA for the water transporting plasmalemma protein CHIP28 was injected into oocytes, there was a similar effect (Fig. 3.10, see Preston *et al.*, 1992). A membrane protein for glycerol transport (GlpF) did not have an effect on water transport (Fig. 3.10). Guerrero *et al.* (1990) and Ludevid *et al.* (1992) found evidence for variation in the amount of water transport proteins in plant membranes. Chrispeels and Maurel (1994) have also reviewed this area.

These demonstrations of protein channels in the plasmalemma and tonoplast verify the concepts of Ray (1960) that water moves primarily through membranes by bulk pressure-driven flow and explains why the flows are so fast, reversible, and affected by pressure and solute in an equivalent manner. The membrane pores appear to be discrete structures in the membrane. As a consequence, we should not expect diffusion to play much part, and diffusion experiments with labeled water will not give an accurate view of how water moves through a membrane. In the latter case, the water moves slowly by diffusion along concentration gradients, and pressure does not change the diffusion direction in contrast to the behavior actually observed with cells.

The presence of water-transmitting pores implies that water transport should vary according to the number of pores present in the membranes. Transport also might be affected by the kinds of pores or regulatory properties of the pores. Nevertheless, at equilibrium where there is no net water flow, the water status would not be altered by the number or nature of the pores. Changes in water status would occur rapidly or slowly depending on the number and size of pores but the equilibrium finally achieved would be the same.

CHANGES IN WATER STATUS

When a cell is dehydrated, its water potential decreases because the cell contents become more concentrated and there is a smaller volume of water to extend the walls. These changes can be represented by

$$d\Psi_{w(p)} = d\Psi_{s(p)} + d\Psi_{p(p)}, \tag{3.8}$$

which shows that the change in water potential is simply the sum of the changes in the osmotic potential and the turgor pressure. This equation does not indicate the rate of change but only the size of the change between the two equilibrium states.

It is useful to know which component causes the most change in the water potential. The answer is simplest if the solute content of the cell remains constant and the $d\Psi_{w(p)}$ is caused only by water. In that case, the change $d\Psi_{s(p)}$ is proportional to the fractional change in water content dV/V (see Appendix 3.2):

$$d\Psi_{s(p)} = -\Psi_{s(p)}\frac{dV}{V}. \tag{3.9}$$

Similarly, the change $d\Psi_{p(p)}$ can be found from the tensile properties of the cell wall. These properties are described by the bulk modulus of elasticity ϵ (MPa) that relates the internal pressure to the fractional change in water content of the cell:

$$d\Psi_{p(p)} = \epsilon\frac{dV}{V}. \tag{3.10}$$

Equations (3.9) and (3.10) have a similar form and show that the effect of a change in water content depends on whether $\Psi_{s(p)}$ or ϵ is numerically larger: the larger the $\Psi_{s(p)}$ or ϵ the larger the effect of dV/V on $d\Psi_{s(p)}$ or $d\Psi_{p(p)}$.

Substituting Eqs. (3.9) and (3.10) in Eq. (3.8) gives the total effect on the water potential

$$d\Psi_{w(p)} = -\Psi_{s(p)}\frac{dV}{V} + \epsilon\frac{dV}{V}, \qquad (3.11)$$

which we can rearrange to give

$$\frac{dV}{d\Psi_{w(p)}} = \frac{V}{\epsilon - \Psi_{s(p)}}, \qquad (3.12)$$

which has been called the capacitance C of the cell (Molz and Ferrier, 1982; Steudle, 1989). This is a useful expression for predicting how much the cell must dehydrate to cause a change in the water potential and also how much of the change is caused by $\Psi_{s(p)}$ or ϵ. Thus, for a cell with $\Psi_{s(p)}$ of -1 MPa and a rigid wall having ϵ of 49 MPa, the ϵ is numerically larger than $\Psi_{s(p)}$ and dehydration will cause mostly a turgor change. Equation (3.12) indicates that a decrease in water content of 2% ($dV/V = 0.02$) causes the turgor to decrease enough to decrease water potential 1 MPa in such a cell. On the other hand, the same cell with an elastic wall having ϵ of 4.9 MPa will still be dominated by the effects of turgor but the water potential decreases only 0.12 MPa for the same dehydration. Clearly, changes in water potential are caused more by changes in turgor than by changes in osmotic potential and are larger when the wall is rigid than when it is elastic.

This conclusion holds whenever there is turgor in a cell and can be demonstrated with a pressure chamber. The pressure is raised around a leaf until it is overpressured and water exudes. The new balancing pressure is measured at the new water content. A comparison of a rhododendron leaf having relatively rigid cell walls ($\epsilon = 97$ MPa) and a sunflower leaf having relatively elastic walls ($\epsilon = 6.4$ MPa) shows that the water potential decreases much more in rhododendron than in sunflower when water is lost from the leaves (Fig. 3.11). The larger decrease in rhododendron allows water to be extracted from the soil with only a slight dehydration of the leaf whereas sunflower requires a large dehydration before it can exert the same force on the soil water. Expressed in terms of the capacitance [Eq. (3.12)], rhododendron leaves having high water contents need to change only 1% in water content per MPa change in water potential whereas the sunflower leaves must change 13%.

Thus cells are affected by water exchange with their surroundings, and the cell water potential changes more dramatically when the wall is more rigid. Plants like rhododendron with evergreen leaves may encounter soils with little water or with frozen water during part of the year, and its rigid walls ensure that large force can be applied to extract water from the soil without excessive

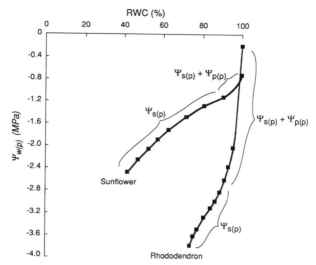

Figure 3.11 Water potential ($\Psi_{w(p)}$) at various relative water contents (RWC) in sunflower and rhododendron leaves measured with a pressure chamber. Both species show a greater decrease in $\Psi_{w(p)}$ when turgor is present ($\Psi_{s(p)} + \Psi_{p(p)}$) than when it is absent ($\Psi_{s(p)}$). However, rhododendron with thick relatively nonelastic cell walls ($\epsilon = 97$ MPa) shows a greater decrease than sunflower with thin elastic walls ($\epsilon = 6.4$ MPa). This results in very low $\Psi_{w(p)}$ in rhododendron with only moderate dehydration compared to sunflower. The Ψ_s was -2.6 and -1.1 MPa in hydrated rhododendron and sunflower respectively. Using ϵ, Ψ_s, and the change in RWC, the capacitance for water can be calculated for these tissues [Eq. (3.12)]. From the calculation, a decrease of 1 MPa in water potential from the fully hydrated state required a 1% decrease in RWC in rhododendron but a 13% decrease in sunflower. From J. S. Boyer (unpublished data).

leaf dehydration. The capacitance of the cells is an important physiological and ecological property.

WATER TRANSPORT

When a potential difference exists across the plasmalemma, the cell changes in water content at a rate determined by the conductivity of the plasmalemma and the size of the potential difference. The pore structure of the plasmalemma probably contributes to the conductivity and the potentials are determined not only by the external conditions but also by the turgor pressure and osmotic potential of the cell. Using the potentials of Eqs. (3.3) and (3.4) for the protoplast and apoplast, the water movement can be described by the transport equation

$$J_v = Lp(\Psi_{m(a)} - \Psi_{p(p)} + \sigma(\Psi_{s(a)} - \Psi_{s(p)})), \tag{3.13}$$

where J_v is the steady rate at which volume crosses the membrane per unit of membrane area ($m^3 \cdot m^{-2} \cdot sec^{-1}$), Lp is the hydraulic conductivity of the membrane ($m \cdot sec^{-1} \cdot MPa^{-1}$), ($\Psi_{m(a)} - \Psi_{p(p)}$) is the pressure difference across the membrane (the matric potential on the outside minus the turgor pressure on the inside of the membrane in MPa, see Fig. 3.4C), ($\Psi_{s(a)} - \Psi_{s(p)}$) is the osmotic potential difference across the membrane (MPa, see Fig. 3.4C), and σ is the reflection coefficient of the membrane (dimensionless, see Appendix 2.2). The Lp indicates the frictional effects encountered by water as it crosses the membrane. A larger Lp shows that water more easily crosses the membrane. According to Table 3.1, Lp ranges between 10^{-6} and 10^{-8} $m \cdot sec^{-1} \cdot MPA^{-1}$ for plant cells. The range of values suggests that the plasmalemma can vary in conductivity.

For most cells and most internal solutes, there also is solute transport across the plasmalemma. Active metabolism usually is required and there is a negligibly small permeability for the passive movement of the solute. The net movement is independent of the movement of water and is much slower. Therefore, for the solutes normally present inside a cell, the plasmalemma can be considered to be an ideal differentially permeable membrane with a reflection coefficient of essentially 1, and the hydraulic conductivity can be considered to be almost entirely for water with little effect of solute transport. Table 3.2 shows that measured values for σ are near 1 for most solutes inside the cell, confirming that the plasmalemma behaves ideally. Under these conditions, Eq. (3.13) becomes simply

$$J_v = Lp(\Delta\Psi_w), \tag{3.14}$$

and water is driven across the plasmalemma by the water potential difference ($\Delta\Psi_w$) between the two sides.

In special situations, this simplification may not hold. Lipophilic solutes that are small molecules such as ethanol or isopropanol have a σ around 0.2 (Table 3.2). Other solutes can alter membrane properties and cause σ to be less than 1 in which case internal solute may leak out. Cells that are suddenly subjected to high concentrations of solutes may shrink enough to cause the plasmalemma to separate from the cell wall (plasmolysis) and disrupt the plasmodesmata. In these situations, it cannot be assumed that $\sigma = 1$.

Significance of Reflection Coefficients

If σ is less than 1, water is not the only molecule crossing the membrane, and Lp also includes the movement of some solute. The solute tends to move in a direction opposite to that of water. While the permeability of the membrane for solute can be described by a solute permeability coefficient analogous to the hydraulic conductivity, the reflection coefficient is not a permeability coefficient but is a key parameter in Eq. (3.13) because it determines how much of the osmotic potential is used in water transport (see Appendix 2.2). When σ is less than 1, the osmotic potential is similarly less than fully effective.

Measuring the osmotic potential inside and outside of cells does not give the reflection coefficient of the membrane and thus does not indicate how much of the measured potential is contributing to the flow. Great care must be taken when placing high concentrations of solute outside of cells for this reason. Depending on how much solute enters the cell, the osmotic effect of the solute can vary dramatically. Moreover, because the reflection coefficient describes a condition of the membrane, its effects are always present and cannot be avoided by making rapid measurements or allowing only small water flows. For this reason, osmotica generally do not simulate the natural dehydration of cells and are no longer used for accurate measurements of cell water status.

The reflection coefficient for a solute can be most simply measured by determining the change in cell water potential that is caused by the solute. In the equilibrium state,

$$\sigma = \frac{d\Psi_w}{d\Psi_s}, \tag{3.15}$$

which indicates that solute supplied externally to change Ψ_s by 1 MPa will change the Ψ_w internally by 1 MPa when $\sigma = 1$. Figure 3.12 shows this kind of measurement using a pressure probe and indicates that the plasmalemma of epidermal cells of *Tradescantia* leaves had $\sigma = 1$ for sucrose but less than 1 for ethanol (Tyerman and Steudle, 1982). When $\sigma = 1$ as for sucrose, the sucrose remained outside and only water moved across the plasmalemma to give a simple shrinkage of the cell (Fig. 3.12A, left). When σ was less than 1 as for ethanol, the shrinkage was less than for sucrose even though the concentration of ethanol was greater. This indicates that the osmotic effectiveness of the ethanol was less than that of sucrose. Because the ethanol could enter the cell, Fig. 3.12 shows that the cell contracted initially as water left the cell but later swelled as ethanol entered. This two-phase contraction followed by swelling is diagnostic for a σ less than 1.

Equation (3.15) has been used to measure reflection coefficients around 1 (e.g., Tyerman and Steudle, 1982) but, for σ less than 1, the two-phase behavior of the cell causes experimental difficulties. As Tyerman and Steudle (1982) point out, permeating solutes can be dragged along by the water moving through the membrane and swept away from the membrane surface. Unstirred layers of water and solute exist next to the membrane and these can limit solute and water transfer. The results depend on how fast the solute penetrates the membrane. Thus, a σ below 1 clearly indicates that the membrane is nonideal but the actual value of σ is usually approximate.

Table 3.1 Half-time of Water Exchange ($t_{1/2}$), Hydraulic Conductivity (Lp), and Tissue Diffusivity for Water (D) in Cells as Determined from Pressure Probe Experiments

Species	Tissue/cell type	Half-time, $t_{1/2}$ (sec)	Hydraulic conductivity, Lp (m·sec⁻¹·MPa⁻¹)	Diffusivity, D (m²·sec⁻¹)	Reference
Chara corallina	Internode cells	1.3–7.5	$(0.8-1.4) \times 10^{-6}$	—	a
Capsicum annuum	Mesophyll of fruit tissue	65–250	$(4-6) \times 10^{-8}$	$(3-6) \times 10^{-11}$	b
	Subepidermal bladder cells of inner pericarp of fruit	1–12	$(2-17) \times 10^{-6}$	—	c
	Tissue cells of inner pericarp	18–54	$(1.2-3.4) \times 10^{-7}$	—	c
Tradescantia virginiana	Leaf epidermis	1–35	$(0.2-11) \times 10^{-7}$	$(0.2-6) \times 10^{-10}$	d,e,f
	Subsidiary cells	3–34	$(2-35) \times 10^{-8}$	10^{-11}–10^{-10}	
	Mesophyll cells	55–95	$(4-6) \times 10^{-8}$	1×10^{-12}	
	Isolated epidermis	9–54	6×10^{-8}	$(0.5-3) \times 10^{-11}$	
Kalanchoe daigremontiana	CAM tissue of the leaf	2–9	$(0.2-1.6) \times 10^{-6}$	6×10^{-10}	g
Pisum sativum	Growing epicotyl	1–27 (epidermis)	$(0.2-2) \times 10^{-7}$	—	h,i
		0.3–1 (cortex)	$(0.4-9) \times 10^{-6}$	3.2×10^{-10}	
Glycine max	Growing hypocotyl	0.3–5.2 (epidermis)	$(0.7-17) \times 10^{-6}$	$(1-9) \times 10^{-11}$	j
		0.4–15.1 (cortex)	$(0.2-10) \times 10^{-6}$	$(1-55) \times 10^{-11}$	
Zea mays	Midrib tissue of leaf	1–8	$(0.3-2.5) \times 10^{-6}$	$(0.4-6.1) \times 10^{-10}$	k
Oxalis carnosa	Epidermal bladder cells	22–213 (adaxial)	4×10^{-7}	—	l
		7–38 (abaxial)	2×10^{-6}	—	
Mesembryanthemum crystallinum	Epidermal bladder cells	200–2000	2×10^{-7}	—	m
Salix exigua	Sieve elements of isolated bark strips	110–480	5×10^{-9} (lateral hydraulic conductivity)	—	n

Species	Tissue			
Hordeum distichon	Root cortex and rhizodermis	1–21	1.2×10^{-7}	$(0.5\text{–}9.5) \times 10^{-11}$ (cortex) $(1\text{–}7) \times 10^{-12}$ (rhizodermis)	[o]
Triticum aestivum	Root hairs, rhizodermis, cortex	8–12	1.2×10^{-7}	—	[p,q]
Z. mays		1–28	$(0.5\text{–}9) \times 10^{-7}$	$(2\text{–}53) \times 10^{-12}$	[r]
	Root cortex, rhizodermis	—	1.2×10^{-7}	—	[q]
Phaseolus coccineus	Root cortex	0.4–2.3	2×10^{-6}	$(0.3\text{–}1.7) \times 10^{-10}$	[s]

Note. the diffusivity D refers to cell transport only.

[a] Steudle and Tyerman (1983).
[b] Hüsken *et al.* (1978).
[c] Rygol and Lüttge (1983).
[d] Tomos *et al.* (1981).
[e] Tyerman and Steudle (1982).
[f] Zimmerman *et al.* (1980).
[g] Steudle *et al.* (1980).
[h] Cosgrove and Cleland (1983b)
[i] Cosgrove and Steudle (1981).
[j] Steudle and Boyer (1985).
[k] Westgate and Steudle (1985).
[l] Steudle *et al.* (1983).
[m] Steudle (1975).
[n] Wright and Fisher (1983).
[o] Steudle and Jeschke (1983).
[p] Jones *et al.* (1983).
[q] Jones *et al.* (1988b).
[r] Steudle *et al.* (1987).
[s] Steudle and Brinckmann (1989).

67

Table 3.2   Reflection Coefficients (σ) of Plant Cell Membranes
for Some Nonelectrolytes

	Reflection coefficients			
Solute	*Chara corallina*[a]	*C. corallina*[b]	*Nitella flexilis*[c]	*Tradescantia virginiana*[d]
Sucrose	0.95	—	0.97	1.04
Mannitol	1.02	—	—	1.06
Urea	—	1	0.91	1.06
Acetamide	—	—	0.91	1.02
Formamide	0.99	1	0.79	0.99
Dimethylformamide	0.76	—	—	—
Glycerol	—	—	0.80	0.93
Ethylene glycol	—	1	0.94	0.99
n-Butanol	0.14	—	—	—
Isobutanol (2-methyl-l-propanol)	0.21	—	—	—
n-Propanol	0.24	0.22	0.17	− 0.58
2-Propanol	0.45	—	0.35	0.26
Ethanol	0.40	0.27	0.34	0.25
Methanol	0.38	0.30	0.31	0.15
Acetone	0.17	—	—	—

[a]Steudle and Tyerman (1983).
[b]Dainty and Ginzberg (1964).
[c]Steudle and Zimmermann (1974).
[d]Tyerman and Steudle (1982).

These examples illustrate the central role of the plasmalemma and its reflection coefficient in the water relations of cells. Water moves at high rates because the plasmalemma allows water to pass readily, and osmotic force is generated by solutes because of the ideal nature of the membrane for the solutes normally in the cell. Without the plasmalemma, the osmotic potential could not be harnessed and water generally would not move rapidly enough to maintain cell hydration.

RATES OF DEHYDRATION AND REHYDRATION

The ease of water movement across the plasmalemma determines how readily cells dehydrate and rehydrate. Hydration changes are frequently seen in cells as algae encounter varying salinities or land plants experience evaporation (transpiration). The rate of dehydration depends on whether a water supply is present or a protective barrier exists to inhibit water loss and also on how fast individual cells lose water. Thus, the rates of dehydration at the cell level are of considerable interest.

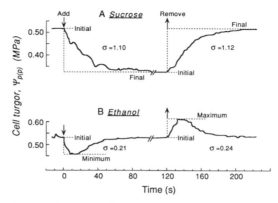

Figure 3.12 Plasmalemma reflection coefficients measured with a pressure probe in leaf epidermal cells of *Tradescantia*. The pressure probe measured the change in cell $\Psi_{p(p)}$. (A) Sucrose having Ψ_s of -0.18 MPa was added to the medium bathing the epidermis and caused water to move out. The turgor decreased by an amount that essentially equaled the Ψ_s of the sucrose, thus giving a reflection coefficient of about $-0.18/-0.18 = 1$ [see Eq. (3.15)]. At the upward arrow, the sucrose was removed and the pattern reversed as water moved in. When the reflection coefficient is 1 as for sucrose, the response is monophasic because only water moves. (B) Ethanol having Ψ_s of about -0.37 MPa caused turgor to decrease about 0.08 MPa to give a reflection coefficient of about $-0.08/-0.37 = 0.2$. Note that the ethanol caused a biphasic response. In the first phase, water moved outward and the cell shrank. In the second phase, ethanol entered and the cell swelled. At the upward arrow, the ethanol was removed and this pattern was reversed. Adapted from Tyerman and Steudle (1982).

The membrane properties in Eq. (3.14) affect the rate of dehydration, and the volume of water lost or more precisely the capacitance in Eq. (3.12) also contributes. By substituting Eq. (3.12) into Eq. (3.14), all of these factors can be combined (Appendix 3.3) for any small change in cell water potential

$$t_{1/2} = \frac{0.693\,V}{LpA(\epsilon - \Psi_s)} = 0.693rC, \tag{3.16}$$

where A is the surface area of the cell (m^2), r is the frictional resistance to water movement through the plasmalemma ($1/LpA$), and $t_{1/2}$ is the time for half the change in water potential.

Equation (3.16) shows that the cell acts much like an electrical circuit with a resistance and capacitance in series. The resistance r is mostly determined by the plasmalemma and controls how fast water enters the cell. The capacitance C [Eq. (3.12)] is determined by the size of the cell, the elasticity of its wall, and the internal osmotic potential, and these control how fast the potential changes for a unit change in the volume of water. The rate of dehydration is the product of the resistance and capacitance, and an increase in either resistance or capacitance makes the dehydration slower.

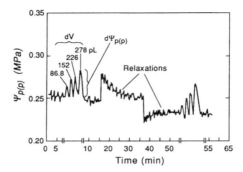

Figure 3.13 Pressure–volume relations measured with a pressure probe when a cell solution is rapidly injected and rapidly removed (left part of trace), injected and allowed to flow naturally out of the cell (relaxation in middle), or removed and allowed to flow into the cell (relaxation on right). The measurements were made in individual cells of a pepper fruit and changes in pressure ($d\Psi_{p(p)}$) and volume (dV) were used to calculate the bulk elastic modulus of the cell wall as described in the text. The relaxations on the right part of the trace measured flow through the plasmalemma and were used to calculate the hydraulic conductivity as described in the text. The small oscillations in the trace were generated to ensure that the microcapillary remained unplugged. Adapted from Hüsken *et al.* (1978).

Figure 3.7 shows the kind of measurement that can be used to determine $t_{1/2}$. A pressure probe injects or removes cell solution and changes the water potential because the turgor changes. Water leaves or enters the cell in response, and the $t_{1/2}$ is a measure of how fast dehydration or rehydration can occur.

From the $t_{1/2}$, Eq. (3.16) can be used to measure Lp as described by Steudle (1989). The method can be most simply explained by considering each term together with the measurement procedures, as shown in Fig. 3.13. The pressure probe is used to raise and lower the turgor rapidly (Fig. 3.13, left) to determine ϵ. By noting the pressure and volume of solution injected into and removed from the cell, the dP and dV are measured and ϵ can be calculated according to Eq. (3.10) (the volume V is determined from the cell dimensions). For the cell in Fig. 3.13, $d\Psi_{p(p)}/dV$ was 1.1×10^{11} MPa·m^{-3} and V was 11×10^{-12} m^3 so that ϵ was 1.2 MPa. On the other hand, if the pressure is raised and water is allowed to move out of the cell at its own pace, the $t_{1/2}$ can be measured (Fig. 3.13, relaxations) and was about 300 sec. The Ψ_s can be measured with extracts of cell solution and was -0.25 MPa. The A can be determined from the cell dimensions and was 2.97×10^{-7} m^2. The Lp can then be calculated from Eq. (3.16) to give 5.8×10^{-8} m·sec^{-1} · MPa^{-1}, which is within the range of values in Table 3.1.

This method of determining Lp, although it requires many measurements, is basically quite rapid and involves observing cell behavior for only short times. Therefore, cell properties should be quite stable while the measurements are being performed. The only disruptive influence is the insertion of the tip of the microcapillary into the cell. This probably causes little effect because other methods that do not penetrate the cell give similar values of Lp (Green et al., 1979; Kamiya and Tazawa, 1956; Levitt et al., 1936).

In general, the Lp measured for plant cells show that rapid water transport occurs across the plasmalemma for small potential differences. Small cells typical of many tissues tend to have water potentials similar to those of nearby cells. Even so, the plasmalemma conductivities vary by over 100-fold (Table 3.1) and thus the plasmalemma must differ widely in its properties depending on the type of cell. Because the rate of water transport is large, the rates of hydration and dehydration tend to be rapid for plant cells. The $t_{1/2}$ are only rarely more than 5 min and then only in cells of rather large dimensions (Table 3.1). As a consequence, the rate of dehydration of plant cells depends to a large extent on other features of the plant in addition to the plasmalemma. Waxy barriers on cell surfaces can decrease evaporation, extensive connections with the soil can supply water, and so on. Also, metabolic activities within the cell can lead to changes in internal solute concentration that delay or prevent dehydration.

OSMOTIC ADJUSTMENT

Changes in the internal solute concentration will occur whenever the water content changes during hydration/dehydration or the solute content changes inside the cell. Changes in water content are passive responses resulting from absorption or loss of water and they dilute or concentrate the solute. Changes in the solute content generally result from metabolic activity and are not passive. Because the solute changes represent a change in solute content per cell and are under the regulatory control of the cell, they are termed osmotic adjustment (Bernstein, 1961). The passive responses are not actively regulated and probably should be unnamed (Munns, 1988), although they are sometimes included with osmotic adjustment and the entire response called osmoregulation (Morgan, 1984). Initially, osmotic adjustment was thought to occur only in plants subjected to high salinities (Bernstein, 1961; Eaton, 1927, 1942; Munns, 1993) but it was later found in plants in drying soils (Meyer and Boyer, 1972) and much work was done to determine the effect on plant growth. Morgan (1984) and Munns (1988) provide useful reviews of the area.

Osmotic adjustment provides a means of maintaining cell water content which is an important cell activity. Because water loss can increase the concentration of solute in the cell, molecules that regulate metabolism can be affected. Some inorganic ions such as K^+, Ca^{2+}, Mg^{2+}, and Cl^- cannot be metabolized or incorporated into cell structure significantly and they are inevitably concentrated by dehydration. Because they play regulatory roles for enzymes, enzyme activities can be affected. For example, photophosphorylation is inhibited by

Mg^{2+} concentrations slightly above the optimum of 1.5 to 3 mM (Pick and Bassilian, 1982; Rao *et al.*, 1987; Shahak, 1986; Younis *et al.*, 1983). Certain K^+-requiring enzymes also can be affected if K^+ concentrations become too high (Evans and Sorger, 1966; Evans and Wildes, 1971; Wilson and Evans, 1968). In addition to the concentrating effects of water loss, exposure of plants to high external salinities adds the extra problem of high concentrations of NaCl. Most enzymes are inhibited by high concentrations of NaCl even in halophytic plants (Flowers *et al.*, 1977; Wyn Jones, 1980).

Osmotic adjustment maintains cell water contents by increasing the osmotic force that can be exerted by cells on their surroundings and thus increasing water uptake. The adjustment results from compatible organic solutes accumulating in the cytoplasm which decreases the osmotic potential of the cytosol. Compatible solutes allow enzyme reactions to occur even though the solutes are in high concentration around the enzymes. Compatible solutes are sugars, glycerol, amino acids such as proline or glycinebetaine, sugar alcohols like mannitol, and other low molecular weight metabolites (Bental *et al.*, 1988b; Flowers *et al.*, 1977; Grumet and Hanson, 1986; Hanson and Hitz, 1982; Meyer and Boyer, 1981; Morgan, 1984; Munns *et al.*, 1979; Voetberg and Sharp, 1991; Wyn Jones, 1980). If large amounts of inorganic salts are present externally, they may be accumulated as well, but are stored in the vacuole which sequesters them and prevents high concentrations from occurring around cytoplasmic enzymes (Hajibagheri and Flowers, 1989). External salts used for osmotic adjustment decrease the amount of compatible solute that needs to be produced in the cytoplasm, and this keeps the energy requirement low.

Good examples of compatible solute production are seen in marine algae such as *Dunaliella* and *Oochromonas* that can withstand saturated solutions of NaCl (Bental *et al.*, 1988a,b; Kauss, 1983; Kauss and Thomson, 1982). A little NaCl enters the cells and is stored in the vacuoles (Hajibagheri *et al.*, 1986). However, the cells mostly produce large quantities of glycerol (*Dunaliella*) or galactosyl glycerol (*Oochromonas*) in the cytoplasm. The solutes are produced from reserves, mostly starch, and are returned to starch under favorable conditions (Gimmler and Möller, 1981). Figure 3.14 shows that the glycerol content nearly doubled in 4 hr in *Dunaliella* after the external salinity was increased to 3.0 *M*. There was a comparable depletion of starch. Thus, the solute was simply converted from an insoluble polymeric form to soluble small molecules. This allowed rapid osmotic adjustment and conserved carbon compounds inside the cells.

When dehydration occurs without high external salinities, similarly rapid increases in solute content can occur in cells. Typically, the growing tissues adjust throughout the plant when the soil dehydrates (Westgate and Boyer, 1985b) and concentrations of solutes can increase markedly in only a few hours. Figure 3.15B shows that cells in the growing regions of soybean stems increased in solute content sufficiently to decrease the osmotic potential by 85% in 12 hr

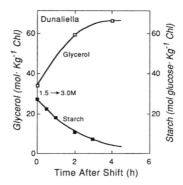

Figure 3.14   Osmotic adjustment in the marine alga *Dunaliella*. At zero time, the cells were shifted from a solution containing 1.5 *M* NaCl to a solution containing 3 *M* NaCl. The increase in cell glycerol came at the expense of cell starch (note that each glucose molecule released from starch produces two glycerol molecules). Adapted from Gimmler and Möller (1981).

after the roots were transplanted to dehydrated vermiculite (one-eighth of the water content of hydrated vermiculite). The accumulating solute was mostly glucose, fructose, sucrose, and amino acids (Meyer and Boyer, 1981). The cell water content changed only slightly (Fig. 3.15D) and turgor was maintained in these cells (Fig. 3.15C). Growth was inhibited but recovered somewhat after 48 hr (Fig. 3.15A). The mature tissues adjusted less osmotically and lost water as a result (Nonami and Boyer, 1989).

Thus, in both algae and land plants, salinity and dehydration cause metabolism to generate osmotica rapidly enough to keep pace with changes in external conditions. Most compatible solutes serve other functions in the cell and normally are produced in small quantities. Accumulation can occur simply by slowing their use in the normal reactions of the cell. In soybean stems growing with limited water, for example, Meyer and Boyer (1981) found that growth became slower and solute normally acting as substrate for growth was used less rapidly. With the slowdown in use, the solute accumulated. The unused solute accounted for most of the osmotic adjustment.

In roots, osmotic adjustment may have somewhat different origins. Sharp and his co-workers (1988, 1990) observed that maize roots decreased in diameter upon encountering dehydrated vermiculite. Fewer new cells were produced and they were smaller at maturity (Fraser *et al.*, 1990) which reduced the demand for imported solute. However, the thinner roots continued to extend at significant rates in conditions that completely eliminated stem growth.

Matyssek *et al.* (1991a,b) showed that the strong osmotic adjustment in growing regions could extract water from nearby mature tissues. Indeed, water moved backward from the shoot to the roots as the roots grew on water extracted from the mature stem (Matyssek *et al.*, 1991b). The use of internal

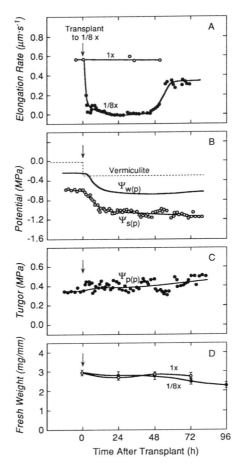

Figure 3.15 Osmotic adjustment in stems of soybean seedlings transplanted to vermiculite containing one-eighth of the water ($1/8\times$) normally present in hydrated vermiculite ($1\times$). (A) Stem elongation rates at $1/8\times$ and $1\times$, (B) water potential of vermiculite (dashed line) and water potential ($\Psi_{w(p)}$) and osmotic potential ($\Psi_{s(p)}$) of the stem-elongating region in $1/8\times$ plants, (C) turgor ($\Psi_{p(p)}$) of the stem-elongating region in $1/8\times$ plants, (D) fresh weight of the stem-elongating region in $1\times$ and $1/8\times$ plants. Osmotic adjustment is seen in (B) as a decrease in $\Psi_{s(p)}$. The adjustment preserves turgor (C) and fresh weight (D) because the water content of the cells remains high. However, in mature cells of the stems of the same plants, osmotic adjustment was less and the turgor and water content decreased (after Bozarth *et al.,* 1987; Nonami and Boyer, 1989).

water to promote root growth is advantageous in dry soil and probably occurs frequently.

Osmotic adjustment solves several problems for cells (Morgan, 1984). Because compatible solutes accumulate in the cytoplasm, enzyme function is maintained. The water content of the cell remains high and regulatory ions do not change concentrations. Turgor is maintained and allows a moderate amount of growth where none would occur otherwise (Meyer and Boyer, 1972; Michelena and Boyer, 1982). Continual root growth brings new water supplies to the plant (Matyssek *et al.*, 1991b; Sharp and Davies, 1985; Sharp *et al.*, 1988).

However, the process is limited by how much solute can be accumulated. Osmotic adjustment depends to a large extent on photosynthesis to supply compatible solute. As dehydration becomes severe, photosynthesis becomes inhibited. With a smaller solute supply, osmotic adjustment is curtailed. Thus, in the face of continued water limitation, osmotic adjustment delays but cannot completely prevent dehydration.

WATER RELATIONS OF CELLS IN TISSUES

Under laboratory conditions, cells may be easily subjected to large pressures or high concentrations of solute. In a tissue, however, these conditions are rare. A Ψ_w difference of 0.5 MPa might occur between leaves or perhaps across an individual leaf but almost never across a single cell. Generally, such a potential difference would be spread across at least 10 cells (20 cell wall/membrane layers) and there would be about 0.025 MPa between the inside and outside of each cell. While these small differences can drive water into the cell [see Eq. (3.14)], they are small enough so that the protoplasts in a tissue are almost always near water potential equilibrium with their own cell walls (Molz and Ikenberry, 1974; Molz *et al.*, 1979). This explains why it is possible to measure the water status of cells in a tissue by determining the vapor pressure of water in the cell walls and why cellular characteristics are distributed uniformly over considerable distances.

A cell in a tissue is not surrounded by unlimited water as an isolated cell would be when bathed in a solution. In a tissue, the cell obtains water mostly from the vascular tissue by way of other cells. Water flows through the apoplast and protoplasts or symplast (Fig. 3.16). The volume of the apoplast usually is

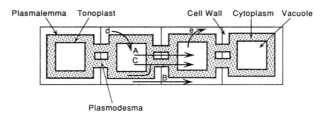

Figure 3.16 Diagrammatic representation of the water pathways in a tissue consisting of four cells. (A) Cell to cell path, (B) cell wall (apoplast) path, (C) cell to cell path through plasmodesmata (symplast), (d) water uptake by cells from apoplast, and (e) water loss by cells to apoplast. Plasmalemma lines plasmodesmata to form continuous membrane between cells. The dimensions of the various compartments are not to scale (after Molz and Ferrier, 1982).

small compared to the volume in the protoplasts or symplast. Changes in cell water potential are rapidly transmitted to the apoplast or through the symplast to affect surrounding cells. Groups of cells tend to act in concert with the immediately surrounding cells. As a consequence, gradients in Ψ_w are detectable only over distances of several cells.

In this situation, macroscopic principles involving groups of cells can be used to help understand cellular water transport. For a time in the early 1900s, water was considered to move from vacuole to vacuole in tissues (path A in Fig. 3.16), traversing the cytoplasm and membranes of each cell (Newman, 1974, 1976). Later, water was thought to move primarily in the porous cell walls (path B in Fig. 3.16), bypassing the cell membranes (Scott and Priestley, 1928). The discovery of plasmodesmata added still another pathway that allowed water to enter the first cell and then move from cell to cell without crossing another plasmalemma (path C in Fig. 3.16). In addition, water moves in and out of the protoplasts from the apoplast (paths d and e in Fig. 3.16).

Philip (1958a–d) was the first to construct a model of tissue water movement based on the characteristics of a group of cells. This seminal work has formed the basis for most of the treatments that followed, although his development did not consider reflection coefficients or water flow through the cell wall path. Molz and Hornberger (1973) extended Philip's theory to include the effects of the reflection coefficient, and Molz and Ikenberry (1974) included the cell wall path in parallel with the vacuole to vacuole path. Steudle and his co-workers conducted detailed studies of water transport in tissues using the miniature pressure probe (e.g., Steudle and Jeschke, 1983; Westgate and Steudle, 1985; Zhu and Steudle, 1991). However, no treatments have yet encompassed the plasmodesmata because their conductivity for water is not well understood (path C in Fig. 3.16). In theory, the plasmodesmata provide a means of connecting the protoplasm of adjacent cells to form a symplasm and, with a pulse of turgor from a pressure probe, one measures water transport out of the cell by both the plasmalemma and the plasmodesmata. Measurements of Lp for cells in a tissue thus include unknown contributions from the plasmodesmata.

Nevertheless, these efforts allow predictions of tissue behavior based on the characteristics of the cells, and some useful information has been obtained. In general, for mathematical purposes, the tissue is considered to be made up of infinitely small units (the cells) that collectively control the macroscopic behavior of the tissue, and equations having a form to describe diffusion are applied. The fundamental equation is

$$\frac{\partial \Psi}{\partial t} = D\frac{\partial^2 \Psi}{\partial x^2}, \qquad (3.17)$$

where D is the diffusivity of water ($m^2 \cdot sec^{-1}$), t is the time, and x is the distance for water movement. This relationship arises from a consideration of the rate of water uptake and loss by the cells and the rate of water storage by the cells applying the concept that the mass of water is conserved. It indicates that the rate at which the water potential changes in a tissue ($\partial\Psi/\partial t$) depends on the difference ($\partial/\partial x$) in the gradients of water potential ($\partial\Psi/\partial x$) along the diffusion path. The difference in the gradients is shown as ($\partial^2\Psi/\partial x^2$) in the equation. The D includes all the cell characteristics that control the development of potential gradients in the tissue.

The D can be determined by monitoring the rate of the change in water potential with time over known distances. Although the mathematics can become quite complex (Molz and Ferrier, 1982), the major goal is to compare individual cell rates with tissue rates of water transfer. If the cell rates are faster than the tissue rates, water movement is from cell to cell (paths A and C in Fig. 3.16). On the other hand, if the cell rates are the same as the tissue rates, all the cells hydrate virtually in unison at a rapid rate and water passes through a highly conductive path between them. This could occur only if water bypassed some of the protoplasts/symplast by way of the apoplast (path B in Fig. 3.16). Thus, cell/tissue kinetics can give information about which flow paths are operating in tissues.

For example, Westgate and Steudle (1985) compared the turgor relaxation of a maize leaf cell when solution was injected directly into the cell or when water was injected into the xylem (Fig. 3.17). The relaxation was much faster for water injected into the cell than into the xylem. Moreover, cells close to the xylem absorbed water faster than cells far from the xylem. This indicates that the cells were not connected to the xylem by a highly conductive path that could bypass the protoplasts. Rather, cells farther away from the xylem had to wait for water that flowed through cells nearer the xylem before they could hydrate fully. This excludes path B in Fig. 3.16 as the main flow path. Rather, water had to move primarily through the intervening cells by paths A and C with at most a modest contribution from B. It is worthwhile noting that the cell-to-cell path allows water to flow quite rapidly since the $t_{1/2}$ was only 15–18 sec (Fig. 3.17). This explains why wilted flowers recover so rapidly when their stems are cut under water and why whole plants regain turgidity soon after the soil is watered.

In several of these studies (Steudle and Boyer, 1985; Steudle and Jeschke, 1983; Westgate and Steudle, 1985; Zhu and Steudle, 1991), the tissue was pressurized on a cut surface. Steudle and Boyer (1985) pointed out the possibility of flooding the intercellular gas spaces in these kinds of experiments, which would make path B in Fig. 3.16 appear more conductive than it would ordinarily be. Therefore, methods need to be sought that do not have this complication, and studies of water transport using noninvasive methods employing nuclear magnetic resonance (Brown *et al.*, 1986, 1990; G. A. Johnson *et al.*, 1987; Kramer

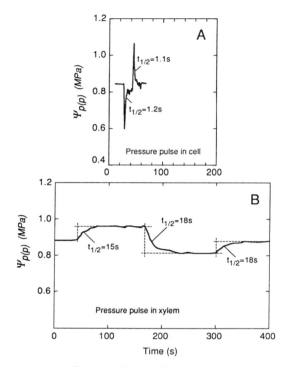

Figure 3.17 Turgor in maize cells situated in a leaf. (A) Turgor relaxation caused by a pressure pulse inside a cell using a pressure probe. (B) Turgor relaxation in the same cell when a pressure pulse was applied to water in the xylem. Relaxations are rapid in (A) where the path involves only the plasmalemma of the cell (d and e in Fig. 3.16) but slow in (B) where the path includes other cells. Note that if the apoplast was the main flow path between the cells, it would need to have a low resistance to flow. Water supplied by the xylem would be able to bypass intervening cells because of the low resistance and would reach each cell readily resulting in similar rates of relaxation in (A) and (B). However, because relaxations in (B) were much slower than in (A) for the same cell, the apoplast path was not the main contributor and water flowed mostly from cell to cell (paths A and C in Fig. 3.16). Adapted from Westgate and Steudle (1985).

et al. in Hashimoto *et al.*, 1990; MacFall *et al.*, 1990; Veres *et al.*, 1991) or mass spectrometry (Yakir *et al.*, 1989, 1990) look promising.

Molz and Boyer (1978) and Silk and Wagner (1980) used diffusion equations to model water movement in growing tissues in comparison with single cells. The models predicted local equilibrium in individual cells but significant water potential gradients over the whole growing tissue. Subsequently, measurements of the potentials of the individual cells (Nonami and Boyer, 1993) confirmed that the predicted gradients exist in growing tissue (see Chapter 11).

This work is of a fundamental nature and is necessary to understand growth processes and water use by plants. Because it involves directly measurable parameters at the cell and tissue levels, it allows understanding to be built up from first principles that can be highly predictive and useful for tissues. Further research could be directed toward identifying the specific paths used by water and the quantitative contributions of these paths as water moves through the tissue. Particularly useful will be understanding of the paths into growing tissues and roots, and the paths leading to the evaporating surfaces in leaves.

For further information, Molz and Ferrier (1982) and Steudle (1989) provide extensive reviews of cell and tissue water transport. Background can also be gained from Kedem and Katchalsky (1958), Dainty (1963), Nobel (1974, 1983, 1991), and Zimmermann and Steudle (1978).

SUMMARY

The plasmalemma is the principal barrier controlling molecular traffic in and out of cells. For water, it functions primarily to harness the osmotic potential without which water would move into the cells too slowly to replace that which is lost by transpiration. At least in part, dehydration damage to cells can be attributed to a change in the plasmalemma and/or tonoplast.

The water status of cells is most usefully characterized in terms of the chemical potential or water potential because it involves a physically defined reference instead of a biological one and because the potential is the force moving water to and through cells. It is applied to the two major regions of the cell: the inside of the plasmalemma (the protoplast or symplast) and the outside of the plasmalemma (the apoplast), and it has different components in the two regions. The cell interior is affected mostly by the osmotic potential and turgor pressure whereas the external region is affected mostly by the osmotic potential and matric potential. These two regions are usually in equilibrium or near equilibrium with each other.

The potentials can be measured by various methods. The three most commonly used are (1) the thermocouple psychrometer which measures the vapor pressure of water (water potential) in the cell walls, (2) the pressure chamber which measures the tensions (matric potential) of water in the walls, and (3) the pressure probe which measures the turgor pressure inside the cells. By extracting the solution in the wall pores or protoplasm, the osmotic potential also can be measured in those regions.

Osmosis involves water movement across the plasmalemma because of differences in potential. Although the differences are usually small in individual cells, they can become substantial when considered over distances of many cells in tissues. Osmotic water movement appears to be driven by pressure gradients in membrane pores that are water filled. The pores appear to be located in water-transmitting proteins embedded in the membranes. The proteins have been found in the plasmalemma and in the tonoplast. Increasing the amounts of pore-forming proteins in membranes increases rates of water transport through the membranes. Water diffusion through the membranes also occurs but is too slow to account for osmotic flow.

The dehydration of cells decreases cell water potential and thus increases the force causing water uptake. The decrease in water potential is caused more by decreases in turgor than by decreases in osmotic potential, and more turgor is lost in cells with rigid walls than in cells with elastic walls.

The dehydration can be delayed by osmotic adjustment of the cells, which increases solute contents rapidly enough to increase the osmotic force that can be applied by the cells. This allows increased water uptake and protects against changes in water content that otherwise would alter regulatory ion concentrations for the enzymes engaged in cell metabolism. The accumulating solute consists of organic molecules compatible with enzyme function that are produced by photosynthesis, released from reserves, or transported to the cells faster than they are used in metabolism.

The rates of dehydration and rehydration can give information about the conduction of water by the plasmalemma, especially when rates of hydration/dehydration are compared in cells and tissues. From differences in rates between a whole tissue and the individual cells in the tissue, conclusions can be drawn about the chief flow paths in the tissue. In those studied so far, the apoplast is rarely a dominant path and instead flow occurs mostly from cell to cell with perhaps a modest contribution from the apoplast.

SUPPLEMENTARY READING

Boyer, J. S. (1995). "Measuring the Water Status of Plants and Soils." Academic Press, San Diego.

Crowe, J. H., and Crowe, L. M. (1986). Stabilization of membranes in anhydrobiotic organisms. *In* "Membranes, Metabolism and Dry Organisms" (A. C. Leopold, ed.), pp. 188–209. Comstock Publishing Association, Ithaca, NY.

Dainty, J. (1963). Water relations of plant cells. *Adv. Bot. Res.* **1**, 279–326.

Gibbs, J. W. (1931). "The Collected Works of J. Willard Gibbs," Vol. 1. Longmans, Green and Co., New York.

Kedem, O., and Katchalsky, A. (1958). Thermodynamic analysis of the permeability of biological membranes to non-electrolytes. *Biochim. Biophys. Acta* **27**, 229–246.

Molz, F. J., and Ferrier, J. M. (1982). Mathematical treatment of water movement in plant cells and tissues: A review. *Plant Cell Environ.* **5**, 191–206.

Morgan, J. M. (1984). Osmoregulation and water stress in higher plants. *Annu. Rev. Plant Physiol.* **35**, 299–319.

Munns, R. (1988). Why measure osmotic adjustment? *Aust. J. Plant Physiol.* **15**, 717–726.

Nobel, P. S. (1974). "Biophysical Plant Physiology." W. H. Freeman, San Francisco.

Nobel, P. S. (1983). "Biophysical Plant Physiology and Ecology." W. H. Freeman and Co., New York.

Nobel, P. S. (1991). "Physicochemical and Environmental Plant Physiology." Academic Press, London.

Robards, A. W., and Lucas, W. J. (1990). Plasmodesmata. *Annu. Rev. Plant Physiol. Plant Mol. Biol.* **41**, 369–419.

Slatyer, R. O. (1967). "Plant–Water Relationships." Academic Press, New York.

Zimmermann, U., and Steudle, E. (1978). Physical aspects of water relations of plant cells. *Adv. Bot. Res.* **6**, 45–117.

APPENDIX 3.1: Preservation of Cell Ultrastructure for Electron Microscopy

Biological observations with the electron microscope depend on the preservation of structures in the living cell so that they can be observed later in the non-living cell. Fixatives have been developed to kill the cells rapidly so that structures are maintained. One that is frequently used is 1.5% glutaraldehyde in 0.05 M phosphate buffer, pH 7.2, which has an osmotic potential of -0.65 MPa that is similar to the water potential of many hydrated plant tissues. As a consequence, osmotic swelling and shrinking are minimized in the cells, and preservation of organelles is maximized.

The fixation of dehydrated cells is more difficult because the water potential of the cells is often lower than -0.65 MPa. The fixative then has a tendency to hydrate the cells because fixation is not instantaneous. Fellows and Boyer (1976, 1978) showed that abnormal cell structures are induced during this kind of fixation. Differing potentials often cause normally appressed membranes to move apart abnormally, and new vesicles and lipid droplets to appear. Several studies (Alieva *et al.*, 1971; Giles *et al.*, 1974, 1976; Kurkova and Motorina, 1974; Nir *et al.*, 1969; Vieira da Silva *et al.*, 1974) that employed this kind of fixation could have contained artifacts. On the other hand, when the osmotic potential is kept the same for fixative and tissue, most structure is preserved in dehydrated cells, as in Fig. 3.1B. Therefore, for dehydrated tissue, we recommend that fixative be prepared by adding sucrose at several concentrations to give a range of osmotic potentials, as done by Fellows and Boyer (1976, 1978). Samples of the dehydrated tissue are fixed at each osmotic potential and parallel samples are used to measure the tissue water potential. After the water potential has been determined, the fixation that most closely corresponds to the measured water potential of the tissue is chosen for microscopy, and the other fixations are discarded.

APPENDIX 3.2: Osmotic Potential and Dehydration

The effect of dehydration on Ψ_s can be calculated from the van't Hoff equation (Appendix 2.1) by replacing the concentration C with n_s/V where n_s is the number of moles of solute and V is the volume of water. Rearranging gives

$$\Psi_s V = -RT n_s \tag{3.18}$$

and, because RTn_s is constant at constant temperature, the differential gives

$$d(\Psi_s V) = 0,$$ (3.19)

which, when expanded, is

$$\Psi_s dV = -Vd\Psi_s$$ (3.20)

and

$$d\Psi_s = -\Psi_s \frac{dV}{V}.$$ (3.21)

Equation (3.21) can be applied to small changes in cell volume caused by water loss or gain.

APPENDIX 3.3: Rates of Dehydration and Rehydration of Cells

The rates of cell dehydration/rehydration are determined by the conductivity of the plasmalemma and the capacitance of the cell. For a cell having a reflection coefficient of 1 and water potential Ψ_w in a medium having a constant water potential Ψ_o, the conductivity is described by Eq. (3.14), which in expanded form is

$$J_v = \frac{dV}{dt}\frac{1}{A} = Lp(\Psi_o - \Psi_w),$$ (3.22)

where V is the volume of water in the cell, t is the time, and A is the surface area of the cell (i.e., plasmalemma).

Substituting for dV the expression from Eq. (3.12) for the capacitance $dV = Vd\Psi_w/(\epsilon - \Psi_s)$ and rearranging gives the rate of change of Ψ_w

$$\frac{d\Psi_w}{dt} = \frac{-LpA}{V}(\epsilon - \Psi_s)(\Psi_w - \Psi_o),$$ (3.23)

which shows that the rate of change of the water potential is proportional to $-(\Psi_w - \Psi_o)$ when $LpA(\epsilon - \Psi_s)/V$ is constant, which is a reasonable approximation for small changes in Ψ_w. Since Ψ_o also is a constant, Eq. (3.23) can be integrated for $(\Psi_w - \Psi_o)$ at any time t

$$\int_{\Psi_w^o - \Psi_o}^{\Psi_w^t - \Psi_o} \frac{d\Psi_w}{\Psi_w} = -\frac{LpA}{V}(\epsilon - \Psi_s)\int_o^t dt$$ (3.24)

$$ln\frac{\Psi_w^t - \Psi_o}{\Psi_w^o - \Psi_o} = -\frac{LpA}{V}(\epsilon - \Psi_s)t,$$ (3.25)

where the superscripts o and *t* refer to zero time and any subsequent time, respectively. The time taken for $(\Psi_w^t - \Psi_o)/(\Psi_w^o - \Psi_o)$ to change by half $(t_{1/2})$ is

$$\ln 0.5 = -\frac{LpA}{V}(\epsilon - \Psi_s)t_{1/2} \tag{3.26}$$

$$t_{1/2} = \frac{0.693\,V}{LpA(\epsilon - \Psi_s)}, \tag{3.27}$$

which indicates that the half-time for a cell to change its Ψ_w depends on the properties of the wall (ϵ) and plasmalemma (Lp) in addition to the physical dimensions of the cell $(A$ and $V)$ and the osmotic potential (Ψ_s) of the cell solution. Because ϵ, Ψ_s, A, and V may be considered constant for small changes in water content, measuring $t_{1/2}$ for the cell allows Lp to be calculated. Of course, when changes in Ψ_w are caused by small changes in the turgor, this expression also can be used to describe the rates of the turgor change and to calculate Lp.

Soil
and Water

INTRODUCTION

Soil is important for plant growth as a source of water and minerals, as the anchorage for plants, and as a medium for development of the root systems essential for absorption and anchorage. It is a complex system, consisting of varying proportions of rock particles and organic matter forming the solid matrix, and soil solution and air occupying the pore space. In addition, soil usually contains an active population of bacteria, fungi, algae, insects, and small animals that directly or indirectly affect soil characteristics and root growth (see Wild, 1988, Chapters 14–17).

IMPORTANT CHARACTERISTICS OF SOILS

Soil characteristics such as texture and structure have important effects on the suitability of soil as a medium for plant growth.

Composition and Texture

Those characteristics of soil most important for plant growth, such as water and mineral storage capacity and suitability for root growth which is related to aeration and resistance to root penetration, depend largely on texture and structure. Soils usually are classified as sands, loams, or clays, depending on the proportions of large ($>2.0–0.02$ mm), intermediate ($0.02–0.002$ mm), and fine particles (<0.002 mm) present. By definition, sands contain less than 15% of silt and clay, clay contains over 40% of fine particles, and loam soils contain intermediate proportions of sand and clay. Clay soils are compact and cohesive, often poorly drained and aerated, but because of their large internal surface they usually store large amounts of water and minerals. Sandy soils are loose, non-cohesive, well drained, and well aerated, but with a limited storage capacity for water and minerals. Loam soils are intermediate in respect to these character-istics (Fig. 4.1). In general, a high clay content increases the storage capacity of soils for water and minerals (cation-exchange capacity) but decreases the aeration so essential for good root growth and functioning. Thus the clay frac-tion is very important in determining the suitability of a soil for plant growth (Fig. 4.1).

A large amount of organic matter also increases the water-holding capacity and cation-exchange capacity, but decreases the effectiveness of pesticides and decreases injury by toxic substances. According to Wild (1988, pp. 585–588), the more organic matter soil contains the more herbicide, insecticide, or nemato-cide must be applied to be effective. Attempts have been made to increase the wa-ter storage capacity of soil by adding water-absorbing polyacrylamide polymers to the soil. This would seem to be promising for sandy soils. However, Letey *et al.* (1992) reported that although the water held by the additives is available

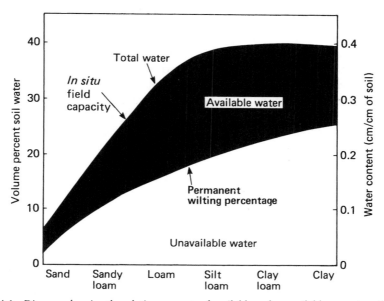

Figure 4.1 Diagram showing the relative amounts of available and unavailable water in soils rang-ing in texture from sands to clay. Amounts are expressed as percentages of soil volume and centi-meters of water per centimeter of soil. After Cassel (1983), from Kramer (1983).

to plants and the interval between irrigations is increased for container-grown plants, little water is conserved over a season.

Numerous attempts have been made to develop systems by which soils with similar properties can be classified together but there is no universally accepted system and readers are referred to Wild (1988, Chapter 24) and to soil text-books for further information on soil classification.

Structure and Pore Space

Soil structure and amount of pore space depend on particle size and the extent to which the basic particles are assembled into stable crumbs or aggregates. Aggregation of clay particles into stable "crumbs" apparently is aided by the presence of root exudates and organic colloids produced by soil organisms. Maintenance of stable aggregates is particularly important in clay soils because it increases pore space, improving infiltration of water and maintenance of good aeration. The structure of clay soil often is damaged by traffic and cultivation when wet, and occasionally in arid regions by flocculation (precipitation) of the clay, caused by an excess of alkali in irrigation water. According to Richter (private communication), growth of tap roots of pine trees increases the bulk density of the soil surrounding them.

The relative amounts of capillary and noncapillary pore space strongly affect soil drainage and aeration and hence its suitability as a medium for plant growth. Capillary pore space consists of small pores (30 to 60 μm or less) that retain water against gravity. This determines the field capacity or the amount of water retained in a soil after a rain or irrigation. Noncapillary pore space is the fraction of soil volume from which water drains by gravity, providing the air space so important for good aeration of roots. About half the volume of most soils consists of pore space, but the proportions of capillary and noncapillary pore space vary widely in different soils, as shown in Figs. 4.2 and 4.3. The past history or treatment of a soil also has important effects on the proportion of capillary pore space, as shown for a forest soil and an adjacent cultivated field soil in Fig. 4.3. Cultivation often damages soil structure and decreases noncapillary pore space. According to Ravina and Magier (1984) the presence of rock fragments decreases compaction and improves aeration of clay soils, but according to Richter *et al.* (1989) it also decreases the water storage capacity, as would be expected. Nobel *et al.* (1992) found more water beneath flat rocks and around boulders than in soil a few centimeters away and stated that this plays an important role for root proliferation of desert succulents and probably for other plants growing in dry soil.

A decrease in pore space increases the bulk density (weight per unit of volume) and resistance to root penetration as measured by a penetrometer. Those changes are accompanied by a decrease in root growth (Barley, 1962; Taylor and Ratliff, 1969). Bengough and Mullins (1991) suggest that some of the early

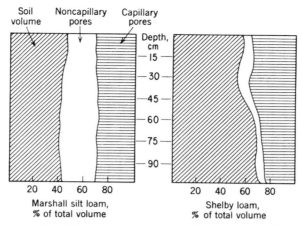

Figure 4.2 Examples of the differences in amount of capillary and noncapillary pore space in two dissimilar soils. A large proportion of noncapillary pore space promotes drainage and improves aeration. After Baver (1948), from Kramer (1983).

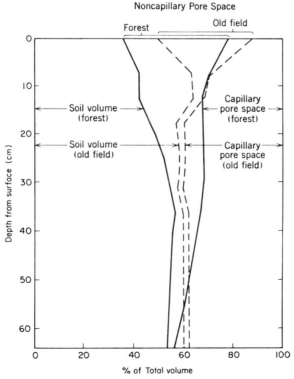

Figure 4.3 Difference between amount of capillary and noncapillary pore space in an old field soil and an adjacent forest soil on the same type of soil. The large volume of noncapillary pore space in the forest soil provides better root aeration and more rapid infiltration of water (Fig. 4.6), reducing surface runoff and erosion during heavy rains. After Hoover (1949), from Kramer (1983).

data on soil resistance and root growth suffer from faulty technology and need to be reinterpreted. However, research indicates that in some instances increasing soil density decreases stomatal conductance, photosynthesis, and shoot growth (Carmi *et al.*, 1983; Masle and Passioura, 1987; Tardieu *et al.*, 1991), but increases water use efficiency, root–shoot ratio, and carbon discrimination in some plants (Masle and Farquhar, 1988). Hamblin (1985) discussed the relationship among soil structure, root growth, water movement through soil, and water absorption in detail.

Soil Profiles

Although alluvial and loess soils often are uniform in texture and structure to considerable depths, most soils show changes with depth that affect their suitability as a medium for root growth and their capacity as a reservoir for water and minerals. Figure 4.3 shows the differences in capillary and noncapillary pore space at various depths in a forest soil and in an adjacent cultivated field on the same soil type, and Fig. 4.4 shows the horizons that might occur in a well-developed forest soil profile.

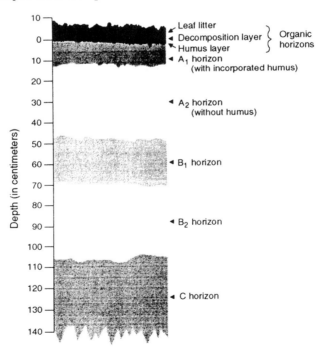

Figure 4.4   A soil profile under an old shortleaf pine stand in North Carolina. Modified from Billings (1978), from Kozlowski *et al.* (1991).

Soil profiles often are interrupted by rock or hard pan layers of natural occurrence (fragipans) or resulting from the passage of heavy machinery (tillage pans) that restrict root growth. Sometimes layers of rock or chemical barriers such as low pH and high concentrations of toxic elements such as aluminum or magnesium in the subsoil, or a high water table, limit root penetration and reduce the volume of soil available to plants as a reservoir for water and minerals. In regions with limited rainfall, a layer of carbonate often accumulates at the deepest point wetted by rain, forming a hardpan or caliche which hinders root penetration. Cassel (in Raper and Kramer, 1983) discussed chemical and physical barriers to root penetration in more detail and it is discussed further in Chapter 5. Perhaps more effort should be made to find genotypes of plants with roots able to penetrate dense soils and to tolerate high acidity and high concentrations of aluminum and other toxic elements. According to O'Toole and Bland (1987) there are extensive genotypic variations in root systems of various crop plants, providing opportunities for selection of root systems with desirable characteristics in special environments.

The effects of restricted root penetration caused by a shallow soil are particularly noticeable during droughts when plants on shallow soils suffer injury sooner than those on deeper soil. Coile (1937) found that height growth of loblolly and shortleaf pine in the Carolina Piedmont was correlated with depth of the A horizon and the characteristics of the B horizon that govern its suitability for root growth. Martin *et al.* (1979) found that subsoiling to break a tillage pan at 25 cm had unidentified beneficial effects on soybean yield in addition to increased availability of water.

SOIL WATER TERMINOLOGY

The water content of soil usually is expressed as a percentage of oven dry weight, or of soil volume. The percentage of soil volume probably is most informative concerning the amount of water available for plants but it is difficult to determine in undisturbed soil. Unfortunately, water content on a percentage basis tells little about the amount of water available to plants because a sand may be saturated at a water content that is near the wilting point for a loam soil, as shown in Table 4.1.

Water Potential

The meaning of water potential and its importance with respect to cell and tissue water relations were discussed in Chapters 2 and 3. The soil water potential depends on four components of varying importance:

$$\Psi_{soil} = \Psi_m + \Psi_s + \Psi_g + \Psi_p. \tag{4.1}$$

Table 4.1 Water Content of Soils of Various Textures at Matric Potentials of -0.03 and -1.5 MPa and at First Permanent Wilting

	Water content as percentage of dry weight	
Name of soil	-0.03 MPa	-1.5 MPa
Hanford sand	4.5	2.2
Indio loam	4.6	1.6
Yolo loam	12.6	7.1
Yolo fine sandy loam	12.6	5.5
Chino loam	19.7	8.0
Chino silty clay	40.8	21.9
Chino silty clay loam	48.9	15.0
Yolo clay	45.1	26.2

Note. From Kramer (1983), based on data of Furr and Reeve (1945) and Richards and Weaver (1944).

In this equation, Ψ_m represents the matric potential produced by capillary and surface-binding forces, Ψ_s represents the osmotic potential produced by solutes in the soil water, and Ψ_g represents the gravitational forces operating on soil water. Ψ_p refers to external pressure and can often be disregarded because the pressure is near atmospheric in the root zone. The exact significance of the various terms, especially the matric term, are discussed in Appendix 2.3 and by Passioura (1980b).

It sometimes is stated that water always moves toward regions of lower total water potential, but this is not always true (Corey and Klute, 1985). The various forces affecting the free energy status of soil water [Eq. (4.1)] are not equally important under all conditions with respect to water movement. For example, although the osmotic potential is an important part of the total water potential with respect to plants growing in saline soils, it has little effect on water movement within the soil. However, it has an important effect on the movement of water from soil to roots because the soil solution is separated from the plant solution by differentially permeable membranes.

Generally, the major forces affecting water movement in soil are matric and gravitational. After the soil is saturated, a fraction of the water moves downward because gravity causes water to drain out of the larger pores. After equilibrium is attained against gravity the soil can be regarded as at field capacity. However, local dehydration by capillary flow, caused by surface evaporation or root absorption, reduces the water potential and causes internal movement. In order to absorb water, roots must generate water potentials low enough to overcome the matric potential plus any osmotic potential of the soil solution. It also should be noted that liquid water and water vapor can move independently of one another in soils (Gurr *et al.,* 1952; Kramer, 1969, p. 69).

Field Capacity

The *in situ* field capacity of a soil refers to the water content after downward drainage has become negligible and water content has become relatively stable. This situation usually is attained several days after a soil has been thoroughly wetted by rain or irrigation, but may require a month or more (Fig. 4.5). Sandy soils usually attain equilibrium much sooner than clay soils, but the presence of a shallow water table or an impermeable layer will slow the process in any type of soil. Also, the water content or apparent field capacity of a soil allowed to drain in the field may be different from that of a column or pot of the same soil allowed to drain over sand in the greenhouse because of the shorter capillary column in the pot than in the field. The problems peculiar to soil in containers were discussed briefly by Hershey (1990) and by White and Mastalerz (1966). The latter termed the upper limit of water content of soil in containers the "con-

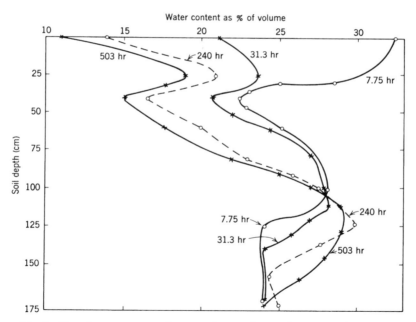

Figure 4.5 Profiles of water content of a nonuniform soil at various times after irrigation. The surface horizon is a sandy loam, changing to a fine sandy loam at about 25 cm, and at 75–100 cm to a clay which holds much more water. The original water content at 125 to 175 cm was about 24%. Over a period of 21 days (503 hr) there was a decrease in water content near the surface and an increase at 100 to 150 cm below the surface. From Kramer (1983) after Rose *et al.* (1965).

tainer capacity" to distinguish it from the field capacity of natural soils. Sometimes laboratory measurements of field capacity are made by subjecting the soil to a pressure of 0.03 MPa in a pressure chamber or an equivalent tension on a tension table. Hillel (1980b, Chapter 3) discussed in detail the problem of measuring field capacity. As it is not really a soil constant, but depends on the conditions under which it is measured, some soil scientists have recommended that it be abandoned. However, it has considerable practical utility if users are aware of its limitations.

Permanent Wilting Percentage

The permanent wilting percentage (PWP) is the soil water content at which plants remain wilted overnight or in a humid chamber unless they are rewatered. Sachs (1882a) observed that plants wilted sooner in sandy than in clay soil, and Briggs and Shantz (1912) developed a standardized method by means of which they found that plants of many species wilted at approximately the same water content in a given soil. They termed this the wilting coefficient, but it is now known as the permanent wilting percentage. Richards and Wadleigh (1952) found that the soil water potential ranged from -1.5 to -2.0 MPa at permanent wilting for many herbaceous plants, with most values near -1.5 MPa, which is now generally used as the approximate soil water potential at permanent wilting. Slatyer (1957) pointed out that this is not a soil constant because wilting really depends on the potential at which leaf cells lose their turgor. However, most crop plants have osmotic potentials in the range of -1.5 to -2.0 MPa so -1.5 is near the point at which wilting can be expected. Although it is convenient to take -1.5 MPa as the soil water potential at permanent wilting, there is no sharply defined lower limit for water availability (Gardner and Nieman, 1964), despite the claim by Veihmeyer and Hendrickson (1950) that soil water is either available or unavailable.

Readily Available Water

Readily available water usually refers to the soil water in the range between field capacity and the PWP. Generally, finer-textured soils contain more readily available water than coarse-textured soils, as shown in Figs. 4.1 and 4.14 and in Table 4.1. Considerable differences occur in the amount of readily available water present at various depths in nonuniform soils, as shown in Fig. 4.5. This suggests that deep rooting can sometimes compensate for a limited supply of available water in the surface soil. An increase in the amount of organic matter, either from root growth or by addition of manure, usually increases the water-holding capacity of coarse-textured soils and improves the aeration of fine-textured soils. The complex role of organic matter in soil is discussed in Chapter 18 of Wild (1988).

Water Demand versus Supply Rate.   The definition of available water as that in the range between field capacity and the PWP is too arbitrary and static to describe accurately the actual situation in the field. From the standpoint of plants, soil water availability depends on the rate at which water can be supplied to roots relative to the plant demand for water. Both supply and demand are variable. Plant demand for water depends primarily on the rate of transpiration which varies widely, depending on the kind and size of plants and meteorological conditions (see Chapter 7). Water supply depends on root length density (root length per volume of soil), root efficiency as an absorbing surface, i.e., their hydraulic conductance, and the hydraulic conductance of the soil which varies with soil type and water content. Thus a water content adequate to meet the plant demand in cool, cloudy weather may become quite inadequate in hot, sunny weather when transpiration is rapid, as shown in experiments of Denmead and Shaw (1962). They found that the average soil water potential in the root zone when the actual transpiration rate of corn rose above the potential rate varied from -1.2 MPa when the rate of transpiration was only 1.4 mm per day to -0.03 MPa when the rate was 6 to 7 mm per day. Thus the need for frequent irrigation is much greater during hot, sunny weather than during cool weather. Some of the implications of this dynamic situation are discussed in more detail in Chapter 7 of Stewart and Nielsen (1990).

WATER MOVEMENT WITHIN SOILS

Movement of water within soils controls the rate of infiltration; the flow to springs, streams, and underground aquifers; and the supply to roots of transpiring plants. The bulk soil solution moves downward under the influence of gravity through the noncapillary pore space, and more slowly through capillary pore space and in films on surfaces of soil particles mostly under the influence of surface or matric forces. Pure water also diffuses as vapor through pore space along gradients of water vapor pressure.

Two important types of water movement are its saturated downward flow (infiltration) after rain or irrigation and its unsaturated flow horizontally toward roots and upward toward the evaporating surfaces of the soil. Sometimes both gravitational and matric forces are involved as in water movement down slopes (Beasley, 1976; Hewlett, 1961). When there are large temperature differences between the surface and deeper horizons there also may be considerable movement of water in the form of vapor from warmer to cooler regions.

Infiltration

The rate of infiltration is important in the recharge of soil water by rain and irrigation. If infiltration is slow, surface runoff is likely to cause erosion, as in the clay soils of the Piedmont of the southeastern United States and in some tropical soils. Downward movement is largely by gravity and depends on the rate at which water is supplied to the wetting front rather than to the difference in Ψ_w between wet and dry soil. Water moves rapidly through the moist soil behind the wetting front because of gravity, but very slowly into the dry soil in front of it. This explains the sharp boundary between wet and dry soil after a dry surface soil has been rewetted by a summer shower. It also explains why soil cannot be rewetted part way up to field capacity by addition of a limited amount of water, because part is wetted to field capacity and part is not wetted at all.

Infiltration into some clays (montmorillonites) is hindered because they swell when wetted, reducing the noncapillary pore space. Infiltration into some sands is hindered because the particles are covered with a hydrophobic coating that prevents wetting (Jamison, 1946). This also sometimes occurs in surface soil after fire. Adams *et al.* (1970) reported that annual vegetation sometimes is excluded from beneath desert scrub because the soil is water repellent. Attempts have been made to improve water infiltration on burned slopes and on watersheds by applying wetting agents (Letey *et al.*, 1962; Mustafa and Letey, 1970). McCauley (1993) reported that the addition of a proprietary compound to irrigation waters increased infiltration and soybean yield on a soil subject to surface crusting. The presence of mulch on the soil surface and the moderate incorporation of organic matter into soil also improve infiltration by preventing puddling and closure of pores. However, incorporation of excessive amounts of peat into potting soils makes wetting slow and difficult. Infiltration usually is more rapid into forest soils than into cultivated soil of similar texture, as shown in Fig. 4.6, because forest soils usually contain more noncapillary pore space (Fig. 4.3). Cultivation tends to destroy noncapillary or macropore space. Channels left by decaying roots probably are important pathways for infiltration in forest soils (Gaiser, 1952) and earthworm burrows also increase the infiltration of water (Wild, 1988, pp. 512–513). Wang *et al.* (1986) concluded that channels made in the soil by roots, insects, and worms have important effects on root growth and movement of gas and water. Infiltration is reduced by compaction caused by human and vehicular traffic and becomes a serious problem in parks, on golf courses, and in some cultivated soils. Various authors in Emerson *et al.* (1978) discuss infiltration in more detail. Bouma (1991) suggested that rapid downward flow of water through macropores is undesirable because it can result in pollution of the groundwater.

Horizontal and upward Movement

In nature there are large diurnal and seasonal variations in soil temperature. For instance, during summer the surface soil often is warmer during the day than at night and it cools more rapidly during the autumn and winter, relative

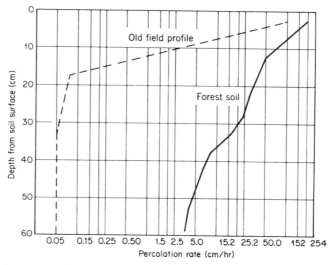

Figure 4.6 Comparison of rates of infiltration into a forest soil and an adjacent old field on the same soil type. Figure 4.3 shows the differences in capillary pore space in the two soils. From Kramer (1983), after Hoover (1949).

to deeper horizons that undergo less change in temperature, causing significant movement of water vapor. This was observed in California by Edlefsen and Bodman (1941), and Lebedeff (1928) reported that in winter in Russia significant amounts of water moved upward from the warm subsoil and condensed in the cooler surface soil. This includes movement both as vapor and as liquid. Movement of water as liquid and vapor was discussed by Slatyer (1967, pp. 109–118), in Gurr *et al.* (1952), and in Hillel (1980b, Chapter 5).

Horizontal movement on a macroscopic scale, as from irrigation ditches, and movement toward roots on a microscopic scale are very important. Soil water in the vicinity of roots of rapidly transpiring plants sometimes tends to become depleted during the day because water absorption exceeds the rate of movement of water through the soil toward roots, resulting in a water-depleted zone around them (MacFall *et al.*, 1990, 1991a, Fig. 5.9, and Chapter 6). Fortunately, the soil in this zone usually is rewetted overnight, but the rate at which this occurs depends on the hydraulic conductance of the soil. This decreases rapidly as the soil dries because the larger pores are emptied first, decreasing the cross-sectional area available for water movement (see Fig. 4.7). Also, contact between soil and roots may decrease as soil and roots dehydrate and shrink (Huck *et al.*, 1970; Faiz and Weatherly, 1982). However, Taylor and Willatt (1983) suggest that the importance of root shrinkage may have been overemphasized because shrinking roots seldom completely lose contact with the soil.

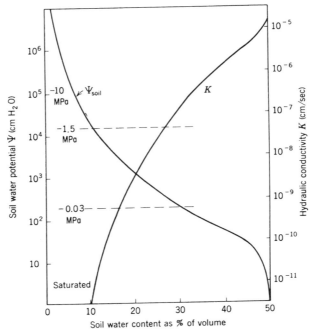

Figure 4.7   Diagram showing approximate decrease in hydraulic conductivity (K) and soil water potential (ψ_w soil) with decreasing soil water content. From Kramer (1983), after Philip (1957).

The movement of water to roots is discussed further in Chapter 6 in connection with water absorption.

Upward movement of water (capillary rise) toward the soil surface is caused chiefly by evaporation from the surface and removal of water by roots of transpiring plants. Upward flow to an evaporating surface from a water table is about twice as rapid in a fine-textured as in a coarse-textured soil (Gardner, 1958). If evaporation is very rapid, loss of water may exceed the rate at which water reaches the evaporating surface which then dries, causing the rate of evaporation to decrease significantly because water movement as vapor in the dry surface soil is much slower than movement as liquid in moist soil. This fact led to emphasis on maintaining a thin surface layer of dry cultivated surface soil as a mulch to reduce the loss of water by evaporation. Gardner and Fireman (1958) reported that lowering the water table from 90 to 180 cm below the soil surface reduced evaporative loss to about 12% of the loss at 90 cm. However, further lowering of the water table to 3 or 4 m had little additional effect on the rate of evaporation. In nature, measurable upward movement sometimes occurs from a depth of several meters (Patric *et al.*, 1965).

There also is evidence that significant amounts of water are transferred from deeper horizons to drier surface soil by roots, where it becomes available to plants (Caldwell and Richards, 1989; Corak *et al.*, 1987; Dawson, 1993, and Chapter 6). In some areas, dew and fog may supply significant amounts of water, and the possible importance of this is discussed in Chapter 6. Vertical movement must be taken into account in estimating the amount of water removed from the root zone. If there is significant upward movement of water and the amount used is calculated from periodic measurements of changes in soil water content in one horizon, water usage will be underestimated unless it is corrected for the amount supplied by upward movement.

In addition to causing loss of water, surface evaporation often results in an undesirable concentration of salt in the surface soil in areas where precipitation is too limited to leach it out. Considerable attention has been given to the reduction of soil surface evaporation by shallow cultivation and by the use of mulches of organic matter and even of pebbles. This is discussed in Hillel (1980a, Chapter 5).

MEASUREMENT OF SOIL WATER

Recognition of the importance of the available water content of soil with respect to plant growth has resulted in development of several methods of measuring soil water content in the field and in the laboratory. Various methods were discussed by Rawlins (in Kozlowski, 1976) and in Stewart and Nielsen (1990, Chapters 6 and 7) and in Pearcy *et al.* (1989, Chapter 3). The variability in soil over area creates serious sampling problems in the field that are discussed in Stewart and Nielsen (1990, Chapter 7).

Soil Water Balance

The chief features of the hydrologic cycle are shown in Fig. 4.8, and a simple equation for the soil water balance can be written as

$$\Delta W = P - (O + U + E_t), \qquad (4.2)$$

where ΔW is the change in water content between samplings, P is the precipitation, O is the runoff, U is deep drainage, and E_t is evapotranspiration from soil and plants between samplings. Precipitation is easily measured, but measurement of the other terms is more difficult. Whitehead and Kelliher (1991a,b) discussed the problems involved in estimating the water balance of a Monterey pine stand in New Zealand for a year. Transpiration from the trees accounted for 50% of annual precipitation, evaporation from wet foliage 15%, evaporation from the understory 7%, deep drainage 24%, and increase in water content of soil in the root zone 4%. Of course these values vary with stand, soil, weather conditions, and the time period between measurements.

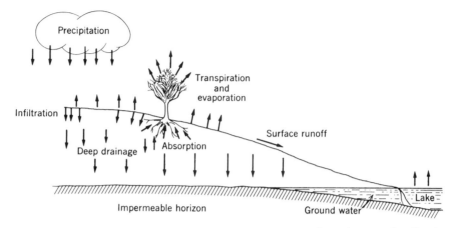

Figure 4.8   The hydrologic cycle, showing disposal of precipitation by surface runoff, infiltration, and deep drainage, and its removal from the soil by evaporation and transpiration. From Kramer (1983).

The most accurate measurements are made with lysimeters. These are large containers filled with soil and equipped with weighing devices which can be buried in a field and planted with whatever crop surrounds them. The use of lysimeters is discussed in Hagan *et al.*, (1967, pp. 536–544) and in Stewart and Nielsen (1990, Chapter 15), and the essential components are shown in Fig. 4.9. Grimmond *et al.* (1992) described a small lysimeter which is said to give as good results as those with a soil area 10 times as great. Good estimates of the soil water balance can also be made from stands of plants growing in small catchment basins where runoff can be measured (Whitehead and Kelliher, 1991b).

Direct Measurement of Soil Water Content

The basic measurement of soil water content is made on samples of known weight or volume dried at 105°C in an oven. Field samples usually are obtained with an auger or sampling tube. Use of a forced draft or microwave oven speeds drying, and other methods requiring less time than oven drying are discussed by Rawlins (in Kozlowski, 1976). Water content as a percentage of dry weight is not very useful unless the permanent wilting percentage and field capacity are known because, as mentioned earlier, a water content representing field capacity in one soil might be below the wilting point in another (Figs. 4.1 and 4.14 and Table 4.1). Sometimes it is useful to convert water per unit of weight into content per unit of volume so the water content can be expressed as millimeters per meter or in inches per foot. Also, additions of water to soil by precipitation are usually given in millimeters or inches. Water content can be converted from

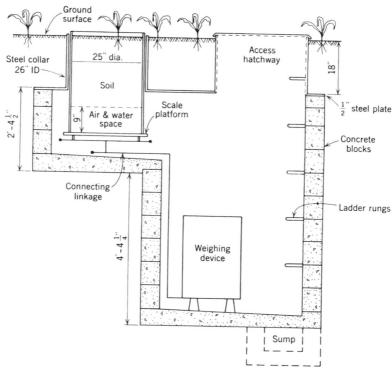

Figure 4.9 The principal components of a weighing lysimeter, modified from England and Lesesne (1962). It consists of a large container filled with soil, mounted on a weighing device. Electronic weighing mechanisms are often used. The lysimeter must be surrounded by a border of similar vegetation if the results are to be applicable to crops or stands of plants. Some lysimetry problems are discussed by Hagan *et al.* (1967, pp. 536–544) and in Stewart and Nielsen (1990). From Kramer (1983).

weight to volume units by multiplying the weight percentage by the bulk density of the soil.

Indirect Measurement of Soil Water

Direct measurement requires many samples obtained at various depths, resulting in considerable disturbance of the soil. It also is very labor intensive. Baarstad *et al.* (1993) describe a method of obtaining soil samples with little disruption of field plots. Most of the measurements are now made by indirect methods which must be related to water content by some kind of calibration procedure.

Neutron Scattering. The neutron probe is used extensively to make repeated measurements of water content at several depths with minimum disturbance. It is based on the fact that hydrogen atoms have a high capacity to slow down and scatter fast neutrons and water is the chief source of hydrogen atoms in most soils. Thus counting slow neutrons in the vicinity of a fast neutron source gives a good measure of the water content of a soil. A neutron probe consists of a source of fast neutrons and a detector for slow neutrons connected to an amplifier and counter (see Fig. 4.10). The probe is lowered in a tube inserted in the soil and measurements are made at various depths. It measures the hydrogen content of a sphere about 20 cm in radius, thus averaging the water content of a fairly large volume of soil. The results can be affected by other sources of hydrogen atoms, such as a high content of organic matter, or by the

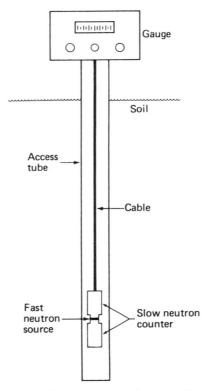

Figure 4.10 The essential features of a neutron meter. A source of fast neutrons and a counter for slow neutrons are lowered to any desired depth in an access tube installed in the soil. The water content of a spherical mass about 20 cm in radius is measured, the size of the mass increasing with decreasing water content. From Kramer (1983).

presence of high concentrations of Cl, Fe, or B. However, some recent models of neutron probes can be corrected and calibrated for most sources of error. The effects of variation in root density on apparent soil water content usually is neglected, but it can be significant if large roots or tubers occur within the soil volume measured by a neutron probe (Faroud *et al.*, 1993). The use of neutron probes was discussed in detail by Hodnett (1986), and Hanson and Dickey (1993) warn of some sources of error.

Gamma Ray Attenuation.   The amount of radiation from a standard source of gamma radiation that passes through the soil is decreased in proportion to an increase in water content. Thus if a source of gamma radiation such as cesium is placed on one side of a column of soil and a detector on the other side in a jig to keep them in alignment, changes in the amount of radiation passing through the soil column can be observed. Of course the readings must be calibrated against gravimetric measurements to convert them into water content. The method was discussed by Ferguson and Gardner (1962), Gurr (1962), and Stewart and Nielsen (1990, pp. 130–131).

Other Methods

Attempts have been made to measure changes in soil water content by changes in electrical capacitance and electrical and heat conductance.

Electrical Capacitance.   The capacitance method depends on the fact that water has a much higher dielectric constant than air or dry materials (water approximately 80, dry soil 5, air 1), hence changes in water content produce measurable changes in capacitance. The method is used to monitor the water content of grain, flour, dehydrated food, and other materials in which samples of fixed volume are easily prepared. Although theoretically attractive, technical problems make it difficult to use on undisturbed soil. However, development of time domain reflectometry has made the method more useful. Parallel rods or wires are inserted in the soil about 5 cm apart to measure the transit time of pulses of microwave energy. This depends on the dielectric constant of the soil, which varies with the water content. Dasberg and Dalton (1985) and Topp and Davis (1985) reported accurate measurements of soil water content by this method, and the former reported satisfactory measurements of electrical conductivity of soil. The method is rapid, but expensive, and thus far is unsatisfactory in wet, saline, or stratified soil. It is discussed in more detail in Chapter 6 of Stewart and Nielsen (1990), and various versions of the method are discussed by Vegelin *et al.* (1990) and Heimovaara (1993). Richardson *et al.* (1992) found that it is reliable for soil in pots and other small containers. Holbrook *et al.* (1992) used the change in the dielectric constant to measure changes in the water content of palm stems.

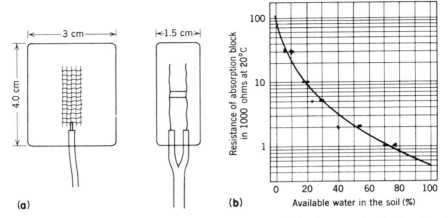

Figure 4.11   Resistance or conductance blocks. (a) Surface and edge diagrams showing location of electrodes in a plaster of paris block. The electrodes are two pieces of stainless-steel screen, separated by a piece of plastic. (b) Resistance in ohms of a plaster of paris resistance block plotted over percentage by weight of available soil water in a silt loam soil in which it was buried. From Kramer, 1983, after Bouyoucos (1954).

Electrical Conductance.   Early attempts to measure changes in soil water content based on changes in electrical conductance or resistance were unsuccessful because of poor contact with the soil, variations in salt content, and temperature variations. The salt and contact errors were reduced by enclosing the electrodes in gypsum blocks (Fig. 4.11) or wrapping them in nylon. The latter last longer in the soil, but gypsum blocks are buffered against salt by the dissolved $CaSO_4$ in the blocks. Gypsum conductance blocks function better than tensiometers in dry soil. A vertical array of conductance blocks is effective in observing the progress of wetting and drying fronts in soil. Other developments in measuring electrical conductance of the soil are described by Kano (in Hashimoto *et al.*, 1990) and Seyfried (1993).

Heat Conductance.   The heat conductance of soil decreases with decreasing water content, and if a heating element is buried in the soil with a detector nearby, changes in water content can be estimated from the rate of heat conduction to the sensor. Unfortunately, the method is not very sensitive in dry soils.

Measurement of Soil Water Potential

Several methods used to measure soil water content really measure the matric potential or pressure which can then be related to the gravimetric water content.

Tensiometers.   Direct field measurements of the matric potential often are made with tensiometers, which consist of porous porcelain cups filled with wa-

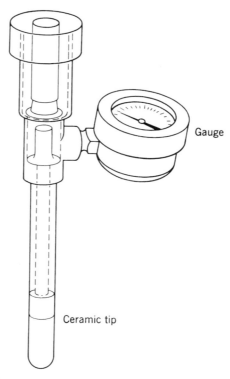

Gauge

Ceramic tip

Figure 4.12   Essential features of a tensiometer, consisting of a plastic tube with a porous porcelain cup attached to the lower end, a screw cap at the upper end for refilling, and a vacuum gauge attached at the side. From Kramer (1983).

ter and buried in the soil at various depths and connected by water filled tubes with vacuum gauges, or pressure transducer systems (Fig. 4.12). As the soil water content decreases the pressure in the water in the porcelain cup decreases in proportion and the decrease is shown on the attached pressure gauge. They work well in moist soil, but at soil water potentials below about -0.08 MPa bubbles of air and water vapor form by cavitation and they often become useless. Methods of dealing with air bubbles have been discussed by Miller and Salehzadeh (1993). Tensiometers are discussed by Rawlins (in Kozlowski, 1976, pp. 31–37), Cassel and Klute (1986), in Chapter 6 of Stewart and Nielsen (1990), and in Chapter 3 of Pearcy *et al.* (1989).

Pressure Plates.   Laboratory measurements of capillary or matric potential are usually made on the pressure plate apparatus developed by Richards (1949, 1954) and others. An example is shown in Fig. 4.13. Moistened soil samples

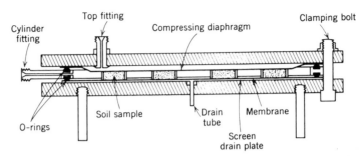

Figure 4.13 A pressure plate apparatus for measuring the matric potential of soil at various water contents. The soil samples are contained in metal rings 5.0 cm in diameter and 1.2 cm deep placed on a cellulose acetate membrane. Compressed air is supplied through the cylinder fitting and air at a slightly higher pressure is supplied through the top fitting to keep the soil samples pressed firmly against the membrane. Sometimes the cellulose acetate membrane is replaced by a porous ceramic plate. The displaced water leaving through the drain tube is collected and measured. Data obtained with this type of equipment are shown in Fig. 4.14. From Kramer (1983).

contained in rings about 5 cm in diameter and 0.5 cm thick are placed on plastic or ceramic membranes permeable to water and solutes and are subjected to a specified pressure until drainage ceases, then removed and the water content determined gravimetrically. By determining the water content at several pressures, instructive water release curves can be constructed such as those shown in Fig. 4.14. The water content of a nonsaline soil at -0.03 MPa is approxi-

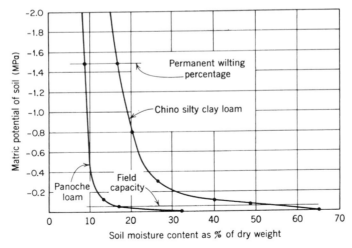

Figure 4.14 Matric potentials of sandy loam and clay loam soils plotted over water content. The curve for the Panoche sandy loam is from Wadleigh *et al.* (1946) and the chino clay loam is from data of Richards and Weaver (1944). From Kramer (1983).

mately the field capacity, whereas that at -1.5 MPa often is taken as the permanent wilting percentage.

Measurement of Osmotic Potential. Measurement of the osmotic or solute potential must be made on samples of soil solution such as those removed in a pressure membrane apparatus or squeezed out of a soil sample in a syringe. The osmotic potential can then be measured in a vapor pressure psychrometer or estimated from its electrical conductivity (Richards, 1954). Sands and Reid (1980) concluded that the most reliable measurements are made on displaced soil solutions.

Measurement of Total Soil Water Potential. The total soil water potential can be measured by burying thermocouple psychrometers protected by enclosure in porous porcelain or stainless-steel screen cylinders at various depths in the soil (Rawlins and Dalton, 1967). Fonteyn *et al.* (1987) used such installations successfully in desert soils. They and others also found that the measurement of predawn plant xylem sap potential usually gives a good indication of soil water potential, although this is questioned by Johnson *et al.* (1991). Measurements also can be made on soil samples removed and placed in thermocouple psychrometers in the laboratory. Boyer (1995) and Brown and Oosterhuis (1992) discussed some of the problems encountered in use of thermocouple psychrometers.

CONTROL OF SOIL WATER

Irrigation

There are two important aspects of soil moisture control, the practical but approximate method used in farming and gardening and the more precise methods used in research. Various aspects of irrigation at the farm and crop level are discussed in Teare and Peet (1983) and Stewart and Nielsen (1990).

Prehistoric farmers evidently realized that rainfall is often inadequate for good crop production because irrigation was being practiced at least 5000 BC in Egypt. It was well developed in Babylonia and China by 2000 BC and in the Americas in pre-Columbian times (Hagan *et al.*, 1967; Masse, 1981 and Chapters 1 and 2). There has been a great increase in the area of land irrigated from about 95 million ha in 1950 to 250 million ha in 1980 (Stewart and Nielsen, 1990, p. 1), but the increase is slowing because of lack of suitable land, competition for other uses of water, and increasing costs. Perhaps the most economical use of water for irrigation is in humid areas as a supplement during droughts (Sneed and Patterson in Raper and Kramer, 1983), where drought often reduces yield (Boyer, 1982). Trends in irrigation practices are discussed in Chapter 3 of Stewart and Nielsen (1990).

Basin and Furrow Irrigation.   The earliest method seems to have been basin irrigation where a small area was enclosed in low dikes of earth and the enclosed surface flooded. This method was used widely in orchards until recently. Row crops often are irrigated by running water down furrows between the rows, but this requires careful grading and leveling of the land and considerable labor to control the flow. There often are differences in the amount of water supplied along the rows, and considerable losses by seepage and evaporation occur by both methods.

Sprinkler Irrigation.   As sprinklers and power sources have improved, sprinklers have increased in size from those that covered lawns or single trees to systems that cover several acres. In addition to saving water and labor they can be used on sloping land and for the addition of fertilizers (fertigation). Occasionally, sprinkler systems also are used for frost control, especially with citrus (Parsons *et al.,* 1985, 1991). The cooling effect of intermittent sprinkling is beneficial in some situations, especially during hot sunny weather (Bible *et al.,* 1968; Gilbert *et al.,* 1970; Southwick *et al.,* 1991; Unrath, 1972). Several types of sprinkler irrigation are discussed by Sneed and Patterson (in Raper and Kramer, 1983) and in Stewart and Nielsen (1990, Chapter 16). A high salt content in the water used for sprinkler irrigation often results in injury to leaves and reduced crop yields (Maas, 1985; Stewart and Nielsen, 1990, Chapter 36). Prolonged sprinkler irrigation may create conditions favorable for leaf and twig disease, and reducing sprinkling periods to 12 hr significantly reduced the incidence of *Botryosphaeria* blight of pistachio in California (Michailides *et al.,* 1992).

Drip and Trickle Irrigation.   This type of irrigation has come into favor because it conserves water and the availability of plastic tubing makes installation convenient. Water is distributed in tubing to individual trees and rows of plants and allowed to drip out through calibrated nozzles, improving control of the amount and the precision of placement. In some instances the tubing is placed in the soil, decreasing the losses from evaporation. One difficulty with the underground system is occasional undetected clogging of nozzles. Sometimes trickle and sprinkler irrigation systems are controlled by tensiometers buried in the root zone, as in the experiments of Adamsen (1992), who found that overhead sprinklers and buried trickle systems produced equal increases in corn yield on a loamy sand soil, but trickle irrigation saved water. Unfortunately, trickle systems wet only limited areas of soil and trees do not always grow as well as with more thorough wetting (Zekri and Parsons, 1989). Many aspects of drip and trickle irrigation are discussed by Bresler (1977) and in Chapter 16 of Stewart and Nielsen (1990).

Secondary Uses of Irrigation Equipment.   Although the primary use of irrigation equipment is to supply water, sprinkler and trickle systems can be used to supply fertilizer (fertigation), pesticides, herbicides, and even growth regulators, and sprinkler systems are used to reduce frost injury to fruit trees (Parsons *et al.,* 1991). Such specialized uses present special problems and require careful management, description of which is beyond the scope of this book. The use of urban wastewater for irrigation also is being explored (Wood, 1988; Mancino and Pepper, 1992). Readers are referred to Stewart and Nielsen (1990, pp. 958– 963 and Chapter 33) for numerous references to such uses on citrus and other fruit crops.

Irrigation in Humid Regions.   Although the need for irrigation usually is associated with arid regions, there is increasing evidence that supplementary irrigation is profitable in humid regions to mitigate the effects of droughts. Sneed and Patterson (in Raper and Kramer, 1983) give numerous examples of improvement of crop yield and quality produced by supplemental irrigation, and its potential may be as great in humid as in arid regions because water is required for a shorter time, it is more available, the quality is higher, and there is less likelihood of damage from salt accumulation. However, the timing of irrigation is more difficult in humid regions with erratic rainfall and this hinders its use as a means of supplying fertilizer. There are several papers on irrigation in humid climates in Raper and Kramer (1983), and Fiscus *et al.* (1984, 1991) discuss methods of increasing the efficiency of irrigation.

Irrigation Scheduling

The area of land irrigated has increased greatly in recent years, creating shortages of good quality water and increasing the use of water high in salt. This creates the need for methods that use water as efficiently as possible. Efficient use of water depends on applying enough to prevent serious plant water deficits without creating a surplus that drains below the root zone. If water is scarce and expensive it sometimes is economically preferable to permit a small reduction in yield to increase the area irrigated. For example, the irrigated area in Israel was increased 30% from 1958 to 1969 by increasing the efficiency with which water was used, without any increase in the total water supply (Shmueli, 1971). Some irrigation scheduling problems are discussed in Taylor *et al.* (1983) and in Stewart and Nielsen (1990, Chap. 17).

Methods of timing irrigations include measurement of soil water content, estimation of loss by evapotranspiration, and measurement of plant water status. The availability of programs for personal computers aids farmers and investigators to store and manipulate the soil, plant, and meteorological data needed to maintain cost efficient irrigation programs.

Soil Water Measurements.   Traditionally, the farmer or gardener dug into the soil and decided from its appearance and "feel" whether water should be applied. The determination of water content by gravimetric methods is useful only if it can be related to soil water content in different horizons, which vary widely as shown in Fig. 4.5. The successful use of neutron meters, tensiometers, and resistance blocks depends on having them installed in zones of maximum root concentration where changes in soil water occur most rapidly.

Use of Evaporation Data.   Because of the difficulty in measuring soil water content over large areas, there is interest in estimating irrigation needs from evaporation losses and the water storage capacity of the soil (Van Bavel and Verlinden, 1956; Blake *et al.*, 1960; Fereres *et al.*, 1981). If the depth of rooting, the water-holding capacity in the root zone, the allowable amount of depletion, and the rate of evaporation are known, the timing of irrigation can be calculated quite accurately. Evaporation data can be obtained from atmometers or evaporation pans or calculated from meteorological data. Tanner (in Hagan *et al.*, 1967) and Hatfield (in Stewart and Nielsen, 1990, Chapter 15) describe several methods of calculating evapotranspiration. Burger *et al.* (1987) discussed irrigation timing for ornamental nurseries based on evapotranspiration from potted plants and from a grass plot. They also calculated crop coefficients based on the ratio of evaporation from various crop plants to that from the grass plot. These varied widely among species, indicating that some kinds of plants use much more water than others.

Use of Plant Water Status.   At least in theory, plants are the best indicators of water availability because they automatically integrate the atmospheric and soil factors that affect plant water status. Many plants show temporary midday wilting and partial closure of stomata on sunny days, even when growing in moist soil. However, if they show evidences of stress at dawn, soil water probably is becoming limiting. Other early indicators of plant water deficit are changes in leaf color, leaf angle, and leaf rolling, and some of these have been used successfully as indicators of the need for irrigation (Oosterhuis *et al.*, 1985; O'Toole and Cruz, 1980; Wenkert, 1980). Blum (1979) used infrared photography to monitor changes in leaf color of sorghum as an indicator of plant water stress. Development of a water deficit usually results in daytime leaf temperatures rising above that found in normally transpiring plants, and remote sensing of leaf temperature seems to be a useful method of evaluating the water status over large areas of crop plants. The measurement of leaf temperature was facilitated by the development of portable infrared thermometers. The use of canopy temperature was reviewed by Jackson (1982) and by Idso *et al.* (1986). Plant water status can be used as a guide to the need for irrigation as discussed by Hsiao (pp. 269–274) and others in Stewart and Nielsen (1990).

The use of plants as indicators for irrigation should take into account differences in tolerance of water deficits among crop plants at various stages of development (Salter and Goode, 1967). For example, corn is very susceptible to water deficit injury during silking and pollination, soybeans during pod formation and filling (Sionit and Kramer, 1977), and wheat during early anthesis (Sionit *et al.*, 1980). Radin *et al.* (1992) found that frequent irrigation of cotton during fruiting increased yield significantly. Hiler and his colleagues attempted to use information concerning susceptibility at various stages of development to develop a stress day factor as a guide to more efficient irrigation (Hiler and Clark, 1971). One of the more successful uses of plants to control irrigation was described by Fiscus *et al.* (1984, 1991). They used a computerized system that included a mass flow porometer to monitor plant water stress, as indicated by reduced stomatal conductance, that turned on an irrigation system when a decrease in stomatal conductance indicated that irrigation was needed. This saved water without reducing yield. Bordovsky *et al.* (1974) reported that timing irrigation of cotton based on plant water stress saved a significant amount of water. Irrigation of various crops is discussed further in Chapter 12, in Stewart and Nielsen (1990), and in Teare and Peet (1983).

Deficit Irrigation.   Where water supply is limited growers sometimes practice deficit irrigation, applying less water to the soil than is removed by evapotranspiration, resulting in increasing water stress late in the season. This reduces the yield of some, but not all, crops. For example, Miller and Hang (1980) found that irrigation of sugar beets in Washington could be reduced to 35–50% of the loss by evaporation on a loam soil without reducing the sugar yield. This was possible because although the fresh weight of roots was reduced the sugar concentration was increased by moderate water stress. According to Snyder (1992), deficit irrigation in California does not necessarily decrease the yield of cotton, the sugar content of sugar beets, or the soluble solids in processing tomatoes under the conditions of their experiments. However, successful deficit irrigation requires careful monitoring of available soil water and the rate of evapotranspiration, and is most successful if the entire root zone of the soil is at field capacity at the beginning of the growing season. Snyder (1992) described a computer program to calculate the acreage to be planted when water supply is limited.

Irrigation Problems

Unfortunately, while irrigation reduces or eliminates the water stress problem, it often creates other problems. At the crop level these chiefly are associated with waterlogging of the soil and those caused by salt accumulation. At the political and economic level they are related to depletion of aquifers and competition with urban and industrial users for water. Unfortunately, wasteful use of water has been encouraged by government subsidies in some regions and is exhausting aquifers and threatening to limit the water available for irrigation in the future. The complicated history of providing water for irrigation and urban development is told in detail by Reisner in "Cadillac Desert" (1986).

Unexpected side effects sometimes occur. For example, sprinkler irrigation of onions in California, especially after bulb formation begins, is said to increase damage from sour skin disease (Teviotdale *et al.*, 1990), and irrigation may either increase or decrease the incidence of other diseases. Sprinkler irrigation can also cause leaching injury to leaves, or if the water is high in salt it sometimes causes injury such as tipburn and discoloration of leaves (Eaton and Harding, 1959; Harding *et al.*, 1958). Some of the technological and social problems associated with the expanding use of irrigation are discussed in Stewart and Nielsen (1990, Chapter 3 and Section VIII).

Soil Waterlogging.   The importance of good soil aeration is mentioned repeatedly in this book (e.g., see Fig. 10.1) and the irrigator must operate between the danger of growth-limiting water deficits and the danger of waterlogging, especially in heavy soils. Basin irrigation is most likely to result in poor aeration, but furrow irrigation also reduces the oxygen supply temporarily. Sometimes poorly drained subsoil becomes saturated because irrigation is based on measurement of the surface soil. Good drainage to remove surplus water is second only to an adequate supply of good quality water in designing irrigation systems. The effects of flooding and deficient soil aeration on plant growth are discussed in Chapter 5, in Kozlowski (1984), and in Iwata *et al.* (1988). Experiments of Sojka and Stolzy (1980) suggest that some forms of irrigation may cause sufficient deficiency in soil oxygen to result in stomatal closure.

Salt Accumulation.   Irrigation is most common in arid regions where the water often is high in salt and the evaporation from surface soil is rapid. As a result, salt accumulation in the soil is a major problem in almost all irrigated areas. The productivity of one-fourth to one-third of the irrigated land in the United States is said to have been reduced by salt accumulation and large areas in the Middle East have been rendered unproductive by it. Several writers address these problems in books edited by Hagan *et al.* (1967), Poljakoff-Mayber and Gale (1975), and Stewart and Nielsen (1990).

Experimental Control of Soil Water Content

It often would be useful in research if plants could be grown in soil kept at various levels of soil water potential below field capacity. Unfortunately, one cannot half wet a soil, as should be evident to any one who has dug into a dry soil following a summer shower and observed the sharp demarcation between wet and dry soil. Since field capacity is the water content held against gravity, a soil cannot be wetted to less than field capacity. If a container is filled with dry soil having a field capacity of 30% and enough water is added to wet the soil to 15%, the upper half of the soil will be wetted to 30% and the lower half will remain dry, a fact pointed out long ago by Shantz (1925) and Veihmeyer (1927). Sometimes in the older literature it was stated that plants were grown in soil maintained at an arbitrary water content such as 20 or 30% of dry weight or field capacity. Strictly speaking, this was impossible as part of the soil volume was wetted to field capacity while part remained dry.

This limitation poses a problem to the investigator who wishes to observe the effect on plant growth of soil water contents intermediate between field capacity and permanent wilting. Attempts to accomplish this by growing plants in pots on top of columns of sand of various heights standing in pans of water (Moinat, 1943) or by growing them in autoirrigated pots at various distances above the water supply (Livingston, 1918; Read *et al.*, 1962) have not been successful. Another approach has been to vary the intervals between irrigation so plants are subjected to various degrees of water stress before rewatering (Richards and Marsh, 1961). However, the plants are not really subjected to a constant degree of water deficit. Wadleigh (1946) discussed calculation of what he termed the integrated soil moisture stress for cycles of wetting and drying.

One way to grow plants with a fixed water supply is to grow them in nutrient solution to which additional solute is added to lower the water potential. Inorganic solutes such as NaCl are absorbed by plants (Boyer, 1965; Eaton, 1942) as are some organic solutes. The latter also are often attacked by microorganisms. Polyethylene glycol is often used because it is not attacked by microorganisms, but it can be toxic if absorbed through broken roots (Lawlor, 1970).

Solutes can have direct effects in addition to lowering the water potential and they do not necessarily exactly reproduce the effects of soil water deficits. Zur (1967) and others have eliminated the direct effects of solutes by growing plants in thin layers of soil in containers with side walls made of a differentially permeable membrane such as cellulose acetate (Fig. 4.15). The containers can then be immersed in solutions maintained at any desired water potential. However, plant size is limited because the soil mass can be only a few centimeters thick in order to maintain a uniform water potential in it and the membranes often are attacked by microorganisms.

Another approach is to supply water to only part of the root system, either by dividing the root system between containers holding moist and dry soil (Davies and Zhang, 1991, pp. 58–59; Gowing *et al.*, 1990) or by supplying water to only part of the total soil volume (Boyer and McPherson, 1975; Mc-

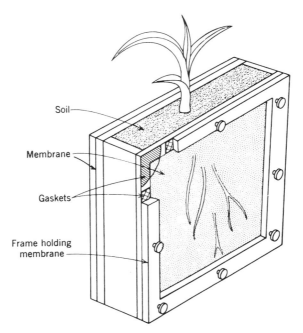

Soil

Membrane

Gaskets

Frame holding
membrane

Figure 4.15   A plastic chamber in which plants can be grown in soil maintained at a fixed water potential by immersion in a solution of polyethylene glycol kept at the desired Ψ_w. The soil mass can be only a few centimeters thick because of slow water conduction at low soil water potentials. Designed by Dr. David Lawlor. From Kramer (1983).

Pherson and Boyer, 1977; Westgate and Boyer, 1985a). The limited root surface in moist soil reduces the water supply to the shoots which can be maintained at a reduced water potential for a long period. The roots in the dry soil are kept alive by water transported from the roots in moist soil. Vaclavik (1966) wetted small areas of soil throughout pots by injecting water through a long needle. All of these methods are compromises.

Actually, growing plants with a constant level of water stress is an artificial situation because in nature the soil water content continually changes. In humid climates the soil water content and water potential decrease steadily after a rain until the soil is rewetted by another rain, as shown in Fig. 4.16. In arid regions plants often must survive through the entire growing season on water stored in the soil during the preceding wet season. According to Passioura (1972), under such conditions it may be advantageous to crops such as wheat for roots to have a high hydraulic resistance that conserves water early in the season to be used later during seed filling. However, this is effective only if no other kinds of plants are present to compete for the water.

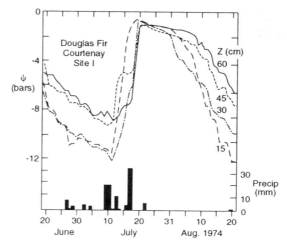

Figure 4.16   Soil water potentials at four depths beneath a Douglas fir forest on Vancouver Island in British Columbia during two drying cycles, before and after midsummer rains. Notice that water was removed most rapidly from the upper soil horizons. From Nnyamah and Black (1977).

SUMMARY

The success of plants is related closely to the properties of the soil in which they grow because it is the source of the water and mineral nutrients essential for growth. It also must constitute a suitable medium for growth of the roots necessary for anchorage and for absorption of water and minerals used in growth. This depends largely on soil texture and structure. Sandy soils are well drained and well aerated and are a favorable medium for root growth, but have limited storage capacity for water and minerals. Clay soils often are poorly drained and aerated and are a less favorable medium for root growth, but because of their large internal surface they store more water and minerals. Loam soils are intermediate in respect to these properties. Occasionally an excess of toxic elements such as aluminum, low pH, excess salinity, or hard pan layers limit root growth.

The amount of soil water available to plants is very important to their success and it varies widely between sands and clays during the growing season. The soil water content can be expressed as a percentage of dry weight of oven-dried samples or measured *in situ* by use of calibrated neutron probes, gamma ray attenuation, or change in electrical resistance. The soil water content is meaningful for plant growth only if it is considered in relation to the field capacity and permanent wilting point of the soil under study. Field capacity is the water content of a soil in which drainage has essentially ceased after thorough wetting (Ψ_w about -0.03 MPa), and the permanent wilting point is the water content at which plants remain wilted overnight unless rewatered (Ψ_w about -1.5 MPa). The water content of a sandy soil at field capacity can be lower than that of a heavy clay soil at the permanent wilting point. The *in situ* water potential of a soil can be measured with tensiometers or psychrometers buried in the root zone.

Control of soil water by irrigation has been practiced since before the beginning of recorded history. There has been a great increase in the area of land irrigated during the 20th century, but the increase is slowing because of lack of suitable land and water and increasing costs. Probably the most economical use of water for irrigation is as a supplement to rainfall during droughts in humid areas. Formerly, irrigation was done chiefly by running water in furrows or basins, but this has become supplanted by sprinklers and trickle systems in which water is supplied directly to plants. Irrigation equipment sometimes is used to supply fertilizer (fertigation), pesticides, and herbicides, and sprinkler systems are used to reduce frost injury to plants. Methods of timing irrigation include measuring the soil water content or rate of evaporation and monitoring the plant water status. Irrigation often creates problems such as waterlogging of the soil and salt accumulation, and good drainage is as important as an adequate water supply for successful irrigation.

SUPPLEMENTARY READING

Box, J. E., Jr. and Hammond, L. C., eds. (1990). "Rhizosphere Dynamics." Westview Press, Boulder, CO.

Brady, N. C. (1990). "Nature and Properties of Soil," 10th Ed. Macmillan, New York.

Carter, M. R., ed. (1993). "Soil Sampling and Methods of Analysis." Lewis Publishers, Boca Raton, FL.

Hillel, T. (1980a). "Fundamentals of Soil Physics." Academic Press, New York.

Hillel, T. (1980b). "Applications of Soil Physics." Academic Press, New York.

Iwata, S., Tabuchi, T., and Warkentin, B. P. (1988). "Soil-Water Interactions: Mechanisms." Dekker, New York.

Marshall, T. J., and Holmes, J. W. (1988). "Soil Physics." Cambridge University Press, Cambridge.

McWilliam, J. R. (1986). The national and international importance of drought and salinity effects on agricultural production. *Aust. J. Plant Physiol.* **13**, 1–13.

Raper, C. D., and Kramer, P. J., eds. (1983). "Crop Reactions to Water and Temperature Stresses in Humid, Temperate Climates." Westview Press, Boulder, CO.

Rendig, V. V., and Taylor, H. M. (1989). "Principles of Soil-Plant Interrelationships." McGraw-Hill, New York.

Stewart, D. A., and Nielsen, D. R., eds. (1990). "Irrigation of Agricultural Crops." Agron. Mon. 30, ASA-CSSA-SSSA, Madison, WI.

Teare, I. D., and Peet, M. M., eds. (1983). "Crop-Water Relations." Wiley, New York.

Wild, A. D., ed. (1988). "Russell's Soil Conditions and Plant Growth." 11th Ed. Longman Group, UK/Wiley, New York.

Wild, A. (1993). "Soils and the Environment." Cambridge University Press, New York.

5

Roots and
Root Systems

INTRODUCTION

The size of root systems has been studied extensively and described by Weaver (1920, 1926), Weaver and Bruner (1927), and others in the United States and by Kutschera (1960) in Europe. However, the physiology of roots received less attention until recently, at least partly because they are usually underground and more difficult to study. Nevertheless, even casual consideration of their functions indicates that physiologically vigorous root systems are as essential as vigorous shoots for successful plant growth because root and shoot growth are so interdependent that one cannot succeed without the other. We will discuss briefly the functions of roots, root growth and structure, and development of root systems in relation to water and mineral absorption in this chapter. Knowledge of root structure is important because it affects the pathway and resistance to water and solute movement, while the extent of root systems affects the volume of soil available as a source of water and mineral nutrients. Much research on factors affecting root growth and root functioning is summarized in books edited by McMichael and Persson (1991) and Waisel *et al.* (1991), in papers by Epstein (in Hashimoto *et al.*, 1990), Feldman (1984), and Zimmermann *et al.* (1992), and in a review by Aesbacher *et al.* (1994), and some of the more important and interesting topics are discussed in this chapter.

FUNCTIONS OF ROOTS

The functions of roots include anchorage, the absorption of water and mineral nutrients, synthesis of various essential compounds such as growth regula-

115

tors, and the storage of food in root crops such as sugar beet and cassava (*Manihot,* spp.).

Anchorage

The role of roots in anchorage often is taken for granted, but it actually is very important because the success of most land plants depends on their ability to stand upright. For example, part of the success of the Green Revolution resulted from development of cereal crops with short, stiff stems that resist blowing over by wind and rain (lodging), but stiff stems are useless unless they are firmly anchored in the soil by vigorous root systems. The mechanical strength of roots also is important in preventing overthrow of trees by wind and winter injury to crops such as winter wheat by frost heaving. Resistance to uprooting by grazing animals may also be important for small herbaceous pasture plants. Roots also increase the stability of soil on slopes (Hellmers *et al.,* 1955). Coutts (1983) discussed the relationship between tree stability and their root systems, and Ennos *et al.* (1993) and Crook and Ennos (1993) studied the mechanics of root anchorage and pointed out that resistance to lodging in maize and wheat is improved by increased spread and bending strength of roots.

Roots as Absorbing Organs

The importance of deep, wide spreading root systems for absorption of water and minerals cannot be overemphasized and is discussed later in this chapter and in Chapter 6.

Synthetic Functions

Root cells possess many of the synthetic functions of shoot cells and some aerial roots even produce functional chloroplasts. Flores *et al.* (1993) cited examples of photosynthesis in aerial roots of orchids and mangroves and reported that roots of several genera of *Asteraceae* and *Orchidaceae* can become adapted to photoautotrophy in solution culture by exposure to light and high concentrations of CO_2. According to Johnson *et al.* (1994), phosphorus deficiency stimulates dark fixation of CO_2 in proteoid roots of lupine and increases production of citrate and its secretion into the rhizosphere where it increases the availability of phosphorus. Bialzyk and Lechowski (1992) observed significant absorption of CO_2 from the root medium and transport of carbon compounds from roots to shoots in tomato.

Most roots are dependent on shoots for thiamin and sometimes for niacin and pyridoxine, and receive auxin from the shoots. The nitrogen-fixing role of *Rhizobium* bacteria in root nodules is important and the activities of microorganisms in the rhizosphere may also be important (Box and Hammond, 1990; Curl and Truelove, 1986; Wild, 1988, pp. 526–530). The role of fungi causing mycorrhizal infection and the effects of pathogenic fungi are discussed later.

Nicotine is synthesized in the roots of tobacco and is translocated to the shoots, and tobacco shoots grafted on tomato root systems contain no nicotine (Dawson, 1942). Ammonia also is converted to organic compounds in roots, but in most species much of the nitrate is translocated to the shoots and reduced. According to a review by Oaks (1992), nitrate reduction in roots differs in legumes and cereals because legume roots export asparagine to their shoots, but cereals do not. Biles and Abeles (1991) found peroxidases and other proteins synthesized in the roots in the xylem sap of several species of plants. According to Rao (1990), flavonoids produced in roots protect them against pests such as fungi and nematodes and have allelopathic effects.

It was suggested long ago that roots probably synthesize hormones essential for shoots (e.g., Went, 1943). Among these are cytokinins (Kende, 1965; Skene in Torrey and Clarkson, 1975), gibberellins (Skene, 1967), and abscisic acid (Davies and Zhang, 1991). It has been suggested that reduced shoot growth of plants whose roots have been subjected to stress such as deficient soil water, deficient aeration, high salinity, or low temperature is caused at least in part by a change in the amount and kind of growth regulators supplied from the roots (Blum *et al.*, 1991; Davies and Zhang, 1991; Itai and Vaadia, 1965; O'Leary and Prisco, 1970; Skene in Torrey and Clarkson, (1975). However, Jackson *et al.* (1988) claimed that stressed pea roots are not a source of excess abscisic acid (ABA) and Munns (1990) argued against ABA as a chemical signal. In contrast, Khalil and Grace (1993) reported that when one-half of a split root system of sycamore maple was water stressed there was a large increase in ABA in the drying roots and the xylem sap, and a decrease in stomatal conductance in the shoots. Of course these stresses are likely to also reduce the supply of water and minerals, which likewise reduces shoot growth. The relative importance under field conditions of decreased absorption of water and minerals versus changes in root metabolism and in the supply of hormones to the shoots deserves further study (McIntyre, 1987). Gowing *et al.* (1993) discuss the current state of knowledge concerning the production and response to ABA.

Roots as Sensors of Water Stress

In recent years considerable attention has been given to the possibility that roots of plants in drying soil function as primary sensors of water stress. According to this view, as the soil dries changes in root metabolism such as a decrease in cytokinin production, an increase in ABA production, and a disturbance of nitrogen metabolism send biochemical signals to the shoots that produce physiological changes such as a decrease in growth, stomatal conductance, and rate of photosynthesis, regardless of the water status of the leaves. The root sensor effect has been demonstrated in pot experiments with split root systems (Khalil and Grace, 1993), with roots in pressure chambers (Davies *et al.*, 1986; Passioura, 1988a; Schulze, 1986a; Turner, 1986 and others), and with maize under field conditions (Tardieu *et al.*, 1991). Tardieu *et al.* (1991) reported that in the field the entire root system of maize must be in dry soil before the ABA concentration in the xylem sap increases.

Under field conditions in areas with adequate rainfall and many sunny days shoots of plants often are subjected to water stress even though the roots are in moist soil. Examples are the curling of leaves of corn plants in moist soil and the midday water stress in flooded rice reported by Tazaki *et al.* (in Turner and Kramer, 1980). It therefore seems doubtful if roots usually are primary sensors of water stress under those conditions (Kramer, 1988), although opposing arguments are presented by Zhang and Davies (1990), Davies *et al.* (1990), and Tardieu *et al.* (1991). Davies and Zhang (1991) and Davies *et al.* (1994) have good reviews of the role of roots as sensors.

Experiments of Kitano and Eguchi (1992a,b) seem to demonstrate that a change in shoot water status can directly affect stomatal conductance. Cramer and Bowman (1991) approached this problem by comparing leaf elongation of intact maize plants and shoots from which the roots have been removed, when placed in a saline solution. The short-term response of leaves on shoots without roots was similar to that on shoots with roots. This led them to conclude that signals from roots are not necessary for the occurrence of short-term reduction of leaf elongation by high salinity in maize. Day *et al.* (1991) reported that chilling one-half of a split root system of *Pinus taeda* did not decrease photosynthesis, suggesting that no nonhydraulic signal originated in the chilled roots.

Of course as the soil dries both roots and shoots dehydrate, altering the biochemistry in both and changing the biochemical and hydraulic communications between them. This makes it difficult to determine the relative importance of chemical and hydraulic messages. This topic is discussed again in Chapters 6 and 9.

ROOT GROWTH

Root growth results from cell division and the pressure developed by enlargement of newly formed cells. Cell enlargement is discussed in Chapter 11 and root growth at the cellular level is discussed by Barlow in Gregory *et al.* (1987). Bret-Harte and Silk (1994) question how sufficient carbon for growth reaches root meristems, which are several millimeters beyond the termination of the phloem. The older portions of roots are anchored in the soil and the tips are pushed forward through the soil by cell enlargement at rates of a few millimeters to a few centimeters daily, often following a tortuous path of least resistance through crevices and around pebbles and other obstructions. Generally, root tips tend to return to their original direction of growth after passing around

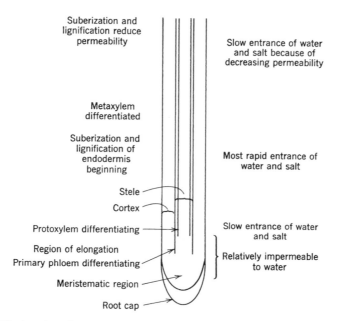

Figure 5.1   The location of primary tissues in an elongating root and relative amounts of absorption at various distances behind the apex. The distance from the apex at which various stages of maturation occurs depends on the species and the rate of root elongation. According to McCully and Canny (1988) the metaxylem of maize and soybean roots becomes capable of significant water conduction 15 to 30 cm behind the root apex. From Kramer (1983).

obstacles, a characteristic observed in the 19th century by Darwin and others, which is known as exotropy (Wilson, 1967). Despite numerous temporary deflections, branch roots of many plants tend to grow outward for a time before turning downward (Wild, 1988, p. 121 and Fig. 5.11). The cause of this change in sensitivity of roots to gravity deserves more study.

During growth and maturation roots undergo changes in anatomy at various distances behind the apex that affect the permeability to water and solutes. The approximate order of maturation of tissues is indicated in Fig. 5.1, but the length of the various zones varies widely, depending chiefly on the species and the rate of growth. In slowly growing roots, differentiation of new tissues occurs much closer to the root tips than in rapidly growing roots (Peterson and Perumalla, 1984). A diagram of a cross section through a fully differentiated dicot root is shown in Fig. 5.2, and a scanning electron micrograph of a cross section of a young barley root is shown in Fig. 5.3.

Enlarging roots often develop pressures sufficient to lift sidewalks and crack masonry walls, as many readers must have noted. Clark (1875) discussed a

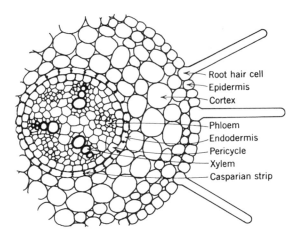

Figure 5.2   Cross section through a squash root in the region where salt and water absorption occur most rapidly. The endodermal cell walls usually become much thickened, except for passage cells opposite the xylem which often remain unthickened. The pericycle usually gives rise to branch roots and the cork cambium found in older roots. From Kramer (1983), after Crafts and Broyer (1938).

number of examples of the pressure developed by growing roots and described an experiment in which a growing squash fruit supported 5000 lb. Gill and Bolt (1955) also presented some data on the pressures developed by growing roots. Dr. Richter (Private communication) found that enlargement of tap roots and large branch roots of pine compresses the soil in their immediate vicinity sufficiently to increase its bulk density. This might decrease infiltration of water around the base of tree trunks. It seems possible that the large number of roots produced beneath grasses (Dittmer, 1937; Pavlychenko, 1937) might temporarily reduce the pore space, but their death and decay would soon increase it. The effect of root growth on soil bulk density deserves further investigation.

Epidermis and Root Hairs

The epidermis and associated root hairs have received considerable attention because they make direct contact with the soil and are the surfaces through which most of the water and minerals usually enter roots. The epidermis is composed of relatively thin-walled, elongated cells, which produce protrusions from the epidermal cells, or from cells of the hypodermis lying beneath it, termed root hairs. The development of root hairs is discussed by Hofer in Waisel *et al.* (1991, Chapter 7) and by Schiefelbein and Somerville (1990).

Root hairs greatly increase the root surface in contact with the soil and decrease the distance that ions and water must travel to reach root surfaces. This presumably facilitates absorption of water and minerals, at least in some situa-

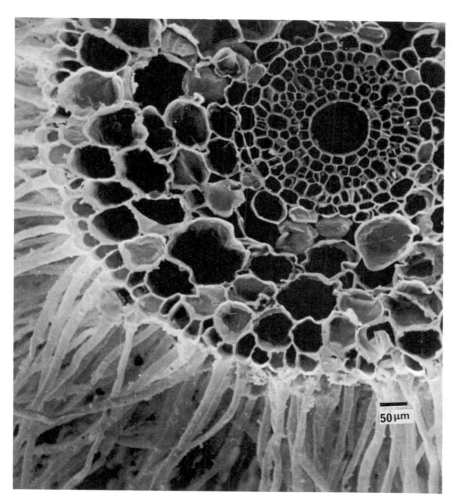

Figure 5.3 Scanning election micrograph of a young barley root showing numerous root hairs, a large central xylem vessel, and a few small vessels. The endodermal cell walls are not yet thickened From Kramer (1983), courtesy of Prof. A. W. Robards and A. Wilson, Department of Biology, University of York, England.

tions (Itoh and Barber, 1983). Because of their small diameter they can penetrate soil pores too small to be penetrated easily by roots. On growing roots the root hair zone moves forward as the roots elongate and the older root hairs usually are destroyed by suberization of epidermal cells. However, they persist for months on some herbaceous plants (Dittmer, 1937; Scott, 1963; Weaver, 1925) and indefinitely on some woody plants (Hayward and Long, 1942). The number of root hairs varies widely among species and with soil conditions, generally being most abundant in moist, well-drained, well-aerated, loamy soils (Itoh and Barber, 1983; Kramer, 1983, pp. 125–127; Meisner and Karnok, 1991). Their formation usually is inhibited by oxygen deficiency and by the presence of ectotrophic mycorrhizae. The number of root hairs reported varies from 20 to 500 per cm^2 of root surface on roots of trees to 2500 on roots of winter rye (Kramer, 1983, p. 126). The large number of root hairs on roots of a cereal is shown in Fig. 5.3. However, observations of McCully and Canny (1988) and Kevekordes *et al.* (1988) question whether extensive absorption of water occurs in the root hair zone of maize and other grasses and some dicots because in their observations the late metaxylem elements were alive and incapable of significant longitudinal water conduction in the apical 15 to 30 cm of maize and soybean roots. This deserves more attention.

The importance of root hairs in the absorption of water and minerals seems to vary among species and cultivars, and perhaps with the growth stage of the plant. For example, Bole (1973), using cultivars of wheat known to differ in root hair frequency, found that the amount of root hair development did not significantly affect phosphorus uptake from a clay loam soil with either a low or high phosphorus content. He also found that rape and flax roots, which bear few or no root hairs, absorbed more phosphorus per unit of length than wheat roots with numerous root hairs. In contrast, Itoh and Barber (1983) reported that root hairs clearly contribute to the phosphorus uptake of Russian thistle and tomato and to a lesser extent lettuce, but not of wheat. Perhaps some of the discrepancies in the literature result from differences in plant demand for minerals at the time of the experiment, which depends on rapidity of growth (Raper *et al.*, 1977; Vessey *et al.*, 1990). Also, if root density is high roots may compete with one another for water and minerals, even if no root hairs are present. It seems that the contribution of root hairs to water and mineral absorption is variable and deserves more study.

The epidermal cells and root hairs are covered by thin films of cutin wherever they are exposed to air (Scott, 1963, 1964), and the root cap and adjacent epidermal cells are covered by a layer of mucigel consisting of polysaccharides secreted by root cells and perhaps partly by associated microorganisms. The mucigel acts as a lubricant to elongating roots and improves contact with the soil (Foster, 1981; Jenny and Grossenbacher, 1963; Oades, 1978). It may also protect roots from aluminum toxicity (Hecht-Buchholz and Foy, 1981; Horst *et al.*, 1982), but Delhaize *et al.* (1993) attribute aluminum tolerance of wheat to the excretion of malic acid. Roots release large amounts of organic matter (Barber and Martin, 1976) which stimulate microorganisms and result in formation of a unique layer high in organic matter in the soil around them, called the rhizosphere (Box and Hammond, 1990; Curl and Truelove, 1986; Wild, 1988, pp. 526–530). Root exudates may use 2 to 20% of the carbon fixed by photosynthesis (Clarkson, 1985, p. 102), and Lambers in Gregory *et al.* (1987, Table 9) gives an average of 5%.

Endodermis

A conspicuous part of the primary structure of many roots is the endodermis (see Fig. 5.2), the innermost layer of the cortex, the walls of which often become conspicuously thickened, plus strips of suberized tissue on the radial walls, the Casparian strips. This is shown in Fig. 5.4. A similarly thickened layer of cells, the exodermis, sometimes develops beneath the epidermis of roots (Esau, 1965; Peterson, 1988). This decreases their permeability and presents a barrier to inward apoplastic movement of water and solutes in the cell walls. However, there is considerable evidence that some water and minerals enter roots even where the endodermis is suberized (Fig. 5.5). Thickening of the endodermis often is quite uneven, with pits and plasmodesmata in the walls which permit symplastic movement of materials from cell to cell (Clarkson and Robards in Torrey and Clarkson, 1975), and the endodermis often is pierced by branch roots that may provide openings for apoplastic radial water movement (Dumbroff and Peirson, 1971; McCully in Gregory *et al.*, 1987; McCully and Canny, 1988; Peterson and Perumalla, 1984; Peterson *et al.*, 1981; Queen, 1967). The development and functioning of the endodermis are discussed in Torrey and Clarkson (1975) and in Clarkson (1993), and there is much interesting information on tissue development during root growth in books on plant anatomy. Benfey and Schiefelbein (1994) stated that a number of mutants have been identified in roots of

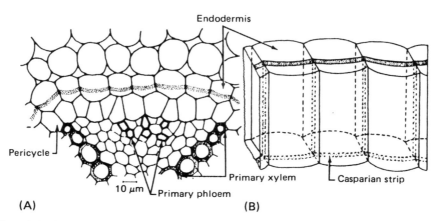

Figure 5.4 Structure of endodermis. (A) Cross section of part of a morning glory root (*Convolvulus arvensis*) showing location of endodermis. (B) Diagram of endodermal cells, showing Casparian strips on radial walls of endodermal cells. From Kramer (1983). Adapted from Esau (1965) by permission of John Wiley and Sons, New York.

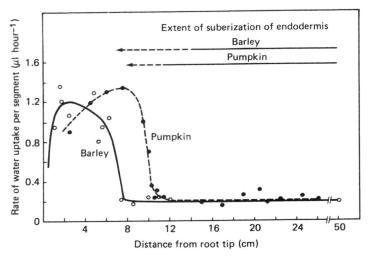

Figure 5.5 Rate of water uptake at various distances behind the apex of barley and pumpkin (*Cucurbita pepo*) roots with varying degrees of suberization of the endodermis. Note that there is measurable uptake even where suberization is complete. From Kramer (1983), after Agricultural Research Council Letcombe Laboratory Annual Report (1973, p. 10).

Arabidopsis that affect growth and reaction to environmental stimuli. The pathway of radial movement of water in roots is discussed in Chapter 6.

Secondary Growth

Secondary growth of roots, resulting from cambial activity, increases root diameter, but causes loss of primary tissues such as the epidermis and cortex, including the hypodermis and endodermis, and development of a suberized outer layer of bark which must significantly modify the pathway of and increase resistance to water and solute entrance. According to Addoms (1946), the entrance of water into suberized roots occurs through lenticels, gaps around branches, and wounds in the bark. The variation in root structure with age and stage of development means that there are large variations in permeability and uptake of water and ions along roots (Fig. 5.5), making it difficult to estimate average uptake per unit of root length.

Root Contraction

A seldom mentioned aspect of root growth that has been observed in many species is their contraction (Esau, 1965, pp. 519–521). It is said to occur in over 450 species, including alfalfa, sugar beet, carrot, and various bulbous monocots. In at least some instances it is brought about by longitudinal contraction and radial expansion of parenchyma cells, resulting in distortion of the vascular tissue. Its importance is uncertain, although Esau speculated that it might pull the shoot apex into the soil surface where the environment is more favorable for growth and the development of adventitious roots.

Rate and Periodicity of Root Growth

The rate of root elongation varies widely among species, with the season, with variation in such soil conditions as water content, aeration, and temperature, and with variations in the shoot environment that affect the supply of carbohydrates. The principal roots of maize were observed to grow 5 or 6 cm per day for 3 or 4 weeks (Weaver, 1925) and the expansion of a maize root system over time is shown in Fig. 5.6. A rate of 10 or 12 mm per day is said to be common in grasses, but rates of 3 to 5 mm seem to be more common in tree roots (Barney, 1951; Reed, 1939; Wilcox, 1962). A few studies indicate that roots sometimes elongate more rapidly at night than during the day (Lyr and Hoffman, 1967; Reed, 1939). Such behavior is most likely to occur when high rates of transpiration produce daytime water stress.

Reich *et al.* (1980) reported that in a constant environment flushes of oak root growth occurred between flushes of shoot growth. There also are seasonal cycles in root growth of perennial plants at least partly related to soil temperature (Lyr and Hoffman, 1967; Romberger, 1963). Turner (1936) and Reed (1939) observed root growth every month of the year in loblolly and shortleaf pine, with the most growth occurring in the spring and summer and the least in the winter. Periods of slow root growth in the summer coincided with periods of low soil moisture (Fig. 5.7). According to Teskey and Hinckley (1981), physiologically optimum soil temperatures and water potentials never occurred simultaneously in the Missouri oak–hickory forest that they studied.

Root extension into previously unoccupied soil is important because it makes additional water and minerals available. Thus the ability of roots to resume growth promptly after transplanting, known as the root growth potential (RGP), is very important to the success of transplants. Conditions in the nursery such as water supply, fertilization, density of seedlings, and time of lifting affect the capacity of seedlings to generate new roots when outplanted. Evidence that roots can grow on water mobilized from stem tissue after transplanting exists and this may aid in establishment (Matyssek *et al.*, 1991a,b). Occasionally, warm winter weather reduces the root growth potential of seedlings of some cooler climate species (Stone and Norberg, 1979). The factors involved in root growth potential of forest tree seedlings are discussed in detail by Kramer and Rose (1986).

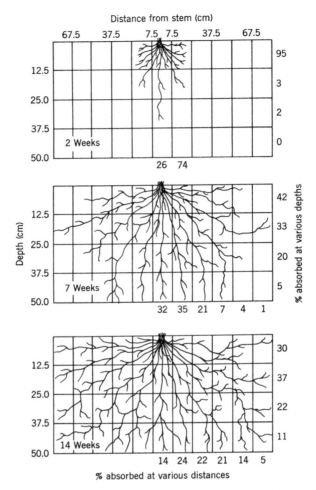

Figure 5.6 Expansion of a maize root system growing in a clay loam soil, based on the uptake of radioactive phosphorus placed in the soil at various depths and distances horizontally from the seedlings. Numbers at right are percentages of total ^{32}P absorbed from various depths, those across the bottom are percentages absorbed at various distances horizontally from the seedlings. From Kramer (1983), after Hall *et al.* (1953).

Depth and Spread of Roots

In deep, well-aerated soil, roots penetrate to great depths and spread widely. Peanut roots reached a depth of 120 cm in 40 to 45 days, but after attaining a certain density showed no further increase (Ketring and Reid, 1993). In prairie soils, corn and sorghum roots regularly penetrate to 2 m, alfalfa roots have been

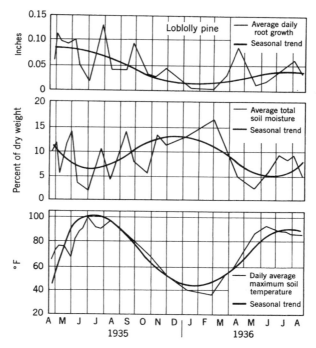

Figure 5.7   Effects of soil temperature and water content on growth of roots of shortleaf pine (*Pinus echinata*) in a forest at Durham, North Carolina. From Kramer (1983), after Reed (1939).

found at depths of 10 m, and Wiggans (1936) found that roots of 18-year-old apple trees penetrated to a depth of at least 10 m and fully occupied the soil between the rows, which were about 10 m apart. Jaafar *et al.* (1993) describe a study of sunflower root development in relation to shoot growth and flowering in a well-watered deep silt loam. The roots penetrated to a depth of 1.8 m at the beginning of disk flowering and 2.0 m by the end of flowering. Roots of several kinds of fruit trees growing in a deep loam soil in California penetrated at least 5 m and the greatest number of roots occurred between 0.6 and 1.5 m (Proebsting, 1943). Hough *et al.* (1965) placed ^{131}I in a forest soil and found that it was absorbed in detectable amounts from as far away as 16 to 17 m by longleaf pine and turkey oak. Hall *et al.* (1953) used uptake of ^{32}P to measure root extension of crop plants, and a study of corn root extension is shown in Fig. 5.6.

The situation is very different in heavy, poorly aerated soils. For example, Coile (1937) found that over 90% of roots less than 2.5 mm in diameter occurred in the top 12.5 cm of soil under pin and oak stands in the heavy clay soil of the North Carolina Piedmont. Pears growing in an adobe soil in Oregon had about 90% of their roots in the upper meter (Aldrich *et al.*, 1935). Even in sandy soils trees often form root mats near the surface because the surface soil contains more mineral nutrients released by decomposition of leaf litter and is wetted by summer showers (Woods, 1957).

The branching and rebranching of root systems often produce phenomenal numbers of roots. Pavlychenko (1937) estimated that a 2-year-old crested wheat grass plant possessed over 500,000 m of roots occupying about 2.5 m^3 of soil. Nutman (1934) estimated that a 3-year-old coffee tree growing in the open bore about 28,000 m of roots, 80% of which occurred in a cylinder 1.5 m deep and 2.1 m in diameter. Kalela (1954) estimated that a 100-year-old pine bore about 50,000 m of roots with 5,000,000 root tips. Considerable further information on the extent of root systems has been summarized by Miller (1938, pp. 137–148) and by Weaver (1926). Much information on conifer root systems has been summarized by Sutton (1969).

Longevity of Roots

Although the larger roots of perennial plants are perennial and some are approximately as old as the plants, mortality is heavy among the smaller roots. The short mycorrhizal roots of pine often die over winter and some of the small roots on apple and other fruit trees live only a week or two (Kinman, 1932; Rogers, 1929). Grier *et al.* (1981) reported a high turnover of roots in a mature forest of *Abies amabilis* in the Washington Cascades, and Caldwell (in Lange *et al.,* 1976) found high rates of root replacement in mixed deciduous forests and cool desert shrub communities. According to Reid *et al.* (1993), most of the fine roots on *Actinidia deliciosa,* a woody vine, survive less than 60 days.

A considerable variation in longevity also exists among the roots of herbaceous plants. It often is stated that the primary roots of grasses live only a few weeks and are succeeded by adventitious secondary roots, but this is not always true. The primary roots of barley, rye, wheat, and various wild grasses are the only roots present and maintain the plants for an entire season. Weaver and Zink (1946) found that the seminal roots of several species of prairie grasses survived two seasons and that some were alive after three seasons.

THE ABSORBING ZONE OF ROOTS

Consideration of root anatomy suggests that entrance of water and minerals into young roots probably occurs chiefly in a region a few centimeters behind root tips, approximately where root hairs are most abundant. However, this is questioned by observations that the xylem is not yet fully functional in the root hair zone of maize and soybean (McCully and Canny, 1988). This situation deserves study in roots of other kinds of plants. Little water enters through the meristematic regions (Frensch and Steudle, 1989), probably because of a lack of functional xylem to carry it away. Farther back the xylem becomes functional, but suberization and lignification of the hypodermis and endodermis usually reduce the permeability in older regions. Figure 5.1 shows diagramatically the location of the absorbing zone in relation to root tissues, and Fig. 5.5 shows the uptake of water in relation to suberization of the endodermis in barley and pumpkin roots.

It formerly was believed that mineral absorption occurred chiefly near the root apex. However, this was based on mineral accumulation studies, and experiments in which radioactive tracers were supplied to roots of transpiring plants at various distances behind the tips indicate that absorption and translocation often occur far behind the apex (Clarkson *et al.*, 1975; Richter and Marschner, 1973; Wiebe and Kramer, 1954). The difference between the zone of mineral accumulation in root cells and mineral absorption through roots is shown in Fig. 5.8. Research by Lazof *et al.* (1992), using secondary ion mass spectrometry, indicates slow influx of NO_3 into root tips and emphasizes the importance of finely branched lateral roots in absorption.

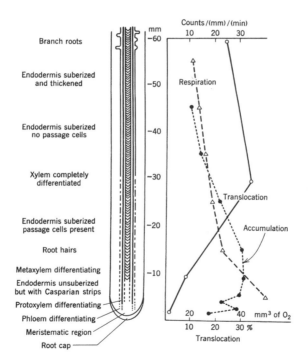

Figure 5.8   Diagram of the apical region of a barley root showing the relationship between the regions where salt is accumulated and where it is absorbed and translocated to the shoot. The curve for respiration is based on data of Machlis (1944), that for translocation on date of Wiebe and Kramer (1954), and that for accumulation on data of Steward and Sutcliffe (1959). The relative positions of those regions probably are similar in most or all growing roots, but the actual distances behind the root tip will vary with the species and the rate of growth. From Kramer (1983).

Absorption through Suberized Roots

Most discussions of absorption deal with young roots and leave the impression that absorption through older, suberized roots is unimportant. Haussling *et al.* (1988), for example, quite properly stressed the importance of growing roots for water and ion uptake of conifers, but absorption through suberized roots must be important in perennial plants where unsuberized roots are often scarce or absent, especially in cold or drying soil. For example, on sunny winter days evergreens such as conifer and citrus trees lose large amounts of water that must be replaced by absorption through suberized roots because few or no unsuberized roots are present (Reed, 1939; Reed and MacDougal, 1937; Roberts, 1948). Kramer and Bullock (1966) found that less than 1% of the root surface in the upper 10 cm of soil under pin and yellow poplar stands in North Carolina was unsuberized in midsummer. Chung and Kramer (1975), Kramer and Bullock (1966), and Queen (1967) all observed a significant uptake of water and phosphorus through suberized roots growing in solution cultures. It seems that the unsuberized root surfaces are too limited in extent, too short lived, and occupy too small a volume of soil to supply all of the water and minerals required by many perennial plants (Chung and Kramer, 1975) and that older, suberized roots must play an important role in absorption, despite their lower permeability. Passioura (1988b) objected to conclusions based on uptake from solution cultures, but MacFall *et al.* (1990, 1991a), using NMR imaging, observed depletion of soil moisture around suberized pine roots growing in fine sand (see Fig. 5.9). This occurs because water sometimes is absorbed more rapidly than it moves toward roots. This is discussed briefly in Chapter 6 in the section on availability of soil water.

Mycorrhizae

The roots of many plants are invaded by fungi that form symbiotic associations called mycorrhizae. These are of two types: endotrophic (VAM, for vesicular arbuscular mycorrhizae) in which the fungus penetrates the root cells, but has little effect on external appearance, and ectotrophic in which the fungus covers the external surface and causes marked hypertrophy and extensive branching of roots, as shown in Fig. 5.10. Mycorrhizae seem to increase the rate of mineralization and solubilization, increasing the supply of minerals, especially phosphorus, available to roots (MacFall, 1994). The presence of mycorrhizal roots also appears to increase the absorption of water and increase drought tolerance, at least in some circumstances (Dixon *et al.*, 1983; Duddridge *et al.*, 1980; Huang *et al.*, 1985; Lamhamedi *et al.*, 1992; MacFall, 1994; Sylvia *et al.*, 1993). Fungal hyphae extend out into the soil and increase the absorbing surface. However, Graham *et al.* (1987) reported that the inoculation of citrus trees with a VAM fungus did not improve their water status, and Harmond *et al.* (1987) found no increase in tolerance of flooding or salinity. Daniels

A

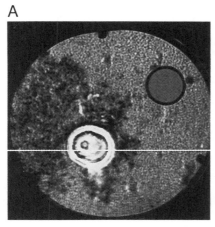

B

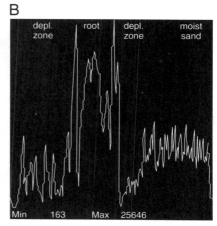

Figure 5.9 (A) A magnetic resonance image showing water depletion (the dark region) around the taproot of a *Pinus taeda* seedling in sand initially at field capacity, after 4.5 hr of transpiration. The circular object at the upper right is a reference tube containing a solution of $CuSO_4$ and H_2O which has a density similar to sand containing 25% water. On the right (B) is a plot of the relative water content along the white line shown in (A). The diameter of the taproot was about 5 mm. From MacFall *et al.* (1991a).

et al. (1987) found that although inoculation with a mycorrhizal-forming fungus improved the growth of well-watered corn, sudan grass, and big bluestem, only the growth of big bluestem was improved by mycorrhizae under water stress.

Endotrophic or VA Mycorrhizae. The role of VA (vesicular–arbuscular) or endotrophic mycorrhizae is discussed by Safir (1987), in McMichael and Persson (1991), and in Waisel *et al.* (1991, Chapters 33 and 34). Some factors controlling VAM symbiosis are discussed by Koide and Schreiner (1992) and Volpin *et al.* (1994), and McArthur and Knowles (1992) suggest that phosphorus deficiency makes plants more susceptible to VAM fungi. Johansen *et al.* (1993) reported that colonization of subterranean clover with VAM fungi increased the uptake of ^{32}P and ^{15}N. According to Syvertsen and Graham (1990), VAM colonization of roots of citrus seedlings did not affect gas exchange, stomatal conductance, or water-use efficiency. They cited work indicating that VAM infection increased the hydraulic conductance of roots of citrus and green ash seedlings only if the plants were deficient in phosphorus (also see Safir *et al.,* 1971, 1972). According to Anderson *et al.* (1988), the colonization of roots of green ash seedlings with VAM fungi did not increase root hydraulic conductance. Of course beyond a certain root density, a further increase in absorbing surface does not increase overall absorption because the new root surface merely competes with the existing surface. Although endotrophic mycorrhizae

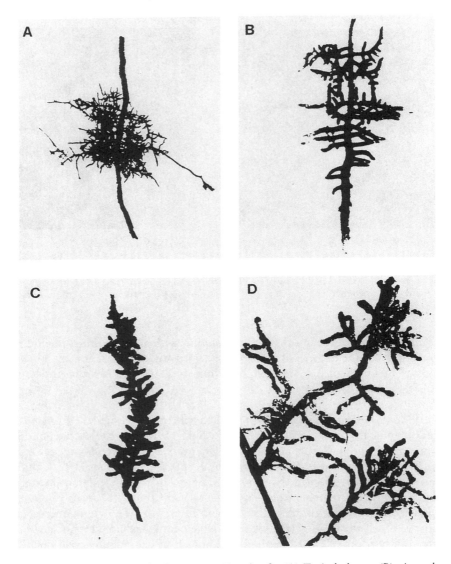

Figure 5.10 Mycorrhizal root development on Douglas fir. (A) Typical cluster, (B) pinnately branched cluster, (C) orange mycorrhiza, and (D) yellow mycorrhiza with rhizomorphs extending into the soil. From Kozlowski *et al.* (1991).

are more common on herbaceous plants, Janos (1980) reported that they also are important on many species of tropical lowland trees. Occasionally, both types of mycorrhizae occur on the same plant.

Ectotrophic Mycorrhizae.   According to Bowen (1984, p. 170) and others, ectotrophic mycorrhizae increase the competitive capacity of trees (and presumably some other plants) and compensate for unfavorable soil conditions such as high pH, excess salt, toxic elements such as aluminum, and deficient aeration. For example, inoculation with mycorrhizal-forming fungi improves tree seedling growth on mine dumps (Marx, 1980; Walker *et al.*, 1989) and increases tolerance of aluminum (Cumming and Weinstein, 1990; Kasuga *et al.*, 1990). Mycorrhizal roots are said to be less susceptible to disease and live longer than the average nonmycorrhizal short root. The beneficial effects of inoculating tree seedlings with mycorrhizal-forming fungi, especially *Pisolithus tinctorius*, are discussed by Marx *et al.* (1984, 1985). MacFall *et al.* (1991b) reported that the growth of red pine seedlings was greatly increased in phosphorus-deficient soil by inoculation with the mycorrhizal-forming fungus *Hebeloma arenosa*, but that the effect on growth decreased as the phosphorus concentration of the soil was increased. Iron was accumulated and copper, calcium, cobalt, boron, and sodium were excluded in mycorrhizal roots in low phosphorus soil as compared with nonmycorrhizal seedlings. Apparently the fungus increases the supply of phosphorus to tree seedlings in phosphorus-deficient soil and regulates compartmentalization of poly phosphorus in mycorrhizal roots (MacFall *et al.*, 1992). Augé and Duan (1991) reported that the presence of mycorrhizae on rose roots hastened stomatal closure when the mycorrhizal-infected half of a split root system was water stressed. They suggested that mycorrhizal roots supply a nonhydraulic signal that affects stomatal aperture.

Although most attention has been given to mycorrhizae in connection with trees and other woody plants (Perry *et al.*, 1987), they may be important in improving water and mineral absorption of herbaceous plants, including cultivated crops, especially on infertile soil (Gerdemann in Torrey and Clarkson, 1975; Safir, 1987; Safir *et al.*, 1972). For example, Bethlenfalvay *et al.* (1987) concluded that mycorrhizal infection increased legume nodule activity and decreased water stress in soybean. However, mycorrhizae require considerable carbohydrate, estimated at 7–10% of that translocated to the roots, and that might reduce yield (Gregory *et al.*, 1987, pp. 140–141, 161–162). Björkman (1942) and Wenger (1955) found that girdling and shading decrease the development of mycorrhizal roots, presumably by reducing the supply of carbohydrate. Mycorrhizae are discussed in more detail in Harley and Smith (1983), by Marks and Kozlowski (1973), and by Mukerji *et al.* in McMichael and Persson (1991). The latter state that there is evidence that the presence of mycorrhizae increases nodulation and nitrogen-fixation by legumes. In contrast, Johnson *et al.* (1992) suggest that the yield decrease accompanying continuous cropping of corn and soybean may be caused by an increase in detrimental VAM fungi. MacFall (1994) has an interesting review of ideas concerning the role of mycorrhizae in forestry and agriculture. She suggests that they play a role in accelerated mineralization and soil biogeochemistry as well as in nutrient uptake through increased absorbing surface. Norris *et al.* (1994) provide a summary of methods for research on mycorrhizae.

DEVELOPMENT OF ROOT SYSTEMS

The amount of water and mineral nutrients available to plants depends on the volume of soil occupied by their roots, and it is well established that plants with deep root systems are more tolerant of drought than shallow-rooted plants. Coile (1940) concluded that the inability of loblolly pine seedlings to compete with hardwood seedlings under closed canopies results from their reduced photosynthesis in the shade (Kramer and Decker, 1944), resulting in failure to produce the deep taproots characteristic of oak and hickory seedlings. Thus pine seedlings are more dependent than hardwood seedlings on water in the surface soil, which is quickly depleted during summer droughts. Fayle (1978) observed that poor growth of red pine plantations was associated with poor vertical root development, and Meyer and Alston (1978) stated that wheat yield in semiarid regions depends on the geometry of the root system in relation to the distribution of soil water at various depths. The development of a maize root system over time is shown in Fig. 5.6. Boyer *et al.* (1980) concluded that newer soybean cultivars yield better than older cultivars because they have a higher root density and therefore are less subject to afternoon shoot water deficits than older cultivars. Frederick *et al.* (1990) questioned this but used only leaves stored for various times to measure water potentials, which may have obscured the differences.

There are some limitations to the concept that very high root density is always favorable. For example, Andrews and Newman (1968) removed 60% of the roots from densely rooted wheat plants without reducing transpiration, even in drying soil, and Eavis and Taylor (1979) reported that treatments increasing the ratio of root length to leaf area did not significantly increase transpiration or leaf water potential. Also, Newman and Andrews (1973) found that although uptake of phosphorus by wheat was well correlated with root growth, uptake of the more mobile potassium was not. Raper and Barber (1970) observed an increased uptake of potassium per unit of root surface in a soybean genotype with lower root density than in one with higher root density. Presumably, as root density increases, competition between roots of the same plant for water and minerals also increases as well as competition between roots of adjacent plants, decreasing the uptake per unit of root surface and the benefits from further increase in root length density.

The depth and lateral spread of root systems depend on both heredity and environment. Figure 5.11 shows the variety of root systems produced by various species of plants growing in the same deep, well-aerated prairie soil and Fig. 5.12 shows the hereditary and environmental limitations on root growth of

Figure 5.11 Differences in depth and spread of root systems of several species of prairie plants growing in a dry, well-aerated prairie soil: *h, Hieracium scouleri; k, Koeleria cristata; b, Balsamina sagittata; f, Festuca ovina ingrata; g, Geranium viscosissimum; p, Poa sandbergii; ho, Hoorebekia racemosa; po, Potentilla blaschkeana.* (From Kramer (1983), after Weaver (1919).

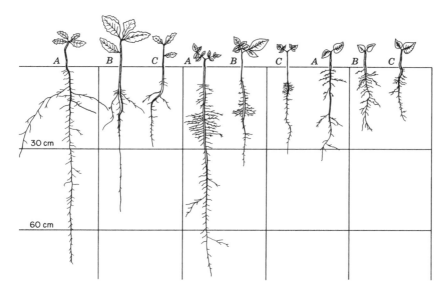

Figure 5.12 Interaction of heredity and environment on amount of root growth produced by three tree species growing in three environments. Left, *Quercus rubra;* center, *Hicoria ovata;* right, *Tilia americana.* Seedlings *A* were growing in open prairie, *B* in an oak forest, and *C* in the deep shade of a moist linden forest. It was necessary to water the linden seedlings in the prairie to prevent death from desiccation. Oak developed the deepest and linden the shallowest root systems in all three habitats. From Kramer (1983), after Holch (1931).

seedlings of three tree species growing in three different environments. Taylor and Terrell (1982) give extensive data describing the spread, depth, and density of root systems in various plant species.

Root–Shoot Interrelationships

The optimum growth of plants depends on maintenance of an efficient balance of functions between roots and shoots, such that neither suffers serious deficiencies in supplies of essential substances contributed by the other. Borchert (1973) suggested that rhythmic shoot growth in a uniform environment may result from cyclic feedback between root and shoot growth, tending to maintain a constant root–shoot ratio. However, it is doubtful if a constant root–shoot ratio is generally maintained in most growing plants. Roberts and Struckmeyer (1946) concluded that the composition and reserve conditions in the shoot were a large and perhaps controlling factor in the production of roots. Sachs *et al.* (1993) proposed that a plant can be regarded as a colony of shoots and roots competing for vascular connections with the remainder of the plant, and their success in this is important for their development.

Effects of Shoots on Roots.   Roots are dependent on shoots for carbohydrates, growth regulators, and some other organic compounds, and severe reduction in leaf area by pruning, insect defoliation, grazing, or diversion of food into fruit and seed production is likely to reduce root growth.

The development of fruits and seeds sometimes reduces root growth significantly, and Fig. 5.13 shows the effect of seed filling on root growth of maize. There was steady increase in root density at all depths until pollination, after which roots began to die more rapidly than they were produced, resulting in a decrease in root density, especially of the older roots in the surface soil, and a decrease in total root weight (Mengel and Barber, 1974). In another experiment, Loomis (1935) found that if the ears were removed corn root growth continued until frost. Figure 5.14 shows that as apple shoot growth increased, root growth decreased and vice versa and that pruning stimulated shoot growth, but reduced root growth (Head, 1967). Buwalda (in McMichael and Persson, 1991), pp. 431–441) found that most of the root growth of kiwi vines (*Actinidia deliciosa*) occurs after shoot and fruit growth is nearly completed. Also, partial defoliation reduces root growth more than fruit growth, suggesting that, as might be expected, kiwi fruits are stronger sinks for photosynthate than roots.

A heavy crop of coffee is said to sometimes reduce the carbohydrate supply to the roots so severely that some die, resulting in injury to the trees (Nutman, 1933), and fruiting of a tropical palm also is said to reduce root growth (Piñero *et al.*, 1982). Root growth of tomato is reduced during fruiting (Hudson, 1960, and others), and Eaton (1931) reported that both root dry weight and root–shoot ratio of cotton were nearly tripled by preventing boll and branch forma-

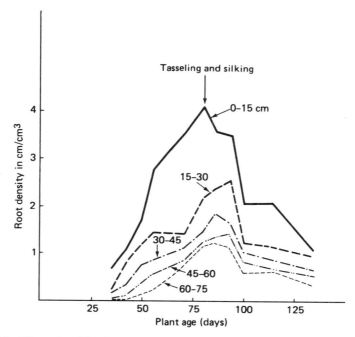

Figure 5.13 Effects of seed development and filling on root growth of maize. Note that root density began to decrease at all depths in the soil after tasseling and silking because old roots died more rapidly than new roots were formed. From Kramer (1983), after Mengel and Barber (1974).

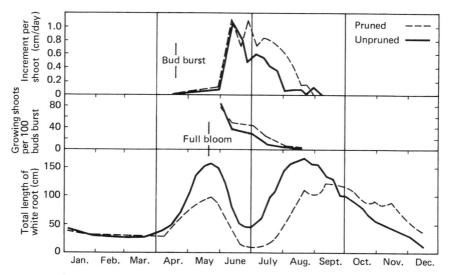

Figure 5.14 Relationship between root and shoot growth of apple. As shoot growth (upper curves) increased, root production (lower curve) decreased. Pruning stimulated shoot growth and reduced midsummer root growth. From Kramer (1983), after Head (1967).

tion. These effects on root growth usually are explained in terms of diversion of carbohydrates from roots to shoot growth or to fruit and seed development. For example, Tripp *et al.* (1991) reported that the increased yield of tomato fruit on plants supplied with a high concentration of CO_2 results from the diversion of photosynthate from roots to fruits because high CO_2 increases seed number in tomato fruits and makes them stronger sinks. However, Van der Post (1968) reported that the appearance of flowers on cucumbers stopped root growth before the fruits were large enough to be important sinks for carbohydrates, suggesting hormonal controls. Wilson (1988) reviewed the extensive literature on root–shoot ratios and concluded that Thornley's (1972) model explains reactions of the root–shoot ratio to environmental factors such as water deficits, light, CO_2, and mineral supply, also to defoliation and root pruning. This model was simplified by Johnson and Thornley (1987) who put more emphasis on the carbon and nitrogen status of plants in the vegetative stage than on hormones. However, as Bingham and Stevenson (1993) state, the carbohydrate supply is only one component of the complex of factors controlling root growth.

Effects of Roots on Shoots.   It is not surprising to learn that damage to root systems severe enough to reduce water and mineral absorption inhibits shoot growth. In addition, shoots are dependent on roots for growth regulators such as abscisic acid, cytokinins, and gibberellins. However, it is somewhat surprising to find that the mineral content of the leaves and the quality of citrus fruit are affected by the kind of rootstock on which the trees are growing, yet those effects have been observed worldwide (Haas, 1948; Sinclair and Bartholomew, 1944). For example, more soluble sugar and total acids are found in fruits from orange trees grown on citrange and trifoliate orange roots than in fruits from trees grown on rough lemon roots. Also, the juice from the Washington navel orange is less bitter in fruit grown on trifoliate orange roots than in fruit grown on other rootstocks such as sour orange, sweet orange, or its own roots. Gregoriou and Economides (1993) found that rootstocks affected fruit size and composition of ortanique tangor, but the differences were too small to be of practical importance. The reasons for these differences are not fully understood.

Horticulturists know that rootstocks differ in disease resistance and tolerance of flooding, salinity, and low temperature, which affect the success of the shoots growing on them. For example, trifoliate orange rootstocks tend to exclude sodium from the shoots grafted on them (Walker, 1986), and Lloyd *et al.* (1987) state that the uptake of sodium and chloride by Valencia oranges varies with the rootstocks on which they are grown. Apparently trifoliate orange rootstocks sequester sodium at the root–shoot transition zone. Maas (1993) surveyed the recent literature on the effects of salinity on citrus. Reciprocal root and shoot grafts between bean genotypes differing in drought tolerance indicated that the differences in tolerance were in the roots (White and Castillo, 1989). In contrast, Delves *et al.* (1987) concluded from reciprocal grafts that the shoot controls supernodulation in soybeans.

It has been known for centuries that the kind of rootstock affects the size and vigor of trees grafted on it. Pears are dwarfed by grafting them on quince roots, and apples are not only dwarfed to one-third of their normal size, but begin to fruit at a younger age when grafted on a dwarfing root system such as M9. Also, on the M9 rootstock about 70% of the photosynthate goes into fruits compared to 40 or 50% in normal trees. However, in California, apple trees on M9 root systems grow too small and M7a and M106 are better for early bearing (Micke *et al.*, 1992). (Numbers with the prefix M indicate that the root systems originated at the horticultural research station at East Malling, England.) The physiology of these root–shoot interactions are not well understood, but some information has been provided in reviews by Lockard and Schneider (1981) and Tubbs (1973). The role of roots as sensors of water stress was mentioned earlier in this chapter and will come up again in later chapters. The importance of chemical signals from roots in controlling shoot processes was reviewed by Davies and Zhang (1991), Gowing *et al.* (1993), and others.

There seems to be renewed interest in electrical potentials as coordinating signals in plants. They received considerable attention a few decades ago, and their role in relation to leaf movement in plants sensitive to touch is well known. Lund (1931) and Rosene (1935) discussed their possible role in coordinating plant growth, and some more recent work is discussed by Fromm and Spanswick (1993). Fromm and Eschrich (1993) claim that electrical signals from roots affect photosynthesis and transpiration in the shoots of willow trees. Hamada *et al.* (1992) reported that there is a decrease in electrical potential on the surface of roots at the point where branch roots will emerge, about 10 hr before emergence. While the role of electrical potentials over distances of a few centimeters seems well established (see references in Fromm and Spanswick), their effectiveness over long distances is more speculative. Perhaps the use of modern technology will result in clarification of their importance. Malone (1993) regards hydraulic signals as important.

The increase in wood production of trees accompanying fertilization usually is attributed to an increase in leaf area and photosynthesis. However, Axelsson and Axelsson (1986) state that there is increasing evidence that the decreased allocation of photosynthate to fine root production by well-fertilized trees is an important factor contributing to increased shoot growth. King (1993) supported the view that increasing the supply of nitrogen decreases root production relative to shoot growth.

Root–Shoot Ratios.   The preceding discussion of the interdependence of roots and shoots suggests that there might be some optimum ratio of roots to

Table 5.1   Amount of Dry Matter in Metric Tons per Hectare Incorporated Annually into Roots and Shoots of Various Plant Species[a]

Species	Roots	Shoots	Root–shoot ratio
Zizania aquatica (wild rice)	0.6	4.0	0.15
Hordeum	3.0	12.0	0.25
Andropogon scoparium (first year)	3.5	14.2	0.25
Triticum (average)	2.0	6.8	0.29
Medicago sativa (average)	3.2	7.4	0.43
Zea mays (average)	4.5	8.7	0.52
Solanum tuberosum (average)	4.0	2.6	1.54
Beta (average)	9.5	3.1	3.06
Pinus sylvestris (average)	1.6	8.9	0.17
Picea abies (average)	2.1	11.9	0.18
Fagus sylvatica (average)	1.6	8.2	0.19
Ghana rain forest	2.6	21.7	0.12

[a]From Bray (1963).

shoots. However, root–shoot ratios vary widely among species, with age, and with environmental conditions. Table 5.1 summarizes some data from Bray (1963) giving ratios varying from 0.15 to 0.20 for various trees, 0.5 for maize, and 3.0 for the storage roots of beets. Such data are not very accurate because of differences in methods and amounts of roots recovered, but they indicate the wide range of root–shoot ratios found among plants in the field. This variation results in part from the wide variations in water supply and other environmental factors to which plants often are subjected during a growing season, as well as to genetic variations among plants such as grasses and root crops.

Perhaps the root–shoot ratio should be considered in terms of root and leaf surface but it is difficult to measure root surface. Fiscus (1981) found that there was a linear relationship between root and leaf surface in growing bean plants. A correlation exists between the sapwood area and leaf area in trees (Chapter 7), and Kaufmann and Fiscus (1985, p. 83) state that the amounts of conducting tissue in roots, stems, and leaves are strongly correlated. Various aspects of root–shoot relations have been reviewed by Klepper (in Waisel *et al.*, 1991).

Root Grafting

The extent of the root system of plants, especially trees, sometimes is increased by natural grafting to the roots of adjacent trees. Bormann and Graham (1959) found so many root grafts in stands of white pine that they regarded the entire stand as a physiological unit, and Kozlowski and Cooly (1961) found a similar situation in stands of angiosperm and gymnosperm trees. Other examples of extensive root grafting include Monterey pine in New Zealand (Will, 1966), slash pine in Florida (Schultz, 1972), red pine in New England (Stone and Stone, 1975b), and quaking aspen in the western United States (Grant, 1993).

Root grafting also is common in tropical trees (LaRue, 1952) and probably will be found wherever a search is made for it. Root grafts provide pathways for transfer of water, solutes, and even fungal spores from one root system to another (Epstein, 1978; Kuntz and Riker, 1955). Bormann and Graham (1960) found that 43% of the untreated trees in a 30-year-old white pine stand were killed by "backflash" to untreated trees through root grafts from treated trees when the plantation was thinned by injecting ammonium sulfate into the trunks of trees to be removed.

Stumps and their attached roots sometimes survive on carbohydrates supplied from intact trees to which they are connected by root grafts. An example of root grafting involving stumps is shown in Fig. 5.15. Apparently root grafting occurs rarely in herbaceous plants, probably because their roots are too short lived for grafts to develop. However, Bormann (1957) found that if roots of tomato plants became firmly intertwined, water moved from one plant to the other even though no grafts occurred. It appears that materials can be transferred between roots by grafting, through fungal hyphae, or between adjoining roots by diffusion through the soil. Woods and Brock (1970) found that radioactive calcium and phosphorus supplied to stumps of red maple were found in the foliage of 19 other species occurring up to 8 m from the donor stumps. They suggested that the root mass of an ecosystem should be regarded as a functional unit rather than a group of separate root systems, a view also held by Bormann

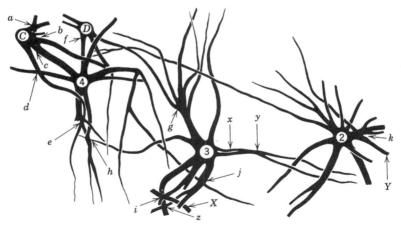

Figure 5.15   Root grafting among roots of three 18-year-old trees of *Pinus radiata* and roots from living stumps of two trees removed 9 years previously. Grafts *a* through *g* were between trees 3 and 4 and roots of stumps C and D. Grafts *h* through *k* were between roots of trees 3 and 4 or between trees 3 and 4 and roots X and Y of trees removed during root excavation. Grafts *x*, *y*, and *z* are between two roots of the same tree. From Kramer (1983), after Will (1966).

and Graham (1959). Another example is the large stands of quaking aspen trees formed by vegetative reproduction (Grant, 1993). One stand consists of 47,000 trees covering 106 acres and can be regarded as a single organism because the tree roots are interconnected and the trees are genetically identical.

Metabolic Cost of Root Systems

A large amount of photosynthate goes into root growth. Some is used in the production of new tissue, some in the respiration supplying energy for the metabolic processes involved in growth, and some in the maintenance respiration of existing tissue (Amthor, 1989; Lambers in Gregory *et al.*, 1987). Caldwell (in Lange *et al.*, 1976) reported that 50% of the annual net primary dry matter production of a deciduous forest and a fescue meadow and 75% of the annual production of short grass prairie and shrub steppe communities goes into the production of new roots. Woods (1980) concluded that the turnover in root biomass of a New England deciduous forest exceeds that of leaves, and Harris *et al.* (1977) reported that root dry matter production is 2.8 times that of aboveground wood production in pine and hardwood forests of the southeastern United States. The economics of root systems are discussed in detail in Givnish (1986).

There has been considerable discussion of why plants often produce more roots than seems necessary and Caldwell (in Lange *et al.*, 1976) was unable to find a satisfactory explanation. It has been argued that death and replacement of roots are efficient because they reduce root respiration at times when they are not needed, but this is doubtful. Part of the problem is the difficulty in determining what constitutes an adequate root system. For example, Teskey *et al.* (1985) reported that they could remove one-fourth of the roots from forest trees in the Pacific Northwest without increasing tree water stress, but Carlson *et al.* (1988) found that removal of any roots from 5-year-old loblolly pine seedlings in Oklahoma increased tree water stress. The benefits of large root systems seem to depend on soil water storage capacity in the root zone and on rainfall patterns, and may vary from year to year.

The role of root density and the extent of root systems in the success of plants have been discussed by Kummerow and by Taylor in Turner and Kramer (1980), by Barley (1970), by Fitter (1987), and also in Waisel *et al.* (1991, Chapter 1). Passioura (1972) discussed the configuration, i.e., branching patterns and depth of rooting, in relation to water and mineral absorption and plant success in competition. The data of Newman and Andrews (1973) indicate that at high densities wheat roots compete with each other for potassium; they probably also compete for water. In their experiments the distance between roots ranged from 4 to 1.5 mm at root length densities of 4 to 16 cm/cm^3 of soil.

Natural selection probably favored survival of plants with large root systems because they are most likely to survive occasional severe droughts. They also are more likely to encounter the nutrients that are distributed irregularly in many soils. This probably resulted in evolution of root systems that are larger than necessary for most cultivated crop plants in humid regions. It therefore seems probable that plant breeders should consider the possibility that selection for smaller root systems would be practical for crops grown in fertile soil in regions with dependable rainfall or with irrigation. According to O'Toole and Bland (1987) the extensive genotypic variation in root systems of various crop plants provides opportunities for selecting root systems with characteristics suitable for special situations.

ENVIRONMENTAL FACTORS AFFECTING ROOT GROWTH

Root growth is greatly affected by environmental factors such as soil texture and structure; aeration; moisture; temperature; pH; salinity; the presence of toxic elements such as aluminum, lead, and copper; competition with other plants; and the presence of bacteria, fungi, and soil-inhabiting animals such as nematodes. A few of these factors are discussed briefly and more extensive discussions can be found in Waisel *et al.* (1991), in Wild (1988), and in a review by Feldman (1984).

Soil Texture and Structure

The physical properties of soil affect root growth directly by restricting root penetration and indirectly by effects on aeration and water content. For example, hardpan layers of natural occurrence (fragipans) and those resulting from tillage operations (tillage pans) often seriously restrict root system development, as shown in Fig. 5.16. Plant growth in such soils often benefits from deep tillage. Figure 5.17 shows the inhibitory effect of various degrees of soil compaction on root penetration and growth of maize in a clay soil, and compaction by heavy machinery reduces the yield of wheat (Oussible *et al.*, 1992). Tardieu (1988) reported that wheel compaction of soil reduced maize root density and water extraction. Sarquis *et al.* (1992) reported that high atmospheric pressure applied to corn plants had little effect on root growth, but a mechanical pressure of 100 kPa on the soil increased ethylene production four-fold and root diameter seven-fold, but decreased root elongation 75%. Further information on the effect of changes in soil structure can be found in Emerson *et al.* (1978) and in Glinski and Lipiec (1990, pp. 75–96); both have references dealing with the effects of soil bulk density on root growth. The term "impedance," borrowed from physics, often is used to refer to the physical resistance of the soil to root extension, but this is undesirable because impedance originally referred to the apparent resistance in an alternating current circuit rather than the physical resistance of the soil.

Figure 5.16   Effect of a compacted layer of soil on root penetration by 11-week-old oat plants. (Left) Undisturbed soil with dense mass of roots above the compacted layer, but few below it. (Right) Uniform penetration of roots into soil loosened by tillage to a depth of 50 cm. The restriction of root penetration was caused by mechanical resistance, as aeration was not limiting below the compacted layer. From Kramer (1983). Courtesy of H. D. DeRoo, Connecticut Agricultural Experiment Station.

In addition to the physical restriction on root penetration caused by various kinds of hardpan layers, there also are chemical barriers. Most important is the effect of a pH below 5.0 which increases the concentration of soluble aluminum

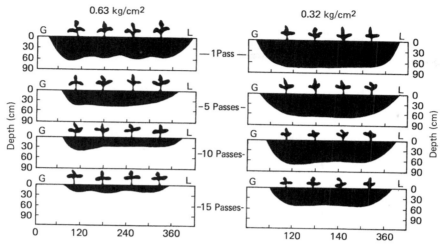

Figure 5.17   Effect on root penetration of soil compaction at 0.63 and 0.32 kg cm⁻² for 1, 5, 10, and 15 passes over a clay soil prior to seeding. From Kramer (1983), after Cassel in Raper and Kramer (1983).

to a toxic level. This condition is common in the acid soils to the southeastern United States and in various tropical soils. The importance of physical and chemical soil barriers to root extension has been discussed by Cassel in Raper and Kramer (1983) and the effects of soil pressure on roots have been discussed by Dexter (1987).

Soil Moisture

Either a deficiency or an excess of soil water limits root growth. Water itself is not directly injurious to roots, as shown by their vigorous growth in well-aerated nutrient solutions, but an excess of water in the soil displaces air from the noncapillary pore space and produces oxygen deficiency that may reduce growth and functioning and cause death of roots. This is discussed further in the section on flooding.

A severe deficiency of soil water usually brings about a reduction in or cessation of root growth, and little or no root growth occurs in soil dried to the permanent wilting percentage. This inhibits water and mineral absorption. The effect of seasonal variation in soil water content on root growth of shortleaf pine in the field is shown in Fig. 5.7. Kaufmann (1968) reported that the root growth of loblolly and scotch pine seedlings in slowly drying soil was reduced to about 25% of the rate at field capacity at a soil water potential of -0.6 or -0.7 MPa, and shoot growth was reduced much more than root growth in drying soil. Teskey and Hinckley (1981) reported that root elongation in a Missouri oak–hickory forest was greatest at a soil water potential of -0.1 MPa, but the number of growing root tips was greater in somewhat drier soil. Vartanian (1981) reported that drying soil reduced root elongation but increased the number of new lateral roots in *Sinapis alba*. Above a soil temperature of 17°C, soil water potential was the dominant limiting factor for root growth in the experiments of Teskey and Hinckley but below 17°C, temperature was most often limiting. During their study, physiologically optimum soil temperature and soil water potential never occurred simultaneously, with one or the other always being limiting. Waring and Schlesinger (1985) cited several experiments suggesting that tree roots do not grow much at a soil water potential below -0.7 MPa, but Logsdon *et al.* (1987) reported growth of maize roots at a soil matric potential of -1.09 MPa. Newman (1966) found that flax root growth was reduced at a soil water potential of -0.7 MPa, but some growth occurred in soil drier than -2.0 MPa. Caution must be used in field experiments to make certain that roots in dry soil are not being supplied with water from other parts of the root systems growing in moist soil (Portas and Taylor, 1976) or by roots of other plants with deeper root systems (Dawson, 1993).

Soil water deficits often reduce shoot growth before root growth is reduced, resulting in increased root–shoot ratios in moderately water-stressed plants. Mild plant water stress also may reduce leaf growth before photosynthesis is reduced (Boyer, 1970), resulting in a surplus of carbohydrates which are available for root growth. Osmotic adjustment also occurs in root tips (Sharp *et al.,* 1990), prolonging root cell expansion, and the combined result is that the absolute size of root systems of mildly stressed plants sometimes exceeds that of well-watered plants (Jupp and Newman, 1987; Sharp and Davies, 1979). However, Westgate and Boyer (1985b) found similar osmotic adjustment in leaves and roots of maize, yet roots grew more than leaves during a period of water stress. The difference in growth of the two organs was attributed to internal factors other than water stress, since the difference remained when the water potential was the same in roots and leaves. Steinberg *et al.* (1990b) reported a 50% greater increase in root biomass than in shoot biomass in young peach trees subjected to water stress.

If root elongation is stopped by soil water stress, roots tend to become suberized to their tips. This is sometimes regarded as a protective adaptation, especially in desert plants, because it decreases water loss from roots to drying soil, but it also reduces their capacity to absorb water when the soil is rewetted. Svenningsson and Liljenberg (1986) found that roots exposed to repeated dehydration contained much less membrane lipid than controls and that the lipid composition was different (Svenningsson and Liljenberg, 1986; Norberg and Liljenberg, 1991). Perhaps as a result plants subjected to severe water stress usually do not regain their full capacity to absorb until several days after the soil is rewetted (Brix, 1962; Kramer, 1950; Leshem, 1965; Loustalot, 1945).

Cruz *et al.* (1992) discussed some of the effects of water stress on the hydraulic conductance of sorghum roots.

Hydrotropism.   It has been claimed, at least since the time of Darwin (1880), that roots can detect water at a distance and grow toward moist soil. This seems to be based largely on observations of the large masses of roots developed around leaky drains, under leaking water taps, and wherever roots encounter moist soil (Oppenheimer, 1941), but this is not necessarily evidence for hydrotropism. Roots extend in all directions and if a randomly growing root encounters an area high in moisture or minerals it is likely to branch profusely and grow. Greacen and Oh (1972) found that wet soil had less resistance to root growth than the same soil at a lower water content and that root growth was more rapid in the wetter soil.

Hydrotropism is difficult to separate from geotropism (gravitropism), and Wareing and Phillips (1981, p. 179) concluded that it is unlikely that true hydrotropism exists. However, research with mutant plants having roots insensitive to gravity indicates that at least some roots show a hydrotropic response (Takahashi and Scott, 1991).

Soil Aeration.

Soil aeration is important physiologically and therefore has effects both on natural vegetation, such as in marshes and swamps, and on cultivated crops. The difference in tolerance of poor aeration between rice and tobacco or cypress and dogwood is striking. Insufficient oxygen often limits root growth in soils which lack sufficient noncapillary pore space for good gas exchange. Poor penetration of nitrogen and oxygen can limit nitrogen fixation by roots of legumes (see Chapter 10). On wet sites roots often tend to be concentrated on hummocks where aeration is best (Lieffers and Rothwell, 1987). Aeration is seldom a problem in sandy soils, but is often a serious limitation in fine-textured soils if less than 10% of the volume consists of noncapillary pore space. Examples are the Shelby loam shown in Fig. 4.2 and the deeper horizon of the old field soil shown in Fig. 4.3. Oxygen diffuses nearly 10,000 times as rapidly in air as in water, and since the concentration of oxygen at 15°C is 30 times greater in air than in water, the actual transport of oxygen to roots is about 300,000 times greater through air-filled pore spaces than when they are filled with water. Even in unflooded soil the presence of water films on the roots reduces the oxygen supply, and flooding of the capillary pore space of soil is likely to result in roots suffering severe oxygen deficiency. Thus two aspects of the aeration problem exist: chronic inadequate aeration in soils deficient in noncapillary pore space and acute episodic deficiency in aeration caused by flooding the soil.

Table 5.2   Oxygen Consumption and Carbon Dioxide Production from a Bare Soil and under Kale in Summer and Winter at Rothamsted, England (Rates are in g · m^{-2} · day^{-1}

	July (17°C)		January (3°C)	
Soil temperature at 10 cm	Cropped	Bare	Cropped	Bare
Oxygen used	24	12	2.0	0.7
Carbon dioxide produced	35	16	3.0	1.2

[a]From Wild (1988, p. 300).

Roots and soil organisms use large amounts of oxygen and produce large quantities of carbon dioxide during the growing season, as indicated in Table 5.2. Unless there is a rapid exchange of gas between the soil and the air the oxygen supply soon becomes limiting for plant growth, as indicated by the data in Fig. 5.18. According to Wild (1988, p. 300), the rate of respiration observed in soil might exhaust the entire oxygen supply to a depth of 25 cm in 2 days in a soil containing 20% of air space, unless it is replenished. The rate of exchange between soil and aboveground air decreases with decreasing soil porosity and increasing water content, and the effects can be easily observed in legumes that require oxygen and nitrogen for nitrogen fixation by the roots.

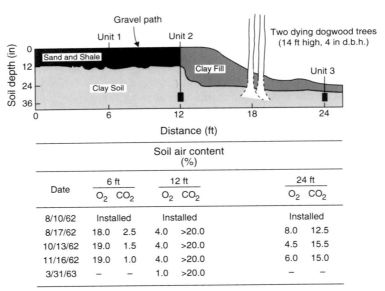

Figure 5.18   Changes in oxygen and carbon dioxide concentration of soil beneath sand and clay fills. From Kozlowski *et al.* (1991), after Yelenosky (1964).

Figure 10.1 shows that the activity for nitrogen fixation can increase initially as water is removed from soil, although severe water deficiency causes lower activity for metabolic reasons. Aeration often becomes seriously limiting in soils compacted by traffic or saturated with water. Raising the soil level by filling or sealing the surface with pavement often creates aeration problems for tree roots (see Fig. 5.18), and trampling by pedestrian traffic around picnic sites and on golf greens also creates aeration problems for trees and grass. Urban trees often are short lived because of the unfavorable environment to which their roots are subjected (Kramer, 1987).

It seems likely that the growth of both cultivated crops and native vegetation is often reduced by undetected deficiencies in root aeration. McComb and Loomis (1944) suggested that root respiration and decomposition of organic matter produce sufficiently anaerobic conditions in prairie soils to hinder tree root growth and exclude trees from grasslands. Howard (1925) stated that in India several species of trees were killed whenever a dense stand of grass developed over their roots, producing anaerobic conditions, and Richardson (1953) reported that grass cover reduced root and shoot growth of *Acer pseudo-platanus*. Karsten (1939) demonstrated in pot experiments that the addition of starch to soil high in nitrogen resulted in anaerobic conditions severe enough to kill wheat seedlings.

There is some uncertainty concerning the relative importance of low oxygen versus high carbon dioxide in inhibiting root growth and function in poorly aerated media, but it seems probable that under field conditions the inhibitory effects of low oxygen are more important than the effects of high carbon dioxide. It appears that bulk air oxygen concentrations above 10% are adequate for roots of most plants. However, it is difficult to determine the actual oxygen concentration at root surfaces from measurements of the oxygen concentration of the bulk air. Methods of measuring soil aeration are discussed by Stolzy *et al.* (1981) and in Glinski and Stepniewski (1985, Chapter 6). The oxygen diffusion rate to root surfaces is the critical factor in soil aeration and this is often measured with platinum electrodes simulating roots. In general it appears that oxygen diffusion rates of $0.4 \text{ mg} \cdot \text{cm}^{-2} \cdot \text{min}^{-1}$ measured with a platinum electrode are adequate and that rates of 0.2 mg are limiting, while the adequacy of intermediate rates depend on the temperature and the plant species. Berry (1949) found oxygen deficiency in the meristematic region of onion roots, and according to Fiscus and Kramer (1970) the interiors of roots usually are subjected to a low concentration of oxygen, as measured with platinum electrodes inserted into them. These observations suggest that root tissues are better adapted than stem tissues to function with low oxygen concentrations. Although Bowling (1973) doubted if the deficiency normally occurring in roots is large enough to affect ion transport significantly, Thomson and Greenway (1991) reported that in a low oxygen concentration the stele of maize roots becomes anoxic. However, the cortex remains aerobic because oxygen is supplied to it from the shoots through the aerenchyma. Ion transport is significantly reduced in poorly aerated roots.

Flooding Injury.   The most severe examples of deficient aeration result from flooding the soil with water because this displaces the soil air, and the low solubility and relatively slow diffusion rate of oxygen in water drastically reduces the supply to roots. Cannell *et al.* (1984) reported that winter flooding of wheat and barley in England caused more injury than summer drought, and Orchard and Jessop (1984) found that the stage of growth at which flooding of sorghum and sunflower occurred affected the amount of injury.

There seem to be multiple causes of injury to shoots of plants growing in flooded soil. The first symptom of flooding injury often is wilting (see Fig. 6.10), and an extreme example is the "flopping" of tobacco when the soil in low areas of a field is saturated by rain, followed by bright sun. This is caused by increased resistance to water flow through roots (Kramer, 1940a, 1951; Smit and Stachowiak, 1988). Similar effects can be produced by saturating the soil with CO_2 or more slowly by displacing the soil air with nitrogen (Kramer, 1940a). Prolonged flooding reduces growth and causes epinasty, leaf chlorosis and death, and development of adventitious roots near the water line. Root elongation ceases and mineral absorption is reduced (Huck, 1970). Some of these effects are incompatible with water stress (Kramer, 1951) and usually water stress is soon eliminated by stomatal closure, probably related to increase in ABA in the shoots (Jackson, 1991). The yellowing and death of leaves have been attributed to a decrease in the supply of cytokinins from roots (Burrows and Carr, 1969). Drew *et al.* (1979a) claimed that the disturbance of nitrogen metabolism is a major cause of poor growth of flooded plants and that the addition of nitrate reduces injury from flooding. There usually is a reduction in the amount of gibberellins in the shoots of flooded plants (Reid and Crozier, 1971) and an increase in ABA. The ABA usually is assumed to come from the roots (Zhang and Davies, 1990), but this is questioned by Jackson *et al.* (1988).

Stem hypertrophy often occurs near the water line on flooded plants and adventitious roots develop on some kinds of plants (Kramer, 1940a). This may result from an accumulation of auxin near the water line (Phillips, 1964), but Drew *et al.* (1979b) attributed the formation of adventitious roots to ethylene. A precursor of ethylene is formed in the roots of flooded plants and moves upward in the xylem (Bradford and Yang, 1980). The new adventitious roots, which usually contain more air space (aerenchyma) than existing roots, seem to take over the functions of the dying roots; plants that develop adventitious roots usually survive flooding better than those that do not (Jackson, 1955; Tsukahara and Kozlowski, 1985; Yu *et al.*, 1969). Roots formed in a poorly aerated environment often contain large air spaces in the cortical parenchyma, called aerenchyma, produced by the breakdown of cells (Fig. 5.19). Formation of

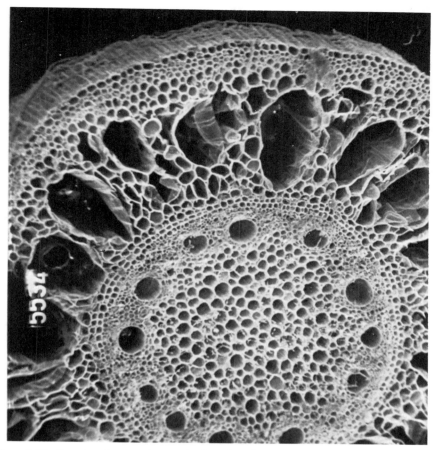

Figure 5.19   Scanning electron micrograph of an adventitious root of corn grown in an unaerated nutrient solution. The section was made 8 to 10 cm behind the root tip. The air spaces in the cortex are formed lysigenously by the breakdown of cells. From Kramer (1983), from Agricultural Research Council Letcombe Laboratory Annual Report (1978, p. 42); courtesy of M. C. Drew.

aerenchyma may result from the high concentration of ethylene stimulating cellulase activity, digestion of the middle lamella, and separation of cells (Kawase, 1979), but this problem deserves more study. Drew *et al.* (1980) reported that the strands of living tissue between air spaces in aerenchyma seem healthy and function effectively in the transport of water and ions to the xylem. It is not clear why some cells of the cortex die while strips of adjacent cells remain uninjured. As aerenchyma seldom occurs nearer than 2 or 3 cm behind the root tip, the apical region does not benefit from it. Crawford (1976) suggested that deep roots often suffer from oxygen deficiency, and Van Noordwijk and Brouwer (1988) thought that the depth of root penetration depends partly on the amount of oxygen supplied through internal air spaces from above ground.

It often is assumed that injury to roots in flooded soil results from accumulation of products of anaerobic respiration such as aldehydes, organic acids, and alcohol (Glinski and Stepniewski, 1985, pp. 141–144). However, Jackson *et al.* (1982) concluded that the concentration of ethanol occurring in roots of flooded plants is not toxic and probably seldom causes injury. Barta (1984) also questioned whether ethanol toxicity is an important factor in root injury. Furthermore, ethanol is found in the cambial region of tree trunks, but causes no injury (Kimmerer and Stringer, 1988). Roberts *et al.* (1985) found by use of nuclear magnetic resonance spectroscopy that flooding injury in root tip cells is correlated with a decrease in pH of the cytoplasm. If acidosis was delayed by treatment with $Ca(NO_3)_2$, root tips survived much longer. Acidosis results from leakage of H^+ from the vacuole into the cytoplasm through the tonoplast, suggesting that properties of the tonoplast may be a factor in the difference between some flooding tolerant and intolerant species. They concluded that acidification of the cytoplasm is an important cause of injury under anaerobic conditions.

Compounds such as methane, sulfides, and reduced iron sometimes accumulate in flooded soil and cause injury to roots. Mendelssohn *et al.* (1981) reported metabolic adaptation to anoxia in *Spartina,* which grows in soil frequently inundated with salt water. The interesting problems of flooded soil were discussed by Ponnamperuma in Kozlowski (1984) and in Chapter 26 of Wild (1988). The book "Flooding and Plant Growth" edited by Kozlowski (1984), contains useful information on effects of flooding, including a chapter on its relation to plant diseases.

Differences in Flooding Tolerance.   It is obvious that there are significant differences among species in tolerance of flooding. Cattails, rushes, rice, mangroves, bald cypress, and tupelo gum thrive in saturated soil, whereas tobacco, tomato, dogwood, longleaf pine, and plants of many other species are killed by flooding the soil. Among shade trees, American elm and honey locust are more tolerant than most other species (Yelenosky, 1964). According to Lin and Lin (1992), orchards of waxapple (*Syzygium samarangense*) survive the 30 or 40 days of flooding, used to induce flowering, without injury.

There are two principal reasons for differences in flooding tolerance: morphological differences resulting in differences in the transport of oxygen to roots and biochemical differences in response to anaerobic conditions such as an increase in alcohol dehydrogenase activity. In some plants, swamp tupelo (*Nyssa sylvatica, V. biflora*), waxapple, and *Spartina,* for example, both mechanisms seem to operate (Hook *et al.,* 1971; Huang and Morris, 1991; Lin and Lin, 1992). According to Laan *et al.* (1990) there was little difference in respiratory pathways between flooding tolerant and intolerant species of *Rumex,* but roots of tolerant species were more permeable to gas. Laan and Blom (1990) stress the importance of photosynthesis in submerged leaves and reserve carbohydrates in increasing flood tolerance of *Rumex*. There is so much literature on flooding tolerance that only a few papers can be cited and readers are referred to books edited by Glinski and Lipiec (1990), Glinski and Stepniewski, (1985), and Kozlowski (1984) for additional information. There are a number of papers in Volume 39 of Aquatic Botany (1991) on aeration problems, and they also are discussed in the book edited by Jackson *et al.* (1991).

Andrews *et al.* (1993, 1994) reviewed some work on effects of deficient aeration at the molecular level. Induction of alcohol dehydrogenase (ADH) formation in corn root tips was small and transient in anoxic conditions (zero O_2) compared to induction by hypoxic conditions (4% O_2 by vol.), and root tips died sooner than older portions of roots. According to Mujer *et al.* (1993), rice and several species of *Echinochloa*, which are tolerant of anaerobic conditions, produce stress proteins which intolerant species of *Echinochloa* do not produce. They cite other relevant research, and more work at the molecular level is likely in the future.

The development of shallow root systems in surface soil and an increase in the amount of internal air space (aerenchyma) are important for survival of plants in waterlogged soil (Armstrong *et al.*, 1991). It is well known that trees and other plants growing in poorly drained soil often develop shallow root systems concentrated in the better aerated surface soil, and seedlings in swamps often start life on hummocks. The adventitious roots formed on flooded plants usually contain considerable aerenchyma in their cortex which facilitates movement of oxygen to the roots and increases flooding tolerance (Tsukahara and Kozlowski, 1985; Jackson, 1955). Glinski and Lipiec (1990, pp. 119–122) and Luxmoore *et al.* (1970) concluded that at least part of the oxygen requirement of maize roots is supplied by transport from the shoots, but Greenwood (1967) found internal aeration of roots of several herbaceous plants to be effective only to about 5 cm below the soil surface. Yu *et al.* (1969) also reported limited gas penetration of existing roots in several species under anaerobic conditions, although porosity of new roots was increased. Van Noordwijk and Brouwer (1988) concluded that the depth of root penetration into soil depends partly on their porosity to gas movement from shoots and described methods of measuring root porosity. The role of root porosity and internal aeration in flooding tolerance seems to deserve more study.

The mechanism by which oxygen moves from shoots to roots has received considerable attention. Some movement occurs by diffusion along concentration gradients caused by the release of oxygen during photosynthesis and its use in root respiration (Laing, 1940; Raskin and Kende, 1985; Waters *et al.*, 1989). However, Armstrong (1968) found that in some instances oxygen entered stems from the air through lenticels a few centimeters above the soil surface and none came from the leaves, but this may have been an unusual situation. D. Barber *et al.* (1962) observed that $^{15}O_2$ diffused from shoots to roots of flooded barley and rice plants. Rice allowed much faster diffusion than barley. Evidence has been found of pressure-induced flow of air from shoots to submerged roots and rhizomes. According to Grosse *et al.* (1991), this was observed by Pfeffer in 1897 (see Pfeffer, 1900), but the early observations were neglected. Grosse *et al.* (1992) regard gas pressure caused by warming of the stem as important for supplying air to the roots and rhizosphere of several tree species, but not for others. Schroeder (1989) invoked thermo-osmotic transport to explain gas transport to roots of European alder. Dacey (1981, 1987) reported that in the water lily, *Nuphar,* and in Nelumbo, which have floating or emergent leaves, warming of the leaves by the sun causes a downward flow of gas. Raskin and Kende (1985) stated that pressure flow in rice results from the use of oxygen in respiration and the solution of CO_2 produced in respiration, whereas Huang and Morris (1991) concluded that pressurization in *Spartina* is caused by water vapor accumulation in intercellular spaces. Sorrell (1991) pointed out that there can be no true flow of air through the air spaces unless there are openings to the exterior for its escape. Enough oxygen sometimes leaks out of roots and rhizomes to oxidize substances in the rhizosphere (Armstrong, 1968; Mendelssohn and Postek, 1982), and this oxidation of the rhizosphere is considered to be important by some investigators, for example, Hook in Kozlowski (1984). Armstrong and Armstrong (1991) and Armstrong *et al.* (1992) reported convective or mass flow of gas to the roots in *Phragmites,* caused by the venturi effect of wind blowing over hollow, broken culms.

Constable *et al.* (1992) found that the concentration of CO_2 in the air spaces of cattail stems was high at dawn, but that it decreased to a near atmospheric concentration by midday, then rose again in the late afternoon. These changes were attributed to the use of CO_2 in photosynthesis, which might contribute to the high productivity of some wetland plants. Brix (1990) suggested that CO_2 from the sediment in which many aquatic plants grow might move to the shoots and be used in photosynthesis. Such movement must be by diffusion along concentration gradients in the intercellular space or in solution in the transpiration stream. Early work on the uptake of CO_2 through roots was reviewed by Livingston and Beall (1934) who thought it reached the leaves chiefly in the transpiration stream, although some might escape into the air from the soil. Their experiments in which the CO_2 concentration of the soil was enriched showed variable results. Billings and Godfrey (1967) concluded that photosynthesis in some plants of wet alpine meadows uses CO_2 from their hollow stems before their leaves are fully expanded. Farmer and Adams (in Waisel *et al.,* 1991) reviewed the recent literature and regarded the uptake of CO_2 from sediments through roots as important for some aquatic plants. Bialzyk and Lechowski (1992) reported that CO_2 is fixed in tomato roots and carbon compounds are transported from roots to shoots.

The metabolic effects of deficient aeration were mentioned in the section on causes of injury, but will be discussed briefly here. Injury often has been ascribed to the accumulation in roots of incompletely oxidized compounds, especially ethanol. In fact, tolerance of flooding often has been attributed to the ability to control ethanol accumulation by increased alcohol dehydrogenase activity, but the importance of this is uncertain. However, Hole *et al.* (1992) and Andrews *et al.* (1993) reported that when maize roots were grown with low oxygen, production of a number of enzymes involved in glycolysis and fermentation was induced and tolerance of anoxia was increased. Kimmerer and MacDonald (1987) reported that there is high alcohol dehydrogenase activity in leaves of many woody plants, but it is not related to flooding tolerance. As mentioned previously, Jackson *et al.* (1982) and others concluded that the concentration of ethanol rarely reaches a toxic level in roots, and Roberts *et al.* (1985) found that acidification of the cytoplasm is an important cause of injury. Thus there is considerable uncertainty concerning the relative importance of various metabolic effects of inadequate aeration. Waters *et al.* (1991) point out that it is difficult to generalize concerning reasons for differences in tolerance among species because of variations in prior treatment and in criteria used by different investigators that may obscure real differences. It therefore seems impossible to generalize concerning the relative importance of structural and metabolic factors in respect to flooding tolerance.

Formation of Intercellular Spaces and Aerenchyma.   There has been frequent mention of the importance of aerenchyma and intercellular spaces in roots. Cells are closely packed in meristematic regions, but as they enlarge they often tend to separate, leaving intercellular spaces. These were termed schizogenous in origin because they were supposed to develop by splitting of the middle lamella. Another and usually larger type of intercellular space (lysigenous) develops by disintegration of entire cells (see Fig. 5.19). The large intercellular spaces formed in poorly aerated roots and in water plants are termed aerenchyma. According to Kawase (1979) the high concentration of ethylene in poorly aerated tissue stimulates cellulase activity and separation of cells, but this needs more study. Some intercellular spaces function as secretory ducts, such as resin and latex ducts. Further discussion of intercellular spaces can be found in plant anatomy texts such as Esau (1965, pp. 62–63).

Under ordinary conditions intercellular spaces in roots are filled with gas, but Canny and Huang (1993) reported that they are sometimes filled with fluid which usually is similar in composition to the vacuolar sap of adjacent cells. According to Burström (1959) the internal atmosphere usually is 79% nitrogen, 2 to 15% CO_2, and the remainder is oxygen. However, according to Burström (1959), intercellular spaces near root tips sometimes are filled with CO_2 which dissolves and they then become filled with water, resulting in dwarf roots.

Zimmermann *et al.* (1992) reported the presence of air-filled, radial, intercellular spaces in NMR images of cross sections of roots. Ethylene often accumulates in the intercellular spaces of submerged tissue because it is relatively insoluble in water (Voesenek *et al.,* 1993).

Root and Earthworm Channels.   Death and decay of old roots often provide channels through which new roots grow downward and they also improve soil aeration. This is very noticeable in forest soils where root channels also increase the infiltration of water (Gaiser, 1952). Research by Hasegawa and Sato (1987) indicated that water absorption from the subsoil by roots growing downward in deep soil cracks is important for crop plants in soils that develop cracks. The role of earthworm activity in loosening soil and improving plant growth was noted by Gilbert White in 1777 and was publicized by Charles Darwin (1881). Earthworm channels not only provide pathways for root growth, but also improve aeration and increase the infiltration of water. Bouma *et al.* (1982) and Hartenstein (1986) discuss earthworm activity, and Wild (1988) also discussed the numerous insects and other animals occurring in soil. Wang *et al.* (1986) discussed the importance of channels formed by decaying roots and soil-inhabiting organisms on root growth and soil aeration. Passioura (1988b, p. 247) pointed out that the tendency of roots to clump in earthworm and old root channels often invalidates the assumption that roots are uniformly distributed in undisturbed soil. The importance of the feeding activity of soil-inhabiting animals on roots is seldom mentioned but probably deserves more study than it has received (Brown and Gange, 1991). For example, nematodes often cause serious injury to roots.

Aeration and Root Diseases.   Important interactions exist between soil aeration and root diseases. For example, the root systems of citrus trees growing in soil frequently saturated by irrigation are prone to attack by *Phytopthora* and pine trees growing in poorly aerated clay soil in the southeastern United States suffer a slow decline (little leaf) because fungal attacks on their root systems cause nitrogen deficiency (Campbell and Copeland, 1954). "Damping off" or death of seedlings caused by attack of the lower stem by fungi sometimes is a serious problem in wet soil. Soil saturation is favorable for the release, spread, and germination of the spores of some pathogenic fungi. It also is favorable for at least some kinds of nematodes. This topic deserves more attention than can be given to it in this book and interested readers are referred to Kozlowski (1984, Chapter 7) and Ayres and Boddy (1986) for more detailed discussion.

Oxygen Toxicity.   Although oxygen is essential for aerobic organisms, an excess can be toxic. Loehwing (1934) reported that a continuous aeration of root systems with normal air containing 20% oxygen can inhibit growth. More recent research indicates that metabolic processes such as photosynthesis give rise to such reactive toxic compounds as superoxide ($O_2^-\cdot$), H_2O_2, and OH^-. These compounds injure some enzyme systems and react with unsaturated fatty acids of membrane lipids, causing damage to membranes and the cells and organelles that they enclose. Environmental stresses such as bright light, water and temperature stress, and some herbicides cause excessive production of superoxide, resulting in damage by photoinhibition and photooxidation. Aerobic organisms possess several protective mechanisms, including catalase and superoxide dismutase. Some of the literature on oxygen toxicity has been reviewed by Scandalios (1993).

Soil Temperature

Root growth often is limited by low soil temperatures, and occasionally exposed surface soil becomes hot enough to limit root growth. Overheating of the surface soil can be reduced by mulching, but a thick mulch slows rewarming of the soil in the spring, which often is undesirable (Wild, 1988, Chapter 8). An example of the effect of temperature on root growth is shown in Fig. 5.20, and the effect of seasonal change in temperature on pine root growth is shown in Fig. 5.7. There are wide differences in the optimum temperature for root growth of different species, and this has important ecological and economic effects. For example, roots of bluegrass are injured at the high midsummer temperatures favorable for Bermuda grass (Brown, 1939) and a low soil temperature often limits growth of cotton roots early in the season (Arndt, 1945). Roots of plants growing in containers sometimes are injured by low temperatures, and unusually cold weather that freezes the surface soil sometimes injures shallow roots such as those of nursery seedlings.

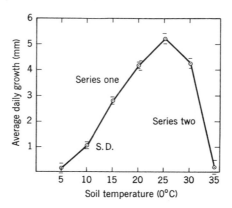

Figure 5.20   Relationship between soil temperature and root elongation of *Pinus taeda* seedlings under controlled conditions. From Kramer (1983), after Barney (1951).

The effects of low soil temperature on water absorption are discussed in Chapter 6. In addition to a decrease in water and mineral supply, a decrease in synthetic activity reduces the supply of hormones to the shoots and results in decreased shoot growth (Duke *et al.*, 1979). The use of cold water for irrigation is said to be a limiting factor for rice yields in such varied locations as northern Italy, Korea, on the island of Hokkaido, and in the Sacramento Valley of California. Occasionally, the use of cold water in the winter on greenhouse crops such as cucumbers has proven injurious (Schroeder, 1939). However, chilling root systems of young soybean plants to 10°C for a week had only temporary effects and did not reduce yield (Musser *et al.*, 1983). Effects of soil temperature are discussed in more detail in Glinski and Lipiec (1990, pp. 152–161) and in Wild (1988, Chapter 8).

Root Competition

The size of root systems usually is reduced when they are grown in competition with other plants. For example, Pavlychenko (1937) reported that root systems of barley and wheat were nearly 100 times larger when grown without competition than when grown in rows 15 cm apart. Although roots often seem to be intertwined in the soil, there is evidence of a mechanism that prevents roots of some kinds of plants from growing very close to one another. This was reported for roots of trees by Lyford and Wilson (1964) and for soybean by Raper and Barber (1970a), as shown in Fig. 5.21, but seems not to occur in tomato (Bormann, 1957). Considerable information on root competition has bene summarized by Caldwell in Gregory *et al.* (1987). The detrimental effects of one plant or crop on another can be attributed to the (1) depletion of water

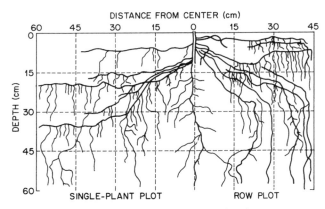

Figure 5.21   Difference in lateral root extension of an isolated soybean root system (left side) and that of a plant growing in a row. From Kramer (1983), after Raper and Barber (1970a).

or nutrients, especially nitrogen; (2) release of toxic substances from roots or leaves; or (3) production of toxic substances during decomposition of plant remains.

Early in this century there was considerable interest in the effects of decomposition of the residue of one crop on the growth of subsequent crops (see Miller, 1938, pp. 164–174), and examples of more recent research were summarized by Putnam and Duke (1978). Although in certain experiments some sequences of crops seemed better than others the results were often contradictory and were so frustrating that research was largely abandoned.

Allelopathy

It has been known for more than 2000 years that some plants seem to inhibit the growth of other plants growing in their vicinity, a phenomenon known as allelopathy. Pliny (died AD 79) reported that walnut trees are injurious to some other plants and the injury is now attributed to a compound called juglone, excreted from their roots (Gries, 1943). Colton and Einhellig (1980) found that extracts from the leaves of velvetleaf (*Abutilon theophrasti*) are injurious to soybean, and Booker *et al.* (1992) found that ferulic acid, an allelochem occurring in soil, decreases ion and water uptake of cucumber seedlings. Putnam (1983) describes chemicals involved in allelopathy. Einhellig *et al.* in Thompson (1985) found that treatment of sorghum root systems with coumaric and ferulic acid or extracts from certain allelopathic weeds reduced water absorption and suggested that one factor in allelopathic reactions is increased plant water stress.

Alfalfa is said to be both allelopathic to other plants such as sorghum and autotoxic, with its residue inhibiting alfalfa seed germination and seedling growth (Hegde and Miller, 1990, 1992). According to Tesar (1993), delaying reseeding alfalfa at least 2 weeks after plowing it under reduces autotoxic effects. Keever (1950), Muller (1969), Webb *et al.* (1967), and others attempted to explain success or failure of plants in natural succession to toxic effects of substances released by competing vegetation, but this is difficult to prove or disprove. In contrast to the examples of inhibition just mentioned, black locust is said to stimulate the growth of neighboring trees because nitrogen fixation in its roots increases the nitrogen supply for neighboring plants (Chapman, 1935). Other possible interactions are discussed in Kozlowski *et al.* (1991, pp. 102–104). There seems to be increasing interest in allelopathy in forestry, chiefly as a factor in seedling establishment, and the increasing interest in agroforestry also has stimulated interest in root interactions (Huck, 1983; Sureshi and Rai, 1988). Possible interactions between shade trees and various kinds of ornamental plant ground cover deserve more attention (Shoup and Whitcomb, 1981). Rice (1984) and Harborne (1988) reviewed the literature on allelopathy, and allelopathy in woody plants was reviewed by Norby and Kozlowski (1980).

Thompson (1985) edited a collection of papers on allelopathy, Putman and Duke (1978) discussed allelopathy in agriculture, and Putnam and Tang (1986) edited a book on allelopathy.

The Replant Problem

When old orchards and vineyards are replanted the new plants sometimes grow poorly or even die. Although this has been attributed to injurious products of root decay (Israel *et al.*, 1973; Proebsting and Gilmore, 1941), there seem to be other possible causes. These include an increase in nematodes and various fungi, poor drainage, and inadequate aeration. Survival of replants often is improved by deep plowing and soil fumigation. The success of soil fumigation suggests that nematodes and fungi are more important than allelopathic compounds. Unfortunately, fumigation also often kills beneficial organisms as well as harmful ones, necessitating reinoculation of the soil with beneficial organisms to restore full productivity. Catska *et al.* (1982) reported that changes in soil microflora have contributed to the decline of apple trees in Czechoslovakia. Yadava and Doud (1980) reviewed the replant problem in horticulture. Although it has not yet been recognized widely in forestry, the increasing use of short rotations is likely to result in its appearance.

Crop rotation is a related problem. For many years rotation was regarded as essential in improving the control of weeds, diseases, and insects; decreasing injury from toxic products of root decay; and increasing the nitrogen supply by including a legume in the rotation. However, an increased use of fertilizers and pesticides and an increased disease resistance of newer cultivars of crop plants has resulted in a decreased dependence on crop rotation. Nevertheless, the long-term effects of rotation on soil, root growth, and crop yield need further study. An extensive review of the numerous effects and benefits of crop rotation has been done by Sumner (1982).

Biochemistry of Competition and Infection

It seems that root competition, allelopathy and autotoxicity, symbiotic relationships such as those with mycorrhizal fungi and nitrogen-fixing bacteria and fungi, and attacks by parasitic organisms involve complex biochemical interactions involving host recognition and defense mechanisms. Research at the cellular and molecular level is beginning to contribute to a better understanding of these interactions. An example is the change in root metabolism associated with the development of mycorrhizal roots on Eucalyptus (Hilbert *et al.*, 1991), although this is questioned by Guttenberger and Hampp (1992). Kapolnik *et al.* (in Schultz and Raskin, 1994) discuss signals between roots and fungi in the formation of VA mycorrhizae. Another example is germination of the seed of witchweed (*Striga*) in response to phenolic compounds synthesized in roots of potential hosts (Lynn and Chang, 1990; Lynn and Boone in Schultz and Raskin, 1994). Lynn and Chang (1990) also discussed the role of specific phenolic compounds believed to be involved in root infection by *Rhizobium* and the general nature of the signals promoting host recognition and infection. There probably will be much more research in this area in the near future.

Atmospheric Conditions

Attention has thus far been on the soil atmosphere, but the atmospheric conditions to which the tops are exposed obviously can have important indirect effects on root growth. For example, low light intensity usually limits root growth and reduces the root–shoot ratio because it reduces the supply of carbohydrates available to the roots. An increase in atmospheric CO_2 often is accompanied by an increase in the ratio of roots to shoots in herbaceous plants. Rogers *et al.* (1992) reported that high atmospheric CO_2 more than doubled the root weight of soybeans, but the effects on roots of woody plants are variable (Sionit and Kramer, 1986). The variable results might have resulted from root restriction in some experiments (Thomas and Strain, 1991). According to Norby (1987), an increase in atmospheric CO_2 increases the nitrogen fixation of some woody species and it also increases mycorrhizal development (O'Neill *et al.*, 1987). Bottomley *et al.* (1993) claimed that high atmospheric CO_2 caused increased water content of the upper roots and lower stems of water-stressed beans. Some effects of CO_2 were reviewed by Bowes (1993), but he gives no data on root growth.

Miscellaneous Effects

Other interesting and sometimes puzzling results of environmental effects on root growth have been reported. For example, Carmi *et al.* (1983) found that restricting the root systems of beans in very small pots kept well watered and fertilized depressed shoot growth, but not the rate of photosynthesis per unit of leaf area. In another study (Carmi, 1993), root restriction reduced the accumulation of nitrogen compounds in the leaves, but this was not regarded as the major cause of reduced growth. The inhibitory effect of small container size and reduced soil volume on growth of well-watered and fertilized ornamental plants also may be important (Krizek and Dubik, 1987). One can speculate that reduced growth might be related to the smaller number of growing root tips and the reduced synthesis of growth regulators. A reduction in sink size may also reduce photosynthesis by feedback inhibition. Thomas and Strain (1991) found that the reduction of root size by growing cotton in small containers eliminated the increase in plant biomass observed in plants grown in larger containers when both were exposed to high atmospheric CO_2. According to Fuleky and Nooman in McMichael and Persson (1991), growing maize in small pots increased root density and uptake of phosphorus per unit of root. Masle and Passioura (1987) reported that an increased bulk density of soil reduced leaf area and root and shoot dry weight of wheat, possibly by changes in hormones exported to the shoots from the roots. However, Tardieu *et al.* (1991) questioned if mechanical restraint on roots caused by soil compaction sends signals to the shoots in maize. Much remains to be learned about root–shoot interactions and the effects of environmental factors on root growth and indirectly on shoot growth.

METHODS OF STUDYING ROOT SYSTEMS

Observation of the development of root systems seems to have begun with work of Duhamel du Monceau about 1764–1765 on root systems of trees and reached a peak in the first half of this century with the observations of Weaver (1925) in the United States and Kutschera (1960) in Europe. The early work was summarized by Böhm (1979) and some later work by Glinski and Lipiec (1990) and Caldwell and Virginia (in Pearcy *et al.*, 1989), along with Vogt and Persson (in Lassoie and Hinckley, 1991), to whose bibliographies the reader is referred. Epstein (in Hashimoto *et al.*, 1990) has a good review of methods useful for studying roots and root systems, with special reference to the mineral nutrition of plants.

The oldest method of studying root systems was to excavate them, a method that requires much time, energy, and a disregard for getting dirty. One form of excavation requires cutting a trench and then removing soil by hand or by a stream of water or air, while mapping the roots on the face of the trench. This was used effectively on fruit trees by Oskamp and Batjer (1932) and on forest trees by Coile (1937) and many others (see Böhm, 1979). Occasionally soil is removed in the horizontal plane to expose the spread of tree root systems. Another approach, the monolith method, is to remove large blocks of soil and wash out the roots. Sometimes these blocks are enclosed in steel boxes driven into the soil to hold the soil mass together while removing it. Boards covered with spikes, called pinboards, are sometimes driven into the soil mass to preserve the root arrangement while the soil is washed away. Root frequency sometimes is sampled in cores of soil removed with various kinds of soil augers, which frequently are driven into the soil and removed by tractor power (Jaafar *et al.*, 1993). W. S. Clark (1875), who made the first measurements of root growth in the United States, known to the authors, grew plants in greenhouse benches and washed out the roots.

Another method, apparently introduced by Sachs in 1873, is to install glass plates in the soil and observe root growth against them. From this has developed the elaborate rhizotrons at East Malling, Auburn, Alabama, Ames, Iowa, and elsewhere which have large underground observation tunnels and observation windows large enough to follow the development of tree root systems. Mini-rhizotrons also have come into use, consisting of plastic tubes about 5 cm in diameter and 2 or 3 m long which are driven into the ground, usually at an angle. A camera and a fiber optic illumination system can be lowered down the tube to record root growth over time. An example is described by Upchurch and Ritchie (1988) and another by Box and Ramseur (1993). Buckland *et al.* (1993) discussed the problems involved in converting minirhizotron observations into root length density data. The advantages and disadvantages of various methods are discussed by Böhm (1979, pp. 75–76) and by Glinski and Lipiec, (1990, Chapter 7). Heeramon and Juma (1993) concluded that destructive sampling is still the best method to study root growth.

Couchat *et al.* (1980) used neutron radiography to study root growth in sand, and nuclear magnetic resonance imaging shows promise for some purposes (Brown *et al.*, 1986, 1991; Omasa *et al.*, 1985b; Rogers and Bottomley, 1987). MacFall *et al.* (1990, 1991a) showed development of a water depletion zone around suberized pine roots by NMR imaging (Fig. 5.9) and Omasa *et al.* (1985b) used it to show changes in root and soil water content. The use of nuclear magnetic resonance technology for research on plants has been discussed by Kramer *et al.* (in Hashimoto *et al.*, 1990) and for plants and soil by MacFall and Johnson (1994). An example of its use is described in a paper by Zimmermann *et al.* (1992). Hegde and Miller (1992) reported that scanning electron microscopy is useful for studying root anatomy and morphology. However, these methods cannot provide images of large root systems.

Several indirect methods have been used to estimate root density and the rate of root extension in the soil. One method is to measure the decrease in soil water content, assuming that a decrease is closely related to root density. This is most useful if rainfall is infrequent or if rain shelters are available. Root extension also has been followed by injecting small quantities of a radioactive isotope of phosphorus, sulfur, or rubidium into the soil at various depths below and distances from seedlings and noting when it appears in the plant. An example is shown in Fig. 5.6. Sometimes the tracer is injected into the plant and is recovered from root samples in soil cores (Fusseder, 1983). Some success has been attained in using ^{14}C as a tracer of carbon transport to the roots and to distinguish between living and dead roots. These methods are discussed in Chapter 8 of Böhm (1979).

Measurement of the ratio of deuterium to hydrogen (D/H ratio) in xylem sap has proven useful in studying plant–soil water relations (White in Rundel *et al.*, 1988). For example, White *et al.* (1985) found that bald cypress growing in a swamp showed no isotopic change in response to rain because most of its roots were below the surface of the water table, whereas trees on a dry site used rainfall water exclusively for several days after a rain. Trees on intermediate sites used both at first, but a few days after a rain they were using groundwater exclusively. In another study, Flanagan *et al.* (1992) used the difference in the D/H ratio between rain water and soil water to determine the relative uptake of rain and groundwater by woody plants in a pinyon–juniper woodland.

SUMMARY

The principal functions of roots are anchorage, absorption of water rand minerals, and synthesis of nitrogen compounds and growth regulators such as abscisic acid, cytokinins, and gibberellins which have essential roles in shoot growth and functioning. Roots may also play a role as sensors of water stress which causes them to send biochemical signals to shoots that reduce leaf growth and stomatal conductance even before there is any significant reduction in leaf turgor.

Growing roots typically show several well-defined regions, including the root tip and root cap, a light colored zone covered with root hairs, and a darker colored region where suberization of epidermal cells occurs. Root hairs increase root contact with the soil and presumably increase the absorbing surface for water and minerals, although the importance of this seems to vary among species and with various ions. The effective absorbing surface also is increased by the presence of mycorrhizal fungi. Secondary growth causes loss of epidermis and cortex and the development of bark which decreases root permeability to water. Nevertheless, considerable absorption of water and minerals occurs through suberized roots, especially when root elongation is slowed by cold or dry soil.

Root elongation varies from a few millimeters to a few centimeters per day, and it sometimes results in root systems hundreds of kilometers in length and occupying large volumes of soil. The depth and the spread of root systems are controlled by heredity and environment, varying widely among species and with water content, temperature, and aeration of the soil. Flooding soil injures the roots of most plants, but is tolerated by a few such as rice, cattails, and cypress trees. Tolerance of flooding depends on the presence of aerenchyma which facilitates movement of oxygen from shoots to roots, on physiological adaptations that permit metabolism at low oxygen concentrations, or on a combination of the two.

Successful plant growth depends on maintenance of a balance between root and shoot growth, but there are wide differences among species with respect to successful root–shoot ratios. Severe defoliation usually drastically reduces root growth, chiefly because it reduces the supply of carbohydrates. The development of fruits and seeds also reduces root growth, but hormonal controls may be involved in addition to competition for carbohydrates. Flushes of root and shoot growth often alternate, probably because of competition for food. There is a strong interdependence between roots and shoots and anything that inhibits one is likely to affect the other.

Roots of trees often form so many root grafts that the trees of a stand are interconnected and form essentially one organism. In general, plants tend to produce more roots than are necessary for survival, except during very severe droughts. However, root growth often is inhibited by soil resistant to root penetration; by deficient aeration, low pH, or an excess of toxic elements such as aluminum; and by competition with other roots. Roots of a few species of plants produce substances toxic to other plants (allelopathy), and growth sometimes is retarded when orchards and vineyards are replanted to the same species. Small containers that restrict root growth often inhibit shoot growth, even when well watered and fertilized.

In conclusion, conditions favorable for root growth are just as important to the success of plants as conditions favorable for shoot growth.

SUPPLEMENTARY READING

Atkinson, D., ed. (1991). "Plant Root Growth." Blackwell, London.

Böhm, W. (1979). "Methods of Studying Root Systems." Springer-Verlag, Berlin.

Box, J. E., Jr., and Hammond, L. C. eds. (1990). "Rhizosphere Dynamics." Westview Press, Boulder, CO.

Brouwer, R., Gasparikova, O., Kolek, J., and Loughman, B. G. eds. (1982). "Structure and Function of Plant Roots." Junk, The Hague.

Caldwell, M. M., and Virginia, R. A. (1989). Hydraulic life. In "Plant Physiological Ecology" (R. W. Pearcy, J. Ehleringer, H. A. Mooney, and P. W. Rundel, eds.), pp. 367–398. Chapman Hall, New York.

Carson, E. W., ed. (1974). "The Plant Root and Its Environment." University Press of Virginia, Charlottesville, VA.

Clarkson, D. T. (1985). Factors affecting mineral nutrient acquisition by plants. *Annu. Rev. Plant Physiol.* 36, 77–115.

Epstein, E. (1990). Roots: New ways to study their function in plant nutrition. In "Measurement Techniques in Plant Science" (Y. Hashimoto, P. J. Kramer, H. Nonami, and B. R. Strain, eds.), pp. 291–318. Academic Press, San Diego.

Esau, K. (1965). "Anatomy of Seed Plants." 2nd Ed. Wiley, New York.

Feldman, L. J. (1984). Regulation of root development. *Annu. Rev. Plant Physiol.* 35, 223–242.

Glinski, J., and Lipiec, J. (1990). "Soil Physical Conditions and Plant Roots." CRC Press, Boca Raton, FL.

Glinski, J., and Stepniewski, W. (1985). "Soil Aeration and its Role for Plants." CRC Press, Boca Raton, FL.

Gregory, P. J., Lake, J. V., and Rose, D. A., eds. (1987). "Root Development and Function." Cambridge University Press, Cambridge.

Harley, J. L., and Russell, R. S., eds. (1979). "The Soil–Root Interface." Academic Press, New York.

Hook, D. D., and Crawford, R. M. M., eds. (1978). "Plant Life in Anaerobic Environments." Ann Arbor Scientific Press, Ann Arbor, MI.

Jackson, M. B., Davies, D. D., and Lambers, H., eds. (1991). "Plant Life Under Oxygen Deprivation." SPB Publ., The Hague.

Kolek, J., and Kozinka, V. (1992). "Physiology of the Plant Root System." Kluwer Academic Press, Norwell, MA.

Kozlowski, T. T., ed. (1984). "Flooding and Plant Growth." Academic Press, Orlando, FL.

Marks, G. C., and Kozlowski, T. T., eds. (1973). "Ectomycorrhizae: Their Ecology and Physiology." Academic Press, New York.

MacFall, J. S. (1994). Effects of ectomycorrhizae on biogeochemistry and soil structures. *In* "Mycorrhizae and Plant Health" (F. L. Pfleger and R. O. Linderman, eds.), pp. 213–238. APS Press, St. Paul, MN.

McMichael, D. G., and Persson, H., eds. (1991). "Plant Roots and Their Environment." Elsevier, Amsterdam.

Rice, E. G. (1984). "Allelopathy." 2nd Ed. Academic Press, Orlando, FL.

Russell, R. S. (1977). "Plant Root Systems." McGraw-Hill, New York.

Torrey, J. G., and Clarkson, D. T., eds. (1975). "The Development and Function of Roots." Academic Press, New York.

Waisel, Y., Eshel, A., and Kafkafi, U., eds. (1991). "Roots." Dekker, New York.

Wild, A., ed. (1988). "Russell's Soil Conditions and Plant Growth," 11th Ed. Longmans Group, UK Wiley, US.

6

The Absorption
of Water
and Root
and Stem
Pressures

INTRODUCTION

The continuous absorption of water is essential for the growth and survival of most plants because they lose large amounts of water daily, often even more than their own weight on hot, sunny days. Unless this is replaced immediately, they will die from dehydration. Only a few xeromorphs, such as cacti, with low rates of transpiration and a large water storage capacity, and plants very tolerant of dehydration can survive without immediate replacement of the water lost by transpiration. Thus, the mechanism of and factors affecting water absorption deserve careful attention. Furthermore, well-watered plants often develop root pressure, resulting in exudation of sap from wounds ("bleeding") and guttation. Other interesting exudation phenomena such as the flow of maple sap, latex, and oleoresin are discussed briefly, although they are not related directly to the absorption of water.

ABSORPTION MECHANISMS

Absorption of water occurs along gradients of decreasing potential from the substrate to the roots. However, the gradient is produced differently in slowly and rapidly transpiring plants, resulting in two absorption mechanisms. Active or osmotic absorption occurs in slowly transpiring plants where the roots behave as osmometers whereas passive absorption occurs in rapidly transpiring plants where water is pulled in through the roots, which act merely as absorbing surfaces (Renner, 1912; Kramer, 1932). When the soil is warm and moist and transpiration is slow, as at night and on cloudy days, water in the xylem often is under positive pressure (root pressure) as indicated by the occurrence of guttation and exudation of sap from wounds. This difference can be demonstrated by bending the stem of a plant below the surface of a container of a dye such as acid fuchsin and making a cut into it. If the plant is transpiring even moderately rapidly, dye will rush in and stain the xylem above and below the cut, but if the plant has been in moist soil and transpiring very slowly, the dye will not enter but sap is likely to exude from the cut. In moist soil, absorption at night and early in the morning is largely by the osmotic mechanism, but as daytime transpiration increases the demand for water in the leaves, absorption increasingly occurs by the passive mechanism and the osmotic mechanism becomes less and less important. The osmotic water absorption causing root pressure occurs only in healthy, well-aerated roots of slowly transpiring plants growing in moist soil, but passive intake of water can occur through anesthetized or dead roots, or in the absence of roots as in cut flowers or branches (Kramer, 1933). Osmotic absorption and some alternative explanations are discussed later in the section on root pressure.

Passive Absorption by Transpiring Plants

The force bringing about absorption of water by transpiring plants originates in the leaves and is transmitted to the roots or the lower end of cut stems through the sap stream in the xylem. Evaporation of water from leaf cells decreases their water potential, causing water to move into them from the xylem of the leaf veins. This reduces the potential in the xylem sap, and the reduction is transmitted through the cohesive water columns to the roots where the reduced water potential causes inflow from the soil (see Chapter 7 for details). In this situation water can be regarded as moving through the plant in a continuous, cohesive column, pulled by the matric or imbibitional forces developed in the evaporating surfaces of stem and leaf cell walls. Evaporation of water from the twigs and branches of bare deciduous trees and from cut flowers also causes upward flow in the xylem.

As shown in Fig. 6.1, passive absorption often lags quantitatively behind transpiration during the day, indicating the existence of resistances to water flow and capacitance or water storage in the system. Resistance may exist at several points, but the most important site usually is in the living cells of the roots. This is indicated by the fact that the removal of roots from shoots exposed to sun is followed by an increase in the rate of water absorption in a variety of herbaceous and woody plants ranging from sunflowers (Kramer, 1938) to pine trees (Running, 1980) to paddy rice growing in water (Hirasawa *et al.*, 1992). An example is shown in Fig. 6.2. Water storage in the parenchyma cells of various organs is responsible for most of the capacitance effect and the absorption lag.

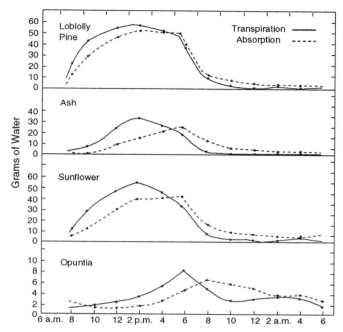

Figure 6.1 The lag of absorption behind transpiration on a hot summer day for four different kinds of plants growing in soil supplied with water by an autoirrigator system that permitted measurement of both water loss and water uptake. Note the evening maximum in transpiration of *Opuntia*, which is a CAM plant, and the midday decrease in sunflower, probably caused by a temporary water deficit and partial closure of stomata. From Kramer (1983), after Kramer (1937).

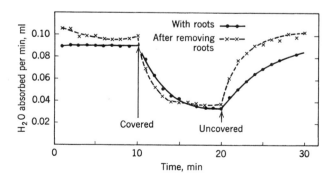

Figure 6.2 Removal of roots decreases the lag of absorption behind transpiration for sunflower plants in a dilute nutrient solution. Absorption of intact plants was measured for 10 min. The shoots were then covered for 10 min to reduce transpiration and uncovered for another 10 min. The root systems were then removed and the experiment was repeated. From Kramer (1983), after Kramer (1938).

Osmotic Absorption and Root Pressure

It was stated earlier that positive pressure in slowly transpiring plants often develops in the xylem sap, resulting in exudation (often termed "bleeding") from cut or broken stems. Hales (1727) measured this pressure and there has been interest in the problem from the 18th century to the present. Some confusion has resulted from failure to distinguish between such diverse phenomena as root pressure, stem pressure, guttation, and exudation from nectaries. It also is necessary to distinguish between exudation caused by root pressure as in birch, grape, and many herbaceous plants and that caused by stem pressure as in maple or by wounding as in agave and palm. The older literature was reviewed by Kramer (1945, 1949, Chapter 7).

There are three groups of explanations for root pressure: secretory activity of the root cells, electroosmosis, and the assumption that the roots behave as osmometers.

Secretion or Nonosmotic Theories.   Ursprung (1929) and others suggested that the water potential is lower on the inner than on the outer side of root cells, resulting in an inward movement of water. This explanation appears to have been revived by Borisova (in McMichael and Persson, 1991). However, it seems very unlikely that significant differences in water potential can be maintained on opposite sides of cells in slowly transpiring plants, and interest in this theory has waned. In the 1940s and 1950s there was renewed interest in the possibility of nonosmotic movement of water in cells and roots. This resulted from observations, such as those of Van Overbeek (1942) and Ginsburg and Ginzburg (1970), that the osmotic potential of exudate from detopped root systems often is higher than the osmotic potential of the solution required to stop exudation (see Kramer, 1983, p. 221, and Mozhaeva and Pilschikova, 1979, for references). This difference was attributed to some kind of nonosmotic force causing water movement. However, other investigators pointed out that roots are not completely impermeable to solutes, also that considerable salt is removed from the xylem sap as it moves up through the roots and the lower part of the stem. This can result in the concentration of solutes in the xylem sap being significantly higher in the roots than at the stem stump where it usually is collected (Klepper and Kaufmann, 1966; Oertli, 1966). This explanation of the discrepancy was disregarded by Schwenke and Wagner (1992) who proposed a complex explanation of exudation based on turgor-regulated changes in permeability of cell membranes. They also reported strong short-term pulses in exudation previously unobserved. Fiscus (private communication) suggested that if ion uptake is greater toward the root apex than toward the base and water enters along the entire length of the root, the concentration in the xylem solution will become increasingly dilute toward the base of the root. This gradient in concentration might explain some of the discrepancies in the literature.

Electroosmotic Theories.   Various investigators have attributed root pressure to electroosmotic transport of water into the xylem (Keller, 1930; Heyl, 1933). Water can be moved across a membrane under an applied electric current, with the direction being toward the pole with the same charge as the membrane. The interior of roots are electrically negative to the exterior and cellulose membranes are negatively charged so water should move inward. Interest in electroosmosis was renewed by Fensom (1958) who observed correlations between cycles of root pressure exudation and bioelectric currents, and Tyree (1973) suggested it as an alternative or supplement to the osmotic theory. However, it is doubtful if electroosmosis can cause significant net water movement in plant tissue because the high permeability of plant cells allows water to leak out almost as rapidly as it is moved inward (Ordin and Kramer, 1956; Dainty, 1963; Slatyer, 1967, pp. 174–175).

Osmotic Theories.   The most satisfactory explanation of root pressure assumes that it is an osmotic process in which roots function as osmometers because of accumulation of solutes in the xylem sap. This view is supported by the pressure probe experiments of Steudle and Frensch (1989) and data in Chapter 3. Most of the solutes are moved into the root xylem by an active ion transport system, but some may be released by the disintegration of protoplasts in maturing xylem vessels (Hylmö, 1953; Kevekordes *et al.*, 1988) and some are carried into roots by the transpiration stream. Cortes (1992) speculated that ions are released in the root interior because of the decreased membrane potential across the cortex.

It is well established that the occurrence of root pressure is correlated with the accumulation of salt in the root xylem. It develops only if root systems are healthy, provided with a supply of minerals, and kept well aerated and at a moderate temperature. Root pressure exudation ceases when root systems are subjected to low temperature, inadequate aeration, dry soil, or an inadequate supply of minerals, as when they are immersed in distilled water.

The principal problem is the location of the differentially permeable membrane permitting solute accumulation in the stele and the development of osmotic pressure. It generally is assumed that in roots which have not undergone secondary growth the endodermis functions as the differentially permeable membrane. This usually may be true, but root pressure was demonstrated in maize roots from which the cortex had been removed, destroying the endodermis (Yu, personal communication). Steudle *et al.* (1993) found that penetrating the endodermis caused a large, but temporary, decrease in root pressure. Ginsburg and Ginzburg (1970) removed cylinders of cortical tissue from maize roots and used them as osmometers to produce osmotic pressure, and Shone and Clarkson (1988) used hypodermal sleeves from maize roots to study the radial flow of water. Thus, it appears that the cortical parenchyma or the pericycle and stelar parenchyma can function as a multicellular osmotic membrane, although probably not as effectively as in an intact root (Atkins, 1916). Clarkson (1993) discussed root structure with respect to the inward transport of solutes and water and considered plasmodesmata in the symplast pathway, but Rygol *et al.* (1993) questioned the conventional view of inward flow.

Little attention has been given to the location of the differentially permeable membrane in woody roots that have lost their cortical parenchyma and endodermis during secondary growth. It must be present because fully suberized grape (Queen, 1967) and conifer root systems (Lopushinsky, 1980) exhibit root pressure. Also, application of pressure increases water flow through suberized pine roots more than salt movement (Chung and Kramer, 1975). It seems possible that the cambial region serves as a salt barrier in woody roots because its cell walls are too thin to permit significant leakage into the apoplast. However, this problem needs more study.

Relative Importance of Osmotic and Passive Absorption

There are differences in opinion concerning the importance of osmotic absorption of water and the accompanying root pressure. Fensom (1958), Minshall (1964), Rufelt (1956), and others claimed that it is quite important. Rufelt (1956) and Brouwer (1965) claimed that osmotic absorption operates in series with passive absorption, but it seems more reasonable to regard it as operating in parallel. Furthermore, positive pressure disappears as transpiration increases and increased water flow lowers the solute concentration in the root xylem. Others suggest that it plays an essential role in refilling xylem vessels with water that have become filled with gas by cavitation during periods of rapid transpiration or freezing weather. This is discussed in Chapter 7. Palzkill and Tibbits (1977) claimed that root pressure flow is essential to supply calcium to slowly transpiring tissues such as leaves in the interior of cabbage heads.

Our view is that root pressure results from the accumulation of solutes in the stele of roots and that any beneficial results are largely incidental. Intact transpiring plants can absorb water from drier soil and more concentrated solutions than can detopped root systems (Jäntti and Kramer, 1957; McDermott, 1945). Also, the volume of exudate usually is only a small percentage of the volume of water lost by transpiration, and no root pressure exists in the roots of rapidly transpiring plants (Fig. 6.3). The early history of root pressure phenomena has been discussed in more detail in Kramer (1969, Chapter 5; 1983, Chapter 8). Some of the problems resulting from the simultaneous occurrence of both osmotic and pressure-driven water flow into roots have been discussed by Fiscus (1975) and Dalton *et al.* (1975). The amount of exudation, its composition, and the pressure developed vary widely among plants and environmental conditions (Canny and McCully, 1988).

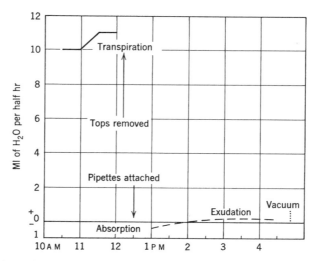

Figure 6.3 The gradual change from absorption to exudation of root systems following the removal of shoots of tomato plants growing in soil at field capacity. Because of a water deficit in the transpiring plants and possibly in the soil immediately surrounding the roots, the roots absorbed water through the stumps for over an hour after the removal of shoots. The maximum rate of exudation was much lower than the rate of transpiration and was increased by application of a vacuum of 64 cm Hg. From Kramer (1983), after Kramer (1939).

CHARACTERISTICS OF ROOT PRESSURE EXUDATION

Although root pressure seems to play a minor role in overall plant water relations, it is so interesting physiologically and has received so much attention that we will discuss it in some detail.

Species Differences

It seems that any plant that can accumulate salt in the root stele should exhibit root pressure exudation. However, there are wide differences among species. Root pressure is seldom seen in woody plants, especially conifers, and the volume of exudate is much greater in plants such as maize, sugarcane, tomato, and sunflower than in legumes. Yu (1966) attributed the larger volume of exudate from corn root systems than from broad bean to the presence of more and larger vessels in corn and less ion leakage. Root pressure was rarely reported in conifers until recently, but it is now well documented. For example, White *et al.* (1958) observed exudation from roots of white pine and white and red spruce. O'Leary and Kramer (1964) observed exudation from detached roots of loblolly pine and white spruce, but not from stumps of detopped seedlings.

Lopushinsky (1980) reported exudation from stumps of detopped seedlings of eight species of conifers, and exudation persisted for many days after detopping. The total volume of exudate varied from 0.1 to 8.0 ml per root system and its osmotic potential varied from 0.024 to 0.078 MPa, which is lower than the potential usually reported for exudate from herbaceous species. Some of these seedlings had new roots, but other root systems were completely suberized. It seems possible that measurable exudation was consistently present in Lopushinsky's experiment because the seedlings were stored for 2 months at $1-2°C$ before detopping, and this probably favored conversion of starch to sugar and accumulation of solutes in the xylem (Clarkson, 1976). Lopushinsky's measurements of exudation were made at room temperature. Thut (1932) observed root pressure exudation from several kinds of rooted, submerged aquatic plants, and Pedersen (1993) observed guttation of tritium-labeled water from leaf tips of aquatic plants.

Magnitude of Pressure.   Hales (1727), who made the first recorded measurements of root pressure, observed a pressure of about 0.1 MPa (0.1 MPa = 1 bar) in grape, and pressures of 0.2 to 0.3 MPa were reported in birch in New England by Merwin and Lyon (1909). Pressures ranging from 0.05 to 0.19 MPa have been reported for root systems of herbaceous species, but the highest pressure recorded, so far as the authors know, was 0.42 MPa on maize root systems growing in solution culture (Miller, 1985). White (1938) reported pressures of over 0.6 MPa on excised tomato roots in culture. Higher pressures have been reported in stems of woody plants, but they are believed to be local pressures caused by wounding or microbial activity (see section on stem pressures).

Volume and Composition of Exudate.   The volume of exudate usually is larger from large root systems, but it often varies among root systems of similar size and past treatment for no identifiable reason. Sugarcane stools exude up to a liter of sap in a week, and maize 100 ml/day for up to 15 days. Crafts (1936) reported a daily volume of exudate from squash root systems greater than the volume of the root systems, and G. H. Yu (unpublished) estimated that exudation from apical segments of corn roots represented a turnover of xylem vessel contents of three times per hour. Minshall (1964, 1968) reported that when abundant urea or KNO_3 was supplied to tomatoes, exudation was increased up to 80 ml/day per plant for tomatoes in the 16 to 18 leaf stage, a rate much greater than usually observed. Birch trees often yield $20-100$ liters or more of sap in a spring (Ganns *et al.,* 1981). The chief sugar in birch sap is fructose, with lesser amounts of glucose and sucrose. Birch sap sometimes is concentrated to make syrup, and in the Ukraine, sugar and citric acid are added to make a soft drink (Sendak, 1978).

The solutes in the exudate from root systems of herbaceous plants usually are chiefly minerals, but the exudate from root systems of legumes and woody plants often contain appreciable amounts of organic substances, chiefly sugar and nitrogen compounds. Xylem sap also contains growth regulators, especially cytokinins and gibberellins, and in stressed plants, abscisic acid. Bollard (1960) reviewed considerable literature on the composition of xylem sap and its role in translocation. Its salt content is increased by fertilization, and attempts have been made to use the composition of root pressure exudate as a guide to fertilizer needs (Lowry *et al.*, 1936; Pierre and Pohlman, 1934). Stark *et al.* (1985) reported that sap extracted from conifer branches by use of a pressure chamber gave a good indication of the relative mineral composition of various trees if they were of similar age and sampled at the same time of day, season, and height. Osonubi *et al.* (1988) reported that the concentration of minerals in the xylem sap was related to the concentration in the foliage. Contrary to Stark *et al.* (1985), they found that the concentration was not sensitive to diurnal variations in rate of water movement. Canny and McCully (1988) discussed the problems encountered in collecting representative samples of xylem sap.

Periodicity.   Over a century ago Hofmeister (1862) observed a diurnal periodicity in root pressure exudation from detopped root systems, and this has been observed by several investigators in recent decades. The maximum pressure and volume usually are observed during the day. This has been attributed to greater translocation of salt into the xylem during the day than at night. However, Parsons and Kramer (1974) found the highest salt concentration in the exudate from cotton root systems at night, possibly because the volume of exudate per hour was much lower at night. There appeared to be little difference in the total amount of salt translocated into the xylem per hour during the day and night. Raper (private communication) reported that a lower uptake of NO_3^- occurs at night than during the day.

Diurnal fluctuations in apparent root resistance or permeability also occur, with the lowest resistance at midday and the highest at night, as shown in Fig. 6.4 and in papers by Skidmore and Stone (1964), Barrs and Klepper (1968), and Parsons and Kramer (1974). The latter found that these cycles could be reset by reversing the light–dark cycle under which the plants were grown and disappeared after 8 or more days in continuous light, suggesting that they are controlled by signals from the shoots. Hagan (1949) found a similar periodicity in movement of water out of roots into dry soil when water was supplied to the stumps of detopped plants.

An intensive investigation by Fiscus (1986) indicated that diurnal fluctuations in the volume of exudate obtained from roots of *Phaseolus* under various pressures is controlled by complex interactions among several transport coeffi-

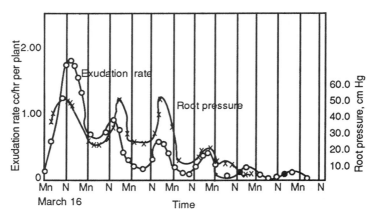

Figure 6.4   Diurnal variation in volume of exudation and root pressure of sunflower plants kept at a constant temperature in half-strength Hoagland solution. From Kramer (1969), after Vaadia (1960).

cients whose importance varies with the volume flux. These include hydraulic conductance, solute flux, reflection coefficient, and the osmotic potential of a solute compartment that lies between the exterior of the root and the xylem (Fiscus, 1977). The osmotic potential of this compartment undergoes dramatic diurnal changes and appears to be particularly important in controlling diurnal rhythms.

According to Ivanov (1980), Russian investigators suggested that the daily periodicity in root pressure exudation is related to daily periodicity in cell division and root growth. This view seems to be supported by work of Bunce (1978) who reported that an increase in apparent root resistance was correlated with decreasing root elongation. More research is needed on this topic. For example, do the completely mature suberized root systems of dormant tree seedlings show periodicity in root pressure exudation?

Guttation

One of the most common results of root pressure is guttation, the exudation of liquid water from leaves and occasionally from leaf scars. Examples are the droplets of water on the margins of leaves in the morning, and much of the water on grass leaves in the morning is the result of guttation (see Fig. 6.5). It usually occurs from hydathodes, which are stomate-like pores in the epidermis located over intercellular spaces where small veins terminate. Figure 6.6 shows a diagram of a hydathode. When sufficient root pressure develops, water is forced out from the xylem into the intercellular spaces and flows out through the hydathodes and occasionally through stomata. Friesner (1940) reported

Figure 6.5 Droplets of guttated water in early morning on leaves. Also note the thin layer of dew on the leaves. Photograph by J. S. Boyer.

guttation through lenticels on stump sprouts of red maple in the early spring, and Raber (1937) observed sap flow from leaf scars of deciduous trees after leaf fall in the autumn. Guttation is said to be common at night in tropical rain forests. Exudation of liquid from young roots was reported by Breazeale and McGeorge (1953), Head (1964), and Schwenke and Wagner (1992), and might be termed root guttation.

The quantity of guttation liquid exuded varies from a few drops to many milliliters, and the composition varies from almost pure water to a sufficiently concentrated solution of organic and inorganic solutes to leave a visible coating of solute on the leaves when the water evaporates (Curtis, 1944; Duell and Markus, 1977). Guttation seldom occurs during the day because transpiration

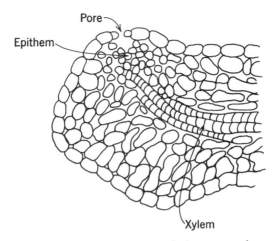

Figure 6.6   Diagram of a hydathode showing a pore, the loose mass of parenchyma cells termed the epithem, and the termination of the xylem. Hydathodes resemble stomata with nonfunctional guard cells and usually occur along margins and at tips of leaves. From Kramer (1983).

reduces or eliminates the pressure in the xylem sap and it ceases in dry, cold, or poorly aerated soil. Guttation is said to be common in submerged aquatic plants, where it may be important in transporting mineral nutrients to growing stem tips (Pedersen, 1993).

Although it is an interesting physiological phenomenon, guttation usually is of little practical importance to plant growth. Occasionally, injury occurs by accumulation of salts on leaf margins as guttated water evaporates, and pathologists suggest that it produces conditions favorable for invasion of leaves by fungi and bacteria (Robeson *et al.,* 1989). Injection of intercellular spaces in leaves of greenhouse plants is said to occasionally cause injury. Some additional information is given in Kramer (1969, pp. 165–167).

STEM PRESSURES

Although exudation from herbaceous plants and some woody plants such as birch and grape is caused by root pressure, exudation from sugar maple, palm, agave, and a few other plants is caused by stem pressures that are independent of root pressure.

Maple Sap Flow

The best known example of stem exudation in North America is the flow of sap from holes bored in maple trees, chiefly *Acer saccharum* Marsh and *A. ni-*

grum, Michx. The sap is collected and concentrated by evaporation in pans to produce maple syrup. Because of the commercial value of maple syrup, maple sap flow has received much attention and has been discussed by W. S. Clark (1874, 1875), Jones *et al.* (1903), Wiegand (1906), Stevens and Eggert (1945), Johnson (1945), Marvin (1958), and others. It can occur from late autumn to spring, whenever freezing nights are followed by warm days with temperatures above freezing, but the largest flows are obtained in the spring. Over 60% of the flow usually occurs before noon and it often ceases during sunny afternoons because transpiration from the branches and twigs eliminates pressure in the xylem sap. The yield varies widely, usually ranging from 35 to 70 liters per tree in a season, but occasionally it is twice that amount. The sap usually contains 2–3% sucrose and certain nitrogen compounds that when heated produce the distinctive flavor of maple syrup (Pollard and Sproston, 1954).

That maple sap flow is caused by stem pressure is indicated by the fact that sap flow can be obtained from sections of tree trunks placed in tubs of water and subjected to alternating freezing and thawing (Stevens and Eggert, 1945). Maple stems are unique in absorbing water while freezing and exuding water while thawing, but this occurs only if sucrose is present in the sap (R. Johnson *et al.,* 1987). Also, simultaneous measurements of root and stem pressures indicate that in maple trees stem pressure often exceeds root pressure, whereas in birch trees root pressure always exceeds stem pressure, as shown in Fig. 6.7 from data of Kramer (1940c).

Maple sap is obtained by drilling holes into the sap wood ("tapping") and installing spouts ("spiles") through which sap drains by gravity into containers. The yield sometimes is increased by applying vacuum to the spouts. The sap is concentrated by boiling to produce commercial maple syrup. There is some uncertainty about the mechanism causing sap flow. According to Sauter (1971), carbon dioxide produced by respiration during the day collects in the intercellular space and forces the sap out. At night, lower temperature reduces carbon dioxide production and that present dissolves, reducing the pressure, causing upward movement of water from the roots and refilling of the xylem vessels. However, the process deserves further investigation (R. Johnson *et al.,* 1987).

Other Stem Pressures

In India and tropical Asia, large amounts of sap are obtained from palms, chiefly coconut, date, and Palmyra. The young inflorescence is cut out, leaving a cavity in which sap collects at a rate of 6–8 or more liters per day. This flow can be maintained for months by rewounding. In some instances sap is obtained by making incisions into palm stems, but in both methods the sap comes from the phloem, and the sugar was originally mobilized for use in inflorescence and growing stem tips (Milburn and Zimmermann, 1977). The sap is used as a source of sugar or is fermented to make palm wine.

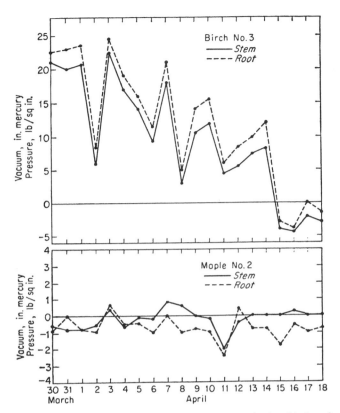

Figure 6.7   Simultaneous measurements of root and stem pressure in river birch and red maple. In birch, root pressure always exceeded stem pressure and the two change simultaneously. In maple, root pressure was usually negative even when positive pressure existed in the stems. Measurements were made on trees in a bottomland forest. From Kramer and Kozlowski (1979).

In Mexico, large quantities of sap containing sucrose are obtained from agaves by cutting out the young inflorescence. This leaves a cavity in which sap collects at the rate of a liter or more per day for 10 to 15 days, and it is removed to be fermented into pulque. Sap flow in palms and agave originally was attributed to root pressure, but it actually is caused by local phloem pressure. Exudation from agave and palm is discussed in more detail by Van Die in Zimmermann and Milburn (1975) and by Milburn and Zimmermann (1977). The latter measured phloem sap pressures of up to 0.76 MPa in *Cocas nucifera* L.

Wounding of tree stems sometimes results in the development of local high pressure. MacDougal (1926) reported local exudation pressures in stems of *Pinus radiata, Juglans regia,* various oaks, and large cacti. Occasionally this sap is fermented by bacteria or yeasts as it flows out of cracks or other wounds, causing slime flux (Carter, 1945). Pressures high enough to blow the cores out of increment borers have been observed in trees containing decaying heartwood (Abell and Hursh, 1931) and sometimes the gas contains enough methane to be flammable. The production of methane in tree trunks has been discussed by Zeikus and Ward (1974).

LATEX AND OLEORESINS

In addition to the exudation of xylem sap, some plants exude complex organic compounds from special duct systems when wounded. The best known of these compounds are the latex from which rubber is produced, the oleoresins from which turpentine and rosin are produced, and the chicle formerly used in chewing gum. In the rubber tree, *Hevea brasiliensis,* latex accumulates in a complex system of vertically oriented ducts in the outer bark, from which it flows when wounded. Oleoresin is produced in the xylem and bark parenchyma of some coniferous trees and accumulates in resin ducts from which it flows when wounded.

These examples of exudation of complex compounds from wounded tissue are quite distinct from the exudation of xylem sap from wounded plants. However, they are related to plant water status because latex and oleoresin flow are reduced when plant water stress develops. According to Milburn *et al.* (1990), the latex pressure of banana is a good indicator of plant water potential. More details and references concerning oleoresins and latex can be found in Kramer and Kozlowski (1979, Chapter 8). Lorio (1993, pp. 91–94; 1994) discussed the rather complex relationship between tree water and resin flow in conifers.

ABSORPTION OF DEW AND FOG THROUGH LEAVES

There is a considerable difference of opinion concerning the importance of absorption of water through leaves and stems. The early literature, reviewed by Miller (1938, pp. 188–190), indicated that significant amounts of water and solutes can enter plants through the leaves because the cuticle is moderately permeable when wet. This makes foliar fertilization possible, but also results in leaching of solutes from leaves by rain and sprinkler irrigation (Tukey *et al.,* 1965). The intake and the loss of water and solutes by twigs of trees have been discussed by Katz *et al.* (1989). According to Anderson and Bourdeau (1955), liquid water from rain, fog, and dew is the only important source of water for mosses as their lack of xylem limits conduction from the soil. This must generally be true for epiphytes, but according to Martin and Schmitt (1989), absorption of water from the humid air of its native habitat is insufficient to maintain the normal water content of Spanish moss (*Tillandsia usneoides,* L.) and occasional rain is essential for its survival.

The importance of dew and fog is somewhat controversial. Duvdevani (1953), Gindel (1973), and others in Israel claimed that absorption of dew is important for survival and growth of both herbaceous and woody plants in arid regions. Stone and Fowells (1955) showed in greenhouse experiments that dew prolonged the survival of unwatered ponderosa pine seedlings, and Went (1975) suggested that *Prosopis tamarugo* growing in the Chilean desert depends chiefly on water absorbed from fog. However, the latter was questioned by Mooney *et al.* (1980), who concluded that *Prosopis* depends on water deep in the soil.

Monteith (1963) claimed that the amount of dew deposited on leaves is too small to be important, and Slatyer (1967, pp. 231–236) pointed out that the vapor pressure deficit even in wilted leaves is so small that the inward movement of water vapor must be very slow. However, the effects of condensed fog and fog drip in reducing water loss can be important enough to affect the species composition of some forests, as in the fog belt of the Pacific Coast (Oosting, 1956, pp. 125–126) and several other areas mentioned by Chaney (1981). The importance of dew and fog probably depends on local conditions (Chaney, 1981; Rundel, 1982).

A new aspect of the fog problem has developed recently with increasing air pollution because various toxic substances become concentrated to a high degree in fog and clouds. Thus foliage at high elevations, which often is exposed to clouds and fog, is more subject to injury (Fowler *et al.*, 1989). Azevedo and Morgan (1974) stated that water and minerals deposited on foliage in the California coastal fog belt significantly affect the water and mineral nutrient balance of the vegetation. Lovett *et al.* (1982) reported that the water input to subalpine forests in the northern Appalachian mountains from fog and clouds amounted to 46% of the total precipitation. Thus, the high concentration of pollutants in a fog seems to explain the severe injury to foliage of trees sometimes observed at high elevations.

FACTORS AFFECTING WATER ABSORPTION THROUGH ROOTS

The numerous factors affecting the absorption of water can be classified in two groups: (1) those affecting the driving force or gradient in water potential from soil to and through roots; and (2) those affecting the resistance to water movement through the soil and roots. Some, such as low temperature, may do both. They can be described by the following equation:

$$\text{Absorption} = \frac{\Psi_{soil} - \Psi_{root\ surface}}{r_{soil}} = \frac{\Psi_{root\ surface} - \Psi_{root\ xylem}}{r_{root}}, \quad (6.1)$$

where Ψ is the water potential, r is the resistance to water movement, and the subscripts soil, root surface, and root xylem indicate positions in the soil–root system. Nnyamah *et al.* (1978) reported that the rate of water absorption by trees in a Douglas fir forest was linearly related to the difference between soil and root xylem water potential and that the difference remained relatively constant in drying soil. One would expect the difference in the Ψ_w required to move water through the soil to increase with decreasing soil water content, but perhaps the resistance was low in that soil.

Absorption of water by plants in moist, warm soil is controlled chiefly by the rate of transpiration, which provides most of the driving force, and by the efficiency of the root system, which depends on its extent and permeability or hydraulic conductance. Low temperature and the deficient aeration common in wet soils decrease root permeability, increasing the resistance to entry of water. As the soil dries, its water potential decreases, reducing the driving force from soil to roots, and soil resistance increases because the larger pores are emptied first (Fig. 4.7 gives an example expressed as decreasing hydraulic conductivity of the soil). Root and soil shrinkage tend to decrease root–soil contact (see Chapter 5). Increased root suberization in drying soil also decreases root permeability, increasing resistance to water absorption.

The limiting effect of low soil water potential increases as atmospheric conditions favor high potential rates of transpiration. Denmead and Shaw (1962) found that a soil water potential of only -0.1 MPa limited absorption at high rates of transpiration, but -1.0 MPa was not limiting at low rates. The practical significance of this is that on sunny days rapidly transpiring plants often develop water deficits in moist soil, but in cool, cloudy weather plants may show little stress in relatively dry soil.

Efficiency of Root Systems in Absorption

The efficiency of root systems in absorption of water and minerals depends on their depth and spread, their density, often expressed as root length density in centimeters of roots per cubic centimeter of soil, and their permeability or hydraulic conductance (Lp). Frequently the inverse of the conductance, the resistance, is used [Eq. (6.1)]. Deep, profusely branched root systems are advantageous because they occupy a larger volume of soil, containing more water. Thus, plants with deep root systems usually survive droughts with less injury than those with shallow root systems (Chapter 5; Hurd, 1974; Kramer, 1983, pp. 236–240). For example, Bremner et al. (1986) found that in the field sunflower extracted soil water more rapidly and to a greater depth (2m) than sorghum. Thus, it is more successful than sorghum in southern Australia when summer droughts occur.

The benefit from deep, wide spreading root systems depends on the resistance to longitudinal water movement in roots being lower than that to movement through the soil. In general, this is true despite occasional exceptions under special conditions, such as those reported for some grasses and cereals by Wind (1955) and Passioura (1972). Passioura (1972) suggested that if plants such as wheat growing in regions with low summer rainfall mature on stored soil water, restriction of absorption early in the growing season by a limited root system may be advantageous because it conserves water that is needed later during grain filling. However, this is effective only where no other kinds of plants are competing for the stored water. Klepper has a useful discussion of root growth in relation to water absorption in Stewart and Nielsen (1990). Landsberg and Fowkes (1978) published a mathematical analysis of the effect of root system geometry and water potential in relation to water absorption.

Meyer and Ritchie (1980) reported that resistance per unit of root length decreases toward root tips in sorghum, compensating for increased distance to the shoot, and a similar situation was found in roots of red pine by Stone and Stone (1975a). Apparently, water is absorbed as readily by roots at distances of several meters from plants as from nearby (Veihmeyer and Hendrickson, 1938; Hough *et al.*, 1965). Apple trees on well-aerated loess soil absorbed water from a depth of 10 m (Wiggans, 1936), and corn grew well after the available water in the upper meter of soil was exhausted (Reimann *et al.*, 1946). McWilliam and Kramer (1968) demonstrated that plants of *Phalaris tuberosa* survived after the water potential in the upper meter of soil was reduced to -1.5 MPa because some of the roots had penetrated to a deeper horizon, containing readily available water.

A root system consists of roots of various ages, ranging from young unsuberized root tips to roots that have lost their cortex during secondary growth and are enclosed in a layer of suberized tissue (bark) (Fiscus, 1977; Sands *et al.*, 1982). Obviously there are wide variations in permeability; young unsuberized roots are more permeable than those which have undergone secondary growth and suberization. Differences in the permeability of grape roots of various ages and conditions are shown in Table 6.1. Considerable water and salt must be absorbed through suberized roots because unsuberized roots often constitute less than 1% of the root surface under forests (Kramer and Bullock, 1966). Van Rees and Comerford (1990) found an uptake of water, potassium, and bromine through woody roots of slash pine and suggested that they play an important role in mineral nutrition. Sanderson (1983) reported that about 50% of the water absorbed by the main axis of barley roots entered through the suberized region. Using NMR imaging, MacFall *et al.* (1991a) showed that suberized roots of pine absorb significant amounts of water from moist sand (Fig. 5.9). Fiscus and Markhart (1979) and Fiscus (1981) studied changes in absorptive capacity of growing bean root systems. The hydraulic conductance (Lp) of the entire root system changes in a complex manner with age and changing proportion of suberized surface, as is shown in Fig. 6.8. Slow maturation of large metaxylem vessels near the tips of some kinds of roots probably hinders absorption through root hairs (Kevekordes *et al.*, 1988; McCully and Canny, 1988; Cruz *et al.*, 1992). The development of root systems is discussed in more detail in

Table 6.1 Relative Permeabilities of Grape Roots of Various Ages to Water and ^{32}P

Zone and condition of roots	Relative permeabilities	
	Water	^{32}P
Roots of current season (growing)		
A. Terminal 8 cm, elongating, unbranched, unsuberized	1	1
B. Unsuberized, bearing elongating branches (dormant)	155	75
C. Main axis and branches dormant and partially suberized before elongation is completed	545	320
D. Main axis and branches dormant and partially suberized	65	35
Roots of preceding seasons (segments bearing branches)		
E. Heavily suberized main axis with many short suberized branches	0.2	0.4
F. Heavily suberized, thick bark, and relatively small xylem cylinder		
Intact	0.2	0.02
Decorticated	290.0	140.00

Note. Measurements were taken under a pressure gradient of about 66 KPa.
[a]From Queen (1967).

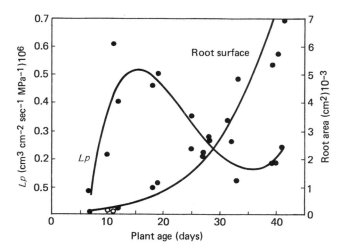

Figure 6.8 Changes in surface area and average hydraulic conductance (Lp) of the root system of growing seedlings of bean (*Phaseolus vulgaris*). While root surface increased steadily with time, the average hydraulic conductance first increased, then decreased over time, probably because of changes in the proportion of the root system in various stages of maturation. From Kramer (1983), after Fiscus and Markhart (1979).

Chapter 5, in Wild (1988, Chapter 4), and in Gregory *et al.* (1987). Methods of studying root growth are discussed in Chapter 5, in books by Böhm (1979) and by Glinski and Lipiec (1990, Chapter 7), and in a paper by Epstein in Hashimoto *et al.* (1990).

Radial Pathway of Water in Roots.   There has been considerable discussion concerning the relative importance of various pathways for radial water movement into roots. There are three possible pathways for water movement from the epidermis across the cortex to the xylem: (1) through the cell walls, termed the apoplastic pathway; (2) through the symplast, which consists of the cytoplasm of many cells connected through plasmodesmata; (3) from cell to cell across the cell walls, plasmalemmas, vacuoles, and tonoplasts; or a combination of these pathways (see Chapter 3 and Fig. 3.16). Pathway 2 is termed symplastic because it involves movement through protoplasm. There also may be limited movement on the outer surfaces of cell walls bordering intercellular spaces. The extensive literature on this topic has been reviewed by Boyer (1985), Davies *et al.* (1986), Drew in Gregory *et al.* (1987), Jones *et al.* (1988a), Passioura (1988b), Weatherley (1982), and others. It is difficult to reach positive conclusions concerning the most important pathway because of uncertainty concerning their relative permeabilities. Perhaps this will be resolved by the use of pressure probes, as described by Steudle in Hashimoto *et al.* (1990), Zhu and Steudle (1991), and Zimmermann et al. (1992). Also, as transpiration increases, the driving force changes from chiefly osmotic to pressure flow (Fiscus, 1975; Fiscus and Kramer, 1975). Boyer (1985, p. 483) concluded that the transport of water through tissue is slower than for single cells, indicating that symplastic movement is more important than apoplastic movement through cell walls. However, Radin and Matthews (1989) reported that root conductivity was about twice that of individual cells, supporting the apoplastic pathway. Zhu and Steudle (1991) also concluded that the apoplastic pathway is more important than the cell-to-cell pathway in maize roots. Hanson *et al.* (1985) found negligible apoplastic flow of a fluorescent dye in red pine roots, but Peterson *et al.* (1981) reported significant movement of apoplastic dye through the endodermis of corn and broad bean, where secondary roots passed out. It is doubtful if tracers are always reliable indicators of the path followed by water (McCully and Canny, 1988) and more research is needed on this problem.

Magnetic resonance images of root cortical parenchyma sometimes show regions resembling spokes in a wheel extending across the cortex that appear to be much higher in water content than the average of the cortex (Brown *et al.*, 1986). These patterns suggest that not all of the cortical parenchyma cells are equally involved in radial water transport. However, Zimmermann *et al.* (1992) question this interpretation. This may be related physiologically to the breakdown of cells in specific regions of poorly aerated roots (see Fig. 5.19). According to Drew *et al.* (1980), the strands of cortical tissue left in the aerenchyma of poorly aerated roots function effectively in water and ion transport. No satisfactory explanation of why some cortical tissue breaks down, leaving adjacent strands of healthy tissue, is available, and this problem deserves more study.

The movement of water into roots is decreased by treatments such as chilling and deficient aeration and by respiration inhibitors that decrease the permeability of protoplasmic membranes, but it is increased by killing the roots. This indicates that at some point much of the water passes through living cells, probably at the epidermis, exodermis, and endodermis, although the protoplasts of cortical parenchyma cells might also be involved. It seems probable that all possible pathways are utilized to varying degrees, depending on their relative conductances and carrying capacities. A similar problem exists with respect to the path of water movement through the mesophyll cells of leaves (Boyer, 1985). The effects of environmental factors on root resistance or conductance and water absorption are discussed later in this chapter.

Resistances to Water Movement in the Soil–Plant System

The existence of substantial resistances to water flow through the soil–plant system is indicated by the development of significant reduction in water potential in leaves of rapidly transpiring plants growing in moist soil or dilute nutrient solution and by the lag of absorption behind transpiration (Fig. 6.1) when plants are exposed to sun. However, there has been some uncertainty concerning the location of the principal resistance which might be in the soil or soil–root interface, in the roots, or elsewhere in the plant water conducting system.

Soil versus Root Resistance. There has been considerable discussion concerning the relative importance of soil and root resistances with respect to water absorption. Early investigators such as Gardner (1960) and Gardner and Ehlig (1962) concluded that resistance in the soil would exceed resistance in the roots at a soil water potential of -0.1 to -0.2 MPa, and this view was supported by Cowan (1965) and others. However, Newman (1969) concluded that the root densities used by previous investigators were much lower than those usually found in nature, and if normal root densities were used, soil resistance in the vicinity of roots would not be limiting until the soil water content approached the permanent wilting percentage. Later, Newman (1973) reported large differences among species in root permeability (or resistance) which he regarded as supporting his view. Blizzard and Boyer (1980) found the plant resistance to water flow in soybean is greater than the soil resistance over the range of soil water potential from -0.025 to -1.1 MPa and other investigators observed root resistances greater than soil resistances in corn and sorghum (Reicosky and Ritchie, 1976) and in cotton (Taylor and Klepper, 1975). Passioura (1980a) found no evidence of the large resistance at the soil–root interface reported by some investigators, but Hulugalle and Willatt (1983) claim that root resistance is important in soils with low hydraulic conductance and where root density is low. Diurnal variations in root resistance to water were discussed earlier in connection with periodicity in root pressure exudation. Harris (1971) observed strong cycling in water absorption by cotton plants growing in a constant environment, and Passioura and Tanner (1985) reported apparent oscillations in root conductivity of cotton when the rate of transpiration was changed, but this may be related to the stomatal oscillations discussed in Chapter 8.

It has been suggested that as the soil dries, water absorption is decreased by root and soil shrinkage which decreases contact at the root–soil interface (Herkelrath *et al.*, 1977; Faiz and Weatherly, 1978; Huck *et al.*, 1970). Nobel and Cui (1992) found that the roots of three species of desert succulents shrank an average of 19% during soil drying. The resulting air gap between roots and soil not only reduced absorption measurably, but also reduced loss of water from roots to soil. Loss of water from roots is discussed later in the section on efflux of water. However, Taylor and Willatt (1983) found negligible shrinkage of roots on wilting soybean plants in moist soil, and the importance of root shrinkage may have been overstated, as pointed out by Tinker (1976). The latter has a good discussion of the factors involved in water absorption, and Landsberg and Fowkes (1978) present a mathematical treatment of absorption, including consideration of soil water movement, root system geometry, and the ratio of radial to axial resistance to water flow. The evidence seems to indicate that over a considerable range of soil water content and with average root density, root resistance exceeds soil resistance to water movement. However, drying soil decreases radial root conductance by increasing lignification and suberization (North and Nobel, 1991) and axial conductance by slowing metaxylem maturation and decreasing the number and size of xylem elements (Cruz *et al.*, 1992) and Lopez and Nobel (1991) also observed decreased conductance in cactus roots in drying soil. Kramer (1950) found that wilting plants overnight decreased root permeability when they were rewatered the next day, and Xu and Bland (1993) reported that exposure of sorghum roots to drying soil decreased water uptake.

It usually is assumed that all water movement to roots occurs as liquid, but as air spaces enlarge in drying soil some movement might occur as vapor. This was proposed by Bonner (1959), but questioned by Bernstein *et al.* (1959). However, a more recent study by Dalton (1988), using deuterium as a tracer, indicates that there might be a small amount of vapor movement to roots, especially with rapidly transpiring plants.

Resistances in Plants.   The resistance to water movement from root surfaces to root xylem was discussed in the preceding section and resistances to radial and axial or longitudinal flow are discussed in Chapter 7 in connection with the ascent of sap. Melchior and Steudle (1993) have interesting data on changes in radial and axial conductivity of growing onion roots. It will suffice here to state that resistance to longitudinal flow in the root xylem generally is relatively low as compared to resistance to radial flow (Frensch and Hsiao, 1993; Frensch and Steudle, 1989; Sands *et al.*, 1982). In general it seems improbable that resistance to longitudinal water flow in the xylem becomes limiting unless it is blocked by injury or cavitation (see Chapter 7). The stomatal control of water movement and the importance of boundary layer resistances for transpiration are discussed in Chapters 7 and 8.

Change in Resistance with Change in Rate of Water Flow. It has been reported by several investigators that the rate of water flow through roots increases more rapidly than the pressure applied (Lopushinsky, 1964; Weatherley, 1982). Also, in some experiments the leaf water potential appeared to decrease with increasing transpiration, but in others it did not (see Kramer, 1983, pp. 208– 210). Some of these observations suggest that the resistance to water flow through roots may decrease with an increasing rate of flow. However, the cause of this change in apparent root resistance is uncertain. Bunce (1978) suggested that in some instances it might be an artifact resulting from failure to allow enough time for equilibration in leaf water potential after a change in the rate of transpiration. In some instances it might result from the transition in driving force from predominantly osmotic to nonosmotic movement. Steudle *et al.* (1987) concluded that water movement caused by an osmotic gradient occurs primarily in the symplast, but that the much larger movement in transpiring plants, caused by hydrostatic gradients, occurs primarily in the apoplast. According to Boyer (1974) the total plant resistance may be 30 times greater at a low flux when stomata are partially closed than at a high flux when they are fully open. He established steady conditions for each measurement so that the leaf water potential reflected the measured flow. Over the entire flow range, the root resistance changed only 2.5 times, and the 30-fold change could be seen in the leaves alone without the roots. Hence he concluded that most of the change in resistance occurs in the leaves and suggested that the apparent decrease in resistance with an increase in flow was caused by an increasing proportion of water bypassing the protoplasts of cells in the leaves. Fiscus *et al.* (1983) later showed that this amount of resistance change could be predicted from a model of the flow. Boyer (1974) suggested that the bypassing water was water vapor in the intercellular spaces because the cuticle lines the intercellular spaces close to the stomata and inhibits evaporation there. Chapters 7 and 11 discuss further evidence that the cuticle inside leaves may cause sites of evaporation to be deep in the leaf.

Root Diseases. As mentioned in Chapter 5, the extent and efficiency of root systems for absorption often are decreased by attacks of pathogenic fungi that kill small roots and block up the conducting system. For example, unfavorable soil conditions and damage to root systems by *Phytophthora cinnamoni* cause the littleleaf disease of shortleaf pine in the southeastern United States and damage to citrus trees in California. Damage to root systems and the lower part of stems, termed damping off, often causes the death of seedlings by reducing water absorption and transport. Root damage caused by various species of *Phytophthora* and other fungi often reduces water absorption and increases root and stem resistance, causing reduced shoot growth or death of plants (Duniway, 1977; Grose and Hainsworth, 1992). In addition to interfering with the absorption of water by decreasing root surface and increasing root resistance, a decrease in the supply of plant hormones and production of toxins may be involved (Ristaino and Duniway, 1991). Prolonged periods of irrigation and saturated soil often increase the injury from pathogens (see papers in Ayres and Boddy, 1986, and Chapter 7 in Kozlowski, 1984).

Environmental Factors

Both root development and water and mineral absorption are strongly influenced by environmental factors, especially soil moisture, concentration and composition of the soil solution, soil aeration, and soil temperature. These factors affect the composition of natural plant communities and the success of crop plants. For instance, most species of trees cannot grow in saturated soil, but species such as cypress, white cedar, and tupelo gum are largely restricted to areas with saturated soil where competitors cannot grow. Many crop plants are injured or killed in poorly drained areas of fields during rainy springs, but some varieties of rice thrive in flooded soil. Low soil temperature also often hinders root development of seedlings of warm climate plants such as cotton.

Availability of Soil Water. The availability of soil water was discussed in Chapter 4, and the difference in available water content of various soils was shown in Table 4.1. The soil water potential decreases sharply as the water content decreases (Fig. 4.7) and the hydraulic conductance also decreases as water drains out of the larger pores. As pointed out earlier, the availability of soil water to plants relative to need also decreases with increasing rate of transpiration (Denmead and Shaw, 1962). The relative importance of soil versus root resistance to water movement was discussed earlier in this chapter.

There is evidence that the soil in the immediate vicinity of roots of transpiring plants often tends to become temporarily dry, increasing the soil resistance to water flow toward the root surfaces. This was demonstrated by Dunham and Nye (1973) by direct sampling of thin layers of soil in the vicinity of a root mat (Fig. 6.9) and by MacFall *et al.* (1990) using magnetic resonance imaging, as shown in Fig. 5.9. Dunham and Nye used a sandy soil and MacFall used a fine sand, both at approximate field capacity. Hasegawa (1986) used a technique somewhat similar to that of Dunham and Nye with soybean roots, except that

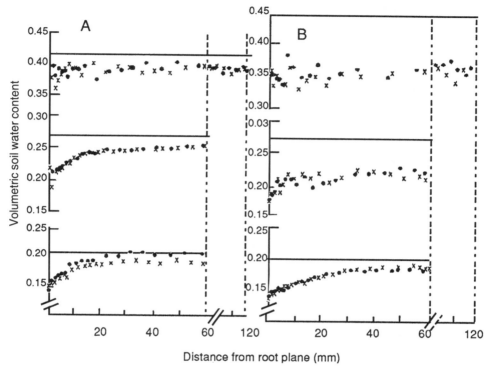

Figure 6.9 A decrease in soil water content near a mat of onion roots attached to transpiring shoots after 2.5 days (A) and 6 days (B). The roots were in sandy soil which initially contained 42–45% (upper curves), 27% (middle curves), or 20% (lower curves) water. The two curves in each part of the figure give the moisture content on opposite sides of the root mat. Readers are referred to the original paper for details. After Dunham and Nye (1973).

the change in soil water content was measured by gamma ray attenuation. He also observed progressive drying of the soil in the vicinity of the roots and absorption of water from a distance of over 10 cm from the root plane. It is possible that under some conditions this water might be at least partly replaced by upward movement through the roots from moist soil (Baker and Van Bavel, 1986; Richards and Caldwell, 1987; Corak *et al.,* 1987), in addition to movement through the soil. However, it was not entirely replaced in the experiments of Dunham and Nye (1973) or Hasegawa (1986). The drying of soil around roots emphasizes the importance of root extension into previously unoccupied soil (Kramer and Coile, 1940). Taylor (in Turner and Kramer, 1980) suggested that an increase in root depth is more important for postponing the onset of water stress in cotton and soybean than increasing root length density (cm of roots per cm^3 of soil) in the surface soil.

Concentration and Composition of Soil Solution.   A high concentration of salt in the soil inhibits the growth of most plants, and large areas of land in arid regions cannot support crop plants because of salt accumulation. Salt accumulation also is a problem on irrigated land because irrigation water often is high in salt which accumulates in the soil as water is removed by evaporation and transpiration. It seldom is a problem in humid regions because rainfall leaches salt out of the surface soil. Osmotic potentials of the soil solution of -0.35 to -0.4 MPa at permanent wilting seriously reduce the growth of most crop plants (Magistad and Reitemeier, 1943) and only a few halophytes survive below -0.4 MPa. Excessive amounts of fertilizer sometimes produce an osmotic potential too low for good plant growth, especially in coarse-textured soils with a low cation-exchange capacity. For example, application of 1300 kg/ha of $3-9-3$ fertilizer to Norfolk sandy loam temporarily decreased the osmotic potential to -1.4 MPa, but a similar application to Cecil clay loam only decreased it to -0.3 MPa (White and Ross, 1939). Excessively high concentrations of salt sometimes develop in greenhouse soils from overfertilization and reduce growth (Davidson, 1945; Merkle and Dunkle, 1944). This can be avoided by occasional overwatering to leach out excess salt. Rhoades and Loveday (in Stewart and Nielson, 1990) discussed the salinity problem in irrigation in more detail and Maas (1993) reviewed the effects of salinity on citrus trees.

Some of the decrease in growth seems to be caused by the osmotic effects on the driving force and root resistance rather than by toxic effects of particular ions (Eaton, 1942). Wadleigh and Ayres (1945) found similar reductions in growth at a given water potential produced by drying soil, adding salt to it, or by combining the two treatments. According to Materechera *et al.* (1992), the root elongation of dicot seedlings was reduced more than the elongation of monocot seedlings in polyethylene glycol. There usually is a considerable reduction in root permeability in concentrated soil solutions (Azaizeh *et al.*, 1992; Hayward and Spurr, 1943, 1944; O'Leary, 1969, and others). However, it is doubtful if the reduction in growth is caused solely by the reduced absorption of water because the uptake of salt by plants tends to maintain similar differences in osmotic potential between plants and substrate at various substrate concentrations, and plants maintain normal turgor over a wide range of substrate concentrations (Bernstein, 1961; Boyer, 1965; Eaton, 1942).

The reduction in growth caused by high salinity probably is related as much to metabolic effects of accumulation of salt in cells as to reduced availability of soil water. In halophytes, much of the salt accumulates in the vacuoles and the cytoplasm is not exposed directly to high concentrations (Wyn Jones, 1980). In many species the cytoplasm accumulates high concentrations of organic compounds such as proline, glycine betaine, and other amino acids and sugars that counterbalance the high salt concentrations in the vacuoles, but are compatible with enzyme and membrane functions (Wyn Jones, 1980; Hanson and Hitz, 1982). Tarczynski *et al.* (1993) reported that transgenic tobacco plants that synthesize mannitol show increased tolerance of high salinity. At some point plant salt tolerance seems to be related to the concentration of salt that can be tolerated by the protoplasm (Munns and Termaat, 1986; Repp *et al.*, 1959; Slatyer, 1967, pp. 301–308).

Although general osmotic effects are emphasized, specific ion effects sometimes are important (Strogonov, 1964). For example, the growth of some kinds of plants is reduced more by sulfates than by chlorides (Hayward *et al.*, 1946), but Maas (1993) stated that citrus is more sensitive to chlorides than to sulfates. It is claimed that chlorides increase the succulence of plant tissue while sulfates decrease it. Van Eijk (1939) attributed the succulence of halophytes to an excess of chlorides and Boyce (1954) reported that ocean spray increases the succulence of vegetation near the beach.

Excesses or deficiencies of specific ions in the soil solution sometimes reduce plant growth. For example, boron is an essential element, but an excess of boron can be toxic. Radin and Boyer (1982) reported that nitrogen deficiency reduces the root permeability of sunflower roots nearly 50%, resulting in a daytime loss of turgor and a reduction in leaf growth; Radin (1983) found this to be true for several other species, although monocots were affected less than dicots. Radin and Matthews (1989) found that a phosphorus deficiency also reduces root permeability.

According to Einhellig *et al.* (in Thompson, 1985), treatment of sorghum roots with ferulic or coumaric acid or extracts from several allelopathic weeds decreases the absorption of water and increases water stress. They suggested one mechanism of allelopathic action may be through an increase in plant water stress.

Root Aeration.   Deficient aeration of the soil not only reduces root growth but also reduces the absorption of water and minerals. The decrease in water absorption is caused chiefly by an increase in the resistance to radial movement into roots, but Everard and Drew (1989) also reported a decrease in the osmotic driving force in sunflower roots. This probably resulted from a decreased uptake of salt. The deficits that develop in flooded soil because of reduced absorption often cause serious injury to plants. For example, tobacco sometimes wilts so rapidly and severely when the sun comes out after the soil is saturated by a rain that farmers term it "flopping." Flooding the soil sometimes causes wilting of other kinds of plants, as shown in Fig. 6.10. There are wide differences among species of plants in respect to the effect of flooding on water absorption. In one series of experiments tobacco was most seriously injured by flooding, sunflower least, and tomato was intermediate (Kramer, 1951). In experiments with woody plants, the water absorption of bald cypress, a native of swamps, was reduced very little, but absorption by flooded oaks, loblolly pine, and red cedar was reduced to less than 50% of the controls (Parker, 1950). Kramer (1940a) and Smit and Stachowiak (1988) agreed that a high concentration of

Figure 6.10 Wilted maize in a flooded field. Water is standing between the rows which were flooded for 3 weeks. Photograph by J. S. Boyer.

CO_2 reduces water absorption more rapidly than a deficiency of O_2, a conclusion reached long ago by investigators cited by Kramer (1940a). However, the importance of CO_2 was questioned by Willey (1970). The effects of deficient aeration on root growth were discussed in Chapter 5.

Although an important effect of deficient root aeration is a reduction in root permeability to water, not all the effects can be explained simply by the development of shoot water deficits (Kramer, 1951; Jackson, 1991; Schildwacht, 1989) because stomatal closure often results in rehydration of the shoots. The development of adventitious roots and hypertrophy of stems near the water line of flooded plants suggest that growth regulators are involved; the occurrence of epinasty suggests the presence of ethylene (Bradford and Yang, 1980). The various roles of ethylene in flooded plants were reviewed by Jackson (1985). Interference with normal root metabolism may affect the supply of minerals, cytokinin, gibberellin, and ABA to the shoots and inhibition of downward translocation of metabolites, and growth regulators may cause some of the symptoms associated with the deficient aeration of root systems. Smit *et al.* (1989) reported that leaf expansion of flooded seedlings of a hybrid poplar was limited by decreased cell wall extensibility. In some instances, stomatal closure has been attributed to an accumulation of ABA in the leaves (Jackson and Hall, 1987), but later experiments of Jackson (1991) did not support roots as the source of ABA in flooded plants.

Temperature.   It was observed by Hales and others in the 18th century that cold soil reduces water absorption, and in about 1860 Sachs observed that tobacco and cucurbits wilted more severely than cabbage and turnips when the soil was cooled to 3–5°C. Numerous other examples of species differences are given in Kramer (1969, Chapter 6) and differences in three species are shown in Fig. 6.11. The water stress caused by chilling roots reduces stomatal conductance and may cause both stomatal and nonstomatal reduction in photosynthesis (DeLucia, 1986). In another study it appeared that the nonstomatal inhibition of photosynthesis in loblolly pine was only important at soil temperatures

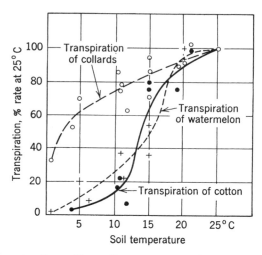

Figure 6.11   Difference in effects of low soil temperature on absorption of water by cool season (collards) and warm season plants (cotton and watermelon), as measured by rate of transpiration. From Kramer (1983), after Kramer (1942).

below 7°C (Day *et al.*, 1991). Chilling 16- or 17-day-old soybean root systems to 10°C for a week temporarily reduced stomatal conductance, CO_2 uptake, water potential, and leaf enlargement, but if rewarmed after a week, no permanent injury or reduction in yield occurred (Musser *et al.*, 1983).

Less is known about the effects of high temperature, although the surface soil often is subjected to fairly high temperatures near midday. BassiriRad *et al.* (1991) found that as the root temperature increased from 15 to 25°C, exudation from barley root systems increased, but then decreased sharply as temperature increased above 25°C. Exudation from detached sorghum roots increased to 35°C, but decreased nearly 50% at 40°C. There also was a large decrease in ion flux into the xylem at high temperatures, which may explain the decrease in the volume of exudation. They suggested that variations in the ion supply to shoots may constitute a messenger from the roots to shoots. Later BassiriRad and Radin (1992) found that the effect of ABA on hydraulic conductance is strongly temperature dependent for barley roots at 15 to 40°C, but for sorghum roots only at low temperatures.

The principal cause of the decrease in water absorption at low temperatures is the increase in root resistance to water movement through the roots, caused partly by the increased viscosity of water, as shown in Fig. 6.12. Much attention

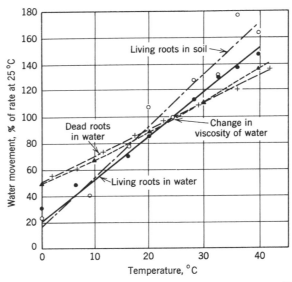

Figure 6.12   Water movement through living and dead root systems in water subjected to a vacuum of about 0.08 MPa for 1 hr. A different set of plants of identical past treatment was used for each temperature and the rates are plotted as a percentage of the rates at 25°C. The curve for the viscosity of water is the reciprocal of viscosity plotted as a percentage of the value at 25°C. Temperature affects water movement through dead roots less than movement through living roots. From Kramer (1983), after Kramer (1940b).

has been given to the effects of temperature on the permeability of cell membranes as a reason for species differences (Lyons *et al.*, 1979). For example, water absorption is reduced much less by cooling broccoli than soybean root systems and when an Arrhenius plot of water flow is made there is a discontinuity in the rate for soybean but not for the cool weather plant broccoli (see Fig. 6.13). Some investigators attribute this break to a phase separation in membrane lipids (Raison *et al.* in Turner and Kramer, 1980; Raison *et al.*, 1982) but this has been questioned by others (Lyons, 1973; Lyons *et al.*, 1979). Markhart *et al.* (1980) found a much greater increase in unsaturated fatty acids in new roots formed at low temperature on the cool weather crop broccoli than in new roots on the warm weather crop soybean, and Osmond *et al.* (1982) found a temporary increase in unsaturated fatty acids in new soybean roots grown at low temperatures. More research is needed on biochemical changes in roots of various species when cooled in order to fully explain the differences among species and cultivars in tolerance of chilling.

The reduced absorption of water through cooled roots usually is accompanied by a reduction in stomatal conductance and photosynthesis, a decrease in

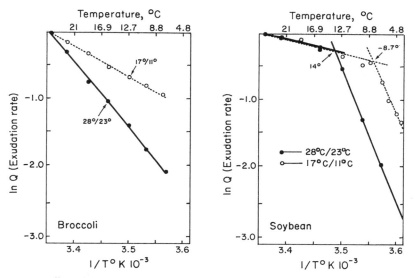

Figure 6.13   Effect of previous growth temperature on the rate of water movement through root systems of broccoli and soybean under a pressure of about 0.05 MPa. Plants were grown at day night temperatures of 28/23°C and 17/11°C, and absorption was reduced much more by cooling in roots grown at the higher temperature. The rates for individual root systems were normalized to the rate at 25°C to permit comparison. The rates are graphed as Arrhenius plots with the natural log of water flow (*Q*) on the ordinate and the reciprocal of the Kelvin temperature (1/*T*) on the abscissa. Celsius temperatures are given at the top for convenience. Plants of both species grown at low temperatures were less affected by cooling than those grown at higher temperatures. Also, there are sharp changes in slope for soybean. From Kramer (1983), after Markhart *et al.* (1979).

leaf water potential, and an inhibition of leaf expansion. Much relevant literature is cited by Day *et al.* (1991), Markhart *et al.* (1979), and Raper and Kramer (1987, pp. 593–597). The effects of root chilling on shoots usually are attributed to shoot water deficit. However, recent experiments with split root systems, discussed in Chapter 5 and in a review by Davies and Zhang (1991), suggest the possibility that some effects of root chilling on shoot physiology may result from chemical signals such as hormones produced in chilled roots. However, Day *et al.* (1991) found that chilling one-half of split root systems of loblolly pine seedlings had no effect on leaf water potential or net photosynthesis, leading them to doubt if nonhydraulic signals are involved in the decreased leaf water potential and photosynthesis they observed in root-chilled seedlings. The possibility that biochemical signals from chilled roots are involved in the shoot reactions probably deserves further investigation under field conditions.

EFFLUX OF WATER FROM ROOTS AND HYDRAULIC LIFT

Although this chapter deals primarily with the absorption of water, there are circumstances in which a measurable efflux of water occurs from roots into the surrounding soil. It was mentioned in Chapter 4 that significant amounts of water sometimes are transferred through roots from deep, moist soil to drier surface soil where it becomes available to shallow rooted plants. This is sometimes termed "hydraulic lift" and has been demonstrated by several investigators, including Baker and Van Bavel (1986), Caldwell and Richards (1989), Corak *et al.* (1987), and Xu and Bland (1993). Bormann (1957) observed water exchange between intertwined roots, Richards and Caldwell (1987) observed a significant transport of water from deep soil to grasses in drier surface soil through roots of *Artemisia tridentata,* and Corak *et al.* (1987) observed sufficient water transport through alfalfa roots to keep maize plants alive. Dawson (1993) claimed that herbaceous plants growing near deeper rooted sugar maple trees benefit from hydraulic lift.

These occurrences require reversal from normal inflow to outflow of water from roots. This probably occurs whenever the soil water potential falls below that of the donor roots, if the roots are permeable to water. However, root permeability probably decreases in dry soil (Cruz *et al.,* 1992; Lopez and Nobel, 1991). Xu and Bland (1993) reported that outflow occurred from sorghum roots when a gradient of about 0.6 MPa developed, and in their experiments, the total outflow into the soil amounted to 5 or 6% of the daily loss in transpiration. The evidence that water efflux from roots sometimes benefits neighboring plants introduces a previously unrecognized factor in plant competition.

SUMMARY

The absorption of water is an essential part of the continuum of processes involving the soil, plant, and atmosphere. Under some circumstances small amounts of water are absorbed through leaves from rain, dew, and fog, but except for epiphytes this is negligible in amount compared to water absorbed from the soil. Two absorption mechanisms exist: osmotic absorption in slowly transpiring plants, growing in moist soil, and passive absorption by rapidly transpiring plants. The roots of slowly transpiring plants growing in moist, well-aerated soil function as osmometers, resulting in the development of root pressure, exudation from wounds, and guttation. The extent of exudation caused by root pressure varies widely among plants of different species and with environmental conditions. Most of the water absorbed by rapidly transpiring plants is "pulled" in by matric forces developed in the evaporating surfaces of cells and is transmitted to the roots through the cohesive water columns of the xylem, where the reduced water potential causes inflow from the soil. The evaporation of water from twigs and branches of leafless deciduous plants on sunny days sometimes creates tension in the xylem and causes the ascent of sap.

Exudation of sap from some plants such as maple, palms, and agave are caused by local stem pressures rather than from root pressures. Occasionally, local pressure develops as the result of microbial activity in wounds resulting in the development of gas pressure and exudation. Some plants exude complex organic compounds such as latex and oleoresin from specialized systems of ducts when wounded. These types of exudation are not directly related to root pressure, but their magnitude is often related to plant water status.

Water absorption depends on the extent and conductance or permeability of the roots and the steepness of the water potential gradient from soil to roots, which provides the driving force. Generally, the resistance to water movement from soil to root surfaces is smaller than the resistance to radial movement in roots, but as the soil dries, resistance to water movement through it increases and the driving force decreases. Water absorption also is affected by soil aeration and temperature, and by the concentration of the soil solution.

SUPPLEMENTARY READING

Ayers, P. G., and Boddy, L., eds. (1986). "Water, Fungi and Plants." Cambridge University Press, Cambridge.

Boyer, J. S. (1985). Water transport. *Annu. Rev. Plant Physiol.* **36**, 473–516.

Glinski, J., and Stepniewski, W. (1985). "Soil Aeration and Its Role for Plants." CRC Press, Boca Raton, FL.

Hook, D. D., and Crawford, R. M. M., eds. (1978). "Plant Life in Anaerobic Environments." Ann Arbor Scientific Press, Ann Arbor, MI.

Jackson, M. B. (1991). Regulation of water relationships in flooded plants by ABA from leaves, roots, and xylem sap. *In* "Abscisic Acid" (W. J. Davies and H. G. Jones, eds.), pp. 217–226. Bio. Sci. Publ., Oxford, UK.

Kozlowski, T. T. (1982). Water supply and tree growth. II. Flooding. *For. Abstr.* **43**, 145–161.

Kramer, P. J. (1969). "Plant and Soil Water Relationships: A Modern Synthesis." McGraw-Hill, New York.

Lyons, J. M., Graham, D., and Raison, J. K., eds. (1979). "Low Temperature Stress in Crop Plants: The Role of the Membrane." Academic Press, New York.

Passioura, J. B. (1988) Water transport in and to roots. *Annu. Rev. Plant Physiol. Plant Mol. Biol.* **39**, 245–265.

Poljakoff-Mayber, A., and Gale, J., eds. (1975). "Plants in Saline Environments." Springer-Verlag, Berlin/New York.

Wild, A., ed. (1988). "Russell's Soil Conditions and Plant Growth." Longman, England.

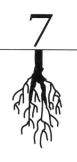

Transpiration and the Ascent of Sap

INTRODUCTION

Fiscus and Kaufmann (in Stewart and Nielsen, 1990) commented that much of what is discussed as plant water relations deals with water movement in the soil–plant–atmosphere continuum (SPAC). It ranges from movement as molecules in diffusion from cell to cell and as vapor in transpiration, to movement of bulk water solution in mass flow of the transpiration stream from roots to shoots through the xylem. Movement of water in cells was discussed in Chapter 3, movement of soil water in Chapter 4, and the absorption of water in Chapter 6. This chapter discusses the loss of water by transpiration and the resulting movement of water and solutes by mass flow through the xylem to transpiring surfaces of shoots.

The Importance of Transpiration

Transpiration is the loss of water from plants in the form of vapor and it is the dominant process in plant water relations because of the large volume of water involved and its controlling influence on plant water status. It also produces the energy gradient that largely controls absorption and the ascent of sap. In warm sunny weather, transpiration often causes transient midday wilting and as the soil dries it causes permanent wilting and finally death by dehydration if the soil moisture is not replenished by rain or irrigation. Worldwide, more plants probably are injured or killed by dehydration caused by excessive transpiration than by any other single cause.

Table 7.1   Relative Water Losses by Transpiration and Evaporation from an Illinois Cornfield during the Period from Mid-June to Early September

Year	Total evapo-transpiration from uncovered plot (cm)	Transpiration from covered plot (cm)	Transpiration as percentage of evapotran-spiration	Total precipitation (cm)	Excess of evapotranspiration over rainfall (cm)
1954	32.25	16.5	51%	18.5	13.75
1955	34.50	17.5	51%	23.0	11.50
1957	33.75	15.1	45%	24.0	9.75

[a]From Peters and Russell (1959).

The quantitative importance of transpiration is indicated by the fact that a well-watered Kansas corn plant loses about 200 liters of water during a growing season or nearly 1300 tons per acre (= 475 metric tons per hectare or about 28 cm of water) (Miller, 1938, p. 412). An Illinois corn field transpired an amount of water equal to 60 to 90% of the precipitation during the growing season (Peters and Russell, 1959) and the combined evaporation from the soil and transpiration from the crop (evapotranspiration) exceeded the precipitation during the growing season, as shown in Table 7.1. In contrast, a deciduous forest in the more humid southern Appalachians transpired 40 to 55 cm per year, which used only 25 to 35% of the annual precipitation (Hoover, 1944), as shown in Table 7.2.

Several hundred grams of water are required to produce a gram of plant dry matter (Table 7.3), but about 95% of this is lost in transpiration. If it were not for the water dissipated by transpiration, a crop could be grown with the water

Table 7.2   Amounts of Water Lost in Various Ways by a North Carolina Watershed Covered with a Deciduous Forest (1940–1941) and the Increase in Runoff Which Followed Cutting of All Woody Vegetation and Elimination of Transpiration (1941–1942)[a]

Process	1940–1941	1941–1942
Precipitation	158.0	158.4
Interception	16.6	9.5
Runoff	53.4	93.0
Soil storage	− 0.4	9.7
Evaporation	39.7	46.0
Transpiration	48.7	00.0

Note. Data in centimeters.
[a]From Hoover (1944).

Table 7.3   Water Requirement or Transpiration Ratio in Grams of Water per Gram
Dry Matter for the Years 1911–1917 at Akron, Colorado, and the
Evaporation from a Free Water Surface from April to September 1[a]

Plant	1911	1912	1913	1914	1915	1916	1917
Alfalfa	1068	657	834	890	695	1047	822
Oats, Burt	639	449	617	615	445	809	636
Barley, Hannchen	527	443	513	501	404	664	522
Wheat, Kubanka	468	394	496	518	405	639	471
Corn, N.W. Dent	368	280	399	368	253	495	346
Millet, Kursk	287	187	286	295	202	367	284
Sorghum, Red Amber	298	239	298	284	303	296	272
Evaporation: April 1 to September 1 (mm)	1239	957	1092	1061	848	1196	1084

[a]From Miller (1938).

supplied by a single rain or irrigation, assuming that evaporation from the soil was controlled by mulching.

It sometimes is argued that transpiration is beneficial because it cools leaves, accelerates the ascent of sap, and increases the absorption of minerals (Clements, 1934; Gates, 1968). Rapidly transpiring leaves usually are cooler than slowly transpiring leaves (Gardner *et al.*, 1981), but leaves in the sun rarely are seriously overheated even when transpiration is reduced by wilting. Water moves to the tops of plants as they grow and transpiration merely increases the quantity and speed of movement. Absorption of minerals probably is increased, but some understory plants thrive in shady, humid habitats where the rate of transpiration is relatively low. Although Winneberger (1958) reported that high humidity reduced plant growth, Hoffman *et al.* (1971), O'Leary and Knecht (1971), and others found that growth generally was better in high than in intermediate or low humidity and Tanner and Beevers (1990) concluded that transpiration is not essential. The numerous harmful effects of water stress caused by rapid transpiration are discussed in later chapters.

Transpiration can be regarded as an unavoidable evil, unavoidable because a leaf structure favorable for uptake of the carbon dioxide necessary for photosynthesis also is favorable for loss of water, and evil because it often causes injury by dehydration. The evolution of a leaf structure favorable for high rates of photosynthesis apparently has had greater survival value in most habitats than one conserving water, but reducing photosynthesis. Thus, the leaf anatomy of most mesophytic plants causes them to live in danger of injury from excessive transpiration. In general, plants adapted to dry environments cannot compete effectively with plants adapted to moist environments when both are well watered (Bunce, 1981; Orians and Solbrig, 1977). The relationship between transpiration and photosynthesis was discussed in detail by Cowan (1982) and is discussed further in Chapters 10 and 12, where research is cited indicating that some varieties of plants that yield well during water deficits also yield well when supplied with water.

THE PROCESS OF TRANSPIRATION

Transpiration involves two steps, the evaporation of water from cell surfaces into intercellular spaces and its diffusion out of plant tissue, chiefly through stomata and the cuticle and to a lesser extent through the lenticels in stem bark of woody plants.

Evaporating Surfaces

It usually is assumed that most of the water evaporates from the surfaces of leaf mesophyll cells into the intercellular spaces (Slatyer, 1967, pp. 215–221; Sheriff, 1984). However, it was argued by some writers (Meidner, 1975; Byott and Sheriff, 1976) that much water evaporates from the inner surfaces of epidermal cells in the vicinity of guard cells, or even from the inner surfaces of guard cells, the peristomatal transpiration of Maercker (1965). However, the exposed surface of mesophyll cells usually is 10 to 15 times greater than the exposed inner epidermal surface. Also, Nonami and Schulze (1989) found the water potential of mesophyll cells in transpiring leaves to be lower than that of epidermal cells. The literature on this interesting question has been summarized by Davies (1986, pp. 65–69). Tyree and Yianoulis (1980) made computer studies indicating that 75% or more of the evaporation should occur from near the guard cells, but their calculations assumed that water evaporates equally freely from all mesophyll cell walls. This assumption probably is incorrect because investigators from von Mohl in 1845 to the present have reported the presence of varying amounts of cutin or suberin on the cell walls bordering intercellular space (Lewis, 1945; Scott, 1964; Sheriff, 1977a, 1984). According to Norris and Bukovac (1968), the cuticle covering the outer surface of the lower epidermis of hypostomatous pear leaves (leaves with stomata only on the lower surface) extends in through the stomatal pores and covers the inner surfaces of the guard cells and the adjacent lower epidermis. It seems that the relative importance of evaporation from various internal surfaces in leaves cannot be decided until more information is available concerning the amount of internal cutinization at various distances from stomata. These problems are discussed further in Chapter 11.

After water vapor has diffused into the intercellular spaces it escapes from the leaves by diffusion through the stomata and to a lesser extent through the cuticle, and through lenticels in the bark of twigs of woody plants (Geurten, 1950; Huber, 1956; Schönherr and Ziegler, 1980).

Driving Forces and Resistances

The rate of transpiration depends on the supply of water at the evaporating surfaces, the supply of energy to vaporize water, the size of the driving forces, and the resistances or conductances in the pathway. The driving force for liquid water is the gradient in water potential, that for water vapor is the difference in vapor concentration or vapor pressure.

Evaporation E can be described by:

$$E = \frac{C_{water} - C_{air}}{r_{air}}, \tag{7.1}$$

where E is given in $g \cdot m^{-2} \cdot sec^{-1}$, C_{water} and C_{air} are given in $g \cdot m^{-3}$ and are vapor concentrations at the evaporating surface and in the bulk air, and r_{air} is in $sec \cdot m^{-1}$ and is the resistance of the air boundary layer to water vapor diffusion. Equation (7.1) indicates that the rate of evaporation is proportional to the concentration difference. Doubling the difference will double the rate, all other factors being equal. Evaporation also is influenced by changes in r_{air} which can vary as the boundary layer next to the leaf or water surface varies. This layer contains unstirred air through which the vapor moves mostly by diffusion which is slow compared to the rate in wind-stirred air. The thickness of the layer is determined mostly by wind speed and the size of the surface: the faster the wind speed, the thinner the boundary layer and the smaller is r_{air} resulting in faster evaporation. Also, the smaller the surface the faster the evaporation.

Equation (7.1) applies at sea level and must be corrected for pressure if it is used elsewhere, as on tall mountains. As a consequence, Eq. (7.1) often is shown with the driving force given as partial pressures ($e_{water} - e_{air}$) in units of kPa/kPa instead of concentrations ($C_{water} - C_{air}$). This has the advantage that the effects of atmospheric pressure are taken into account explicitly by dividing the vapor pressure of water by the total pressure of the atmosphere. The pressures can be calculated from the concentrations and the rate of evaporation usually is given in $mol \cdot m^{-2} \cdot sec^{-1}$ and the resistance of the boundary layer in $sec \cdot m^2 \cdot mol^{-1}$. Otherwise, the equations are fundamentally the same.

In a plant, this process is more complex because water vapor must travel through the leaf. The leaf contains intercellular spaces and stomata, and water vapor moves through them mostly by diffusion before it can enter the boundary layer. This additional resistance makes the rate slower than for a free water surface. The leaf resistance is in series with the boundary layer resistance, and the whole effect can be described by including an additional resistance in the vapor path in which case the rate of transpiration T is

$$T = \frac{C_{leaf} - C_{air}}{r_{leaf} + r_{air}}, \tag{7.2}$$

where C_{leaf} is the vapor concentration at the evaporating surface inside the leaf and r_{leaf} is the diffusive resistance of all the paths for vapor diffusion in the leaf

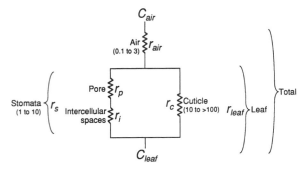

Figure 7.1 Diagram showing approximate resistances in seconds per centimeter to diffusion of water vapor from a leaf. Stomatal and cuticular resistances vary among species and with leaf hydration and atmospheric humidity. The transpiration is proportional to ($C_{leaf} - C_{air}$), the vapor concentration difference between the leaf and air, and inversely proportional to the resistances in the pathway. From Kramer (1983).

in units of sec·m^{-1} or, for the driving force in pressure units, sec · m^2 · mol^{-1}. Because r_{leaf} causes T to be less than E when the other factors are maintained constant, one can evaluate r_{leaf} by comparing the rate of transpiration of a leaf with the rate of evaporation of a comparably shaped piece of wet filter paper. The leaf will always lose water more slowly than the paper and the magnitude of r_{leaf} can be calculated from the difference. Note that a leaf with closed stomata will have a larger r_{leaf} than it will with open stomata and the rate of transpiration will be slower.

The resistance of a leaf can be divided into a number of component resistances as shown in Fig. 7.1, and some species differences in cuticular and stomatal resistance are shown in Table 7.4. If most or all of the stomata are on one surface, r_{leaf} will be quite different for the upper (adaxial) and lower (abaxial)

Table 7.4 Resistance to Movement of Water Vapor through the Boundary layer (r_a), Stomata (r_s), and Cuticle (r_c) in Leaves of Several Species[a]

Species	Resistance to water vapor (sec · cm^{-1})		
	r_a	r_s	r_c
Betula verrucosa	0.80	0.92	83
Quercus robur	0.69	6.7	380
Acer platanoides	0.69	4.7	85
Circaea lutetiana	0.61	16.1	90
Lamium galeobdolon	0.73	10.6	37
Helianthus annuus	0.55	0.38	—

[a]From Holmgren *et al.* (1965).

surfaces. Also the stomata on the upper and lower surfaces often differ in their reaction to light and water stress (Pallardy and Kozlowski, 1979; Sanchez-Diaz and Kramer, 1971, for example), as shown in Fig. 8.8. There are measurable differences in temperature in different parts of transpiring leaves, reflecting local differences in rate of air movement from the margin inward (Cook *et al.*, 1964; Hashimoto *et al.*, 1984).

Energy Relations

The energy required to evaporate water from leaves and plants comes from direct solar radiation, from radiation reflected or reradiated from the soil and surrounding objects, and from the flow of sensible heat between the leaf and the environment. Sensible heat is transferred because of temperature differences between plants and the surrounding air, and it moves by diffusion, convection, and advection. Diffusive heat transfer occurs in unstirred air and is observed when there are temperature differences across the boundary layer of unstirred air next to leaves and other surfaces. Convection and advection refer to heat transfer by the movement of air in bulk due to temperature-induced density differences or horizontal wind movement. Convection can be seen in the warm air rising from a warm surface such as a radiator and advection in the winds that bring new air into a plant canopy. An example of the increase in transpiration caused by increased advection resulting from the removal of surrounding vegetation is shown in Fig. 7.2. Sensible heat transfer is positive or negative,

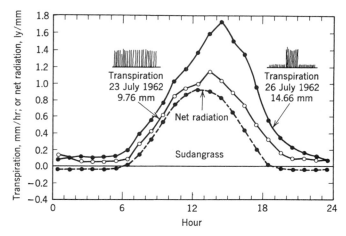

Figure 7.2 The "oasis effect" as shown by the increase in transpiration from a lysimeter in a closed stand of Sudan grass after removal of the surrounding crop, exposing the plants in the lysimeter. Although radiation was similar on the 2 days, additional energy supplied by increased exposure to air movement and advection resulted in a much higher transpiration after the plot was exposed. From Kramer (1983), after Van Bavel *et al.* (1963).

positive when the leaves are cooler than the air, negative when they are warmer, as in the evening or when suddenly shaded. When transpiration occurs, liquid water changes state and energy is required even though no change in the temperature of the evaporated water is involved. In effect, the latent heat for evaporation is extracted from energy in the molecules that stay behind in the liquid.

When there is condensation, the reverse occurs and the latent heat is released to the surrounding molecules. The ratio of heat transferred as sensible heat to that transferred as latent heat by transpiration is called the Bowen ratio (see Nobel, 1991, p. 483).

The energy budget for a leaf can be partitioned as

$$R_n + H + E_\ell = 0, \tag{7.3}$$

where the R_n (net radiation) is the net flux of radiation actually absorbed by the leaves and consists of the total solar radiation plus that from the environment, minus reflected radiation and that reradiated. H is the sensible heat flux between the leaves and the environment, and E_ℓ is the latent heat lost by evaporation of water or gained by condensation. Each term refers to the rate of energy transferred per unit of projected leaf area (the flux) and the fluxes in the equation are large by comparison with the energy for photosynthesis and respiration so these effects are ignored. In this equation, each term can be positive or negative depending on whether energy moves toward or away from the leaf, and the various terms add to zero, that is, the energy budget is in balance whenever leaf temperature is constant. When the energy budget changes, leaf temperature quickly adjusts and stabilizes at the new condition dictated by the equation.

The energy budget is important for understanding the behavior of transpiration. If any two terms in the equation are known, the other can be calculated. For example, during the daytime when the leaves are irradiated by the sun, R_n is positive and E_ℓ is negative because transpiration is occurring. Occasionally the positive R_n just balances the negative E_ℓ. In this situation, the H becomes zero because no sensible heat transfer occurs, and the leaf has the same temperature as the air. Usually, however, R_n is not balanced by E_ℓ and leaf temperatures differ from air temperatures. Even in controlled environments leaves rarely have the same temperature as the air. Thus, it is usual for sensible heat transfer to occur to or from leaves, depending on whether their temperature is lower or higher than that of the air.

Similar principles operate at night. R_n is negative because leaves radiate to the cold night sky and are not receiving input from the sun. E_ℓ also is negative because of transpiration, particularly early in the evening. Both of these negative fluxes tend to decrease leaf temperature. H becomes positive and balances the equation. The air conducts sensible heat to the cool leaf whose temperature stabilizes at a cooler temperature. If the leaf is cold enough, dew condenses on the leaf or frost forms. The condensation releases latent heat that helps to prevent further temperature depression of the leaf. This explains why dew is most likely to be heavy on a cloudless night when the night sky is clear and leaves lose the most energy by radiation. Clouds obscure the distant sky because of the water vapor and the water droplets they contain.

Many other aspects of transpirational behavior can be explained by an analysis of leaf energy budgets and the causes of the variations in each term constitute a field of study in their own right. Readers are referred to Gates (1968, 1980), Knoerr (1967), and Nobel (1991, Chapter 7) for more detailed discussions.

Vapor Pressure Gradients

The driving force causing water vapor to move out of plants or from other evaporating surfaces is the difference between the vapor pressure of the evaporating surface and the humidity or vapor pressure of water in the bulk air. The humidity and vapor pressure of bulk air depend on its water content and its temperature, while those of the evaporating surface of plants depend on the water potential and temperature. The water potential of fully turgid plant tissue is zero, but the tissues of rapidly transpiring plants often have water potentials of -1.0 MPa or less. However, as shown in Table 7.5, a large reduction in cell water potential causes only a small reduction in vapor pressure. For example, at 30°C and a cell water potential of -3.0 MPa the vapor pressure of the cell sap is 98% of the vapor pressure of pure water. Thus, a large reduction in cell water potential has little effect on the rate of evaporation from the cell surface.

Relative and Absolute Humidity.   Weather reports emphasize the relative humidity of the atmosphere, which is the water content as a percentage of that at saturation at the same temperature, instead of the absolute humidity, which

Table 7.5   The Vapor Pressure of Water (e), the Vapor Pressure at Cell Surfaces at Three Cell Water Potentials, and the Difference in Vapor Pressure (Δe) between Cells and Air at Relative Humidities of 80 and 50% at 30°C[a]

Ψ_w of water and of mesophyll cells (MPa)	e at cell surfaces (kPa)	Δe at 80% relative humidity (kPa)	Δe at 50% relative humidity (kPa)
0.0	4.243	0.849	2.121
-1.5	4.200	0.806	2.079
-3.0	4.158	0.764	2.037
-6.0	4.073	0.679	1.952

[a]The relative effect of a low Ψ_w diminishes as the relative humidity decreases. All vapor pressures are for 30°C.

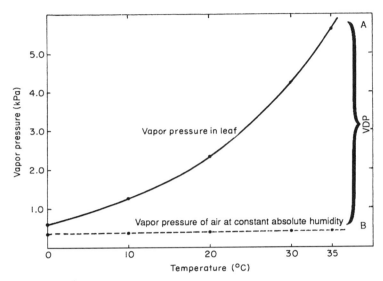

Figure 7.3 Effect of increasing air and leaf temperatures on vapor pressure deficit (VPD) from leaf (A) to air (B), assuming that the air in the leaf is saturated, but the absolute humidity of the atmosphere at 0°C remains unchanged. The saturation vapor pressure at the evaporating surfaces in the leaf increases over 900% over this temperature range, but the vapor pressure of the water in the atmosphere increases only about 13%, in proportion to the increase in absolute temperature. Thus, there is a rapid increase in vapor pressure deficit, evaporation, and transpiration, as shown by the increasing spread between lines A and B.

is the actual water content in g · m^{-3}. Other things being equal, the rates of evaporation and transpiration are proportional to the size of the absolute humidity difference or vapor pressure difference (Bange, 1953; Cole and Decker, 1973), as shown in Fig. 7.3. A relative humidity of 50% at 20°C must be increased to 75% at 30°C to maintain the same vapor pressure difference and the same rate of evaporation. If the relative humidity were kept at 50% at both 20° and 30°C, the rate of evaporation would be about 80% greater at 30°C than at 20°C because the vapor pressure of liquid water increases much more than the vapor pressure of the water in the air (Fig. 7.3). Table 7.6 shows that increasing the temperature from 20° to 30°C increases the vapor pressure difference about 140%, although it decreases the relative humidity over 40%.

A related problem is that air temperature and leaf temperature often are different and the use of relative humidity, which is defined for identical temperatures, does not accurately describe the vapor pressure difference. Because the vapor pressure difference takes the temperature differences into account (Fig. 7.3), it avoids this problem. The adjustment of atmospheric moisture in growth chambers and other closed containers to maintain similar rates of

Table 7.6   Effect of Increasing Temperature of Leaf and Air on Vapor Pressure
Gradient from Leaf to Air, with Relative Humidity of 80% at 10°C
and No Change in Absolute Humidity from that at 10°C

	Leaf and air temperature (°C)		
	10	20	30
Relative humidity of air in %	80	43	30
Vapor pressure at evaporating surface of leaf (kPa)	1.228	2.339	4.246
Vapor pressure in air (kPa)	0.981	1.015	1.051
Vapor pressure gradient from leaf to air (kPa)	0.247	1.324	3.196

evaporation or transpiration at various temperatures requires that the vapor pressure difference between the evaporating surface and the air be the same at all temperatures. Basing the humidity setting of a growth chamber on absolute humidity at air temperature will give a first approximation, but basing the setting on air and leaf temperatures will give a better one.

Estimation of the vapor pressure difference between the interior of leaves and the surrounding air usually assumes that the vapor pressure of water in the cell wall surfaces is that of pure water. This is not exactly correct, as leaf water potential usually is less than zero. However, as indicated earlier in Table 7.5, even fairly low water potentials have little effect on the vapor pressure (e) at cell surfaces. The air in the intercellular spaces presumably is approximately saturated when the stomata are closed, but when they are open it usually is below saturation because water vapor is diffusing out (Nobel, 1991, pp. 419–421; Ward and Bunce, 1986).

Resistances to Diffusion

Two groups of resistances affect transpiration: the internal resistances in plants and the external or boundary layer resistances in their immediate environment.

Leaf Resistances.   The diffusive resistances of a leaf are shown in Fig. 7.1. The total internal resistance (r_{leaf}) is composed of cuticular resistance (r_c) and stomatal resistance (r_s) which consists of resistance in intercellular space (r_i) and resistance in stomatal pores (r_p).

Because stomatal and cuticular resistances are in parallel, an equation for total leaf resistance involves their reciprocals:

$$\frac{1}{r_{leaf}} = \frac{1}{r_c} + \frac{1}{r_s}. \tag{7.4}$$

The components of stomatal resistance are in series and can be described by their sum:

$$r_s = r_i + r_p. \tag{7.5}$$

The cuticular resistance varies widely among species, as indicated in Table 7.4, but when stomata are open, r_s is so much lower than r_c that most water vapor escapes through the stomata, and transpiration is largely controlled by stomatal aperture. Stomatal behavior is discussed in Chapter 8.

There has been considerable discussion of the importance of the mesophyll resistance with respect to transpiration (Kramer, 1969, pp. 306–309; Slatyer, 1967, pp. 256–260). Early in this century Livingston and Brown (1912) observed a decrease in transpiration of rapidly transpiring plants when no decrease in stomatal aperture occurred and attributed it to an increase in mesophyll resistance caused by retreat of the water menisci into the pores of the mesophyll cell walls, reducing the vapor pressure at the evaporating surfaces. This decreases the driving force in addition to increasing the resistance. Evidence for an increase in mesophyll resistance was reported by Shimshi (1963b) and Jarvis and Slatyer (1970), but neither Weatherspoon (1968) nor Jones and Higgs (1980) found significant increases.

Few data are available concerning the relationship between the intercellular space volume of leaves and the rate of transpiration, but a large volume should decrease r_i (Jarvis *et al.*, 1967). Measurement of the resistance in intercellular spaces with pressure flow porometers indicates that r_i can be a significant component of the total leaf resistance when the stomata are open (Jarvis and Slatyer, 1970). It increases as leaves are dehydrated because leaf shrinkage causes a decrease in the size of intercellular spaces, which may account for some of the reports of mesophyll resistance. The r_i can be expected to be larger in thick than in thin leaves. According to Stålfelt (1956b) the intercellular air space varies from 70% of the leaf volume in shade leaves to 20% in sun leaves.

The total resistance for CO_2 is greater than for water vapor (Gaastra, 1959; Holmgren *et al.*, 1965) because of its larger molecules and the longer diffusion path to the reaction site in chloroplasts (Nobel, 1991, pp. 425–428) which includes diffusion through the liquid phase of the cell. This additional mesophyll resistance for CO_2 is the basis for the assumption that partial closure of stomata will reduce transpiration more than it reduces photosynthesis. This is discussed in Chapter 12 in connection with water use efficiency and the use of antitranspirants.

External or Boundary Layer Resistances.   Wind. The boundary layer resistance depends chiefly on wind speed and leaf size and shape for individual

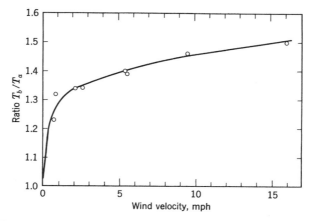

Figure 7.4   Effect of increasing velocity of air movement on the transpiration rate of potted sunflowers growing in a sunny greenhouse. The ordinate is the ratio of transpiration rate of plants exposed to wind (T_b) to the rate of plants in quiet air (T_a). Most of the effect occurred at a velocity of less than 2 mph. A velocity of 1 mph equals 44.69 cm · sec^{-1}. From Kramer (1983), after Martin and Clements (1935).

plants and on density of the stand and "roughness" of the canopy surface in stands of plants. The effects of wind on transpiration are complex because while an increase in velocity decreases the boundary layer resistance, it also cools leaves, decreasing the vapor pressure gradient from leaf to air. Most of the increase in transpiration occurs at low velocities as shown in Fig. 7.4. Knoerr (1967) pointed out that although a breeze should increase transpiration of leaves exposed to low or moderate radiation where leaves tend to be near air temperature, with high radiation water-deficient leaves may be warmer than the air and a breeze might reduce transpiration by cooling them. The behavior of leaves of various species in wind is quite variable, as shown in Fig. 7.5. Davies *et al.* (1974) also reported differences among tree seedlings of various species exposed to varying wind velocity. Wind increased transpiration of white ash in all velocities, decreased transpiration of sugar maple, and had no effect on red pine. The differences probably are related chiefly to differences in stomatal behavior (see Chapter 8). Van Gardingen *et al.* (1991) also reported surface abrasion of conifer needles exposed to wind, probably affecting cuticular resistance. The effects of change in wind velocity on plant water balance have been discussed by Kitano and Eguchi (1992b). Other effects of wind on plants are discussed in Chapter 12 of Kozlowski *et al.* (1991), in books by Grace (1977), and by Vogel (1989).

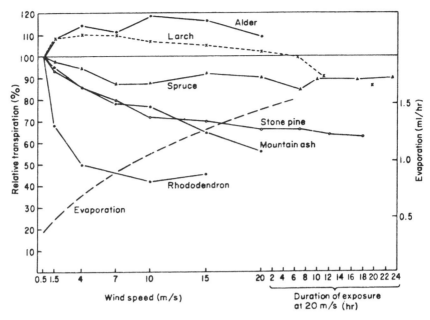

Figure 7.5 Effect of wind velocity on young, potted subalpine plants in a wind tunnel at an air temperature of 20°C, a soil temperature of 15°C, and a light intensity of about 2800 foot candles. Evaporation was measured with a green Piche atmometer. Although evaporation increased steadily with increasing wind velocity, transpiration of most species decreased. From Kramer (1983), after Tranquillini (1969).

OTHER FACTORS AFFECTING TRANSPIRATION

A few of the specific plant and environmental factors affecting transpiration are discussed briefly. Some of these, such as stomata, are discussed in more detail in other chapters.

Leaves

Leaf Temperature and Energy Budget.   The ratio of energy transferred as sensible heat to that dissipated as latent heat by transpiration obviously varies widely and is known as the Bowen ratio. It is said to be about 0.2 for tropical forests, 0.4 to 0.8 for temperate forests, 2 to 6 for semiarid regions, and 10 for deserts (Oke, 1987). According to Idso and Baker (1967), it averages about 0.1 for crops, but can be as high as 6.0 for exposed leaves.

The size, shape, surface characteristics, and exposure of leaves affect the absorption and emission of energy which in turn affect leaf temperature, vapor pressure, and the rate of transpiration. Leaf exposure that is favorable for transpiration also should be favorable for photosynthesis. However, selection for what was believed to be optimum leaf orientation for photosynthesis did not result in any important increase in yield of maize (Loomis *et al.,* 1971). Individual leaf exposure probably decreases in importance in stands of plants.

Leaf temperature affects rates of metabolic processes such as photosynthesis and respiration, but its effects on transpiration are particularly important because it affects the vapor pressure gradient from leaves to the ambient atmosphere, as shown in Fig. 7.3.

In the early part of this century numerous measurements were made of leaf and air temperatures with conflicting results (see Miller, 1938, pp. 472–481), at least partly because of faulty technology. Miller and Saunders (1923), working in Kansas, found that the average leaf temperatures of corn, sorghum, watermelon, and pumpkin were near the temperature of the surrounding air, but that alfalfa leaves were about a degree cooler, probably because of their relatively higher rate of transpiration. Eaton (1930) found transpiring cotton leaves to be about a degree cooler than the air, and young leaves were cooler than older leaves. In contrast, Curtis (1936), working in the cooler, more humid climate of Ithaca, New York, never found leaves cooler than the surrounding air.

In recent decades interest has shifted to canopy temperature as an indicator of plant water stress and as a means of determining differences in development of water stress among cultivars. This was made possible by improvements in infrared thermometry and development of remote sensing methods ranging from observations made from the top of a stepladder to those made by weather satellites. In an early example, Blum *et al.* (1978) used infrared photography to detect differences in drought tolerance of sorghum cultivars because stressed plants were lighter in color than unstressed plants. Jackson (1982) reviewed considerable research on crop canopy temperatures as indicators of water stress and the need for irrigation, and Inoue *et al.* (1990) used leaf temperature measured with an infrared thermometer to estimate the stomatal resistance of cotton. Pinter *et al.* (1990) used midday canopy temperatures to screen wheat cultivars for water use and yield when well watered and when grown with deficit irrigation. It seems likely that leaf temperature will be used more extensively in the future as an indicator of crop water status.

Leaf Area.   Leaf area is an important factor in both transpiration and photosynthesis. The leaf area of individual plants of course varies from a few square centimeters to many square meters. Leaf areas of stands of plants often are expressed as the leaf area index (LAI) which is the ratio of projected leaf area (area of one surface) to land area bearing the plants. This varies widely from less than 2 for open Eucalyptus forests on dry sites in southeastern Australia (Anderson, 1981) to 16 to 45 for hybrid poplars grown under intensive culture (Isebrands *et al.,* 1977); crop plants are usually about 5 or 6.

Table 7.7   Transpiration Rates per Unit of Leaf Surface and per Seedling of Loblolly
Pine and Hardwood Seedlings for the Period August 22–September 2. [a]

	Loblolly pine	Yellow poplar	Northern red oak
Transpiration $(g \cdot day^{-1} \cdot dm^{-2})$	5.08	9.76	12.45
Transpiration $(g \cdot day^{-1} \cdot tree^{-1})$	106.70	59.10	77.00
Average leaf area per tree (dm^2)	21.00	6.06	6.18
Average height of trees (cm)	34.00	34.00	20.00

[a] Average of six seedlings of each species.

Plants and stands of plants with large leaf areas usually transpire more than
those with smaller leaf areas, but the decrease in transpiration accompanying
reduced leaf area may be partially offset by increased exposure of the remaining
leaves, resulting in increased transpiration per unit of leaf area, even though
total transpiration is reduced (Parker, 1949). For example, removal of 56% of
the trees in a young pine stand reduced transpiration about 33% (Whitehead
and Kelliher, 1991a) and removal of 24% of the leaf mass of a *Chamaecyparis*
stand reduced transpiration 21% (Morikawa *et al.*, 1986). However, increased
exposure of the soil after thinning may increase evaporation from the soil. The
large differences in the rate of transpiration per unit of leaf area are sometimes
partially compensated by differences in leaf area per plant, as shown in Table 7.7,
or by leaf arrangement. For example, there is considerable self shading of
needles in clusters of pine needles which reduces photosynthesis and probably
transpiration per unit of area (Kramer and Clark, 1947). Sampson and Smith
(1993) discussed the influence of such features of canopy architecture as leaf
area index, leaf clumping, and canopy gaps on light penetration, and Vogel
(1989) discussed leaf shape and size in relation to behavior in wind.

Some woody perennials, such as creosote bush and buckeye, shed most
of their leaves when water stressed, greatly reducing the transpiring surfaces
(Kozlowski, 1973). The curling or rolling of leaves seen in water-stressed maize,
sorghum, and other grasses is said to reduce transpiration by 35 to 75% (Stålfelt, 1956a,b). Leaf orientation and rolling also were discussed by Begg (in
Turner and Kramer, 1980) and Clarke (1986). Developmental and seasonal
changes in the transpiring surface of leaves were discussed by Killian and Lemée
(1956) and Kozlowski (1973) discussed leaf shedding. According to Swank and
Douglas (1974), conversion of a North Carolina watershed from deciduous
hardwoods to evergreen white pine reduced stream flow.

Ratio of Roots to Leaves.   The relationship between root surface and leaf
surface is particularly important in recently transplanted trees and shrubs where

it often seems desirable to reduce the leaf surface by pruning or defoliation to compensate for the loss of root surface caused by transplanting. However, research by Evans and Klett (1985) indicates that dormant branch pruning of Newport plum (*Prunus cerasifera*) and Sargent crabapple (*Malus sargentii*) at planting time does not affect subsequent overall growth or survival. More research is needed on the effects of root and shoot pruning on trees and shrubs subjected to various degrees of water stress after transplanting. Sometimes the leaves are dipped in or sprayed with an antitranspirant to reduce water loss after transplanting. Use of antitranspirants is discussed in Chapter 12.

Leaf Size and Shape.   A decrease in leaf size decreases the boundary layer resistance (r_a) as shown in Fig. 7.6. Small leaves, the leaflets of compound leaves, and deeply dissected leaves tend to be cooler than large entire leaves because their thinner boundary layers permit more rapid sensible heat transfer. However, the lower boundary layer resistance also is favorable for more rapid water vapor loss so the two effects tend to balance each other (Raschke, 1976).

Parkhurst and Loucks (1972) made an analysis of leaf size in relation to environment based on the assumption that natural selection should have resulted in leaves with maximum water use efficiency in their environment. However, they found many exceptions to their model. Apparently, water use efficiency has not been a major selective force for plants in humid or subhumid environments (Cowan, 1977). Anyway, survival usually depends on possession of a combination of favorable adaptations rather than a single one (Bradshaw, 1965). Furthermore, the importance of leaf size, shape, and orientation tends to decrease in closed stands of plants where the canopy dominates over single leaves.

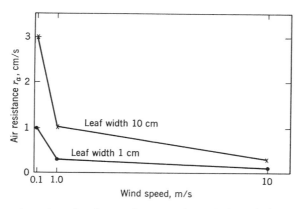

Figure 7.6   Approximate boundary layer resistance at three wind speeds for a cotton leaf 10 cm wide and a grass leaf 1 cm wide. From Kramer (1983), after data of Slatyer (1967).

Leaf Orientation.   Leaves with surfaces perpendicular to the sun's rays are warmer than those more or less parallel, and are likely to have a higher rate of transpiration. Leaves of many plants tend to be more or less perpendicular to midday radiation, but there are exceptions such as some species of *Lactuca* and *Silphium*, turkey oak (*Q. laevis*), and jojoba (*Simmondsia chinensis*). The leaves of *Stylosanthes humilis* normally are perpendicular to the sun's rays but when the pulvini lose turgor they become parallel. Pine needles that occur in bundles, such as those of the three and five needle pines, shade one another and reduce transpiration and photosynthesis per unit of surface (Kramer and Clark, 1947). Leaf orientation may be less important in stands of plants than in individual plants.

Leaf Surface Characteristics.   The outer surfaces of leaves, flower petals, stems, and fruits often are covered by a relatively waterproof layer, the cuticle. This is composed of cutin and wax, anchored to the epidermis by a layer of pectin (Fig. 7.7). When dry, the cuticle is relatively impermeable to water, as shown by its high resistance (Table 7.4), but when wet it is fairly permeable to water vapor and a variety of solutes (Schönherr, 1976). The cuticle generally is thinner on shade than on sun leaves, and thinner on leaves grown with adequate water than on leaves subjected to water deficits (Bengston *et al.*, 1978). Cuticle mutants have been found in sorghum, some of which displayed thin cuticles and high water loss through the cuticle (Jenks *et al.*, 1994). Permeability to water must depend more on the amount of wax than on the thickness of the cuticle because cuticular transpiration often increases severalfold when the wax is re-

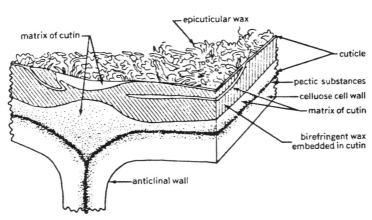

Figure 7.7 Diagram of an upper surface of an epidermal cell of pear leaf, showing cuticle, composed of cutin and wax. The cutin is anchored to the cellulose wall by a layer of pectin. From Kramer (1983), after Norris and Bukovac (1968).

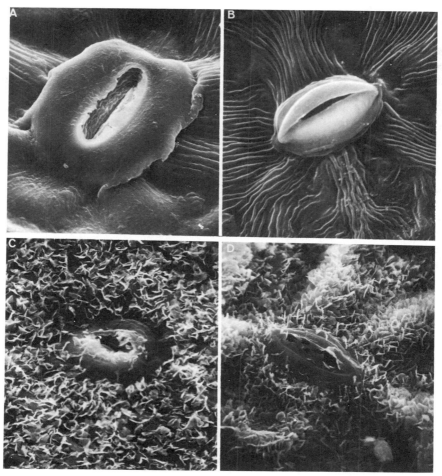

Figure 7.8 Variations in patterns of wax deposited on leaves of four tree species: (A) American elm, (B) white ash, (C) sugar maple, and (D) redbud. Also note differences in appearance of stomata. Photographs by W. J. Davies, from Kramer (1983).

moved (Radler, 1965; Jeffree *et al.*, 1971). However, Chamel *et al.* (1991) claimed that the extraction of soluble lipids did not increase permeability. Examples of wax deposits on leaves are shown in Fig. 7.8. The early work on cuticle was reviewed by Stålfelt (1956b), and Kolattukudy (1981) reviewed the synthesis of cutin and suberin.

Leaves of some plants are covered by thick layers of trichomes or epidermal hairs that reflect light, decrease the driving force by decreasing the leaf temperature, and increase the boundary layer resistance. Ehleringer (in Turner and Kramer, 1980) studied the role of leaf hairs in the genus *Encelia*. The densely pubescent, white leaves of the desert shrub *E. farinosa* reflect more radiation than the green leaves of *E. californica* which reduces their midday temperature about 5°C below that of coastal *E. californica*. Unfortunately, the hairs reflect so much light that the net photosynthesis of *E. farinosa* also is reduced. Overall, Ehleringer concluded that at high air temperatures the gain in net photosynthesis from a reduced leaf temperature is more valuable than the reduction in transpiration. Although thick coats of hairs may have some adaptive advantages, many desert plants thrive without them while some mesophytes such as mullein (*Verbascum thapsus*) have them, although mullein grows in environments where hairs can scarcely be beneficial. Johnson (1975) concluded that pubescence cannot be regarded as a simple adaptation to arid environments.

Hairs sometimes have other functions unrelated to leaf temperature or transpiration. They often secrete chemicals which give plants characteristic odors and other special properties. For example, Turkish-type tobacco has more aroma than other types of tobacco because of the greater density of secretory hairs on its leaves (Wolf, 1962). Other plants such as nettles sting the skin on contact because of substances produced in hairs on their epidermal cells. Epidermal cells on leaves of some plants develop into glands (Sachs and Novoplansky, 1993).

Leaf Structure.   Leaf structure is greatly modified by the environment; sun leaves usually are smaller and thicker, with thicker cuticle than shade leaves of the same species, as shown in Fig. 7.9. Van Volkenburgh and Davies (1977) found that leaves of field-grown cotton and soybean were thicker than leaves on plants grown in greenhouses or growth chambers and that cool nights increased the leaf thickness on plants in growth chambers. Patterson *et al.* (1977) also found more mesophyll tissue and a higher rate of photosynthesis per unit of leaf surface in field-grown cotton than in the thinner leaves of cotton grown in growth chambers. The transpiration rate of well-watered *Ilex glabra* and *Gordonia lasianthus,* which have thick, heavily cutinized leaves, is greater per unit of leaf surface in sunny weather than transpiration of oak, maple, and yellow poplar (Table 7.8). Swanson (1943) reported that on sunny days American holly transpired more per unit of leaf surface than lilac, coleus, or tobacco, but on cloudy days when the stomata were closed it transpired less. The ratios of internal to external surface were 12.9, 7.1, and 4.6 to 1.0 for holly, tobacco, and coleus, and a high ratio of internal to external surface provides more internal evaporating surface (Turrell, 1936; Nobel in Turner and Kramer, 1980).

Stomata.   As mentioned earlier, most of the water lost by plants escapes through the stomata and most of the carbon dioxide used in photosynthesis

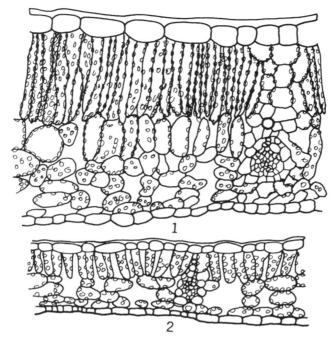

Figure 7.9 Difference in structure of sun leaves from the southern edge (1) and shade leaves from the center of the crown (2) of an isolated sugar maple (*Acer saccharum*). From Kramer (1983), after Weaver and Clements (1938).

enters through them. Thus, stomatal behavior plays a very important role in plant physiology. For example, the cycling in stomatal aperture discussed in Chapter 8 causes cycling in transpiration (Shirazi *et al.*, 1976, and Fig. 8.7). Stomatal behavior is discussed in Chapter 8 and stomatal control of transpiration is discussed by Jarvis and McNaughton (1986).

According to Meinzer and Grantz (1991), an increase in the leaf area of a stand of sugarcane is accompanied by a decrease in stomatal conductance per unit of leaf area, causing transpiration to approach a maximum per unit of ground area instead of increasing linearly with the leaf area. This keeps leaf water status quite constant over a wide range of growing conditions. They suggest that this coordination of canopy conductance with hydraulic conductance may be mediated by chemical regulators of stomata carried from roots to shoots in the transpiration stream (Meinzer *et al.*, 1991).

Leaf Coatings.   It would be expected that substances leaving a light-colored residue on leaves, such as the Bordeaux mixture, or light-colored dust would

Table 7.8 Midsummer Transpiration Rates of Various Species of Trees[a,b]

Species	Location	Season	Duration (days)	Number of Plants	Average transpiration $(g \cdot dm^{-2} \cdot day^{-1})$
Liriodendron tulipifera[c]	Columbus, OH	August	1	7	10.11
Liriodendron tulipifera[d]	Durham, NC	August	3	4	11.78
Quercus alba[d]	Durham, NC	August	3	4	14.21
Quercus rubra[d]	Durham, NC	August	3	4	12.02
Quercus rubra[e]	Fayette, MO	July	14	6	8.1
Acer saccharum[e]	Fayette, MO	July	14	6	12.2
Acer negundo[e]	Fayette, MO	July	14	6	6.4
Platanus occidentalis[e]	Fayette, MO	July	14	6	8.8
Pinus taeda[d]	Durham, NC	August	3	4	4.65
Clethra alnifolia[d]	Durham, NC	August	3	4	9.73
Ilex glabra[d]	Durham, NC	August	3	4	16.10
Myrica cerifera[d]	Durham, NC	August	3	4	10.80
Gordonia lasianthus[d]	Durham, NC	July	23	4	17.77
Liriodendron tulipifera[f]	Durham, NC	August 26– September 2	12	6	9.76
Quercus rubra[f]	Durham, NC	August 26– September 2	12	6	12.45
Pinus taeda[f]	Durham, NC	August 26– September 2	12	6	5.08

[a] Expressed as grams of water lost per square decimeter of leaf surface per day.
[b] All seedlings were grown in soil near field capacity.
[c] From Meyer (1932).
[d] From Caughey (1945).
[e] From Biswell (1935).
[f] From Kramer (unpublished).

lower leaf temperature and reduce transpiration. In fact, there were early reports that the Bordeaux mixture did reduce water loss, but later studies, summarized in Miller (1938, pp. 465–470), indicate that it causes considerable increases in night-time transpiration. Southwick and Childers (1941) found little effect of the Bordeaux mixture on transpiration of apple leaves, but some reduction in photosynthesis. It was reported by Beasley (1942) that application of light-colored dusts such as talc increased the night transpiration of coleus, but only if applied to the lower leaf surfaces when the stomata were open. This suggests that talc operates by mechanically blocking the stomata open. In recent years interest has shifted to waterproof coatings and substances that cause stomatal closure. These are called antitranspirants and are discussed in Chapter 12.

Disease

The effects of disease on plant water relations are often neglected, but can be important. According to Durbin (1967), when sporulation begins, transpiration from bean leaves infected with rust increases as much as 50%. Infection of barley leaves with powdery mildew also increases transpiration and increased transpiration is likely to occur when any organism breaks the cuticle. Pathogens that attack roots decrease water and mineral absorption, and various wilt diseases, such as Fusarium wilt of tomato, and elm and oak blight are caused by pathogens that block the conducting system. Robb *et al.* (1979) reported that *Verticillium* wilt causes partial blocking of the xylem by tyloses and coating on the inner vessel walls. Vascular diseases often result in reduced transpiration, wilting, and death. Grose and Hainsworth (1992) concluded that infection by *Phytophthora* causes more injury to lupine seedlings by blockage of stem xylem than by injury to roots. Zinc toxicity also causes development of an abnormal coating on the inner walls of xylem vessels that may inhibit water movement (Robb *et al.*, 1980). The interaction of plant water status and plant disease is discussed in Kozlowski (1978) and in Ayres and Boddy (1986).

MEASUREMENT OF TRANSPIRATION AND EVAPORATION

Thus far transpiration has been discussed in terms of individual leaves or plants that can be studied easily in the laboratory or greenhouse. However, from the ecological and agricultural standpoint the behavior of plants in stands is much more important. Light intensity and air movement decrease with distance downward in plant stands and the important boundary layer resistance is that of the stand rather than that of individual leaves. Thus, the effects of leaf size and shape and stomatal aperture are likely to be less important in the water relations of stands than for isolated plants. Some information on gas fluxes in stands is given in Nobel (1991, Chapter 9). The effects of evaporation from the soil also can be important in open stands, but it usually decreases as stand density increases (Kaufmann and Kelliher in Lassoie and Hinckley, 1991). The important and difficult problem of extrapolation from individual plants to stands is discussed in several papers in "Tree Physiology," Vol. 9, 1991, by Jarvis and McNaughton (1986), and in a book edited by Ehleringer and Field (1993).

Measurement of Transpiration of Plants and Leaves

The importance of transpiration in the overall water economy of plants has resulted in numerous measurements by several methods, including gravimetric and volumetric measurements, measurement of water vapor, and the velocity of sap flow. Considerable useful information can be found in Chapter 8 of Pearcy *et al.* (1989). We will first discuss measurements on leaves and individual plants and then turn to measurement of combined losses by evaporation from the soil and transpiration from stands of plants (evapotranspiration).

Measurement of transpiration was reviewed by Crafts *et al.* (1949), Franco and Magalhaes (1965), Slavik (1974), Kaufmann and Kelliher (in Lassoie and Hinckley, 1991), and others, and there has been little change in the basic methods for several decades. The best method depends on the kind and size of plant, the objective and duration of the experiment, and the equipment available.

Phytometer or Gravimetric Methods.   Most measurements from the time of Hales to the present have been made on plants growing in containers which are weighed at appropriate intervals to detect water loss. The size of the plants is limited by the size of the containers and the capacity of the weighing equipment. Small plants usually are grown in small metal or plastic containers of soil weighing a few hundred grams, but Nutman (1941) measured transpiration of coffee trees in tanks of soil weighing 225 kg with a sensitivity of 25 g. An even larger lysimeter was constructed by Fritschen *et al.* (1973) containing a Douglas-fir tree 28 m in height, growing in a block of soil 3.7 × 3.7 × 1.2 m. The total weight of the container, soil, and tree was 28,900 kg and the sensitivity of the weighing apparatus was 630 g. Lysimeters were discussed in Chapter 4.

The containers in which plants are grown should be protected from direct sun to prevent overheating, and the ideal arrangement is to have them buried with the tops flush with the surrounding soil in the habitat where they would normally grow. This was done by Biswell (1935) and Holch (1931) in their studies of the transpiration of tree seedlings in sun and shade in Nebraska. The soil surface should be covered to keep out rain and prevent evaporation but if the covers are too tight aeration may become limiting. A good seal is needed around stems to prevent unmeasured wetting of the soil by stem flow. Care must also be taken to prevent soil moisture from becoming limiting, especially as plants grow larger. In long-term experiments, corrections also need to be made for the increasing fresh weight of rapidly growing plants by harvesting samples of plants occasionally and weighing them. Estimates of transpiration over short periods of time may also involve errors caused by a decrease in water content resulting from the lag of absorption behind transpiration shown in Fig. 6.1. A variation of this method is to grow the plants in nutrient solution and measure the volume or weight of solution removed by the roots during a given period of time. Harris (1971), Kitano and Eguchi (1992a), and Simonneau *et al.* (1993) described apparatus for simultaneously measuring water absorption, transpiration, relative water content, and growth. Some data are presented in Fig. 7.20.

Extrapolation of data from potted plants to plants in the field is questionable, but the phytometer method is useful for comparing plants of various species under the same conditions and for measuring the effects of environmental factors and experimental treatments. The desire to estimate the transpiration of plants too large to be easily accommodated in phytometers and the need for rapid measurements resulted in the development of other methods of measuring or estimating the rate of transpiration.

Cut-Shoot Method.   This sometimes is called the rapid weighing method and was formerly used extensively because of its convenience and applicability to large plants in the field. Shoots or single leaves are detached and suspended on a special balance as near their original location as possible and the change in weight over time is recorded. Such measurements can last for only a few minutes because the transpiration rate tends to decrease with decreasing leaf water content. Also there often is a transient increase in the transpiration rate a minute or two after detachment, the Iwanoff effect, caused by the sudden increase in stomatal opening after release of tension by detachment from the plant. Furthermore, the detached leaf or stem no longer is in competition with the remainder of the shoot for water. The method was discussed by Franco and Magalhaes (1965), Slavik (1974) and others, but is seldom used today.

Sometimes stems of plants are detached under water and are placed in containers, and water uptake is measured over time. However, water uptake often is reduced by plugging of the cut stem surface with air or debris or by microbial activity and the stem must be recut frequently. Ladefoged (1963) and Roberts (1977) made useful measurements of transpiration of forest trees up to 16 m in height by cutting them off under a stream of water and placing the bases of the trunks in containers of water while their tops were supported in their normal position in the canopy. Knight *et al.* (1981) also measured transpiration of lodgepole pine trees up to 30 cm in diameter by this method. Although absorption slowed after 2 or 3 days, it could be restored by recutting the base of the trunk. Myers *et al.* (1987) reported that cutting Eucalyptus stems under water increased leaf water potential, but decreased leaf conductance.

Measurement of Water Vapor Loss.   Transpiration can be measured by monitoring the change in humidity of an air stream after it is passed through a container enclosing plant material. The containers usually are made of plastic and vary from tiny cuvettes holding part of a leaf to those enclosing a branch (Kaufmann, 1981) to plastic tents enclosing an entire tree (Decker *et al.*, 1962). Although the method eliminates the errors caused by detaching leaves or branches, it creates a somewhat artificial environment. Differences in leaf and air temperature, humidity, and wind speed inside and outside the chamber can cause important differences in transpiration between the enclosed plants and those in the open. Those errors can be minimized by equipment to control temperature and other factors at the same level within as outside the container (Jarvis and Slatyer, 1966). The cuvettes for single leaves or parts of leaves are essentially porometers, as described in Kramer (1983, pp. 328–330). Kaufmann (1981) and Kaufmann (in Hashimoto *et al.*, 1990) used electronically operated chambers large enough to enclose branches for short intervals to measure leaf conductance and estimate tree canopy transpiration. Larger enclosures can be used to measure evapotranspiration from groups of plants in a stand.

The cobalt chloride method described in Miller (1938, pp. 491–493) measures the escape of water vapor by the time required to produce a given change in color from blue to pink of a piece of filter paper impregnated with cobalt chloride and pressed against the leaf surface. As it replaces the normal leaf environment, rates obtained by this method usually do not agree with those obtained by weighing and are not very useful. Other investigators pressed bottles containing a small amount of $CaCl_2$ against the leaf to absorb water vapor, but that method has some of the disadvantages of the cobalt chloride paper method. This method is seldom used today.

Velocity of Sap Flow.   Several investigators have attempted to estimate the rate of transpiration from measurement of the velocity of sap flow by thermal methods. The technology is discussed later in the section on velocity of sap flow in stems. Decker and Skau (1964) found good agreement between simultaneous measurements of velocity of sap flow and of transpiration. Steinberg *et al.* (1990a) reported that the transpiration rate of 5-year-old pecan trees estimated from their sap flow gauge was similar to measurements made on the same trees in a weighing lysimeter. Granier *et al.* (1990) reported that if correction was made for transpiration from the understory there was good agreement between transpiration of a French maritime pine stand calculated from sap flow and from water vapor flux measurements. Dugas *et al.* (1993) compared estimates of transpiration based on sap flow measured by a thermal method and the injection of deuterium as a tracer, or based on porometer measurements of stomatal conductance, compared with gravimetric measurements on small trees. The difference between gravimetric measurements and heat balance methods was less than between gravimetric measurements and losses calculated from stomatal conductance or the deuterium tracer method. While sap flow measurements provide good nondestructive estimates, reliable measurements of transpiration require calibration against potted plants or plants in lysimeters. Readers will find additional references in Dye and Olbrich (1993), Dugas *et al.* (1993), and in the section on sap flow later in this chapter. According to McKenney and Rosenberg (1993), different methods of estimating evapotranspiration differ in their response to a change in temperature and other climatic conditions. This needs to be taken into account in predicting the effects of climatic change caused by increasing concentrations of atmospheric CO_2 on potential evapotranspiration.

Bases for Calculating Transpiration Rates.   After investigators have settled on a suitable method of measuring transpiration, they must select a method of expressing the rate. Generally, it seems reasonable to express transpiration in terms of leaf surface area, but should it be based on one or both surfaces? We have generally used one surface but use of the stomate-bearing surface seems reasonable. Many European measurements have been expressed as units of water lost per unit of leaf fresh weight, but the fresh weight of leaves varies with time of day, from day to day, and among species. An example of the differences in rates based on fresh weight and on leaf area is shown in Fig. 7.10.

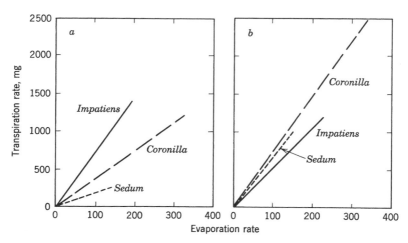

Figure 7.10   Transpiration of three species (*Impatiens noli-tangere, Coronilla varia,* and *Sedum maximum*) expressed in (a) milligrams per gram of leaf fresh weight and (b) milligrams per square decimeter of leaf surface. The rate of transpiration is plotted over the rate of evaporation from a filter paper atmometer. The transpiration of *Sedum* varied most with the method used to express it because of its low ratio of surface to mass. Adapted from Pisek and Cartellieri (1932), from Kramer (1983).

Leaf area seems to be the best basis for calculating rates of transpiration and photosynthesis because the receipt of energy is more closely related to the area than to the weight of leaves. However, as mentioned earlier, Nobel (in Turner and Kramer, 1980) pointed out that leaf thickness and the amount of mesophyll tissue can affect the rates of transpiration and photosynthesis per unit of surface area. Use of the leaf area introduces the necessity of measuring or estimating it. If leaves can be separated sufficiently, their area can be measured on a commercial leaf area machine or the area usually can be estimated from a formula relating area to length or width (Wiersma and Bailey, 1975). Sometimes the relationship between area and weight of a representative sample can be obtained and used to estimate the area of a large mass of leaves, such as all of those on a tree, from their weight. Johnson (1984) and Svenson and Davies (1992) discussed methods for measurement of surface area of pine needles, and leaf area measurement is discussed in Lassoie and Hinckley (1991, pp. 466–475).

Transpiration from stems usually is disregarded, but Gračanin (1963) reported that the rate of water loss from the stems of several herbaceous species constitutes a significant fraction of the total water loss. Water loss from large herbaceous stems probably deserves further study. According to Huber (1956), transpiration from the bark of trees is negligible in comparison to leaf transpiration. However, enough water is lost from branches of leafless deciduous trees during sunny winter days to reduce the pressure in the xylem sap and stop maple sap flow by early afternoon.

Rates of Transpiration. Despite various errors inherent in the measurement of transpiration, some useful information on daily and seasonal differences and differences among species have been obtained. An example of differences in diurnal rate between a C_3 and a CAM species are shown in Fig. 7.11, and the seasonal course of transpiration of an evergreen and a deciduous species is shown in Fig. 7.12. The evergreen transpiration rate was lower than that of the deciduous tree in the summer, but was higher in the winter because it retained leaves. Table 7.8 shows transpiration rates, measured gravimetrically on potted plants, of a number of woody species measured at different locations. It is somewhat reassuring to find the transpiration rates of well-watered yellow poplar (*Liriodendron tulipifera*) seedlings measured in August at Columbus, Ohio, and Durham, North Carolina, so similar (10.11 and 11.78 g dm^{-2} day^{-1}). Also, a comparison of rates of transpiration of six species of European trees measured by two methods, made by four investigators, all indicated that spruce had the lowest rate and birch the highest (Kramer and Kozlowski, 1960, p. 299). Roberts (1983) found that transpiration rates for trees reported from various places in Europe were quite similar, suggesting that the rate of transpiration of European forests is relatively stable and lower than would be expected from potential rates of evaporation.

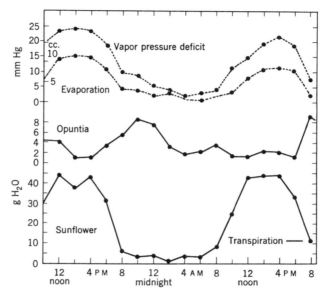

Figure 7.11   Transpiration of sunflower and *Opuntia* plants in soil at field capacity on a hot summer day. The maximum rate of transpiration of *Opuntia* occurred at night, a characteristic of plants with CAM. The early afternoon decrease in the transpiration of sunflower the first day probably resulted from a decrease in turgor and partial closure of stomata. From Kramer (1983), after Kramer (1937).

In contrast, Kaufmann (1985) reported that in the subalpine forests of Colorado there are large differences among tree species. Taking aspen as 1, the ratios of three conifers were lodgepole pine, 1.8; subalpine fir, 2.1; and Engelmann spruce, 3.2. The differences were attributed to differences in leaf conductance,

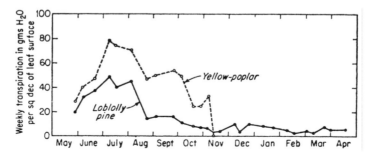

Figure 7.12   Seasonal course of transpiration of potted seedlings of an evergreen and a deciduous species growing outdoors at Durham, North Carolina. From Kramer and Kozlowski (1979).

leaf to air temperature, leaf area index, and the shorter transpiring season of aspen. Such differences have important implications for watershed management.

Roberts (1983) cited studies indicating that moderate variations in soil water content have a negligible effect on forest transpiration. Perhaps this surprising conclusion results from the relatively mild climatic conditions in the British and European forests studied. Low soil moisture certainly often becomes a limiting factor during drought in North America (Lopushinsky and Klock, 1974).

Evapotranspiration from Stands of Plants

The discussion of transpiration has dealt thus far largely with leaves or individual plants, but plants in nature usually occur in stands or communities and water loss occurs both from the transpiring plants and by evaporation from the surface of the soil in which they are growing, which collectively is known as evapotranspiration (ET). In a stand of plants, transpiration occurs from leaves of various ages and physiological conditions, located at various heights in the canopy and exposed to different microenvironments. Chason *et al.* (1991) discussed methods of estimating the leaf area of forest stands and Baldocchi *et al.* (1991) discussed methods of estimating canopy stomatal conductance. The amount of water lost from the soil by evaporation varies widely, depending on the kind and density of plant cover and thickness of the litter layer.

Jarvis and McNaughton (1986) present an interesting discussion of the problem of scaling up from leaves or plants to stands of plants. The difference in scale results in physiologists emphasizing the importance of stomatal behavior and species differences whereas micrometeorologists emphasize atmospheric conditions measured on a large scale for stands of plants. Problems of scale also are discussed in Ehleringer and Field (1993).

Measurement of Evapotranspiration.   Measurement of total water loss by evapotranspiration per unit of land surface is more important in agriculture and forestry than measurement of transpiration of individual plants. Such measurements are important for scheduling of irrigation, estimating yields from watersheds, and comparing water use efficiency of various crops. Methods of measuring water loss from land areas include: (1) determination of changes in soil water content, the water balance method; (2) the energy balance method; (3) eddy diffusion, the determination of net upward flow of water vapor from the soil and vegetation; (4) estimates from meteorological data; and (5) the use of lysimeters and catchment basins. The various methods are discussed by Tanner in Hagan *et al.* (1967) and in Kozlowski (1968), by Slatyer (1967, pp. 57–64), by Whitehead and Kelliher (1991b), by Kaufmann and Kelliher (in Lassoie and Hinckley, 1991), and by several authors in Kaufmann and Landsberg (1991).

The water balance method depends on measuring changes in soil water content between rains or irrigations, neglecting or measuring deep drainage beneath the crop or plant stand under study. Measurement of soil water is often made by neutron probes or an array of electrical resistance blocks installed at various depths in the root zone or by time domain reflectometry (see Chapter 4). The most reliable data are obtained from lysimeters located in the plant stand being studied or from stands in catchment basins (Whitehead and Kelliher, 1991b). Measurement of soil water content was discussed in more detail in Chapter 4.

Estimates of water loss by the energy balance method depend on the fact that the amount of water lost from an area of land or water by evaporation depends chiefly on the energy available to evaporate water, the net radiation of Eq. (7.3). Over a period of time of a few days the only important changes are in the heat exchange with the atmosphere and the latent heat used in evaporating water. If net radiation and heat loss are measured the fraction used in evaporating water can be calculated (Slatyer, 1967, pp. 58–59).

Vapor Flow.   The most direct measurement requires determination of the water vapor flow above a stand of plants, but this requires delicate instrumentation and adequate fetch, and works best on level land. Measurement of the humidity and temperature gradients above the plant stand, and of wind velocity, are necessary to provide data for calculating the vapor flux. Meteorological estimates of evapotranspiration from forest stands have been discussed by Baldocchi *et al.* (1991), Granier *et al.* (1990), Kaufmann and Kelliher (in Lassoie and Hinckley, 1991), and others.

Empirical Formulae.   Many estimates of evapotranspiration have been made from meteorological observations of such factors as bulk air humidity and temperature, wind speed, and length of day by various workers cited in Kramer (1969, p. 344) and Slatyer (1967, pp. 62–64). The equation developed by Penman and modified by Monteith (1965) is used. Whitehead and Hinckley (1991) and Luxmoore *et al.* (1991) discussed the use of several models and the problem of scaling up from leaves to plant stands.

Jarvis (1985) and Jarvis and McNaughton (1986) discussed the extent to which the transpiration of crops is coupled to atmospheric conditions and termed it the "omega factor, Ω." Tree crops seem to be more strongly coupled to the atmosphere than field crops because they are taller and exposed to more turbulence. If plant stands are closely coupled to atmospheric conditions, porometer data on stomatal conductance can be used to make reliable estimates of transpiration rates.

In full sun, 320 to 380 calories of incident energy per day are available per square centimeter of surface, but 570 calories are required to evaporate a cubic centimeter of water so the maximum possible evapotranspiration is about 6 mm per day. However, the actual evaporation usually is only about 80% of the theoretical maximum because evaporation also is affected by the vapor pressure gradient and by resistances in the vapor pathway. If evaporation exceeds the calculated potential rate, it often is because incident energy is supplemented by advection, the oasis effect shown in Fig. 7.2. Although the leaf area of plant stands (leaf area index) usually is several times that of the soil on which they grow, water loss cannot exceed that from an equal area of moist soil or a water surface receiving the same amount of energy. As the soil dries, stomatal closure reduces transpiration and drying of the soil surface reduces evaporation from the soil, hence actual ET often is considerably lower than the potential rate calculated from pan evaporation.

Evaporation and transpiration losses from a field of corn and a forest are given in Tables 7.1 and 7.2, and the effect of removing plant cover in increasing runoff and evaporation is shown in Table 7.2. One might expect water losses to be similar from various types of plant cover receiving similar amounts of radiation, but variations in reflectance (albedo), roughness of the canopy, depth of rooting, and stomatal behavior result in significant differences (Rider, 1957). An extreme example is shown in Fig. 7.13. Pineapple is a CAM plant (Joshi

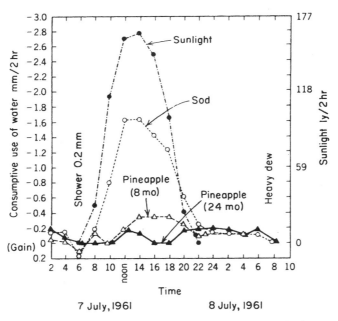

Figure 7.13 Evapotranspiration from a Bermuda grass sod and from stands of pineapple 8 and 24 months after planting. All plants were grown in soil in lysimeters in Hawaii. From Kramer (1983), after Ekern (1965).

et al., 1965) with such a low rate of transpiration that when a mature stand completely shades the ground the rate of evapotranspiration may be lower than that of a younger stand which leaves the soil partly exposed (Ekern, 1965; also see Chapter 12).

Over the entire United States about one-fourth of the total precipitation escapes as stream flow and three-fourths returns to the atmosphere by evapotranspiration from the soil and vegetation. Some precipitation is intercepted and is evaporated directly from the surface of vegetation, but most reaches the soil and is returned either by direct evaporation from the soil surface (49% in an Illinois cornfield, Table 7.1; 45% in a North Carolina deciduous forest, Table 7.2) or by transpiration. The relative losses by evaporation from the soil and from the vegetation are of special interest on watersheds in respect to vegetation management. For example, experiments in humid southwestern North Carolina indicated that conversion from mature, deciduous forest to low growing vegetation increased stream flow 12.5 to 40 cm the first year after cutting; the increase was much greater from north-facing than from south-facing slopes (Hewlett and Hibbert, 1961). According to Bosch and Hewlett (1982) in nearly 100 experiments with catchment basins removal of vegetation always resulted in an increase in runoff. Clear cutting lodgepole pine in Colorado resulted in a 30% increase in stream flow and conversion of chaparral to grass in the central valley of California even doubled runoff. Instances are reported where the removal of forests resulted in a measurable rise in a shallow water table. McNaughton and Jarvis (1983) discussed methods of measuring the effects of vegetation change on transpiration and evaporation, and Kaufmann (1984) developed a canopy model for estimating transpiration of forest stands.

It is fairly common for evapotranspiration to exceed precipitation during the growing season for a year or two, even in an Illinois cornfield, as shown in Table 7.1. The deficit was supplied from stored soil water; the water content of the upper 2.3 m of soil was reduced from 31.0 to 17.7 cm during the growing season. This deficit would be made up during an average winter. An extreme instance was the 18- to 20-year-old apple orchard in eastern Nebraska, studied by Wiggans (1938) during the drought of the 1930s. The trees removed 25 to 38 cm more water per year from the soil than was returned by precipitation, and it was estimated that in 3 more years they would exhaust all the available water to a depth of over 9 m. Presumably the trees eventually would have died from desiccation, but unfortunately they were killed by a severe autumn freeze, prematurely terminating the observations. Many tree plantations in the prairie and Great Plains states flourished for several years on stored soil water, but died when it was exhausted (Bunger and Thomson, 1938).

Another example is the deep-rooted perennial alfalfa, which made excellent growth in virgin soil on stored water in certain areas of Nebraska for the first 3 years, quite independently of current rainfall, after which the yield declined and became dependent on current rainfall. Alfalfa can absorb water from depths of at least 9 m in well-aerated loess soil, but this water may not be replaced for many years. In fact, Kiesselbach *et al.* (1929) reported that it had not been replaced below 2.0 m after 15 years of crops other than alfalfa.

THE ASCENT OF SAP

The success and even the survival of land plants depend on sufficient water moving upward from the roots to replace that lost from the shoots by transpiration. The existence of land plants more than a few centimeters in height depended on evolution of an effective water-conducting system because on a hot sunny day the volume lost may exceed the total water content of the plant. Replacement requires an efficient conducting system and a strong driving force in addition to the absorbing system discussed in Chapters 5 and 6. Mosses lack a water-conducting system and are limited in height to a few centimeters. In this section we will first discuss the structure of the water-conducting system, then the driving force that propels water through it.

The Conducting System

The xylem has been recognized as the principal pathway for the upward movement of water at least since the time of Malpighi and Grew in the 17th century. Anyone can demonstrate this by cutting the stem of a transpiring plant under a dye solution such as acid fuchsin and later making cross sections of the stem above the cut. If the stem is left in dye a few minutes the dye may even become visible in the leaf veins.

There are two kinds of water-conducting elements in xylem: tracheids which are spindle-shaped cells up to 5 mm long and 30 μm in diameter and vessels which are formed by disintegration of end walls of rows of cells, forming tube-like structures ranging from a few centimeters to many meters in length and 20 to 700 μm in diameter. Angiosperm xylem contains both vessels and tracheids, but gymnosperm xylem contains only tracheids. During maturation the protoplasts of vessel elements die and disintegrate and many of the end walls disintegrate, leaving only the lateral cell walls. The end walls of the remaining xylem elements contain perforations of various shapes through which water flows (Schulte and Castle, 1993, and anatomy texts). Most of the water movement in gymnosperms occurs through tracheids, in angiosperms through vessels, though other types of cells such as xylem parenchyma, and fibers may play a minor role. Movement from one xylem element to another is facilitated by pits or thin places in their walls. Detailed descriptions of vessels and tracheids can be found in anatomy texts, and Zimmermann (1983) discussed xylem structure in relation to water movement. Ewers and Cruiziat (in Lassoie and Hinckley, 1991) also discussed the transport and storage of water in stems.

Root–Shoot Transition Zone.   There is a change in the arrangement of the conductive tissues in the transition from roots to stems, especially in herbaceous plants. The xylem usually forms a cylinder in the center of roots, but becomes split into a number of vascular bundles arranged in a ring near the outside of stems in most herbaceous plants. Sometimes tissues in the transition zone appear to have special physiological properties. For example, citrus grafted on certain rootstocks are said to be more salt tolerant because Cl^- and Na^+ ions are intercepted in the transition zone (Lloyd *et al.*, 1987; Walker, 1986).

Woody Plants.   Because of their longer life the conducting systems of woody plants are exposed to greater variations in temperature and water stress than those of herbaceous plants. Because of the greater height of most woody plants they also face special problems. The conducting system often forms a spiral and dye injected near the base of a tree often follows unexpected paths as it moves upward (Kozlowski and Winget, 1963). Thomas (1967) reported that dye in dogwood moves spirally as much as 90° of the circumference per meter of ascent. Rudinsky and Vitè (1959) suggested that a spiral pattern of sap ascent provides a more effective distribution of water to all parts of a tree crown than direct vertical ascent. Kozlowski and Winget (1963) stated that pin oak in which the transpiration stream spreads out in the top suffers more from oak wilt than white oak in which water usually moves straight upward, keeping the fungus more localized.

There seems to be some uncertainty concerning the extent to which particular roots supply particular parts of the shoot. Klepper (in Gregory *et al.*, 1987) stated that in wheat each lateral root usually supplies a particular region of the shoot, but lateral transfer of water and minerals can occur in the xylem. Zimmermann (1983, p. 25) stated that each branch of the crown in trees is supplied with water from many roots, providing a safety factor in case of injury to part of the roots. For example, when Auchter (1923) removed the roots from one side of a peach tree the leaf water content was reduced equally on both sides, indicating free lateral transfer of water in the trunk. Furr and Taylor (1933) also observed free lateral transfer of water in lemon trees irrigated on one side. In view of the interconnections among vascular bundles at nodes in the stems of most herbaceous plants, it seems likely that when the conducting system is damaged, any part of the root system might supply any part of the shoot. However, as indicated in the section on phyllotaxy, considerable resistance to the lateral flow of water may exist.

The stems of woody plants are composed of xylem surrounded by a thin layer of bark containing the phloem and separated from the xylem by the cambium. In young woody plants the wood or xylem consists of sapwood containing many living cells, but as they age those cells begin to die and heartwood forms (Kramer and Kozlowski, 1979, pp. 602–610). Heartwood contains no living cells and is not involved in conduction because the vessels usually have been blocked by gas bubbles, tyloses, or deposits of gum, resin, and other substances. In older stems most of the cross section is composed of dead heartwood, which usually is darker in color and denser than the sapwood. The lack of essential physiological function in heartwood is indicated by the survival of hollow trees in which it has decayed and disappeared.

Xylem of woody angiosperms may be either ring or diffuse porous. In ring porous species such as ash, oak, and elm the first xylem vessels formed each spring (earlywood) are much larger in diameter than those formed late in the preceding summer (latewood), resulting in a sharp demarcation between successive annual rings. In diffuse porous trees such as birch, maple, and poplar the vessels usually are smaller in diameter, and those formed early in the growing season (earlywood) are nearly the same diameter as those formed at the end of the previous growing season (latewood). The difference between early and latewood enables observers to determine the number of annual rings, and differences in their width indicate the occurrence of droughts and other conditions unfavorable to growth. This relationship led to the development of dendrochronology, the study of the width of rings of old trees and timbers of old buildings, both as an indicator of past rainfall and of age of buildings (Fritts, 1976; Kozlowski et al., 1991, p. 66).

Interest in the conducting system of trees developed long ago. According to Zimmermann (1978), Leonardo da Vinci (d. 1519) wrote that when all branches of a tree are put together they are equal in thickness (area) to the trunk below them. This was developed into the pipe model by Shinozaki et al. (1964a,b) who treated the xylem as a collection of tubes. Zimmermann (1978) has an interesting discussion of the varying specific conductivity in various parts of the conducting system, examples of which are shown in Fig. 7.14. The relationship between structure and water movement was reviewed in detail by Fiscus and Kaufmann (in Stewart and Nielsen, 1990) and by Tyree and Ewers (1991), and the structure of woody stems was discussed in Kramer and Kozlowski (1979, pp. 18–38).

In many ring-porous species, water movement is confined to one or a few outer rings because older vessels and tracheids are blocked by air bubbles (embolisms), tyloses caused by extrusion of protoplasts through cell wall pits into the cavities of neighboring cells, and deposits of gum and resin. In diffuse-porous species, several annual rings may function in water movement. The resistance to water flow is much higher in conifer stems where water moves through tracheids with numerous cross walls than in deciduous, broadleaf trees, and the resistance to longitudinal flow is relatively low in vines and roots. The specific conductivity or conductivity per unit of leaf weight in various parts of a birch tree is shown in Fig. 7.14 and the rate of flow in various parts of an oak tree can be seen in Fig. 7.18. According to Tyree and Sperry (1989) the specific conductivity is low in twigs and petioles and the chief resistance to water move-

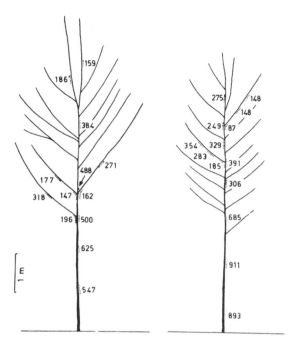

Figure 7.14   Leaf specific conductivity in microliters per hour, per gram fresh weight of leaves supplied, for two paper birch trees (*Betula papyrifera*). Conductivity decreases from lower to upper trunk and in the branches. It also is low when branches are attached to the trunk, as further indicated by the high velocity of sap flow at the point of branch attachment shown in Fig. 7.18. From Kramer (1983), after Zimmermann (1978).

ment is in branches and twigs. It is especially high at the junction of branches with the trunk (Figs. 7.14 and 7.18). It is suggested that growing conditions can modify the conductivity of tree trunks. According to Sellin (1993), the stem xylem of open-grown Norway spruce has greater specific conductivity than that of shade-grown trees because the latter has smaller tracheids and a smaller sapwood area.

Ratio of Sapwood Area to Leaf Area.   A fairly close relationship exists between the leaf weight or leaf area of trees and the cross-sectional area of the sapwood that supplies the leaves with water, but there are large differences among species. For example, Kaufmann and Troendle (1981) found similar ratios between leaf area and sapwood area in the lower and upper part of the crown in each of the four species they studied. However, there were large differences among the species (see Fig. 7.15), presumably because there are large dif-

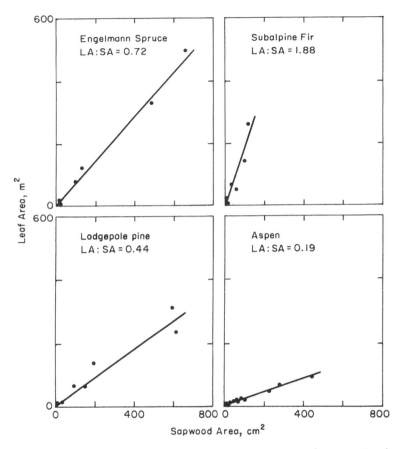

Figure 7.15 Relationship between leaf areas (LA) in square meters and cross-sectional area of sapwood (SA) at 1.37 m above ground for four subalpine forest tree species in Colorado. From Kramer (1983), after Kaufmann and Troendle (1981).

ferences in the efficiency or specific conductivity of their conducting systems. Whitehead *et al.* (1984) also found a linear relationship between leaf area and sapwood cross sections in Sitka spruce and lodgepole pine, but the relationship differed in the two species. No significant difference in the relationship between fertilized and unfertilized trees was found. Gartner (1991) studied the conducting systems of the shrubby and vine forms of *Toxicodendron diversilobum* growing in the same environment. The xylem cross-sectional area per unit of leaf area was smaller in vines, but the sap conductivity was higher in the larger vessels of vines, resulting in a similar conducting capacity per unit of leaf area in shrubs and vines. Gartner *et al.* (1990) found a higher specific hydraulic conductivity in stems of tropical vines than in trees, but there was much variability within species. A relationship probably exists between leaf area and xylem supplying it in herbaceous plants, but this seems not to have been investigated.

It might seem that the lower resistance to water flow through the large vessels of ring-porous species would be advantageous, but some of the tallest trees in the world are conifers in which water moves through tracheids only a few millimeters in length. Perhaps the shorter, smaller tracheids minimize blockage of the conducting system by decreasing cavitation and the spread of air bubbles as compared with ring-porous species. Thus, a trade-off may exist in which higher resistance to flow through tracheids than vessels is compensated by better protection against blockage and increased resistance later in the season, and both systems are successful.

Herbaceous Plants.   The conducting system of herbaceous plants often is quite complex, as indicated by the diagram of the branching of vascular bundles at nodes of a potato stem (Fig. 7.16). Because of the numerous interconnections, a local injury does not necessarily produce a serious blockage to water movement (Dimond, 1966). Monocots often have even more complex conducting systems than dicots (Zimmermann and Tomlinson, 1974; Zimmermann, 1978). There is said to be a high resistance to water flow into leaves because of the reduction in number and size of vessels in leaf traces and petioles and there often is a high resistance to flow where branches join the main stem of woody plants, as shown in Figs. 7.14 and 7.18. The water-conducting system of plants is discussed in detail in Chapter 4 of Zimmermann (1983) and by Tyree and Ewers (1991); Black (1979a,b) made a detailed study of resistance to water movement in sunflower.

Phyllotaxy.   It might be assumed that all leaves on the same side of a stem are supplied with water by the same vascular bundle, but this is not necessarily true. For example, tobacco has 2/5 or 3/8 phyllotaxy, meaning that a vascular bundle supplying leaf 1 goes around the stem twice between leaves 1 and 5 or three times between leaves 1 and 8. Thus, as shown in Fig. 7.17, on a plant with 2/5 phyllotaxy, leaves 4, 5, and 6 are supplied by different vascular strands (Fiscus *et al.*, 1973). Failure to take this into account can lead to misinterpretation of data on leaf water status. Larson (1980) reported that in *Populus deltoides* seedlings the phyllotaxy changes from 1/2 in the cotyledonary stage to 2/5 to 3/8 to 5/13, accompanying increasing leaf production as seedlings grow. In trees with a straight grain, tracers injected at the base rise straight up to leaves above the point of supply but in trees with a spiral grain, tracers may move

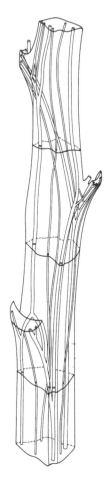

Figure 7.16 Vascular system of a potato stem (*Solanum tuberosum*) showing the complex branching at the nodes. Readers are referred to Dimond (1966) for a study of water flow in tomato stems, which have a similar structure. From Kramer (1983), after Eames and MacDaniels (1947).

around the stem and appear on the side opposite to the injection point. In monocots such as sorghum, maize, and palm trees there is so much interconnection of vascular bundles at lower nodes that a tracer spreads out in all directions.

Injury to the Conducting System.   There is some uncertainty concerning the amount of injury required to seriously restrict the flow of water to the leaves. Jemison (1944) observed that broad-leaved trees in which the sapwood in over half the circumference near the base of the trunk had been killed by fire made

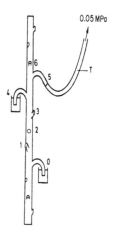

Figure 7.17   Diagram showing leaf attachment on a tobacco plant with 2/5 phyllotaxy. Leaves at positions 0 and 5 are supplied by the same vascular strand whereas leaves at positions 4 and 6 are on different vascular strands. By supplying different dyes to petioles at positions 0 and 4 and removing sap by suction from petiole 5, it was demonstrated that resistance to lateral movement of water from petiole 4 to 5 is much greater than the resistance to longitudinal movement from 0 to 5. From Kramer (1983), after Fiscus *et al.* (1973).

as much growth during the following 10 years as nearby uninjured trees. However, Rundel (1973) reported a strong correlation between fire injury to the base of the trunk and occurrence of dead tops in redwood, a conifer. Postlethwait and Rogers (1958) demonstrated that radioactive phosphorous moved past as many as four horizontal cuts made only 15 cm apart halfway through trunks of pine trees. Thus, it seems that although injury to part of the cross section of woody stems increases resistance to flow, it does not always cause a serious reduction in water and mineral supply to the leaves. Tyree and Sperry (1989, p. 33) suggested that this is possible because the resistance to water flow usually is lower in the main stem than in the branches. Tyree and Sperry (1988) concluded that many water-stressed trees operate near the limit of catastrophic failure when embolisms block part of the xylem, but apparently the limit is seldom exceeded. Yang and Tyree (1993) reported that midday water stress commonly causes a reduction in stomatal conduction and gas exchange of maple leaves, but it is doubtful if excessive stem resistance causes the death of many healthy plants.

Activities of bacteria, fungi, and insects also cause blockage of the xylem. Bark beetles introduce a fungus that blocks water transport in pines and also causes blue stain of the wood. The complex interactions among bark beetles, water stress, and tree physiology were discussed by Lorio (1993). Plugging of the vascular system has often been observed in Dutch elm disease (*Verticillium* wilt) of elm and maple and in mimosa wilt. In some instances, toxins produced by the fungi stimulate the production of gums which block the xylem. Hsu and Goodman (1978) reported that the bacteria causing fire blight produce a toxin causing rapid wilting of apple, and Suhayda and Goodman (1981) reported that they also produce a high molecular weight polysaccharide that blocks xylem vessels. Thus, injury from vascular wilt disease may have multiple causes (Talboys, 1978; Zimmermann and McDonough, 1978). Experiments of Grose and Hainsworth (1992) suggest that blockage of water conduction in stems can be a more important cause of injury to plants infected with *Phytophthora* than damage to the root system.

Various kinds of stresses, including zinc deficiency and attacks by pathogenic organisms, are said to induce deposits of material on and in walls of xylem vessels (Robb *et al.*, 1980; Van Alfen and Turner, 1975). A later study showed that ABA and *Verticillium* infection cause deposition of suberin on the inner walls of the xylem of tomato and more suberin is produced in resistant than in susceptible cultivars (Robb *et al.*, 1991). The increased suberization presumably hinders the growth of hyphae into adjoining cells and slows spread of the disease, but it must also slow movement of water and solutes through and out of the xylem.

Blockage by Cavitation.   In rapidly transpiring plants, blockage often occurs by cavitation, the spontaneous rupture of sap columns when subjected to large negative pressure or tension, accompanied by an infiltration of air. The resulting bubbles or emboli are filled with water vapor and air and block the xylem. There is an extensive literature on cavitation, partly because the sound made by its occurrence can be detected with acoustic equipment, making it easy to study. Examples can be found in papers by Milburn and Johnson (1966) and Tyree *et al.* (1984, 1986) and in a review by Tyree and Sperry (1989). Sperry *et al.* (1988a,b) found that in some plants acoustic data did not predict a decrease in hydraulic conductivity accurately because of interference by embolism formation in xylem fibers. Ritman and Milburn (1991) used a combined detector for audible and ultrasound emissions that enabled them to locate sites of cavitation more accurately. Freezing also causes formation of gas bubbles that block xylem elements (Zimmermann, 1964). According to Sperry and Sullivan (1992), freezing produces fewer embolisms in conifers than in nonconiferous trees. The large earlywood vessels of oak are most susceptible to embolism formation, the small vessels of poplar and birch are intermediate, and the tracheids of conifers are least susceptible. Hammel (1967) suggested that the tiny gas bubbles formed in the tracheids of conifers during freezing are reabsorbed during thawing, but it is unlikely that the large bubbles in the larger vessels of ring-porous trees could be removed in this manner. In plants of a few species such as grape and birch, root pressure may aid in refilling gas-filled xylem vessels with water in the spring (Sperry *et al.*, 1987, 1988a). However, recovery of full capacity to move sap in ring-porous trees probably depends chiefly on production of new xylem each spring (Huber, 1935; Zimmermann, 1964). Tyree *et al.* (1986) concluded that there was enough root pressure in maize, even when water stressed, to refill vessels overnight that have been blocked by cavitation. Borghetti *et al.* (1991) discussed the difficulties in explaining the refilling of embolized tracheids in conifers, and Tyree and Yang (1992) discussed refilling in sugar maple.

One would expect large diameter xylem elements to be more subject to cavitation than elements with a smaller diameter and Dixon *et al.* (1984) found this to be true within a species. However, it is not always true between species, as maple with vessels is less vulnerable to cavitation than cedar and hemlock with only small diameter tracheids. Cochard (1992) reported large differences among species in the xylem water potential required to cause cavitation. Tyree *et al.* (1991) found the stems of the tropical tree *Schefflera morotoni* very susceptible to cavitation, but this was compensated by its large stem capacitance and very high stem conductance per unit of leaf area (leaf specific conductance). Tyree and Sperry (1989) concluded that differences among species in vulnerability to embolism formation depend on differences in the air pressure required to penetrate the intervessel pit membranes instead of on vessel diameter. Cavitation occurs in roots and reduces longitudinal flow (Byrne *et al.*, 1977), and also occurs in cells and may contribute to freezing injury (Ristic and Ashworth, 1993).

Velocity of Sap Flow.   Measurements of sap flow velocity have been made by injecting dyes or radioactive tracers (Fraser and Mawson, 1953; Moreland, 1950), but the rate probably is at least temporarily affected by cutting into the stem. The heat pulse method devised by Huber and his colleagues (Huber, 1932; Huber and Schmidt, 1937) and modified by Sakuratani (1981) and others seems to be the most reliable method of measuring the velocity of sap flow because it is noninvasive. Heat is supplied to the sap stream by small electric heating units attached to or embedded in the stem and sap velocity is determined by the time required for warm sap to reach a thermocouple or thermistor placed on the stem above the heater. The theory and inherent problems of this method have been discussed by Marshall (1958), Pickard (1973), Lassoie *et al.* (1977b), Cohen *et al.* (1988), and Groot and King (1992). Cermák and Kučera (1981) and Steinberg *et al.* (1989) developed heat balance methods that compensate for temperature variations in tree trunks and permit more accurate estimation of volume flow than the original heat pulse method. These require temperature measurements at several points relative to the heat source and good insulation of the trunk from external temperature variations. Ham and Heilman (1990) also proposed modifications for high flow rates. Steinberg *et al.* (1990a) reported that transpiration of 5-year-old pecan trees measured with their sap flow gauge was comparable to measurements made on the same trees in a weighing lysimeter. Schulze *et al.* (1985) has an interesting comparison of canopy transpiration and stem flow in conifers, and Lopushinsky (1986) studied daily and seasonal variations in flow rates in conifers. In the spring while the soil was moist the maximum velocity occurred at midday, but late in the growing season, as the soil dried, it occurred earlier in the day. Jones *et al.* (1988b) used the heat pulse method on apple trees and discussed the difficulty in relating it to actual transpiration rates. Dye *et al.* (1992) claimed that the heat pulse and deuterium tracer methods give similar results if care is taken to minimize sources of error, but Cermák *et al.* (1992) claimed that dye injection gives higher velocities than the heat balance method. According to Groot and King (1992), even the latest version of the heat balance method suffers from errors at very low and high rates of sap flow, and Cohen *et al.* (1993) found that the accuracy of both the heat pulse and heat balance methods varied with sap flow velocity.

Sap movement generally starts first in the upper part of trees in the morning and there may be a lag of 2 to 4 hr before sap movement is measurable in the lower part of the stem (Schulze *et al.*, 1985). Steinberg *et al.* (1990a) found sap flow starting simultaneously near the ground and in the upper part of young pecan trees, but this may have resulted from the short transport distance and low capacitance in young trees. In oak and ash the greatest velocities occur near the base, but in birch the rate increases upward, apparently because it has less conducting capacity per unit of leaf area in its slender branches (Fig. 7.14). The rate of conduction in various parts of an oak tree at midday are shown in Fig. 7.18 and rates of movement in stems of various species, measured by various methods, are given in Table 7.9.

Cermák *et al.* (1992) measured water flow velocity at various distances from the periphery in trunks of pedunculate oak and Norway spruce. In oak, velocity was greatest in the youngest annual ring and decreased inward for about 10 rings. In Norway spruce, velocity was greatest in the center of the conducting xylem and decreased toward both the cambium and the heartwood, and variability around the trunk was greater than in oak. According to Granier *et al.* (1990) the rate of water flow is quite uniform at various points across the xylem in Maritime pine.

Another method of measuring water flow in plants is by use of nuclear magnetic resonance imaging. It has been used to measure movement on a scale ranging from less than a millimeter in wheat grains (Jenner *et al.*, 1988) to flow through intact cucumber stems (Van As and Schaafsma, 1984) and leaf petioles (Millard and Chudek, 1993). It appears to be particularly useful for nondestructively measuring changes in water content and water binding in intact plant tissue (Colire *et al.*, 1988; G. A. Johnson *et al.*, 1987; Veres *et al.*, 1991) and for studies of soil–root water exchange (MacFall *et al.*, 1990). Although the

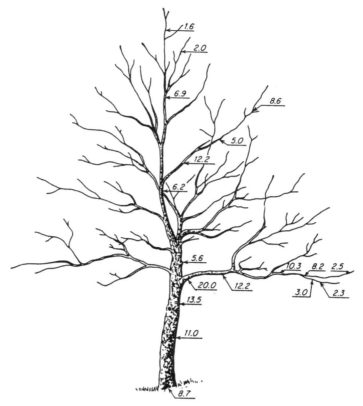

Figure 7.18 Velocity of sap flow in meters per hour in various parts of an oak tree at midday, measured by the thermoelectric method. The rate of flow decreases toward the top because the relative conducting surface (ratio of xylem cross section to leaf area supplied with water) increases toward the top in oak. In birch (see Fig. 7.14) the relative conducting surface decreases toward the top and the velocity of flow increases. From Kramer (1983), after Huber and Schmidt (1937).

method requires expensive equipment and highly trained technicians, it is developing into a promising tool to study the state of water (bound or free) and its movement in living plant tissue, as well as for morphological studies. Brief descriptions of the technology and some research results are given in Kramer *et al.* (in Hashimoto *et al.*, 1990) and in MacFall and Johnson (1994).

Direction of Sap Flow.   Water movement generally is upward because transpiration lowers the water potential in the leaves, producing a gradient in driving force in that direction. However, water can move through the xylem in the reverse direction as demonstrated by John Ray in 1669, Hales, and other early

Table 7.9   Rates of Water Movement in Xylem Measured by Various Methods[a]

Investigator	Method	Material	Velocity (m · hr^{-1})
Bloodworth et al. (1956)	Heat pulse	Cotton	0.8–1.1
Greenidge (1958)	Acid fuchsin	Acer saccharum	1.5–4.5
	Acid fuchsin	Ulmus americana	4.3–15.5
Huber and Schmidt (1937)	Heat pulse	Conifers	0.5
	Heat pulse	Liriodendron tulipifera	2.6
	Heat pulse	Quercus pedunculata	43.6
	Heat pulse	Fraxinus excelsior	25.7
Klemm and Klemm (1964)	^{32}P	Betula verrucosa	± 3.0
Kuntz and Riker (1955)	^{86}Rb	Quercus macrocarpa	27.5–60.0
Moreland (1950)	^{32}P	Pinus taeda	1.2
Owston et al. (1972)	^{32}P	Pinus contorta	0.1–0.8
Decker and Skau (1964)	Heat pulse	Juniperus osteosperma	0.25

[a]Kramer and Kozlowski (1979).

investigators. Williams (1933) reviewed the early work and demonstrated that enough water will move from leaves immersed in water to keep other leaves on the same plant, but exposed to air, turgid for several days. Daum (1967) demonstrated a reversal of flow in two major branches of an ash tree. Sometimes there was upward flow in the exposed branch and downward flow in the other when it was shaded. Also, after the top was wetted by an afternoon shower there was downward movement toward the roots when they were in dry soil. Occasionally, during periods of plant water deficit, reverse water flow occurs from fruits back into the plant on which they are growing. Lang (1990) observed this in apple and other examples are cited in Kramer (1969, pp. 355–356). Apparently there is no more resistance to water movement in one direction than the other and water moves along the prevailing gradient in water potential. This fact has been used to inject solutions into plants to give unusual flavors to fruits and colors to flowers, remedy mineral deficiencies, and control insects and diseases. The early work was reviewed by Miller (1938, pp. 883–885) and Roach (1934). Baxter and West (1977) reported that injection of water through cut branches of apple trees increased fruit size and Boyle et al. (1991b) increased kernel number and size in water-stressed maize by infusion of a culture solution. Rand (1983) discussed movement of water in plants as a problem in biomechanics.

Capacitance Effects.   Figure 6.1 showed that transpiration often quantitatively exceeds water absorption for much of the daylight period, resulting in transpiration and absorption curves that are out of phase. This phase difference results from what usually is termed the absorption lag. It results in a temporary decrease in the water content of leaves and stems and in the diameter of stems and thickness of leaves, the size of which depends on the storage capacity (capacitance) of the plant. An example of midday stem shrinkage is shown in Fig. 7.20 and measurement problems are discussed by Neher (1993). Hellkvist *et al.* (1974) published interesting data on vertical gradients of water potential in trees, some of which are shown in Fig. 7.19. For example, when transpiration increases rapidly on a sunny morning the water content and water potential of

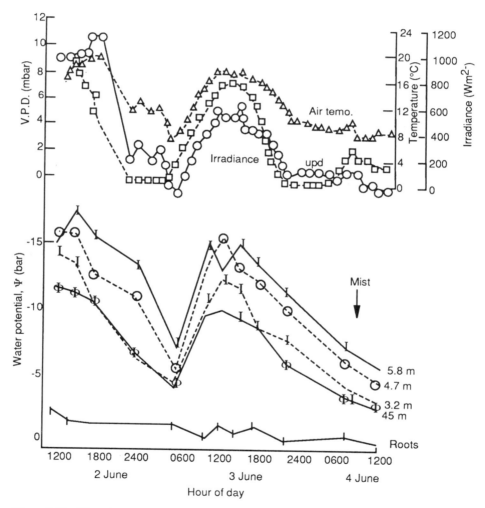

Figure 7.19 Diurnal variation in water potential at four heights on a Sitka spruce tree on a sunny day. The heights are at the point of attachment of the branches on which water potentials were measured with a pressure chamber. From Hellkvist *et al.* (1974), by permission of the authors.

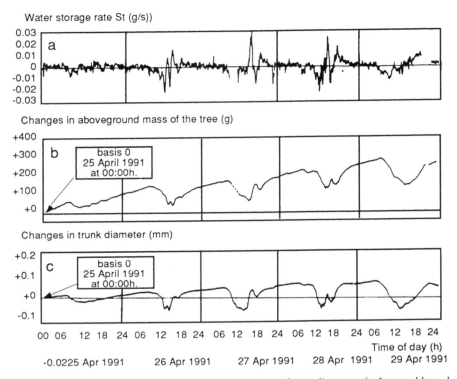

Figure 7.20 Diurnal changes in water content, stem mass, and stem diameter of a 5-year-old peach tree. (a) Gain or loss in stem water in g · sec^{-1}. Changes in shoot mass (b) and in trunk diameter (c). The first day was cloudy, the others were sunny. Note the good correlation among loss of water, afternoon loss of shoot mass, and stem diameter. From Simonneau *et al.* (1993).

the leaves and upper part of the shoot often decrease and transient wilting may take place. This occurs because the resistance to movement of water out of parenchyma cells is lower than the cumulative resistance to flow of water in from the soil through the roots. If the plants are in moist soil when transpiration slows in the afternoon and evening the parenchyma tissues regain the water lost in the morning. This capacitance effect tends to "dampen" or decrease the severity of midday leaf water deficits in exposed regions of both herbaceous and woody plants. Waring and Running (1978) and Waring *et al.* (1979) claimed that water stored in tree trunks significantly postpones water stress in forest trees, but Carlson and Lynn (1991) found that capacitance had little effect in small plants. Kitano and Eguchi (1993) reported that about 10% of the shoot water content of cucumber was lost by the midday water deficit, causing stomatal closure and a reduction in the rate of leaf expansion.

Simonneau *et al.* (1993) obtained interesting data on transpiration, water absorption, water storage in stems, and changes in stem diameter and mass of 5-year-old peach trees. Some of their data are shown in Fig. 7.20. Nobel (1991, pp. 528–532) discussed the theory of capacitance and it is discussed by Koide *et al.* (in Pearcy *et al.*, 1989). The importance of stored water is discussed in more detail in Chapter 12 in connection with drought tolerance.

The Mechanism of Sap Rise

The mechanism by which sap rises to the tops of trees has been a subject of speculation for centuries. Root pressure has often been assigned a role, but Hales observed in 1727 that root pressure does not occur in rapidly transpiring plants. Nevertheless, some writers consider its importance to be underestimated (Fensom, 1957; Minshall, 1964) and it probably plays a role in refilling vessels blocked by embolisms (Sperry *et al.*, 1988a,b). During the 19th century several physiologists thought the living cells of stems played an essential role in the ascent of sap, and the Indian physiologist Bose (1923) claimed that sap rise is caused by rhythmic pulsation in root and stem cells. Kargol (1992) developed a hypothesis involving the osmotic activity of living xylem cells to supplement the cohesion theory. However, investigators from Boucherie (1840) and Strasburger (1891) to Kurtzman (1966) have shown that sap will rise through stems or stem segments killed by heat or poisons. Although living cells do not cause the ascent of sap they may be important for the maintenance of favorable conditions by excluding air.

The Cohesion Theory. Today it is generally agreed that sap is pulled upward to the transpiring surfaces. This was suggested by Hales (1727), and in the 19th century Sachs and Strasburger concluded that transpiration produces the pull causing the ascent of sap. The cohesion theory became more acceptable when Askenasy (1895) and Dixon and Joly (1895) demonstrated that water confined in tubes has considerable tensile strength and Böhm (1893) demonstrated that transpiring branches can lift mercury above the height to which it is supported by atmospheric pressure.

The following are the essential features of the cohesion theory:

1. Water has high cohesive forces and when confined in small tubes such as xylem elements it can be subjected to tensions of many MPa before the columns rupture.

2. Water in plants forms a continuous system through the water-saturated cell walls and xylem elements from the evaporating surfaces in the leaves to the absorbing surfaces of the roots.

3. Evaporation of water from plant cells, chiefly those of leaves, lowers their water potential, causing water to move from the xylem to the evaporating cells

and cell surfaces. This reduces the pressure in the xylem sap and produces a tension or water potential gradient in the cohesive hydraulic system of the plant.

4.   This tension is transmitted to the roots where it causes an inflow of water from the soil. Thus, in transpiring plants in moist soil, water absorption is controlled by transpiration, modified only by the lag caused by water stored in plant tissue (capacitance) which sometimes results in absorption lagging an hour or more behind transpiration.

Problems with the Cohesion Theory.   There was some reluctance among physiologists to accept the cohesion theory (Greenidge, 1957, 1958; Preston, 1961, for example). This was based partly on the instability of water under tension and partly on the blocking of xylem elements by gas embolisms (cavitation). Blockage by gas bubbles (embolisms) formed by cavitation seems to be the principal problem. A large decrease in the water content of many tree trunks exists during the summer, as shown in Fig. 7.21. As mentioned earlier, Tyree

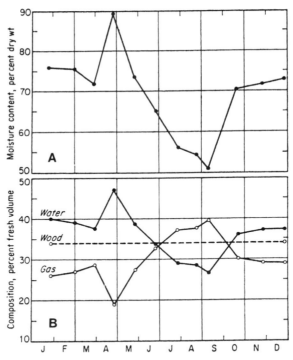

Figure 7.21   (a) Seasonal change in water content of yellow birch tree trunks calculated from disks cut from the base, middle, and top of the trunks. (b) Changes in gas and water content of yellow birch tree trunks calculated as a percentage of total volume. From Kramer (1983), after Clark and Gibbs (1957).

and Sperry (1988) claim trees often operate near the limit of their conducting capacity because of cavitation, but survival after wounding by experiments and accidents suggest that most trees have considerable excess conducting capacity and it seems that blockage by cavitation rarely is lethal.

Other objections to the cohesion theory of the ascent of sap have appeared. Direct measurement, of the negative pressure in the xylem sap with pressure probes by Balling and Zimmermann (1990) and others indicate that they are lower than those obtained by the Scholander pressure chamber method and are too low to support the cohesion theory. The significance of the new data is discussed by Zimmermann *et al.* (1993), who concluded that several processes are involved in the ascent of sap. These include capillary and interfacial forces, and what they term Maragoni convection which is related to the presence of tiny gas bubbles on the walls of the xylem vessels. Apparently in their view, tension is important only at high rates of transpiration. This has caused some controversy (Passioura, 1991; Zimmermann *et al.*, 1991) and a full treatment would require more space than is available. The main difficulty is that the pressure probe must penetrate the xylem water column while it is under tension, which may disrupt the tension. Thus, there probably are as many errors in measurements with pressure probes as with Scholander pressure chambers. More data are needed before final conclusions can be reached. Furthermore, we emphasize the statement of Renner (1912) that plant survival depends on coordination between rates of water absorption and transpiration, and this can best be accomplished by a continuous sap stream connecting the absorbing and transpiring surfaces. At least for the present, the cohesion theory is regarded as the best explanation of the ascent of sap.

CONDUCTION IN LEAVES

At each node a branch of the vascular system, the leaf trace, extends out into the leaf. In most conifers a single vein extends the length of the leaf. In grasses, numerous small veins extend more or less parallel to the midvein to the edges of the leaves, where they often anastomose. Those veins are connected by numerous veinlets extending across the mesophyll tissue which they supply with water. The phloem in the veins also provides the means of removing carbohydrate from the photosynthetic tissue. Altus and Canny (1985) have an interesting discussion of the complex venation of wheat leaves. Some dicots have palmate venation in which a few large veins extend out from the base of the leaf blade and are connected by a complex system of smaller veins. Other dicot leaves are pinnately veined, with numerous small veins extending outward on each side of the midrib. Veins usually extend to the tips of lobes and teeth, and Wilson *et al.* (1991) observed that tracers accumulate quickly in the hydathodes of teeth of transpiring balsam poplar leaves. They suggested that hydathodes retrieve solutes from the transpiration stream. In many species, leaf xylem elements

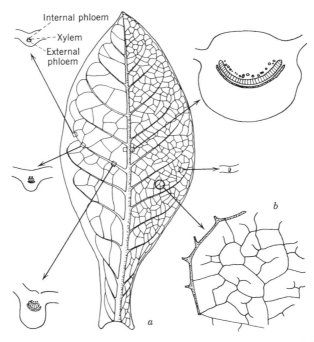

Internal phloem
Xylem
External
phloem

b

a

Figure 7.22 (a) Diagrams of venation of a mature tobacco leaf, showing midrib and principal lateral veins, also cross sections of midvein and lateral veins of various sizes. (b) Enlargement of a small section of a leaf blade to show the network of small veins. There were 543 mm of veins per square centimeter on this leaf. From Kramer (1983), after Avery (1933).

branch, rebranch, and reconnect to form a complex network, as shown in Fig. 7.22. As a result, most cells of a leaf are only a few cells away from a vein. Most leaves have so much excess conductive capacity that they often survive injury to major veins (Plymale and Wylie, 1944).

There is some question concerning the pathway by which water is supplied to the epidermal cells. LaRue (1930) and Williams (1950) pointed out that the epidermis in some kinds of leaves is or can be detached from the underlying mesophyll without injury and must be supplied with water from main veins. In some leaves, bundle sheath extensions reach the epidermis and provide a pathway for movement of water and solutes to the epidermis. It has even been suggested that it might be easier for water to move to the epidermis through bundle sheath extensions and then back to mesophyll cells than to move laterally through the mesophyll cells (Wylie, 1943), but this is speculative. There also is a question in leaves, as in roots, concerning the importance of apoplastic versus symplastic movement of water across masses of cells (Boyer, 1985, pp. 480–483). Canny (1990) reviewed some of the literature on use of tracers to follow water movement in leaves and other tissues and warned that water movement does not always coincide with apparent solute movement. As mentioned earlier, developments in technology such as NMR spectroscopy and imaging, and pressure probe measurements may aid in answering these questions.

USE OF XYLEM SAP BY PARASITES

Thus far, we have dealt with the role of the xylem in supplying water and solutes to the shoots of plants. However, there are many parasitic plants and some insects that depend on xylem sap for water, minerals, and organic substances. Over 3000 species of angiosperms are said to be fully or partly parasitic; the dodders and mistletoes are the best known. These plants are connected to their hosts by specialized tissues known as haustoria. The haustoria form close contact with the host tissue, but there usually is no direct connection between the xylem of host and parasite except in some mistletoe species (e.g., Sallé, 1983). The transpiration rate of at least some parasitic plants is higher than that of their hosts because their stomata remain at least partly open at night. This probably aids in maintaining the water potential gradient necessary to move water from host to parasite.

A number of different kinds of insects, including cicadas and spittle bugs, use xylem sap, although others such as aphids use phloem sap. Aphids can be employed to sample phloem sap by excising their stylets, which penetrate individual sieve elements, and collecting the exudate (Fischer and Frame, 1984). In contrast to most parasitic plants, insects that penetrate the xylem with their stylets establish direct access to the xylem sap. They filter out solutes and secrete large quantities of water in droplets from the anus. Sometimes these insects spread viruses and bacterial diseases. Much information on sap-feeding organisms has been summarized by Press and Whittaker (1993) and Raven (1983).

SUMMARY

This chapter deals with the loss of water by transpiration and the movement of water from roots to the transpiring surfaces of shoots, which may vary from a few centimeters to over 100 m above the soil surface. Transpiration refers to the loss of water from plants in the form of vapor and it largely controls plant water status. In warm sunny climates, plants often lose hundreds of times their fresh weight of water during a growing season and several hundred grams of water often are lost per gram of dry matter produced. Transpiration can be regarded as an unavoidable evil, unavoidable because a leaf structure favorable for the uptake of the CO_2 required for photosynthesis is unavoidably favorable for the loss of water vapor, and evil because it often causes injury and death to plants by dehydration.

Transpiration chiefly involves the evaporation of water from cell surfaces into intercellular spaces and its diffusion out of plant tissue, chiefly through stomata, and to a lesser extent through epidermal cells of leaves and lenticels in the bark. The rate of transpiration depends chiefly on the supply of water to the evaporating surfaces, the supply of energy to vaporize water, and the resistances in the vapor pathway. The driving force is the difference in vapor pressure between the evaporating surfaces and the outside air, which varies with temperature and absolute humidity. The chief resistances are the stomatal and cuticular resistance of leaves and the boundary or air layer resistance around leaves which varies with leaf size and wind velocity. Stomatal resistance is most important because most water escapes from plants through open stomata.

Measurement of water loss from plants by transpiration and by evaporation from the soil in which they grow, collectively known as evapotranspiration, is useful in comparing the water use efficiency of different kinds of plants, for scheduling irrigation, and for studying the effect of plant cover on yield from watersheds. Most measurements of water loss from individual plants involve gravimetric measurement of plants in pots or in lysimeters, or by frequent measurements of changes in soil water content. Measurements also are made by enclosing leaves or twigs in containers and measuring the change in humidity of a stream of air passed through the container. Measurements of loss in weight of detached leaves are not very reliable, but some success has been attained in estimating the rate of transpiration from the velocity of the sap stream measured by a thermoelectric method.

Measurement of water loss or evaporation from stands of plants is more difficult, but very important. It is estimated from measurement of the soil water balance or the energy balance, or from the humidity gradient above the stand, or is calculated from meteorological data. It can be measured more directly with plants in lysimeters or in small catchment basins.

Replenishment of the water lost from shoots by transpiration requires an efficient water conducting system and it was only after evolution of the xylem that land plants could grow more than a few centimeters in height. In herbaceous plants the xylem occurs in vascular bundles that are interconnected and spiral upward in complex patterns so that one leaf is often supplied from a different xylem strand than that above or below it. In woody plants the xylem forms most of the stem but only a few outer rings of the sapwood usually function in conduction. Water ascent in the heartwood is reduced by gum deposits, tyloses, and gas bubbles. Xylem conduction sometimes is blocked by activities of bacteria, fungi, and insects, and frequently by cavitation, the development of gas bubbles in the lumens of the xylem elements.

There has been debate for centuries concerning the mechanism by which water reaches the tops of trees. However, it is generally agreed that it is pulled up by the matric forces developed in the evaporating surfaces and transmitted to the roots through the cohesive water columns in the xylem. Although the cohesion theory is under attack, it appears to be the best explanation available. The minimum tension developed in the sap stream is about equal to that required to overcome the force of gravity on a column of water twice the height of the plant. The velocity of sap flow varies with the rate of transpiration and the conductive capacity of the xylem and is much greater in woody vines than in trees. A good correlation also exists between the cross-sectional area of sapwood and the leaf area supplied, although there are wide variations in the ratio among species. The greatest velocity of sap flow usually occurs near midday and movement is slowest at night and on cloudy and rainy days. Downward movement of water can occur as readily as upward movement if the driving force is reversed.

There is some uncertainty concerning the path water follows through masses of living cells such as the root cortex and mesophyll parenchyma of leaves. In such tissues it is difficult to determine how much water moves through protoplasts (the symplast) and how much through cell walls (the apoplast). Use of modern technology such as the pressure probe and nuclear magnetic resonance may help to solve this problem.

SUPPLEMENTARY READING

Ayres, P. G., and Boddy, L., eds. (1986). "Water, Fungi and Plants." Cambridge University Press, Cambridge.

Boyer, J. S. (1985). Water transport. *Annu. Rev. Plant Physiol.* **36,** 473–516.

Cowan, I. R. (1982). Regulation of water use in relation to carbon gain in higher plants. *Encycl. Plant Physiol. New Ser.* **12B,** 589–613. Springer-Verlag, New York.

Cutler, D. F., Alvin, K. L., and Price, C. E., eds. (1982). "The Plant Cuticle," Academic Press, New York.

Dixon, H. H. (1914). "Transpiration and the Ascent of Sap in Plants." Macmillan, New York.

Davies, W. J. (1986). Transpiration and the water balance of plants. *In* "Plant Physiology" (F. C. Steward, J. F. Sutcliffe, and J. E. Dale, eds.), Vol. 9, pp. 49–154. Academic Press, Orlando, FL.

Ehleringer, J. R., and Field, C. B., eds. (1993). "Scaling Physiological Process." Academic Press, San Diego.

Esau, K. (1965). "Plant Anatomy," 2nd Ed. Wiley, New York.

Fiscus, E. L., and Kaufmann, M. R. (1990). The nature and movement of water in plants. *In* "Irrigation of Agricultural Crops" (D. A. Stewart and D. R. Nielsen, eds.), pp. 193–241. Agric. Mon. 30, ASA-CSSA-SSSA, Madison, WI.

Gates, D. M. (1980). "Biophysical Ecology." Springer-Verlag, Berlin/New York.

Givnish, T. J., ed. (1986). "On the Economy of Plant Form and Function." Cambridge University Press, Cambridge.

Grace, J. (1978). "Plant Response to Wind." Academic Press, London.

Hashimoto, Y., Kramer, P. J., Nonami, H., and Strain, B. R., eds. (1990). "Measurement Techniques in Plant Science." Academic Press, San Diego.

Jarvis, P. G., and Mansfield, T. A., eds. (1981). "Stomatal Physiology." Cambridge University Press, Cambridge.

Lassoie, J. P., and Hinckley, T. M., eds. (1991). "Techniques and Approaches in Forest Tree Ecophysiology." CRC Press, Boca Raton, FL.

Nobel, P. S. (1991). "Physicochemical and Environmental Plant Physiology." Academic Press, San Diego.

Passioura, J. B. (1982). Water in the soil-plant-atmosphere continuum. *Encycl. Plant Physiol. New Ser.* **12B,** 5–33. Springer-Verlag, New York.

Pearcy, R. W., Ehleringer, J., Mooney, H. A., and Rundel, P. W., eds. (1989). "Plant Physiological Ecology." Chapman and Hall, New York.

Slavik, B. (1974). "Methods of Studying Plant Water Relations." Czechoslovak Acad. of Sci., Prague.

Steward, F. C., Sutcliffe, J. F., and Dale, J. E., eds. (1986). "Water and Solutes in Plants." Academic Press, Orlando, FL.

Tyree, M. T., and Ewers, F. W. (1991). The hydraulic architecture of trees and other woody plants. *New Phytol.* **119,** 345–360.

Zimmermann, M. H. (1978). Hydraulic architecture of some diffuse porous trees. *Can. J. Bot.* **56,** 2286–2295.

Zimmermann, M. H. (1983). "Xylem Structure and the Ascent of Sap." Springer-Verlag, Berlin.

8

Stomata and
Gas Exchange

INTRODUCTION

Stomata (singular, stoma), sometimes anglicized as stomates, provide an essential connection between the internal air spaces of plants and the external atmosphere. The external surfaces of most herbaceous plants and the leaves of woody plants are covered with a waxy layer of cutin (see Fig. 7.7) which is relatively impermeable to water vapor and carbon dioxide. This enables plants to conserve water in dry air, but it also hinders the entrance of the carbon dioxide essential for photosynthesis. Stomata are pores in the epidermis and associated cuticle bordered by pairs of structurally and physiologically specialized guard cells and adjacent epidermal cells termed subsidiary cells. This group of cells forms the stomatal complex and facilitates gas movement through the epidermis. Stomatal development and structure are discussed in Jarvis and Mansfield (1981), in Weyers and Meidner (1990), and in anatomy texts. In the absence of stomata the supply of carbon dioxide for photosynthesis would be inadequate for survival of most plants, but at the same time the unavoidable loss of water vapor through them creates the danger of dehydration. Thus, the ability of stomata to adjust their aperture is extremely important to the success of plants (Cowan, 1982; Raschke, 1976).

Historical Review

Malpighi observed the presence of pores in leaves in 1674 and in 1682 Grew pictured them in his plant anatomy. Apparently A. de Candolle applied the term "stomata" in 1827 and the study of stomatal behavior began with von Mohl about the middle of the 19th century. The history of early research on stomata was reviewed briefly by Meidner (1986). The study of diffusion through small pores such as stomata was placed on a sound physical basis by the research of Brown and Escombe (1900). This was followed by the work of Stålfelt (1932, 1956a) and Bange (1953) who showed that in moving air, where the boundary layer resistance is low, transpiration is closely correlated with stomatal aperture. Various investigations (see Mansfield, 1986, p. 202) showed that ABA increases in water-deficient leaves and that an external application of ABA usually causes stomatal closure. This led to the concept that stomatal closure in water-deficient plants often is caused by chemical signals from the roots. This was reviewed by Davies and Zhang (1991) and Davies *et al.* (1994), and is discussed later in this chapter and in Chapter 5 in the section on roots as sensors of water deficits, also in Chapter 10.

Occurrence and Frequency

Stomata occur on stems, leaves, flowers, and fruits, but not on aerial roots. They occur on both surfaces of many leaves (amphistomatous) or on only one surface, usually the lower (hypostomatous), especially in woody plants. Common exceptions among woody plants are poplar and willow which are amphistomatous. The adaptive importance of this is not clear (Parkhurst, 1978). Stomata vary widely in size and frequency, as shown in Table 8.1, and species with smaller stomata usually have a higher frequency. The frequency ranges from 60 to 80 per mm^2 in corn to 150 in alfalfa and clover, 300 in apple, and over 1000 in scarlet oak. There often are variations in number in various parts of a leaf (Smith *et al.*, 1989) and among genotypes of a species (Muchow and Sinclair, 1989). Additional data on frequency can be found in Meyer *et al.* (1973, pp. 74–75), Miller (1938, p. 422), and Weyers and Meidner (1990). In monocots, conifers, and some dicots, stomata occur in parallel rows, but in leaves with netted venation they are scattered. They sometimes are sunken below the surface but occasionally are raised, and usually they open into substomatal cavities in the mesophyll tissue. They are easily visible on leaf surfaces under magnification because of the peculiar shape of the guard cells (Figs. 7.8, 8.1, and 8.2) and the fact that guard cells, unlike other epidermal cells, usually contain chloroplasts. When wide open, stomatal pores usually are 3–12 μm wide and 10–30 μm or more in length. Usually, specialized epidermal cells, called subsidiary cells, are associated with the guard cells and play a role in guard cell functioning. According to Meidner (1990) and Stålfelt (1956a), the full opening of stomata is associated with a slight decrease in turgor of epidermal cells.

Table 8.1 Representative Dimensions and Frequencies of Guard Cells

Plant type	Representative species	Comments	Stomatal frequency (pores · mm^{-2})		Guard cell dimensions (μm) Upper surface		Lower surface		Dimensions of stomatal pore (lower surface, μm)		Pore area as proportion of total leaf area if open to 6 μm (%)
			Upper	Lower	Length	Width	Length	Width	Length of pore	Depth of whole pore	
Moss	*Polytrichum commune*	Stomata present on sporophyte only	16	16	46	15	46	15	15	12	—
Fern	*Osmunda regalis*	Many chloroplasts in guard cells	0	67	—	—	56	19	30	15	0.5
Gymnosperm tree	*Pinus sylvestris*	Sunken stomata, needle-like leaf	120	120	28	7	28	7	20	6	1.2
Dicot tree	*Tilia europea*	Hypostomatous leaf	0	370	—	—	25	9	10	8	0.9
Dicot herb	*Helianthus annuus*	Typical mesophyte	120	175	35	13	32	14	17	—	1.1
Dicot herb (xerophyte)	*Sedum spectabilis*	Succulent leaves	35	56	32	10	33	10	20	18	0.4
Monocot herb	*Allium cepa*	Cylindrical leaves, elliptical guard cells	175	175	42	19	42	19	24	18	2.0
Monocot C$_3$ grass	*Avena sativa*	Graminaceous guard cells	50	45	52	15	56	13	20	10	0.5
Monocot C$_4$ grass	*Zea mays*	Graminaceous guard cells	98	108	38	10	43	12	20	10	0.7

Note. The mean values quoted are derived from Meidner and Mansfield (1968) with additional data estimated by the authors. "Upper" and "lower" surfaces refer to leaves except in *Polytrichum*; note that there is no differentiation of surfaces in *Allium*, *Pinus*, and *Polytrichum*. Dashes indicate that category is not applicable. It should be appreciated that the values of parameter shown depend on cultivars, growth conditions, insertion level of leaf, and other factors. From Weyers and Meidner (1990).

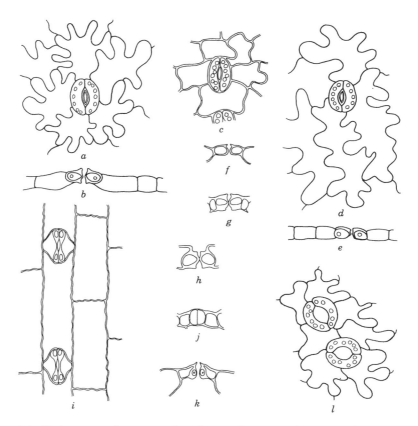

Figure 8.1   Various types of stomata: (a,b) *Solanum tuberosum* in face view and in cross section; (c) apple; (d,e) *Lactuca sativa;* (f) *Medeola virginica;* (g) *Aplectrum hyemale;* (h) *Polygonatum biflorum;* (i,j,k) *Zea mays*. Part (i) is a face view; (j) is a cross section near the ends of guard cells; (k) is a cross section through the center of a stoma; and (l) is a face view of *Cucumis sativus*. From Kramer (1983), after Eames and MacDaniels (1947), by permission of McGraw-Hill.

STOMATAL FUNCTIONING

Guard Cells

The walls of guard cells bordering the pores usually are thickened and sometimes have ledges and projections that extend into the pores, as shown in Figure 7.8A and in Weyers and Meidner (1990, p. 8). Wax filaments also often extend into stomatal pores, especially in conifers (Gambles and Dengler, 1974; Jeffree *et al.*, 1971). The thickening of the inner walls was supposed to play an essential role in causing turgid guard cells to bulge and separate, opening the

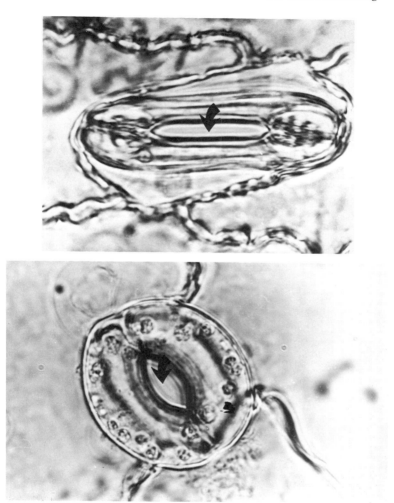

Figure 8.2   (Top) An open stoma of maize, typical of stomata of grasses. (Bottom) An open stoma of bean, typical of most dicots. From Kramer (1983). Courtesy of J. E. Pallas, U.S. Department of Agriculture.

stomatal pores, but Aylor *et al.* (1973) concluded that the micellar structure of the cell wall is more important than the thickening. This is discussed in Mansfield (1986, p. 160). Guard cells are often described as kidney or bean shaped, but those of grasses (Fig. 8.2) are elongated and the ends are enlarged, resembling dumbbells, and various other shapes occur. When wide open the stomatal pores occupy from less than 1 to 2% or more of the leaf surface. Interesting scanning electron micrographs of leaf surfaces, stomata, and leaf interiors can be found in Troughton and Donaldson (1981).

Stomatal Behavior

The most important characteristic of stomata is that they open and close, and the change in size of their aperture regulates gas exchange. In general they are open in the light and closed in darkness, although the stomata of plants with Crassulacean acid metabolism (CAM plants) behave in the opposite manner, being largely closed during the day and open at night. CAM plants have the capacity to fix large amounts of CO_2 in darkness as malic acid. This is decarboxylated during the day, releasing CO_2 that is refixed into carbohydrates in the light by photosynthesis. A comparison of daily cycles of CO_2 exchange and transpiration of the C_3 plant, sunflower, and the CAM plant, *Agave americana*, is shown in Fig. 8.3. This behavior greatly reduces water loss without an equiva-

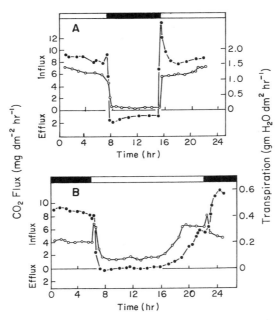

Figure 8.3   Comparison of daily cycles of carbon dioxide exchange (●) and transpiration (○) of the C_3 plant, sunflower (A), and the CAM plant, *Agave americana* (B). Both carbon dioxide uptake and transpiration of sunflower ceased in darkness, and there was some efflux of carbon dioxide released by respiration. The situation was reversed in *Agave* with little transpiration and no carbon dioxide uptake during most of the light period. Note that the units for transpiration are about four times greater for sunflowers than for *Agave* and that the transpiration rate of sunflower is proportionately greater. Adapted from Neales *et al.* (1968), from Kramer (1983).

lent decrease in dry matter production because the rate of transpiration is low at night. It is found in a number of succulents and other plants of dry habitats; pineapple is the best example among crop plants. Its leaves are heavily cutinized and its stomata are at the bottom of deep furrows covered with hairs and do not open until late afternoon or evening. As a result it has a very high water use efficiency, using only 50 or 55 g of water per gram of dry matter produced (Joshi *et al.*, 1965) compared with several hundred grams for most crop plants (Table 7.3). Crassulacean acid metabolism has been discussed in detail by Kluge and Ting (1978), Osmond (1978), and Ting (1985).

Mechanism of Stomatal Opening and Closing

The opening of stomata requires an increase in turgor of guard cells while closing requires a decrease in turgor. Although explanation of the cause of turgor change has been drastically revised in recent years, many questions remain unanswered (Kearns and Assmann, 1993). Originally, changes in turgor were attributed to changes in proportions of starch and sugar in guard cells (Lloyd, 1908; Sayre, 1926). It was believed that in light when photosynthesis removed CO_2 the increase in pH resulted in hydrolysis of starch to sugar, causing a decrease in osmotic potential and an increase in turgor as water entered. Decreasing light intensity and photosynthesis resulted in an accumulation of CO_2, decreasing pH and causing conversion of sugar back to starch. This neat explanation was rendered obsolete by observations in Japan and by the work of Fischer (1968b), Fischer and Hsiao (1968), and Fischer (1971) showing that the transport of K^+ in and out of guard cells is chiefly responsible for changes in turgor (see Mansfield, 1986, p. 164). Actually, Macallum observed in 1905 that the K^+ concentration was much higher in guard cells of open stomata than in those of closed stomata, but the significance of this early observation was neglected for more than half a century in favor of Lloyds' explanation. It now seems to be well established that the concentration of K^+ in guard cells of open stomata is several times greater than that in the surrounding cells, and there appears to be a good correlation between the K^+ content of guard cells and stomatal aperture. In the cell, the K^+ is accompanied by various anions that balance the positive charge on K^+. Some guard cells take up Cl^- as a balancing anion but organic acids also can be synthesized internally and serve the same function. This brings back a possible role for starch as the source of organic compounds, including sugar and organic acids, chiefly malic (Outlaw and Manchester, 1979). There seems to be renewed interest in the carbohydrate metabolism of guard cells (Hite *et al.*, 1993). However, this is complicated by the fact that onion guard cells contain no starch, yet function normally (Schnabl and Ziegler, 1977).

The guard cell chloroplasts exhibit fluorescence transients resembling those of mesophyll chloroplasts (Ogawa *et al.*, 1982; Outlaw *et al.*, 1981; Zeiger *et al.*, 1980), and K^+ and abscisic acid affect the transient as though energy from guard cell chloroplasts is used to accumulate K^+ (Ogawa *et al.*, 1982). Evidence exists that photophosphorylation occurs in these chloroplasts (Grantz *et al.*, 1985a; Shimazaki and Zeiger, 1985) and that CO_2 fixation probably occurs as well, although this has been a controversial area (Outlaw, 1989). According to Cardon and Berry (1992), guard cell chloroplasts probably carry on photosynthesis that is similar to that occurring in mesophyll cells, although it may be slow (Reckmann *et al.*, 1990). It now seems most likely that CO_2 is fixed chiefly by the enzyme phosphoenolpyruvate carboxylase and that the oxaloacetate product is reduced to malate (Scheibe *et al.*, 1990) that balances some of the charge of the incoming K^+. The malate together with incoming Cl^- thus form osmoticum that adds substantially to the osmotic effect of the incoming K^+. It is likely that mitochondrial respiration can supply the energy for opening in the absence of guard cell photophosphorylation and photolysis since opening can occur in the dark under certain conditions, particularly low CO_2 (Fischer, 1968a; Raschke, 1972). The starch of guard cell chloroplasts probably serves as a store of carbon compounds that can be used for energy as well as for organic counterions for K^+ (Raschke, 1975; Zeiger, 1983). The K^+ available in fertile soils appears to be sufficient for guard cell function (Ishihara *et al.*, 1978).

The specialized CO_2 fixation of guard cells can sometimes be seen in the response to CO_2 concentration which would ordinarily be expected to enhance opening at high concentrations. However, the reverse often occurs and stomata open more fully at low CO_2 concentrations, indicating that CO_2 fixation is different from that in leaf mesophyll cells. So far, an explanation of this behavior has not been forthcoming, although Mansfield *et al.* (1990) suggest that there could be more than one process competing for CO_2 one of which is inhibitory and the other stimulatory for opening. Opening would be affected according to whichever process dominates.

The loss of K^+ that results in stomatal closure can be brought about by elevated levels of abscisic acid around the guard cells (Ehret and Boyer, 1979; Mansfield and Jones, 1971) and this probably is the main means of closure (Harris and Outlaw, 1991; Neill and Horgan, 1985). Because the guard cells can metabolize and thus inactivate abscisic acid (Grantz *et al.*, 1985b), they exert considerable local control over the opening and closing process. This suggests that there could be some variability in stomatal aperture across a leaf because of variable rates of local breakdown of the abscisic acid. It is commonly observed that leaves have a statistical distribution of openings as described by Laisk *et al.* (1980) and rarely have all their stomata at the same aperture. The loss of K^+ that also occurs during stomatal closure in water-deficient leaves is found whether the roots are present or not (Ehret and Boyer, 1979) and further indicates that local synthesis and metabolism of abscisic acid probably account for much of the opening and closing response during water deficits.

Readers who wish to learn more about guard cell metabolism are referred to Mansfield (1986), Zeiger *et al.* (1987), Cardon and Berry (1992), and the current literature. Metabolic inhibitors such as sodium azide and the absence of oxygen prevent stomatal opening (Walker and Zelitch, 1963), emphasizing the dependence of opening on metabolic processes. Kearns and Assmann (1993) point out that the process of stomatal closure is not exactly the reverse of stomatal opening.

FACTORS AFFECTING STOMATAL APERTURE

The changes in guard cell turgor that bring about stomatal opening and closing are dependent on a number of environmental factors, including light, carbon dioxide concentration, humidity, and temperature (Schulze and Hall, 1982), and on internal factors such as tissue water status and the level of such plant growth regulators as ABA and cytokinins. Complex interactions often exist among these factors which make it difficult to distinguish the relative importance of individual factors such as light and CO_2 or water status and ABA. Information about these interactions can be found in Burrows and Milthorpe (in Kozlowski, 1976) and Weyers and Meidner (1990, pp. 27–30). Ball and Berry (1982), Ball *et al.* (1987), and Collatz *et al.* (1992) have proposed a simple model to account for the interactions.

The relationship among stomatal conductance, transpiration, and photosynthesis of a larch tree and some environmental factors is shown in Fig. 8.4. However, it is not entirely clear how stomatal conductance responds to these environmental signals. Collatz *et al.* (1991), drawing on earlier work, suggested that responses of stomata to environmental factors be divided into two groups, those dependent on photosynthesis and those independent of photosynthesis, but there are important interactions between the two groups. The role of stomatal conductance with respect to photosynthesis has been discussed in detail by Cowan (1982) and by Farquhar and Sharkey (1982). The latter concluded that although stomatal conductance substantially limits transpiration, it rarely seriously limits photosynthesis because the latter is limited by other factors in addition to those contributing to stomatal closure. Collatz *et al.* (1991, p. 122) state that the primary factor causing a midday decrease in stomatal conductance is a decrease in net photosynthesis, related to rise in leaf temperature above the optimum for photosynthesis but this must be a special case. The temperature rise is said to be caused by low boundary layer conductance.

The Role of Light

Although it has been known for many years that stomata usually open in the light, it has been difficult to determine whether this is a direct effect of light or

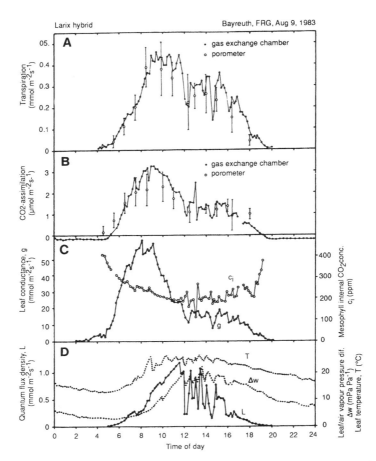

Figure 8.4   Relationships among (A) the daily course of transpiration, (B) CO_2 uptake during the day and efflux at night, and (C) leaf conductance (g) and internal CO_2 concentration (C_i) of a larch tree. Air temperature (T), vapor pressure deficit (Δw), and light (L). From Schulze *et al.* (1985), by permission of the author and *Oecologia*.

whether it occurs because photosynthesis decreases the internal concentration of CO_2. However, the effect of light, independent of its role in photosynthesis, has now been demonstrated (see reviews by Mansfield, 1986, pp. 181–191; Zeiger, 1983, pp. 460–463). It is believed that two photoreceptors are involved; one that is sensitive to red and far red light and another that absorbs in the blue and ultraviolet (Hsiao *et al.*, 1973; Ogawa *et al.*, 1978). These seem similar, or possibly identical, to the systems controlling photomorphogenetic processes such as photoperiod and phototropism. They must operate by affecting the amount and direction of ion transport across guard cell membranes (Serrano *et al., 1988).* Although the relative importance of red and blue light in stomatal opening is of physiological interest, plants are exposed to both wavelengths in the field. Adding to the complexity is the fact that plants often show rhythmic opening and closing of stomata after being transferred to continuous darkness.

A considerable difference seems to exist among plants in the response of stomata to light. Kitano and Eguchi (1992a) found that increasing irradiance from 135 to 540 w/m^2 caused oscillation in stomatal conductance, transpiration, water absorption, plant water balance, and leaf expansion of cucumber seedlings with a period of about 20 min. However, Schulz *et al.* (1993) reported that although *Impatiens pallida* wilted immediately in direct sunlight, stomatal conductance increased at first and only decreased after several hours. Wilting reduced the heat load on the leaves about 40% and reduced transpiration significantly, but caused little reduction in net photosynthesis. This is puzzling and deserves further investigation.

Most research on the effects of light on stomatal behavior has been done with relatively constant light intensity. However, leaves within canopies and on the forest floor are commonly exposed to rapid fluctuations in light intensity, known as sunflecks. Pearcy (1990) reviewed the literature on the effects of sunflecks on photosynthesis. Cardon *et al.* (1994) reported that oscillations in light intensity cause stomatal conductance to vary widely, either above or below that in steady light, depending on the frequency of the oscillations. They also reported that oscillations in the carbon dioxide concentration of the atmosphere, such as those caused by air turbulence, can cause oscillations in stomatal conductance. The ecological significance of such oscillations probably deserves more study.

Carbon Dioxide

The stomatal aperture of many kinds of plants is approximately inversely proportional to the CO_2 concentration in both light and darkness, increase in CO_2 in the intercellular or ambient air causing closure and decrease causing opening. According to Mott (1988), stomata respond only to changes in the intercellular CO_2 concentration, but this is affected by the external concentration. Ball and Berry (1982) suggested that the ratio of internal to external concentration of CO_2 is important in controlling stomatal aperture. Rogers *et al.* (1983) reported that the stomatal conductance of corn, soybean, and sweet gum decreased about 50% when the concentration of CO_2 in open topped outdoor chambers was increased from ambient to 910 ppm, but photosynthesis of soybean and sweet gum increased 65 to 70%. There was no increase in photosynthesis of corn which has the C_4 carbon pathway that is saturated at a low CO_2 concentration. The reaction to CO_2 seems to vary with light intensity, temperature, humidity, and presence of ABA (Raschke, 1986). However, the stomata of some conifers seem less responsive to CO_2 than those of most other plants (Jarvis in Turner and Kramer, 1980). The complex relations of stomatal behavior to CO_2 are discussed in detail by Mansfield *et al.* (1990, pp. 61–67). The effects of CO_2 on photosynthesis and stomatal behavior also were discussed in Chapter 4 of Lemon (1983) and by Raschke (1986). As CO_2 concentration affects photosynthesis and the latter affects stomatal aperture, there may be indirect effects through photosynthesis on stomatal conductance. However, the exact mechanism by which CO_2 affects guard cells remains uncertain (Kearns and Assmann, 1993). Peñuelas and Matamala (1990) reported that the stomatal density and nitrogen content of leaves have decreased since the CO_2 concentration of the atmosphere has increased. This is based on a study of leaves of various ages preserved in herbaria.

Humidity

It was long assumed that the midday closure of stomata was caused by loss of leaf turgor, but it has now been demonstrated that exposure of the epidermis to dry air also causes closure in at least some kinds of plants (Hirasawa *et al.*, 1988; Lange *et al.*, 1971; Schulze *et al.*, 1972; Sheriff, 1977b). Hall and Kaufmann (1975) found closure of stomata of sesame and Kaufmann (1976) found closure of spruce stomata in dry air while Lawlor and Milford (1975) reported that stomata of sugar beets could be kept open in humid air even when a water deficit existed in the leaves. However, some plants show less response to humidity; for example, camellia and privet (Wilson, 1948) and *Atriplex halimus* and *Kochia* (Whiteman and Koller, 1964). It is uncertain how much of the difference in response reported by various investigators is intrinsic and how much is related to previous treatment and differences in experimental methods. Mansfield (1986, pp. 194–202) discussed the effects of humidity in detail and emphasized Meidner's idea that the effect of humidity on evaporation from guard cells is important. However, Nonami and Schulze (1989) found that the water potential of the mesophyll cells in transpiring leaves was lower than that of epidermal cells. This suggests that transpiration from epidermal and guard cells is less important than sometimes claimed.

Aphalo and Jarvis (1991) found that stomatal conductance is better correlated with a vapor pressure deficit than with humidity in *Hedera helix*, and Assmann and Grantz (1990) found a similar situation in sugarcane and sorghum. Of course the vapor pressure gradient is greatly affected by temperature, as shown in Fig. 7.3. Kaufmann (1982) found a good correlation between stomatal conductance and the difference in absolute humidity between air and leaves in subalpine forest trees. In fact, he concluded that temperature effects on stomata in those trees could be explained fully in terms of photosynthetic photon flux density and the difference in absolute humidity between leaves and air, except following freezing nights or during severe water deficiency. However, Mott and Parkhurst (1991) concluded that stomata do not respond directly either to the absolute humidity at the leaf surface or to the difference in vapor pressure between the interior and exterior of leaves, but to the rate of transpiration.

Schulze (1986b) discussed the effects of dry soil and dry air on stomatal behavior; the role of signals from roots is discussed in the section on internal factors later in this chapter. Sojka and Stolzy (1980) found that oxygen deficiency in wet soil causes stomatal closure and speculated that closure caused by deficient soil aeration may be as important as that caused by deficient moisture.

Temperature

The effects of temperature on stomatal aperture and the rate of response to stimuli vary among different kinds of plants (Meyer and Anderson, 1952). It is difficult to separate direct effects of temperature from indirect effects caused by larger vapor pressure deficits associated with increasing temperature (see Chapter 7 and Fig. 7.3). Temperature effects were so small in Kaufmann's (1976, 1982) study of subalpine forest trees that he could disregard them except near freezing. In contrast, Wuenscher and Kozlowski (1971) found that in five Wisconsin tree species stomatal aperture decreased and stomatal resistance increased significantly as the temperature increased from 20 to 40°C; the change was the greatest for trees that normally grow on dry sites. However, the vapor pressure deficit also varied in this study. Pereira and Kozlowski (1977) found considerable interaction between light intensity and temperature on stomatal aperture of sugar maple, as shown in Fig. 8.5. Wilson (1948) found that the responses of

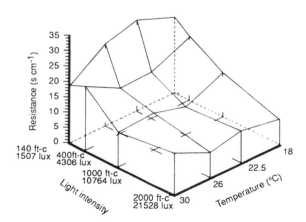

Figure 8.5 Interacting effects of light and temperature on stomatal resistance of sugar maple (*Acer saccharum*). From Pereira and Kozlowski (1977), by permission of the authors and the *Canadian Journal of Forest Research*.

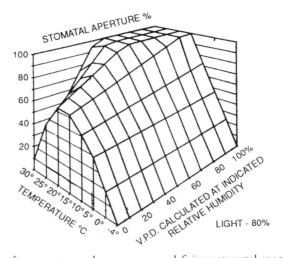

Figure 8.6   Effect of temperature and vapor pressure deficit on stomatal aperture of camellia at 60% of full sun. From Wilson (1948).

stomata of camellia and privet to light and humidity were slowed and that stomatal resistance increased at low temperatures, as shown in Fig. 8.6.

There are important differences in the stomatal reaction to low temperature among plants of various species; there is no increase in the stomatal resistance of collards (*Brassica oleracea,* var. *viridis*) at 5°C, but there is a large increase in cotton and bean (McWilliam *et al.,* 1982). A slow closure at low temperatures was observed in cotton and bean (McWilliam *et al.,* 1982), but not in soybean at 10°C (Musser *et al.,* 1983). An increased opening was observed at low temperatures in garden bean (*Phaseolus vulgaris*) but a pre-exposure to cool temperatures eliminated the effect (Wilson in Raper and Kramer, 1983), indicating that stomata can acclimate to low temperature conditions.

Wind

The effect of wind on transpiration was discussed in Chapter 7, and the varying effects on transpiration of plants of different species are shown in Fig. 7.5. This probably results at least in part from varying effects on stomata and in part from effects on the boundary layer conductance. It also affects the latent heat flux from leaves. Although there have been few studies on the effect of wind on stomatal conductance, it appears that it usually decreases with increasing wind velocity (Burrows and Milthorpe in Kozlowski, 1976, pp. 126–127). Davies *et al.* (1974) reported that stomata of American ash and sugar maple had lower conductances at wind speeds of 0.6 to 2.7 m sec^{-1} than in quiet air, and students sometimes find that when they turn a fan on plants to increase transpiration it actually is decreased by stomatal closure. According to Kitano and Eguchi (1992b), a sudden increase in wind velocity caused strong cycling in absorption, transpiration, and stomatal conductance of cucumber plants in bright light, but less in low light and none in darkness. It also has been reported that shaking plants or branches causes stomatal closure, and both dehydration and shaking may be involved in decreased stomatal conductance in wind. Perhaps this problem deserves more research.

Mineral Nutrition

There seems to be some uncertainty concerning the role of mineral nutrition in respect to stomatal behavior, perhaps partly because of lack of research on the problem. Desai (1937) reported that deficiency of nitrogen, phosphorus, and potassium all reduced stomatal responsiveness in several kinds of plants, and Pleasants (1930) reported that nitrogen deficiency reduced stomatal response to water deficit and resulted in increased transpiration in bean seedlings. In contrast, Radin and Parker (1979b) found that nitrogen deficiency caused stomata of water-deficient cotton plants to close sooner. However, further research by Radin and Boyer (1982) indicated that nitrogen deficiency increases root resistance of cotton and the resulting leaf water deficit probably was the cause of early stomatal closure. Radin (1984) found that the stomata of phosphorus-deficient cotton plants closed before the leaf mesophyll cells lost their turgor and suggested that phosphorus deficiency might affect the balance between ABA and cytokinins. Calcium appears to be involved as a second messenger in signal transduction in various events triggered by red light and involving phytochrome in plant cells (Hepler and Wayne, 1985), and probably is involved in stomatal reactions (Mansfield *et al.*, 1990; McAinsh *et al.*, 1990). There is further discussion of the relationship between mineral nutrition and stomatal behavior in Chapter 10.

Stomata and Air Pollution

The effect of air pollutants such as SO_2, ozone, and fluorides on stomata is important because stomata are the pathway for the entrance of most gaseous pollutants into leaves. There is likely to be much less leaf injury from fumigation with SO_2 or ozone when stomata are closed than when they are open (see references in Omasa *et al.*, 1985a; Kozlowski *et al.*, 1991, pp. 365–366). Low concentrations of SO_2 are said to cause stomatal opening in moist air, but not in dry air (Mudd in Mudd and Kozlowski, 1975), but this may not be true for all species. Omasa *et al.* (1985a) found that 1.5 μl of SO_2 per liter of air caused patchy stomatal closure and tissue injury to sunflower leaves, and there was wide variation in the degree of closure among stomata in various parts of leaves. Olszyk and Tingey (1986) found ozone twice as effective as SO_2 in causing stomatal closure in pea and saw evidence of synergistic effects between the two. In preliminary experiments on young beech trees, Pearson and Mansfield (1993) found that ozone causes stomatal closure on well-watered trees, but keeps them open on water-deficient trees. According to Martin *et al.* (1988), data from both short- and long-term experiments suggest stomatal limitation of photosynthesis in polluted air.

Occasionally, stomata are plugged by particulate material such as dust and soot. This sometimes occurs on foliage near sources of dust such as the volcanic explosion of Mt. St. Helens, cement plants, and along dusty roads. Injurious accumulations of smoke and soot are said to occur on evergreen foliage in cities, but Rhine (1924) reported that stomata of at least some conifers are partially plugged by naturally occurring wax that is not related to smoke injury. This is corroborated by observations of Gambles and Dengler (1974) and Jeffree *et al.* (1971). Some conflicting evidence on the importance of particulate material is presented in Mudd and Kozlowski (1975).

Stomata and Fungi

Arntzen *et al.* (1973) reported that a toxin produced by *Helminthosporium maydis* causes rapid closure of stomata on leaves of corn plants containing the Texas male-sterile gene, possibly by inhibiting the uptake of K^+ by guard cells. However, fusicoccin slows stomatal closure of alfalfa, and Turner (1970) suggested that the drying of alfalfa hay might be hastened by spraying alfalfa with fusicoccin a few hours before cutting. Some effects of pathogenic organisms are discussed by Ayres in Jarvis and Mansfield (1981).

Internal Factors Affecting Stomata

Guard cell behavior and stomatal aperture are affected by internal factors such as leaf water status, internal CO_2 concentration, and growth regulators, especially ABA and cytokinins. As indicated earlier, changes in stomatal conduction seem to be correlated with changes in the rate of photosynthesis or at least in photosynthetic capacity (Wong *et al.*, 1979). According to McCain *et al.* (1988), NMR spectroscopy indicated that chloroplasts in sun leaves contain less water than those of shade leaves, but the significance of this is unknown.

Formerly it was supposed that loss of turgor in leaf cells was the principal cause of the midday closure of stomata observed on dry, sunny days. However, experiments involving split root systems and use of pressure to keep the shoots turgid, although part of the root system is water deficient (Davies and Zhang, 1991; Davies *et al.*, 1994; Gollan *et al.*, 1986), indicate that signals from drying roots can cause stomatal closure even in turgid shoots. The signals might include decreases in amino acids, ions, and cytokinins, but an increase in ABA probably is the chief signal. Generally the ABA concentration increases in roots of water-deficient plants, and Davies and Zhang (1991) and Davies *et al.* (1994) discuss in detail the possible role of roots as detectors of increasing soil water deficit and sources of chemical signals to the shoots that can modify or override the effects of shoot water status. The failure of stomata to close in "wilty" tomatoes (Tal, 1966) was attributed by Livne and Vaadia (in Kozlowski, 1972) to their inability to synthesize ABA. Stomata often fail to reopen immediately after water-deficient plants are rewatered (Fischer, 1970) and this has been attributed to persistence of a high concentration of ABA, although experiments of Beardsell and Cohen (1975) and Harris and Outlaw (1991) led them to question this. On the other hand, cytokinins promote opening and interact with ABA.

It seems possible, as Trewavas (1981) suggested, that too much emphasis is being placed on the role of ABA. Closure of stomata is not well correlated with the ABA content of leaves in all plants (Ackerson, 1980) and stomata sometimes remain open in leaves high in ABA or stay closed or partly closed after the ABA concentration has decreased (Beardsell and Cohen, 1975). Trejo *et al.* (1993a) reported that ABA is rapidly metabolized in mesophyll tissue, and this probably has contributed to the uncertainty concerning the role of ABA in stomatal closure. Stomatal sensitivity to ABA also varies with the water status of the tissue. Munns and King (1988) suggest that some compound in the xylem sap in addition to ABA must be involved in stomatal closure in wheat. It also is difficult to reconcile the midday closure and late afternoon reopening of stomata of plants in moist soil on hot, sunny days with control by ABA from roots (Kramer, 1988).

In their review, Mansfield *et al.* (1990) suggest that calcium ions may also affect stomatal aperture, as Ca^{2+} ions reduce it on epidermal strips. It is even suggested that such varied stimuli as darkness, ABA, and cytokinins use calcium ions as second messengers. Ehret and Boyer (1979) reported that large losses of K^+ occur from guard cells of leaves on slowly dehydrated plants and suggested that this contributes to loss of guard cell turgor in water-deficient plants.

ANOMALOUS BEHAVIOR OF STOMATA

Stomata do not always behave as expected. It was mentioned earlier that opening and closing sometimes continue for several days after plants are placed in continuous darkness and they sometimes cycle during the day. Also, not all of the stomata on a leaf behave in the same manner. A few examples of anomalous behavior will be discussed.

Cycling

Under some circumstances stomata show cycling or oscillation between the open and closed condition, as shown in Fig. 8.7. Cycling occurs most commonly

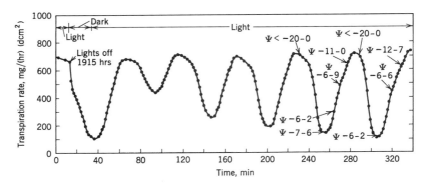

Figure 8.7 Cyclic variation in transpiration and water potential of cotton leaves, caused by stomatal cycling initiated by a 20-min dark period at 3:15 PM. The cycling was attributed to oscillations in water potential caused by high root resistance in rapidly transpiring plants. From Kramer (1983), after Barrs and Klepper (1968).

in water-stressed plants with relatively high root resistance and can be initiated by a sudden shock such as a short period of darkness, cooling the soil, or changes in humidity or temperature. The cycling has a periodicity ranging from minutes to hours, but most often occurs in the range of 15 to 120 min. It has been observed in several kinds of herbaceous plants (Barrs, 1971; Barrs and Klepper, 1968; Kitano and Eguchi, 1992b; Shiraishi *et al.*, 1978), and Levy and Kaufmann (1976) observed it in citrus trees in an orchard. Shiraishi *et al.* (1978) observed cyclic variation in tobacco leaves with a scanning electron microscope. Various leaves on a plant can be at different phases of the oscillation cycle at the same time. Raschke (1975) suggested that it might occur when a negative feedback signal is delayed and reaches the guard cells at a time such that it reenforces the initial response instead of counterbalancing it. Cowan (1972) proposed that cycling optimizes the conflicting requirements for carbon dioxide uptake and control of water loss.

As mentioned earlier, there also are endogenous rhythms that cause stomata to open and close for several days after plants are moved to continuous darkness. These anomalies increase the difficulty of developing a general theory of stomatal behavior. The possible occurrence of short-term cycling and daily and seasonal rhythms should be considered in research on stomatal behavior.

Heterogeneity in Stomatal Response

The stomata on the upper and lower surfaces of leaves sometimes behave differently, as shown in Fig. 8.8. One might expect most or all of the stomata on one surface of a leaf to respond similarly to a given stimulus, but this is not always true. For example, when ABA was supplied to detached leaves through

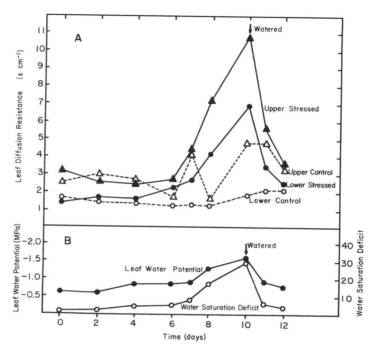

Figure 8.8   (A) Diffusion resistances of upper (adaxial) and lower (abaxial) surfaces of the seventh and eighth leaves of water-deficient and control maize, measured in the morning with a diffusion porometer. The plants were grown in a controlled environment chamber. (B) Leaf water potential and saturation deficit of water-deficient maize plants. From Kramer (1983), after Sanchez-Diaz and Kramer (1971).

the petioles, stomatal closure occurred in patches over the leaf surface and similar patchiness occurred in plants of other species when the endogenous ABA concentration was increased by water deficit (Downton *et al.*, 1988a; Terashima *et al.*, 1988). This heterogeneity in stomatal behavior might result either from an uneven distribution of ABA to different parts of the leaf or from differences in sensitivity of guard cells in various parts of the leaf. The margins of leaves tend to dry out sooner than the central regions (Cook *et al.*, 1964; Hashimoto *et al.*, 1984), causing early closure of stomata in the margins. The relationship between patchiness in stomatal opening and photosynthesis is discussed in more detail in Chapter 10.

The occurrence of heterogeneity or "patchiness" in stomatal behavior suggests the need for observation of leaves under a microscope to supplement porometer measurements of average leaf conductance (Omasa *et al.*, 1985a) to determine if patchiness occurs. Laisk (1983), Terashima *et al.* (1988), and

Mansfield *et al.* (1990, pp. 67–70) suggested that heterogeneity in stomatal closure causes errors in estimating internal CO_2 concentration, explaining observations that carbon fixation sometimes decreases, although the average internal CO_2 concentration (c_i) does not appear to be limiting. However, Cheeseman (1991) claims models show this explanation to be untenable. Also, Wise *et al.* (1992) found that although "patchiness" occurs in unstressed, chamber-grown cotton and sunflower plants, it is uncommon under field conditions. Apparently the importance of stomatal heterogeneity deserves further study.

The reaction of stomata to environmental factors varies with the age of leaves and their past treatment and it often is difficult to determine how much of the difference among experimental results is intrinsic and how much is caused by differences in methods, age, and previous treatment of the experimental plants. As leaves grow older the stomata often become less responsive and may open only partly, even at midday (Brown and Pratt, 1965; Slatyer and Bierhuizen, 1964a), and Tazaki *et al.* (in Turner and Kramer, 1980) reported that stomata on older leaves of mulberry do not close. According to Ackerson and Kreig (1977), although stomata of corn and sorghum close when they are water deficient during the vegetative stage, they do not close in similar conditions in the reproductive stage. However, Longstreth and Kramer (1980) found no significant change in leaf conductance during flower induction or flowering of cocklebur or soybean. It also has been reported that stomata of cotton grown in greenhouses and controlled environment chambers close at higher leaf water potentials than those of field-grown plants (Davies, 1977; Jordan and Ritchie, 1971). Schulze and Hall (1982) reviewed the effects of environmental factors on stomatal behavior and urged that short- and long-term effects be differentiated. Meinzer and Grantz (1991) discussed the coordination between stomatal conductance in sugarcane and the capacity of soil, roots, and stems to supply water, resulting in a relatively constant leaf water status over a wide range of plant sizes and growing conditions. They observed a decrease in stomatal conductance that compensated for the increase in leaf area, in their opinion probably controlled by signals from the roots. The effects of age and past treatment probably account for many of the differences in behavior found in the literature and increase the difficulty in generalizing about stomatal behavior.

OPTIMIZATION

There have been several attempts to develop theories explaining stomatal behavior in terms of optimization of cost to benefit, cost being a reduction in the uptake of CO_2 used in photosynthesis and benefit being a reduction in water loss. Optimum efficiency should occur when the stomatal aperture varies during a day in such a manner that there is minimum transpiration (E) for maximum photosynthesis (A) and dE/dA is constant. Wong *et al.* (1979) claim that in at least some plants there is a feedback mechanism relating stomatal aperture to photosynthetic capacity of the mesophyll tissue and maintaining a fairly constant ratio of internal to external concentration of CO_2. The optimization principle was discussed in detail by Cowan (1982), by Farquhar and Sharkey (1982), and in Givnish (1986). However, Sandford and Jarvis (1986) cited some contradictory findings and reported that dE/dA is not constant in all woody plants. Bunce *et al.* (1977) question if an optimal stomatal response saves much water for plants in stands.

DIFFUSIVE CAPACITY OF STOMATA

The research of Brown and Escombe (1900), Sayre (1926), and Ting and Loomis (1965) establish that *in quiet air* diffusion through small pores such as stomata is better related to their circumference or perimeter than to their area. Because of their more or less elliptical shape, stomata have a high diffusive capacity even when only partly open. This situation is largely academic, however, because it applies only in quiet air where the boundary layer resistance is greater than the stomatal resistance. In nature, leaves usually are exposed to wind, decreasing the boundary layer resistance relative to the stomatal resistance and transpiration increases rapidly with increasing pore area, as shown in Fig. 8.9. Nobel (1991, Chapter 8) has a discussion of stomatal diffusion.

BULK FLOW IN LEAVES

Thus far we have assumed that all movement of gas in and out through stomata is by diffusion, but this may not be true. It is claimed that the bending of leaves in the wind and temperature changes due to passing clouds and intermittent shading cause bulk flow of gas through the stomata. Shive and Brown (1978) reported that the fluttering of eastern cottonwood leaves increases gas exchange. However, Rushin and Anderson (1981) reported that leaf fluttering has little effect on stomatal conductance. It seems possible that wind flowing over the surfaces of leaves or sudden changes in temperature can cause gas flow through the intercellular spaces of amphistomatous leaves. Vogel (1981, p. 83) observed such flow in water-filled leaves injected with a dye solution and immersed in a flow tank.

MEASUREMENT OF STOMATAL APERTURE AND CONDUCTANCE

Because of the importance of stomata in controlling water loss and CO_2 uptake, there has been much interest in the measurement of stomatal aperture or

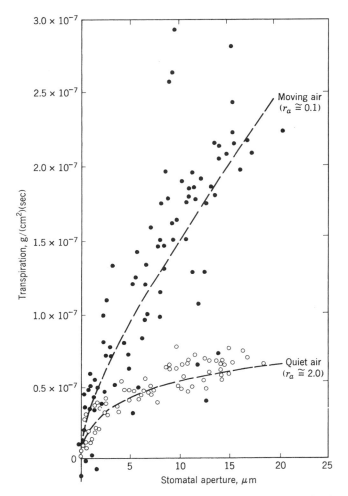

Figure 8.9 The effect of increasing stomatal aperture on transpiration rate of Zebrina leaves in quiet and moving air. The effect of increasing aperture is very large in moving air, but small in quiet air where the boundary layer resistance is relatively large compared to stomatal resistance. From Kramer (1983), after Bange (1953).

the conductance of stomata. Much relevant literature has been reviewed by Weyers and Meidner (1990) and Smith and Hollinger (in Lassoie and Hinckley, 1991), and only a few methods are discussed briefly.

Some investigators have expressed the stomatal opening in terms of conductance, others as resistance which is the reciprocal of conductance, resulting in hyperbolic curves. Burrows and Milthorpe (in Kozlowskki, 1976, pp. 106–107) discussed the problems resulting from the use of resistance, but agreed that it has advantages in dealing with the complex pathway between plants and the atmosphere. The use of resistances and conductances also is discussed in Nobel (1991, Chapter 8).

Visual Observations

Most of the early observations were made by stripping off bits of epidermis and fixing them in absolute alcohol before observing them under the micro-scope. This method seems to have been introduced by Lloyd (1908) and was used for the classical studies of Loftfield (1921), but it is difficult to strip epidermis from some leaves and stripping sometimes causes change in stomatal aperture. Another method is to make impressions of the epidermis of attached leaves in collodion (Clements and Long, 1934), silicone rubber (Zelitch, 1961), or dental paste (Kuraishi et al. in Hashimoto et al., 1990; Weyers and Meidner, 1990, p. 111), or even Duco cement or nail polish, strip them off, and examine them under the microscope. However, Pallardy and Kozlowski (1980) warned that epidermal impressions can be unreliable if cuticular ledges develop on the inner surfaces of guard cells, as shown in Fig. 7.8A. Shiraishi et al. (1978) studied variations in stomatal aperture under the microscope and reported that the aperture of ledges often changes independently of the apparent pore aperture, leading to misinterpretation of effective pore aperture.

Modern technology has made better visual observations of stomata possible. For example, scanning electron microscopy reveals details of stomatal structure not previously observed (Shiraishi et al., 1978). Hashimoto (in Hashimoto et al., 1990), Hashimoto et al. (1984), and Omasa et al. (1985a; and in Hashimoto et al., 1990) described use of a remote-controlled light microscope, television camera, and image processing system to observe the effects of stresses such as water deficit and air pollutants on stomata in different parts of a leaf. This method showed that stomata in different parts of a leaf sometimes responded differently, as mentioned in the section on Heterogeneity in Stomatal Response. They concluded that visual observation under a microscope is desirable to observe the effects of air pollutants on stomata in different parts of a leaf. Chlorophyll fluorescence also is useful in the study of guard cell metabolism (Cardon and Berry, 1992; Ogawa et al., 1982; Outlaw et al., 1981).

Infiltration

This method was popularized by Molisch (1912). It depends on measuring the time required for the infiltration of leaves by liquids of various viscosities. Alvim and Havis (1954) used mixtures of paraffin oil and n-dodecane whereas investigators in Israel used mixtures of paraffin oil and turpentine or benzol, or paraffin oil (kerosene) alone. Fry and Walker (1967) described a pressure infiltration method for use on pine needles. This method clearly indicates whether stomata are open or closed, but the toxic fluids used kill the tissue infiltrated, the infiltration depends on wettability by the solvent, and the method does not permit calculation of stomatal resistances (Lassoie *et al.*, 1977a).

Porometers

Visual methods have been largely supplanted by porometers that measure the movement of gas out of leaves in such a manner that readings can be converted into diffusion resistance of the leaf in sec · m $^{-1}$ or its reciprocal conductance in m · sec $^{-1}$. This method measures total leaf resistance, but if cuticular resistance is high it gives a good approximation of stomatal resistance. There are two types of porometers: pressure or bulk flow, and diffusion.

Pressure or Bulk Flow Porometers.   Francis Darwin, the plant physiologist son of Charles Darwin, is credited with developing the porometer and Darwin and Pertz (1911) described a simple pressure flow porometer. Numerous modifications have been made, including recording porometers (Gregory and Pearse, 1934; Wilson, 1947) and portable models such as that developed by Alvim (1965). Fiscus *et al.* (1984) and Fiscus (in Hashimoto *et al.*, 1990) described a viscous flow porometer that is computer controlled and can be used to schedule irrigation of a crop. Viscous flow porometers are not suitable for use on leaves with stomata on only one surface (many trees) and with bundle extensions that extend to the epidermis and prevent free lateral movement of gases through the intercellular spaces. Also a pressure greater than 10 cm of water can cause alterations in stomatal aperture (Raschke, 1975). Meidner (1992) discussed some problems with bulk flow porometers. However, these types of porometers have the advantage of being highly responsive to changes in aperture and indicate aperture effects rather than diffusion through the leaf.

Diffusion Porometers.   These usually are small cuvettes that can be attached to leaves that measure the time required for a predetermined change in humidity to occur. They have undergone many modifications, but most of the earlier ones used a cuvette containing a humidity sensor connected to a meter to read humidity and a timing device. Many are based on the instrument described by Kanemasu *et al.* (1969) with various modifications to increase speed and accuracy. Beardsell *et al.* (1972) introduced the null point porometer that measures the steady-state rate of transpiration of a leaf enclosed in a cuvette into which dry air is blown to maintain a constant humidity. This eliminates some calibration problems and the lag caused by adsorption of water vapor on cuvette walls. Kaufmann (1981; and in Hashimoto *et al.*, 1990) used ventilated cuvettes, approximately 15 liters in volume and large enough to enclose leafy twigs. They were closed from time to time long enough to measure transpiration. A computer controlled opening and closing and data acquisition and processing. With this apparatus Kaufmann was able to estimate canopy transpiration from measurements of leaf conductance. The use of diffusion porometers has been reviewed by Smith and Hollinger (in Lassoie and Hinckley, 1991).

One of the limitations on use of diffusion porometers is that they average the behavior of stomata in a relatively large area of leaf surface, but provide no information concerning the behavior of individual stomata (Omasa *et al.*, 1985a) or differences in different parts of leaves (Hashimoto *et al.*, 1984). Thus, for some purposes visual observations may be desirable (Omasa *et al.*, 1985a; and in Hashimoto *et al.*, 1990). However, the data provided by larger samples generally are most useful. Diffusion porometers are affected by any factor controlling water loss by leaves; stomatal aperture is only one. Differences in cuticle thickness, leaf thickness, intercellular space dimensions, and other factors can contribute to diffusion differences but do not indicate differences in stomatal aperture. Care needs to be used when interpreting data from diffusion porometers.

SUMMARY

Stomata are pores in the epidermis whose aperture is controlled by pairs of specialized epidermal cells called guard cells. They provide passageways between the ambient air and the air spaces in photosynthetic tissue essential for the entrance of the CO_2 used in photosynthesis. Unfortunately, they also provide pathways for the exit of water vapor, subjecting plants to the danger of excessive dehydration in sunny, dry weather. Thus stomatal aperture plays an important role in controlling both transpiration and the CO_2 supply for photosynthesis.

Stomata usually open in the light and close in darkness. However, there are exceptions to this generalization. For example, the stomata of plants with CAM tend to remain closed in light and open in darkness. Opening of stomata is caused by an increase in turgor of guard cells caused largely by the uptake of K^+ and accumulation of its counterions, and closing results from the loss of K^+ and the consequent loss of turgor. The accumulation of K^+ is driven by energy from metabolism often involving photosynthesis or respiration, but direct effects of particular wavelengths of light indicate that photosynthesis is not the sole source of light effects on the process. Guard cells usually contain chloroplasts and most of the metabolic apparatus of photosynthesis but they can also fix CO_2 into malate which acts as a counterion for K^+. Abscisic acid exerts important control over stomatal opening by controlling the ability of the guard cells to accumulate K^+, and high abscisic acid appears to cause stomatal closure in water-deficient plants by this mechanism.

Changes in guard cell turgor are affected by environmental factors such as light intensity, carbon dioxide concentration, humidity, and temperature and by internal factors such as internal carbon dioxide concentration, leaf water status, and the concentration of growth regulators, especially abscisic acid. The complex interactions among the factors often make it difficult to determine their relative importance. There also is evidence that chemical signals from roots affect stomatal behavior, especially in drying soil.

Occasionally, stomata exhibit anomalous behavior such as cycling in a constant environment. They sometimes show a "patchy" response, with stomata in some areas on a leaf closing while those in other areas remain open. The stomata on the upper and lower surfaces of leaves often behave differently.

The number of stomata per mm^2 of leaf area varies from 60 to 80 on corn leaves to over 1000 on scarlet oak. The size tends to decrease with increasing frequency, but the area of pore space when wide open usually amounts to 1% or more of the leaf area. The stomatal aperture can be observed visually on epidermal strips fixed in alcohol or by applying collodion or silicone rubber to leaves, stripping it off, and examining the impressions under a microscope. Direct visual observations can be made with special cameras and imaging equipment. Rapid estimates of stomatal aperture can be made from the rate of infiltration of liquids of various viscosities. However, most observations are now made with porometers which permit estimates of stomatal conductance.

SUPPLEMENTARY READING

Carlson, T. N., ed. (1991). Modeling stomatal resistance. *Agric. For. Meteorol.* **54**, 103–388.

Cowan, I. R. (1982). Regulation of water use in relation to carbon gain in higher plants. *In* "Encyclopedia of Plant Physiology" (O. L. Lange and J. D. Bewley, eds.), Vol. 12B, pp. 535–562. Springer-Verlag, Berlin.

Givnish, T. J. (1986). Optimal stomatal conductance, allocation of energy between leaves and roots, and the marginal cost of transpiration. *In* "On the Economy of Plant Form and Function" (T. J. Givnish, ed.), pp. 171–213. Cambridge University Press, Cambridge.

Jarvis, P. G., and Mansfield, T. A., eds. (1981). "Stomatal Physiology." Cambridge University Press, London.

Mansfield, T. A. (1986). The physiology of stomata: New insights into old problems. *In* "Plant Physiology" (F. C. Steward, J. E. Sutcliffe, and J. E. Dale, eds.), Vol. 9, pp. 155–224. Academic Press, Orlando, FL.

Mansfield, T. A., Hetherington, A. M., and Atkinson, C. J. (1990). Some current aspects of stomatal physiology. *Annu. Rev. Plant Physiol. Plant Mol. Biol.* **41**, 55–75.

Meidner, H., and Mansfield, T. A. (1968). "Physiology of Stomata." McGraw-Hill, London.

Ting, I. P. (1985). Crassulacean acid metabolism. *Annu. Rev. Plant Physiol.* **36**, 595–622.

Weyers, J. D. B., and Meidner, H. (1990). "Methods of Stomatal Research." Longman, Scientific & Technical, U.K.

Zeiger, E., Farquhar, G. D., and Cowan, I. R., eds. (1987). "Stomatal Function." Stanford University Press, Stanford, CA.

9

Ion Transport
and Nitrogen
Metabolism

INTRODUCTION

Plants require inorganic salts that are absorbed from solutions in contact with the plant. The management of their uptake and use is an important part of agriculture and is treated in several books (Epstein, 1972; Marschner, 1986; Nye and Tinker, 1977). This chapter provides an overview of salt uptake with emphasis on how plant water status affects it and how the mineral nutrients act as signals for enzyme change during dehydration.

The salts are required in growth because they contain chemical elements that together with the products of photosynthesis become part of the structures, enzymes, and metabolites of the cell. The elements are needed in various amounts and those used in greatest quantity are N, P, K, Ca, Mg, and S and are termed macronutrients while others used in minor quantity are Fe, B, Mn, Cu, Zn, Mo, and Cl and are termed micronutrients. Each element plays a particular role that determines its requirement by the plant. Some provide elemental constituents for macromolecules that function as enzymes or part of cell structures. Others provide cofactors for enzyme activity and thus play a regulatory role. In every case, an inadequate supply reduces growth and can prevent reproduction if the deficiency becomes severe.

For primitive plants in the marine environment, the surrounding water provides a ready supply of most salts, and the dissociated ions are simply absorbed from the surrounding solution by crossing the plasmalemma. The surrounding solution generally is kept uniform by the stirring from water currents. On land, salts are in the soil solution, and uptake and transport require specialized organs, the roots, to supply the rest of the plant. In general, the soil restricts ion movement compared with the marine environment because the soil structure prevents stirring. Large gradients in concentration can form next to the root. The ions diffuse from place to place and because roots often absorb large quantities of water, there is some bulk flow of soil solution toward roots. Some ions such as phosphate form insoluble complexes with soil minerals and others such as potassium bind to soil particles.

The concentrations of ions are markedly different in the ocean and in soil (Epstein, 1972). Sodium and chloride dominate in the ocean (nearly 500 mM for each), but Fig. 9.1 shows that the macronutrients calcium, magnesium, potassium, and sulfate also are present in considerable quantity (10–56 mM). In the soil, these nutrients are normally present at concentrations below 10 mM (Fig. 9.1). In contrast, nitrate and phosphate are barely detectable in most marine environments but occur in higher concentrations in soil solutions. Thus, plants in marine environments enjoy higher availabilities of most nutrients than land plants do except for phosphate and nitrate where the reverse is true.

The ability of roots to supply nutrients is enhanced by various symbioses, in particular by the mycorrhizal relationships that form between roots and fungi (see Chapter 5) and by nitrogen-fixing bacteria (see later). Mycorrhizal fungi surround roots or penetrate them and extend out into large soil volumes. The root/fungus system is particularly important for solubilizing phosphate from its insoluble complexes with iron, calcium, and organic matter, allowing transport to the plant (Gerdemann, 1968; Gerdemann in Torrey and Clarkson, 1975; Harley and Smith, 1983). Nitrogen fixation is carried out by free-living bacteria and by symbiotic bacteria such as in legumes. The N$_2$ of the atmosphere serves as the substrate and the bacteria supply reduced N to the plant. Together these symbioses further enhance the supply of phosphorus and nitrogen to land plants.

The supply of salts is more stable in the ocean than in soil solutions. Marine concentrations become variable mostly in coastal waters where they are affected by runoff from the land. Soil concentrations depend on the parent material, degree of weathering, and amount of water and organic matter. This complexity in the soil environment is illustrated by the different soil types occupying the global land area. Each can be classified according to its suitability for growing plants (Dudal, 1976). Table 9.1 shows that 22% of the world land area has soils with mineral nutrient problems. Another 28% is too dry and 24% too shallow (subject to frequent drying). Flooding affects 12% and 3% is too rocky. About 15% of the land area is too cold. Only 10% of the global land area is relatively free of unfavorable characteristics and those soils are our richest farmlands.

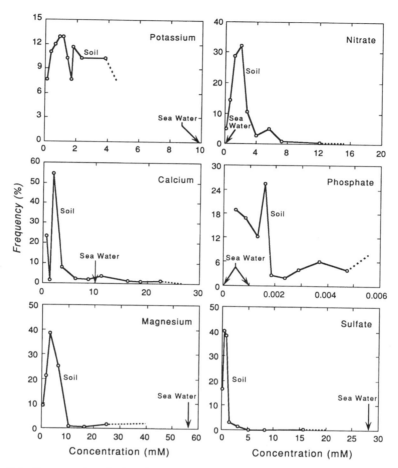

Figure 9.1   Concentration of mineral nutrients in soil solution and seawater. The frequency with which various concentrations occur is shown for each range of concentrations. The number of soils sampled was 149 to 979 depending on the element. The dashed lines indicate that a few soils had concentrations above those shown. Seawater concentrations are stable except for phosphate and nitrate, which vary in the range shown. After Epstein (1972).

ION UPTAKE AND TRANSPORT

Optimum Conditions

Almost any ion can enter plant cells in small quantity, but nutrient ions are accumulated selectively by specific mechanisms, often against an electrical or concentration gradient. Metabolic energy often must be spent for some aspect

Table 9.1   Area of Total World Land Surface
Subject to Environmental Limitations
of Various Types[a]

Environmental limitation	Area of world soil subject to limitation (%)
Drought	27.9
Shallow soil	24.2
Mineral excess or deficiency	22.5
Flooding	12.2
Miscellaneous	3.1
None	10.1
Total	100.0
Temperature	14.8

Note. Area affected by unfavorable temperatures overlaps with other classifications and is shown separately.
[a] From Dudal (1976).

of their uptake, and the expenditure can be large. For example, Pate *et al.* (1979) reported that roots of plants deprived of nitrate salts for 10 days respired at only 71% of the rate for roots supplied with nitrate, presumably because less energy was used for ion uptake in the starved roots. Transport across the plasmalemma and tonoplast often involves specific energy-driven, molecular carriers. The energy is consumed primarily by ATPases that pump protons across the membrane. The resulting proton gradient is then dissipated when the protons return to the original side of the membrane through carriers coupled to the transport of another ion such as potassium. Because the carrier is specific for the other ion, uptake selects for each ion separately. The cells are thus actively involved in accumulating the ions that provide elemental nutrients, and the rate of uptake depends not only on the concentration of the ion in the extracellular solution, but also on the carriers and energy available for transport.

Plants containing a vascular system dump the ions into the root xylem where they are swept along in the transpiration stream at concentrations that depend on the rate of water entry into the xylem. When water enters rapidly, ion concentrations can become quite low. However, the rate of delivery to the shoot is determined strictly by the rate of delivery from the roots, and thus by the ability of the roots to unload ions into the xylem. Shoot tissues absorb the ions from the nearby transpiration stream and the plant experiences a steady supply of mineral nutrients as long as the roots unload ions steadily.

The selective uptake results in a regulated supply of ions to the cells. The cells contain each element in a particular range of concentrations based on dry weight. Table 9.2 shows that a mature maize plant contains 1.46% of its dry mass as nitrogen, 0.20% as phosphorus, 0.92% as potassium, and so on. As a

Table 9.2   Elemental Composition of a Maize
Plant as a Percentage of Dry
Matter

O	44.43	P	0.20
C	43.57	Mg	0.18
H	6.24	S	0.17
N	1.46	Cl	0.14
Si	1.17	Al	0.11
K	0.92	Fe	0.08
Ca	0.23	Mn	0.04

[a]From Miller (1938).

first approximation, the requirement for these elements is proportional to the rate that dry mass accumulates, and inorganic ions generally are absorbed most rapidly during the most rapid phases of growth.

The ions are carried to the root surfaces by bulk flow and diffusion (Barber, 1962). Bulk flow occurs mostly because ions are swept along in the water absorbed by roots. Diffusion occurs because absorption during periods of low transpiration lowers the concentration at the root surface and creates a concentration gradient extending from the soil to the root surfaces, and the ions diffuse down the gradient toward the root. Occasionally, ions are delivered by bulk flow faster than they are taken up and they accumulate at the root surface, forming a gradient for back diffusion away from the root.

As mineral nutrient requirements increase with increased growth rates, transpiration increases and more nutrients move toward the roots by bulk flow. For some nutrients, it is possible to account for all nutrient delivery to the roots by this means (S. A. Barber *et al.*, 1962a,b; Renger *et al.*, 1981). The supply of nutrients can be calculated from the concentration in the soil solution multiplied by the flow of the solution to the root. Table 9.3 shows that the calculated supply was adequate to account for the calcium and magnesium requirement

Table 9.3   Supply of Elements to Maize Roots by Bulk Flow Caused by Transpiration

Element	Percent of plant dry matter	Concentration of soil solution needed if transpiration is 500X dry matter (mM)	Concentration of soil solution in 145 Maize soils (mM)
Ca	0.22	0.11	0.83
Mg	0.18	0.15	1.15
K	2.0	1.02	0.10
P	0.20	0.13	0.002

[a]After S. A. Barber *et al.* (1962a).

but not the potassium or phosphorus requirement in maize. Renger *et al.* (1981) reached a similar conclusion for sugar beet, spring wheat, and spring barley and found that the nitrate requirements often were met as well.

If the inorganic ions are absorbed at a relatively greater rate than bulk flow can provide, as with phosphate and potassium, the concentration in the soil solution will decrease next to the root. In response, ions are released from the soil particles and tend to buffer the concentration. Nevertheless, there is a lowering at the root surface and ions will tend to move into the depletion zone by diffusion in addition to bulk flow. Figure 9.2A shows that the depletion of phosphorus around an onion root decreased the concentration at the root surface to less than half the concentration in the bulk soil solution. The depletion zone extended about 2 mm into the soil. Root hairs can extend 1 to 4 mm from the root and increase the depletion zone to 6 mm or so (Bhat and Nye, 1974). Rhodes and Gerdemann (1975) showed that mycorrhizal fungi extend as much as 10 cm beyond this depletion zone and supply phosphorus to the root from great distances. This appears to be one of the main benefits of mycorrhizae (Gerdemann, 1968; and in Torrey and Clarkson, 1975; MacFall *et al.*, 1991b,c; Sanders and Tinker, 1971).

On the other hand, if the ion is absorbed at a relatively slow rate compared to the bulk flow to the root, the concentration in the soil solution may build up next to the root. Sulfate is an example because its uptake is moderate in plants but the ion is highly mobile in the soil (S. A. Barber *et al.*, 1962b). Figure 9.2B

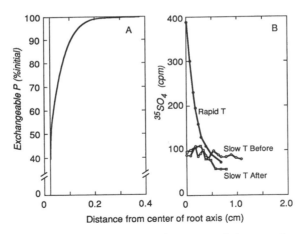

Figure 9.2   (A) Depletion zone for phosphate and (B) accumulation zone for sulfate extending radially from the root. For (B), transpiration (T) was varied, and the accumulation zone was measured before, during, and 16 hr after exposing the plant to rapid T for 18 hr. After Nye and Tinker (1977).

shows an example from Wray (1971) who observed that transpiration caused sulfate to accumulate around onion roots, with the amount depending on how rapidly transpiration occurred. Rapid rates caused an accumulation in a zone extending about 5 mm into the soil. No accumulation was evident when transpiration was slow either before or 16 hr after the rapid transpiration, indicating that the accumulation zone dissipated by back diffusion when transpiration was slow but could quickly reform when transpiration resumed.

Thus, depending on the interaction of the two processes by which soil solutes move—bulk flow and diffusion—solutes can be depleted or accumulated in the soil next to the root. Over the life of the plant, some of the solutes can be supplied to the root sufficiently rapidly by bulk flow associated with transpiration but even for these solutes depletion or accumulation zones may be set up as transpiration and ion uptake vary. A related question is whether transpiration also affects ion transport inside the plant itself. The experiments have been reviewed by Russell and Barber (1960), Viets (1972), and Hsiao (1973) who concluded, essentially, that over long times transpiration had little effect on uptake by roots in low external solution concentrations, but had a significant effect when the external concentrations were high. Ion uptake tended to proceed independently of transpiration, probably because it depended on energy requiring processes quite different from the physical factors driving transpiration.

The action of transpiration on ion transport can be seen by varying the transpiration rate when the plant roots are surrounded by a well-stirred inorganic nutrient solution. When transpiration is rapid, xylem concentrations of the nutrient ions are low because the incoming water dilutes the xylem solution. Root ion uptake is rapid because opposing concentration gradients are small inside the root. When transpiration slows, the concentrations build up to high levels in the xylem because metabolically driven ion uptake proceeds even though water flow is slow. However, as the xylem concentrations build up, root uptake can be inhibited. Shaner and Boyer (1976a) found that NO_3^- uptake was generally independent of transpiration when transpiration was rapid (see also Schulze and Bloom, 1984) and that the concentration in the xylem varied inversely with transpiration, as expected. However, when transpiration became slow, NO_3^- concentrations climbed to such high levels (40 mM) in the xylem that root uptake of NO_3^- also became slow. The effect was reversible when transpiration increased sufficiently to dilute the xylem concentration. Thus, NO_3^- uptake changes from independent to dependent on transpiration in the same plant as the rate of transpiration changes.

High soil concentrations of mineral nutrients promote high xylem concentrations and increase the tendency for transpiration to affect ion uptake. Figure 9.3 shows an example in barley (Hooymans, 1969) where high potassium concentrations outside the roots caused higher potassium concentrations inside the roots. Suppressing transpiration caused root potassium concentrations to be-

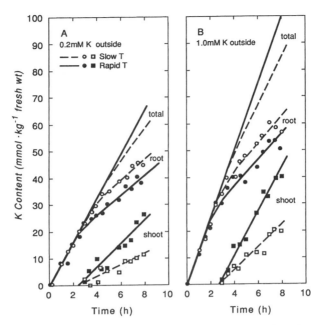

Figure 9.3   Potassium content of roots and shoots of barley plants at various times after exposing the plants to high or low rates of transpiration (T). (A) Roots in nutrient solution with 0.2 mM K. (B) Roots in nutrient solution with 10 mM K. Note that for the total plant, rapid transpiration suppressed uptake more when the external K concentration was high than when it was low. After Hooymans (1969).

come even higher. For the plant as a whole, the suppressive effect of reduced transpiration became greater when the root concentration became higher.

Effects of Dehydrating Conditions

When soil dehydration occurs, mineral nutrient uptake tends to decrease because transpiration is reduced, decreasing the bulk flow of soil solution to the root, but photosynthesis and plant growth also are curtailed. With less increase in dry matter, the decreased nutrient uptake has only a modest effect on the elemental concentration in the tissues. Several long-term field studies (Greaves and Carter, 1923; Janes, 1948; Jenne *et al.*, 1958) showed that drought had only small effects on the elemental composition of grain, in agreement with this idea. Vegetative tissues show a similar response and a typical result with apple plants (Table 9.4) showed only a moderately higher tissue concentration of N and slightly lower concentration of P, K, Ca, and Mg than in plants grown with adequate water (Mason, 1958).

Table 9.4   Effect of Soil Moisture on
Percentage of Elements in Dry
Matter of Apple Plants[a]

| | Shoot | | Root | |
Element	Dry	Wet	Dry	Wet
N	0.97	0.53	1.40	1.21
P	0.09	0.10	0.10	0.21
K	0.46	0.47	0.64	0.86
Ca	0.41	0.39	0.50	0.57
Mg	0.06	0.06	0.08	0.11

[a] From Mason (1958).

For this mechanism to hold, mineral nutrient uptake should decrease as soon as the accumulation of dry mass is reduced. Eck and Musick (1979) showed that in sorghum, nutrient uptake was suppressed as soon as dry matter accumulation was inhibited. Thus, the final concentration of nutrient elements in the tissue depended on the whether the decrease in uptake was greater or less than the decrease in dry matter accumulation. O'Toole and Baldia (1982) also found that both dry matter and nutrient uptake were reduced when rice was subjected to a water deficit. Figure 9.4 shows that the decreased nutrient uptake was apparent as soon as dry matter accumulation decreased. Using the data from Fig. 9.4, calculations in Table 9.5 show that the nutrient concentrations (as a percentage of dry matter gained) were slightly less for N, P, and K in the water-deficient plants than in the controls. Thus, over the long term, the water deficit caused only moderate effects on nutrient concentrations in the tissue. The lack of a wide variation indicates that nutrient delivery remains reasonably well coupled to the demand resulting from dry mass accumulation, regardless of the availability of water. Imsande and Touraine (1994) similarly conclude that NO_3^- uptake is coordinated with biomass production.

In respect to nutrient delivery to the root, O'Toole and Baldia (1982) showed that the mechanism did not change during a water deficit. The bulk flow of water to the roots continued to be capable of supplying a large part of the N and K, and diffusion continued to supply P (Table 9.5).

NITROGEN METABOLISM

Nitrogen is the most common element in plant tissues after carbon, hydrogen, and oxygen. Like the other inorganic elements, its uptake is diminished when the supply of water is curtailed, but the tissue *concentration* shows slight changes depending on the extent of growth inhibition (cf. Tables 9.4 and 9.5). Table 9.4 shows that the direction of N concentration change sometimes can be

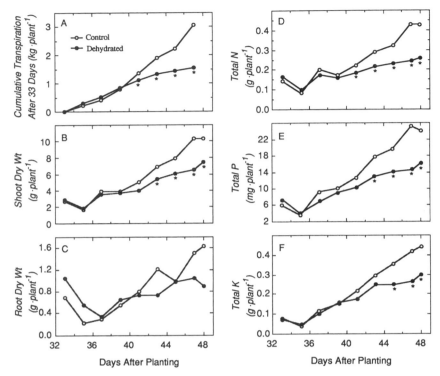

Figure 9.4 (A) Transpiration, (B) shoot dry mass, (C) root dry mass, (D) total N, (E) total P, and (F) total K contents of rice plants from which water was withheld after Day 30. The asterisk shows significant differences at the 5% confidence level. After O'Toole and Baldia (1982). The data in (E) are $10\times$ greater than reported in the original paper because of a decimal error in the original (J. C. O'Toole, personal communication).

counter to that of the other ions during dehydration. Nitrogen acquisition differs from most of the other elements because it undergoes extensive metabolism after uptake by the cells. Nitrogen usually is oxidized in the soil and is taken up as NO_3^-. The NO_3^- must be reduced before it can be incorporated into amino acids and thence to proteins and other cell constituents. The unique changes often observed in tissue concentration of N imply that dehydration affects N metabolism.

Soil NO_3^- is present in the soil solution and, although NH_4^+ sometimes is present in small amounts, it is oxidized rapidly to NO_3^- by soil bacteria. Naturally occurring NO_3^- and NH_4^+ come mostly from the breakdown of pre-existing organic matter, a process that also loses a substantial amount of N by volatilization of NH_3 or production of N_2 from NH_4^+ by some soil bacteria.

Table 9.5 Supply of Elements to Rice Roots Calculated from Bulk Flow Caused by Transpiration[a]

Treatment	Gain (% of DM gain)[b]			Transpiration ratio (g H_2O · g DM gain^{-1})[b]	Concentration needed if transpiration provided nutrients (mM)[d]			Concentration typical in soil solution (mM)[e]		
	N	P[c]	K		N	P[c]	K	N	P	K
Control	2.8	0.18	3.3	256	7.8	0.23	3.3			
H_2O deficient	1.9	0.14	2.9	190	7.1	0.24	3.9	0.1–4	0.0005–0.002	0.1–4

Note. The effect of a water deficit is also shown.
[a] From O'Toole and Baldia (1982).
[b] From Fig. 9.4
[c] Data for phosphate are reported here at 10X the concentration in the original paper because of a misplaced decimal point in the original (J.C. O'Toole, personal communication).
[d] Calculated as:

$$\frac{(\text{Gain in Element})(10^4)}{(\text{Transpiration Ratio})(\text{Atomic Wt})} = \frac{(\% \text{ of DM gain} \cdot 10^{-2})(10^3 \text{ g } H_2O \cdot \text{kg}^{-1})(10^3 \text{ mmol} \cdot \text{g} - \text{atom}^{-1})}{(\text{g } H_2O \cdot \text{g DM gain}^{-1})(\text{g element} \cdot \text{g} - \text{atom}^{-1})}$$

$$= \text{Concentration Needed (mM)}$$

[e] From Fig. 9.1.

Also, plants volatize small amounts of N (Farquhar *et al.*, 1979). As a consequence, N needs to be continually added to the soil–plant system for plant productivity to remain stable. The input of new N comes from the fixation of N_2 gas from the atmosphere, mostly by microorganisms that can reduce N_2 to NH_3. If the microorganisms are free-living, the NH_3 is released to the soil as organic matter after the microorganisms complete their life cycle and begin to break down. The NH_3 is generally converted quickly to NO_3^- by other microorganisms and is absorbed by plants. If the microorganisms are not free-living but instead grow in symbiosis with the plant, the NH_3 is released directly to the plant without being converted first to NO_3^-. In addition to N_2 fixation by microorganisms, rainfall supplies a small amount of NH_3 because of electrical activity in the atmosphere. In agriculture, N depletion is prevented by adding nitrogen-containing fertilizers, some of which are manufactured by the Haber process that reduces N_2 to NH_3 nonenzymatically, using high temperatures and pressures.

Nitrogen Fixation

Nitrogen is fixed chiefly by bacteria growing symbiotically with other plants. The most common form is the rhizobium–legume symbiosis but others occur such as the actinomycetes found in species of alder and casuarina (Tjepkema *et al.*, 1986), and cyanobacteria in cycads (Lindblad *et al.*, 1987). Less important are some free-living bacteria and a few that form loose associations with plant roots (Bothe *et al.*, 1983; Döbereiner, 1983; Quispel, 1983) or leaves (Döbereiner, 1983). Also, it has been reported that N_2-fixing bacteria can be found in the intercellular spaces of sugarcane (Dong *et al.*, 1994). In some instances, significant N_2 fixation occurs in specialized cells of cyanobacteria (Bothe *et al.*, 1983). This is an intriguing situation because no symbiosis is involved and instead photosynthesis and N_2 fixation occur in the same plant, although in different cells.

In these microorganisms, the fixation is carried out by the enzyme nitrogenase that binds N_2 and uses ATP and reductant to form NH_3 and H_2 without releasing any intermediates:

$$N_2 + 8[H] \rightarrow 2NH_3 + H_2, \qquad (9.1)$$

where [H] indicates reduced ferredoxin and the stoichiometry is only approximate. The requirement for ATP is not shown in the equation. The nitrogenase is inhibited by O_2 and thus there must be a discrimination between N_2 and O_2, which are both gases. In legumes, this is accomplished by the formation of layers of host tissue around the nitrogen-fixing bacteroids. The resulting root nodules are conspicuous and contain a modified hemoglobin (leghemoglobin) that binds O_2 making its way into the nodule. The leghemoglobin holds internal O_2 at a low level (King *et al.,* 1988). As N_2 and O_2 diffuse from the soil pores into the nodule, the O_2 also is consumed to produce ATP and reductant and this further decreases the O_2 reaching nitrogenase (Criswell *et al.,* 1976; Dakora and Atkins, 1989; Denison *et al.,* 1992; Hunt *et al.,* 1987; Layzell *et al.,* 1989; MacFall *et al.,* 1992; Sheehy *et al.,* 1985; Weisz and Sinclair, 1987).

The nitrogenase releases NH_3 which serves as the substrate for the production of glutamate and related amino acids. The carbon skeletons for amino acid formation are provided by the host plant. Thus, the symbiosis results in an energy and carbon supply for the bacteroids but an amino acid supply for the host plant. The nodules contain xylem elements that connect to the xylem of the host root, and the amino acids are transported to the shoot in the transpiration stream.

This process requires a supply of photosynthetic products from the host plant and N_2 gas from the atmosphere as well as enough O_2 so that respiration can occur in the host and bacteroid tissues but not so much that nitrogenase becomes inhibited. The delicate balance between the supply of each substrate is dynamically controlled. Removing the source of photosynthetic products by detopping the plant or girdling the stem rapidly inhibits N_2 fixation (Denison *et al.,* 1992; Hartwig *et al.,* 1990; Huang *et al.,* 1975a,b; Vance *et al.,* 1979; Walsh *et al.,* 1987), even though substantial photosynthates (sugars and starches) are stored in the nodule tissues (Hartwig *et al.,* 1990; Vance *et al.,* 1979; Walsh *et al.,* 1987). Thus, a continued flux of recently formed photosynthate seems necessary for N_2 fixation and there is evidence that it also may play a role in the control of the O_2 concentration inside the nodule (Denison *et al.,* 1991, 1992).

The balance also is affected by the availability of soil water. Decreasing soil water at first increases the availability of N_2 and O_2 because the soil pores drain and become filled with air. The effect becomes apparent when the N_2-fixing activity is assayed with the alternate substrate acetylene, which nitrogenase can reduce to ethylene. The substrate and product are both gases and measuring the ethylene production gives a convenient assay of nitrogenase activity *in situ*. The gases can be used in intact soil–plant systems to mimic the effects of soil water on gas diffusion during fixation. Figure 9.5A shows that nitrogenase activity measured as acetylene reduction initially increased when the soil dehydrated after water was withheld, confirming that the gaseous substrates for N_2 fixation became more available. However, as dehydration became more severe, nitrogenase activity was markedly inhibited. The inhibition was accompanied by a similar inhibition in photosynthesis.

Huang *et al.* (1975b) showed that decreasing photosynthesis by decreasing the CO_2 concentration around the shoot of hydrated plants had the same effect on N_2 fixation as decreasing photosynthesis by dehydration (Fig. 9.5B). This similarity indicates that decreased photosynthesis was limiting or near limiting for N_2 fixation in dehydrated soil. Increasing photosynthesis with high CO_2

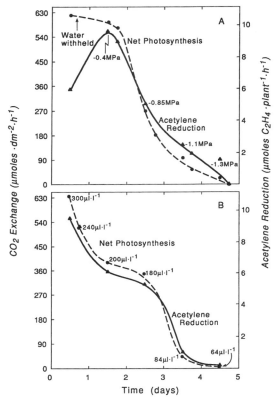

Figure 9.5 Net photosynthesis in shoots and nitrogen fixation (acetylene reduction) in roots of soybean plants at various times after (A) withholding water from the soil or (B) decreasing CO_2 around the shoot without withholding water. The nodule water potentials are shown in (A). The CO_2 concentrations around the shoot are shown in (B). In (B), the low CO_2 was selected to mimic the effects of low water potentials on photosynthesis as in (A). Note that because (B) involves only decreased CO_2 around the shoot, acetylene reduction was inhibited by the lack of products of photosynthesis. Therefore, in (A) the effect on photosynthesis indicates that acetylene reduction must have been similarly inhibited by a lack of photosynthetic products. In (A) and (B), photosynthesis and acetylene reducing activity were measured simultaneously in the same plants *in situ*. After Huang *et al.* (1975a,b).

around the shoot reversed some of the inhibition of N_2 fixation in the dehydrated plants (Huang *et al.*, 1975b). These experiments established that the loss in photosynthetic activity caused the decrease in nitrogen-fixing activity at least in part. Huang *et al.* (1975b) further showed that decreasing the availability of recently formed photosynthate to the nodules inhibited nitrogenase activity perhaps because the production of ATP, reductant, and carbon compounds for the fixation process would have decreased [Eq. 9.1]. In effect, the lack of photosynthate was a signal from the shoot to the root that caused the decreased N_2 fixation.

Others (Davey and Simpson, 1990; Fellows *et al.*, 1987; Irigoyen *et al.*, 1992b; Pararajasingham and Knievel, 1990) showed that there is a significant amount of sugars and starches in nodules of dehydrating plants and argued against a limitation of N_2 fixation by the limited availability of photosynthate. However, in view of the dependency of N_2 fixation on recently formed carbohydrate (Hartwig *et al.*, 1990; Huang *et al.*, 1975b; Vance *et al.*, 1979; Walsh *et al.*, 1987), the flux of carbohydrate appears more important than the amount that is stored (Hunt and Layzell, 1993). Thus, the nitrogenase appears to depend on a small pool of photosynthetic products that is rapidly diminished when the delivery of new products decreases.

Pankhurst and Sprent (1975, 1976) observed that increasing the O_2 concentration around the nodules also could reverse some of the inhibition of nitrogenase activity caused by dehydration. Figure 9.6 shows that the recovery was complete in nodules that had been slightly dehydrated but was not complete when dehydration was more severe. Nodule respiration showed a similar response to high O_2 (Guerin *et al.*, 1990; Pankhurst and Sprent, 1975). The enhancement by O_2 could be overcome by slicing the nodules, indicating that there was an O_2 barrier in intact nodules that was broken by slicing and that the barrier had become more effective when the nodules were dehydrated. Weisz *et al.* (1985) directly demonstrated the barrier by showing that dehydration decreased the gas permeability of nodules. Irigoyen *et al.* (1992a,c) showed that nodule enzyme activity shifted toward hypoxic metabolism. Thus, in addition

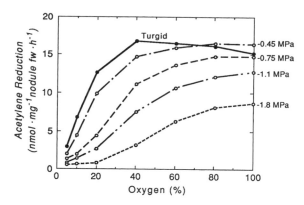

Figure 9.6 Nitrogen fixation (acetylene reduction) at various oxygen partial pressures in detached soybean nodules dehydrated to varying degrees. Dehydration is indicated by nodule water potentials beside each curve. After Pankhurst and Sprent (1975).

to an inhibition by a lack of photosynthate (Huang *et al.*, 1975b), a lack of O_2 also can be limiting in some conditions, presumably because there is an inhibition of conversion of photosynthate to the substrates needed by nitrogenase (Pankhurst and Sprent, 1975, 1976).

These results indicate that water availability has two main effects on N_2 fixation. In one, gas diffusion is increased as water drains from the soil pores and N_2 fixation tends to increase. In the other, photosynthesis and nodule gas diffusion are decreased as dehydration becomes more severe and N_2 fixation decreases. The exact conditions where the shift occurs probably depend on soil conditions, the amount of photosynthate being produced by the plant, and the structure and condition of the nodules. Sanchez-Diaz *et al.* (1990) indicate that mycorrhizal symbiosis also can alter the nodule response.

Nitrate Metabolism

The ability of plants to use NO_3^- depends on the amount of NO_3^- surrounding the roots (Beevers and Hageman, 1969, 1983). When the ion is absent, the uptake capability is negligible (Beevers *et al.*, 1965; Ingle *et al.*, 1966) but, when the ion is supplied, uptake begins rapidly and after about 30 min, the metabolism of NO_3^- increases (Beevers *et al.*, 1965). The NO_3^- is metabolized first by being reduced in the cytosol where the enzyme nitrate reductase catalyzes the reaction. The NO_3^- is reduced to NO_2^- as

$$NO_3^- + 2[H] \rightarrow NO_2^- + H_2O \qquad (9.2)$$

and metabolic energy is consumed to produce the reductant [H] which is $NADH+H^+$ or $NADPH+H^+$, depending on the tissue. The stoichiometry of the reaction has not been fully worked out.

The signal for the induction of reductive activity is the NO_3^- itself that is entering the cell (Beevers and Hageman, 1969, 1983). The activity results from the synthesis of new nitrate reductase in the cytosol and, after more enzyme has been synthesized, the NO_3^- is rapidly reduced so that NO_3^- does not accumulate in the cytosol. The NO_3^- that escapes reduction is transported to the vacuole, further keeping the concentration low in the cytosol.

The NO_2^- produced is kept from accumulating by a second reduction that converts the NO_2^- to NH_3 in the approximate reaction

$$NO_2^- + [H] \rightarrow NH_3, \qquad (9.3)$$

where the reductant is reduced ferredoxin. The enzyme is nitrite reductase which carries out the reductive steps requiring ATP and reductant and, as with nitrogenase, no intermediates are released. The NH_3 is combined with carbon compounds to form amino acids that are utilized in protein synthesis. NO_2^-

is toxic so that the continued presence of nitrite reductase is essential. Nitrite reductase is a constitutive enzyme thought to be in various kinds of plastids (Beevers and Hageman, 1983).

Thus, the uptake of NO_3^- sets in motion a series of reactions that are essential for the growth of the cell. Growth depends on NO_3^- metabolism to supply new protein and nucleic acids but the same metabolism is necessary to synthesize nitrate reductase for the process. As a consequence, there is always a small NO_3^- reduction activity present in plant cells. Depriving the plant of NO_3^- disrupts the flow of nitrogen to protein synthesis and limits the synthesis of nitrate reductase but does not eliminate reductase activity. Because of this, recovery from a nitrogen deficiency occurs slowly at first and more rapidly after new nitrate reductase is synthesized.

The availability of NO_3^- to the cells is thus an important signal, and Shaner and Boyer (1976a) found that the NO_3^- flux to the cytosol was the critical feature of the signal. They used several methods to change the NO_3^- flux to maize shoots and found that the nitrate reductase activity corresponded with the flux even though the NO_3^- content of the shoots (mostly in the vacuoles) did not change. Whenever the activity increased, it was because the synthesis of new enzyme protein increased (Morilla *et al.*, 1973; Shaner and Boyer, 1976b). Therefore, the flux controlled the expression of the nitrate reductase genes (Crawford and Campbell, 1990).

The fact that the enzyme responded to changing fluxes in the transpiration stream rather than NO_3^- in the vacuole indicated that a small cytoplasmic pool of NO_3^- must be involved and, because of its small size, the concentration could change rapidly. In effect, changes in flux were read from changes in concentration of this small pool.

With this in mind, Shaner and Boyer (1976b) decreased the water supply to maize roots and found that the NO_3^- flux to the shoot was decreased together with a lowered activity of nitrate reductase (Fig. 9.7). The NO_3^- flux decreased because of a combination of lower transpiration and a lower NO_3^- concentration in the transpiration stream due to the decreased root uptake (Fig. 9.7A, insets). If the plants were supplied with additional NO_3^- before the dehydration, the NO_3^- flux was elevated and nitrate reductase activity was similarly elevated during the dehydration (Fig. 9.7). This indicated that the NO_3^- flux regulated the activity of nitrate reductase during dehydration.

If water was resupplied, the tissue rehydrated and nitrate reductase activity recovered (Fig. 9.8), but the recovery depended on a high NO_3^- flux to the shoot (Fig. 9.8) and the synthesis of a new enzyme (Shaner and Boyer, 1976b). This indicates that the loss in enzyme activity during dehydration was not caused by a direct inhibition of the enzyme but rather by a decreased synthesis of the enzyme. The synthetic activity was determined simply by the flux of a

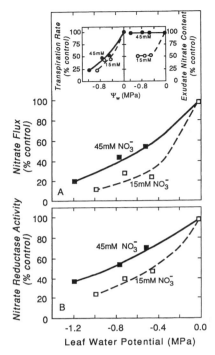

Figure 9.7   (A) Nitrate flux to the shoot and (B) leaf nitrate reductase activity in the shoot at various leaf water potentials in intact maize seedlings. (Insets) Transpiration and nitrate concentration in the xylem solution of the intact seedlings. The nitrate flux in (A) was calculated as transpiration rate × nitrate concentration shown in the insets. Immediately before dehydration, half the plants were provided with supplemental nitrate (45 mM) around the roots. After Shaner and Boyer (1976b).

regulatory molecule. The NO_3^- flux thus represented a dehydration signal from the roots that altered the activity of the shoot enzyme.

The basis for this response is the short half-life of the enzyme in the cell. Morilla *et al.* (1973) found that nitrate reductase has a half-life of only 4 hr in maize and that it would need to be continually synthesized to maintain steady activity. When the NO_3^- flux was diminished during dehydration, the synthesis signal was diminished and, in the absence of synthesis, the activity rapidly died away. Bardzik *et al.* (1971) suggested that similar effects would be expected if a general decrease in protein synthesis occurred in plant tissue; enzymes having short half-lives would show losses in activity before longer-lived enzymes. Therefore, decreases in enzyme activity could be caused by regulators specific for each enzyme (such as NO_3^-) or by regulators of protein synthesis itself.

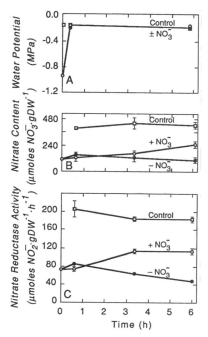

Figure 9.8   (A) Leaf water potential, (B) leaf nitrate content, and (C) leaf nitrate reductase activity in maize shoots recovering from dehydration after cutting under solution with or without nitrate. There was a substantial nitrate flux to the shoot in the $+NO_3^-$ treatment but little if any in the $-NO_3^-$ treatment. Note that the recovery of leaf nitrate reductase activity depended on a NO_3^- flux during rehydration. The recovery could be blocked by inhibitors of enzyme synthesis, indicating that recovery required new enzyme molecules to be synthesized. After Shaner and Boyer (1976b).

Protein Synthesis

Decreased protein synthesis is a central feature of plant dehydration. Synthesis often slows during mild dehydration (Hsiao, 1970; Mason and Matsuda, 1985; Mason *et al.*, 1988a; Morilla *et al.*, 1973; Scott *et al.*, 1979) and can cease entirely in severely desiccated tissues (Bewley, 1979). In part, the decrease can be explained by the decreased NO_3^- flux to the plant cells but synthesis decreases even when the NO_3^- supply does not appear to be involved. Usually, the decreases in synthesis are detected as losses in the polyribosomal content of the tissue because most other methods of measuring protein synthesis require aqueous media that rehydrate the tissue. Because polyribosomes are complexes of messenger RNA and ribosomal RNA actually synthesizing the protein, their disappearance is evidence for a general slowdown in protein synthesis.

Figure 9.9 shows that even 1 hr of dehydration causes substantial losses of

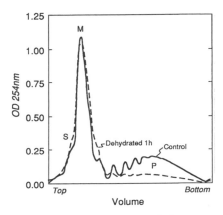

Figure 9.9   Polyribosomes (P), monoribosomes (M), and ribosome subunits (S) isolated from young maize leaves dehydrated for 1 hr to a water potential of −1.27 MPa. The ribosomes were separated from each other on a density gradient so that the largest ones (P) are on the right and the smallest ones (S) are on the left. The ribosomes are detected by their absorbance of ultraviolet radiation (OD 254 nm). Note the loss in P after the dehydration. After Morilla *et al.* (1973).

polyribosomes in maize (Morilla *et al.*, 1973). Growing cells typically have large numbers of polyribosomes and undergo large losses during dehydration whereas mature tissues contain fewer polyribosomes and lose fewer (Bewley and Larsen, 1982; Morilla *et al.*, 1973). The losses in growing tissues take place soon after growth is inhibited by the water deficit (Mason and Matsuda, 1985; Mason *et al.*, 1988a). Figure 9.10 shows that the losses in polyribosomes were apparent about 4 hr after stem growth was inhibited by transplanting soybean seedlings to vermiculite of low water content. The general loss in protein synthesis did not cause the growth inhibition because growth already had slowed, which was also noted by Scott *et al.* (1979) in wheat leaves. After a time, growth resumed in soybean and the polyribosomal content recovered. The loss also did not appear to be caused by a decrease in NO_3^- flux because amino acids were supplied by reserves in these seedlings, which were dependent on stores in the cotyledons, and NO_3^- reduction was not necessary. Importantly, the polyribosomes in the roots were not affected (Creelman *et al.*, 1990). The roots continued to grow after the transplanting and thus there was a relationship between the factors causing growth inhibition and those causing less protein synthesis. The polyribosome loss was not a whole plant response but rather a tissue-specific response.

The possibility that ABA could be a signal regulating growth began to attract attention when it was observed that treating plants with ABA inhibited shoot growth but not root growth much as occurred during dehydration (Davies *et al.*, 1986). Saab *et al.* (1990) found that a maize mutant having less ABA

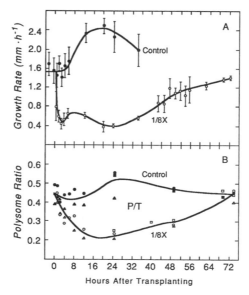

Figure 9.10   (A) Growth and (B) polyribosomal content in the elongating region of stems of intact soybean plants after transplanting to vermiculite of low water content. The dehydrated vermiculite had a water content one-eighth that of the control vermiculite. The response in control vermiculite is shown by dark points and in dehydrated vermiculite by open points. The polyribosome (P) content is shown relative to the total ribosomes (T) in the tissue. After Mason *et al.* (1988a).

production during dehydration also showed less growth inhibition of the stem and less growth enhancement of the roots. They concluded that elevated ABA might be necessary to inhibit stem growth but stimulate root growth during dehydration. Bray (1988) had shown that treating tomato leaves with ABA or dehydration affected protein synthesis in a like manner, supporting the notion that growth changes might be caused by ABA. However, Creelman *et al.* (1990) specifically studied protein synthesis in the growing regions of roots and stems and found that the patterns were different for the two treatments. Moreover, the inhibition of stem growth was followed by a loss in polyribosomes during dehydration but not after exposure to high ABA. They concluded that high ABA levels could simulate the growth responses but not the molecular responses that occur during dehydration, and thus high ABA appeared not to cause the growth responses.

Bewley and Larsen (1982) and Dasgupta and Bewley (1984) further found that the loss in polyribosomes was not due to the loss in synthesis of a particular protein, although some synthetic differences were found. Dhindsa and Bewley (1976) were unable to attribute the polyribosome loss to increases in ribonuclease, and mRNA was conserved (Dhindsa and Bewley, 1978). This indicated that the losses in polyribosomes were not caused by a breakdown of mRNA. There is evidence that the polyribosomes translate some different mRNAs during dehydration (Creelman *et al.*, 1990; Oliver, 1991) and the changes in nitrate reductase activities support this concept. Therefore, two phenomena appear to be occurring. First, synthesis for certain proteins is more affected than for others and, second, the overall rate of protein synthesis tends to decrease. However, while there is some understanding of the molecular regulation of individual protein synthesis such as nitrate reductase, the molecular control of the overall rate remains obscure.

DEHYDRATION AND ROOT/SHOOT SIGNALS

The altered nitrogen metabolism and protein synthesis occurring during dehydration reflect many enzymatic changes that take place under these conditions, raising certain questions. What are the molecular signals that cause the changes? How do plants "sense" that the soil is dehydrating?

The enzymes of nitrogen metabolism can be used to gain molecular insight to these questions. It is apparent that water deficiency inhibits nitrogen acquisition at about the time growth is inhibited. Variations occur depending on the species and environmental conditions, but overall the effects are the same (Eck and Musick, 1979; Janes, 1948; Jenne *et al.*, 1958; Greaves and Carter, 1923; Mason, 1958; O'Toole and Baldia, 1982). As a consequence, the nitrogen concentration of the plant tissues is not changed by large amounts which indicates that the decreased nitrogen acquisition did not cause the decreased growth but somehow was coordinated with it.

The work described in the previous sections of this chapter indicates that the nitrogen acquisition enzymes nitrogenase and nitrate reductase are regulated by root/shoot signals that vary with changes in the root water supply and the growth of the plant. Both signals originate in another part of the plant. For nitrogenase, shoot photosynthesis is inhibited and less photosynthate travels to the root. For nitrate reductase, root NO_3^- uptake is inhibited and less NO_3^- travels to the shoot. Nitrogenase responds to the availability of photosynthate and nitrate reductase to the availability of NO_3^- and both signals are fundamental for the overall growth of the plant, which is inhibited simultaneously.

Therefore, the regulation is specific for each enzyme and represents the way in which the enzyme "senses" that the water supply to the roots is reduced. A key feature is that a pool of regulatory molecules exists that is small in the target cells. For nitrogenase, there is a small pool of useable, recently formed photosynthate (Huang *et al.*, 1975a,b). For the reductase, there is a small pool of cytoplasmic NO_3^- (Shaner and Boyer, 1976a,b). As a consequence, changes in supply are quickly detected and the molecules in the pool function as regulatory "sensors."

Another key feature is that the activities of the enzymes consume the regulator. As the pool size for the regulator increases, the enzyme activity increases, using the regulator at a faster rate. Thus, there is a feedback that stabilizes the pool size. For example, as soon as the photosynthate or NO_3^- supply increases, nitrogenase and nitrate reductase activity increase and consume the extra flux into the pool, thus stabilizing the regulator signal at the new level. Without consumption of the regulator, this stabilization would not occur and the pool would grow indefinitely, preventing its use as a sensor.

These simple elements of regulation are found in most control systems and are shown schematically in Fig. 9.11A. Note that the supply flux for the regulator acts as a signal because the pool size is small and the concentration in the pool quickly responds to changes in supply (Fig. 9.11B). When the flux to the

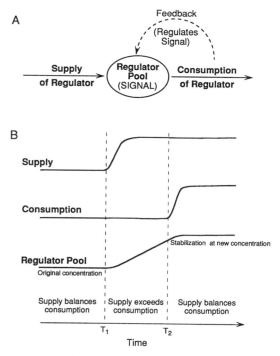

Figure 9.11   (A) Basic elements of a regulating system applicable to the control of enzyme activity in dehydrated plants. (B) Operation of the regulating system. There is a small pool of an enzyme regulator that acts as a signal. The pool is fed by a supply of the regulatory molecule (supply) and is depleted by reactions associated with the enzyme (consumption). Before time T_1, the supply is balanced by consumption. At T_1, the supply increases and the concentration of the regulator pool increases. This causes an increase in the enzyme activity at T_2, which feeds back to stabilize the regulator concentration at the new level. Because the pool size is small, it rapidly reflects changes in supply and thus responds to the flux of the regulator.

pool increases, the signal increases because the supply exceeds use and the pool concentration increases. In response, enzyme use of the pool increases until use is brought back into balance with the supply flux. This stabilizes the concentration of the pool at a new, higher level and thus maintains the enzyme activity at a new higher level.

It seems that all regulatory systems contain at least these simple control elements. For example, ABA shows certain features related to this concept. Most tissues increase in ABA content when they dehydrate because ABA is synthesized in the cells or is transported to them from other plant parts (increased supply). High ABA causes the stomata to close (Beardsell and Cohen, 1975; Kriedemann *et al.*, 1972; Mittelheuser and Van Steveninck, 1969) and can be inhibitory to shoot growth but stimulatory to root growth (Creelman *et al.*, 1990; Saab *et al.*, 1990). It has been proposed that the roots detect soil dehydration (Bates and Hall, 1981; Davies *et al.*, 1990; Saab and Sharp, 1989; Turner *et al.*, 1985) and that an increase in root ABA might be sent to the shoot in the transpiration stream and act as a signal for the shoot to close its stomata and slow its growth (Davies and Zhang, 1991; Gowing *et al.*, 1990; Zhang and Davies, 1989a,b, 1990) but Munns and King (1988) found little evidence for ABA in this role.

The issue is complicated because ABA is produced both in the root and in the shoot (Creelman *et al.*, 1990; Walton, 1980; Zeevaart and Creelman, 1988). It is more difficult to establish a signaling role than for other molecules that are produced or absorbed in one tissue and transported to another. Also, according to the root/shoot signaling described earlier, there needs to be consumption of the molecule in order for it to act as a signal, and the consumption of ABA will need to be better understood. This will require the signaling pool to be located. ABA in the xylem is not likely to be the pool because its concentration changes in passive response to changes in transpiration, and Davies and Zhang (1991) and Trejo *et al.* (1993b) point out that more attention will need to be paid to fluxes instead of concentrations of ABA in the xylem.

Root/shoot signals are important in plants and have large effects on biochemical responses to environmental conditions. Because many biochemical responses can be demonstrated in the absence of roots or shoots, root/shoot communication is not the only way that environmental signals are conveyed to metabolism. However, their presence indicates that many signals will not be understood unless enzyme studies are extended beyond the test tube to the whole plant.

DEHYDRATION AND ENZYME ACTIVITY

Direct Enzyme Effects

For many years, the causes have been sought for the metabolic changes occurring in plants exposed to dehydration. It has been proposed that low water availability may directly cause enzyme activity to change because of changes in the free energy of water within the cells (Kramer, 1969; Slatyer, 1967), changes in spatial relationships of membrane systems, volume changes, and concentrating of enzymes resulting from losses of water, or decreases in the water of hydration surrounding macromolecules (Hsiao, 1973; Walter, 1931). These are direct effects because they involve the water molecules acting directly on the enzyme or cell structure, and some of them are undoubtedly important in some systems. Figure 9.12 shows that dehydrating an isolated enzyme, in this case urease, decreases enzyme activity. The enzyme reaction produces ammonia and CO_2, both of which escape into the gas phase where they are easily detected without disturbing the enzyme mixture. Activity is completely lost when the enzyme is dehydrated at a relative humidity below 60%. Rupley *et al.* (1983) point out that enzyme activities depend on the motion of portions of the peptide molecule and that dehydration of an isolated enzyme acts mostly by restricting this motion. The activity begins to decrease when most of the water has been removed around the peptide and only a monolayer of surface water remains. As further water is removed, the internal motion of the peptide becomes limited

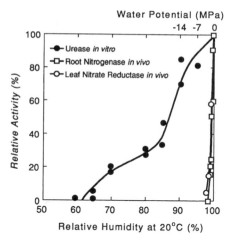

Figure 9.12 Activity of urease in equilibrium with various relative humidities, nitrate reductase exposed to various leaf water potentials, and nitrogenase exposed to various nodule water potentials. Water potentials equivalent to humidities are shown at the top of the graph (calculated according to Chapter 2). The urease catalyzes the reaction converting urea and water to NH_3 and CO_2. The enzyme was isolated, frozen, dehydrated, mixed with dehydrated substrate, and exposed to the humidity shown (*in vitro* dehydration). The rate of reaction was determined as the enzyme mix equilibrated at each humidity. The nitrate reductase was dehydrated in the intact maize leaf (*in vivo* dehydration), then isolated, and assayed in aqueous medium. Nitrogenase was assayed *in vivo* in roots of intact soybean plants. Note the large difference in the enzyme response to *in vitro* and *in vivo* dehydration. After Skujins and McLaren (1967), Morilla *et al.* (1973), and Huang *et al.* (1975a).

and the activity decreases. Eventually, activity disappears when only enough water remains to bind to a few exposed polar groups of the peptide, and the internal motion of the peptide becomes minimal.

As is readily apparent, dehydration directly affects enzyme activity only when too little water remains to cover the peptide surface with a single layer of molecules. This is much less water than is necessary for an effect on many enzymes. As pointed out earlier, nitrate reductase activity is completely lost at a water potential of about -2.5 MPa equivalent to a relative humidity of about 98% (Fig. 9.12). Similarly, nitrogenase activity disappears at a nodule water potential around -2.1 MPa (Fig. 9.12). About half of the cell water remains under these conditions, which is far more than monolayer coverage. The difference seems to be that these enzymes are functioning inside cells and depend on certain regulators to continue their function. The regulators affect the enzyme before dehydration has reduced the water content to only a monolayer.

Regulator Hypothesis of Enzyme Control

The response of nitrate reductase and nitrogenase is controlled by specific molecules different from water. For nitrate reductase the molecule is NO_3^-, and for nitrogenase it is the photosynthetic product or O_2. Other enzyme systems often follow a similar pattern. As described in Chapter 10, enzymes of photosynthesis in spinach or sunflower lose activity during dehydration, and for thylakoid ATP synthetase (coupling factor) there appear to be inhibitory interactions with Mg^{2+} and perhaps other solutes that are concentrated by water loss in the cells. This mechanism is different from that for nitrogenase or nitrate reductase and illustrates that each enzyme has its own response to changed concentrations of regulators in the cell.

To account for these effects, we suggest that enzyme control is based primarily on molecules other than water that act as regulators during dehydration. The effect is shown in Fig. 9.13B as an additional complexity of the molecular environment in the cell (*in vivo*) compared to that of an enzyme without regulators, demonstrated in the isolated enzyme urease (*in vitro*) where only water, substrates, and products are present with the enzyme. For tissues that do not withstand complete dehydration, all the enzymes studied so far show the more sensitive response of *in vivo* dehydration like that of nitrate reductase and nitrogenase (Fig. 9.12) and thus molecular regulators in the cell probably control most of these responses to dehydration. Because the effects are specific for each enzyme system, a complete picture is still being assembled to show how cell metabolism is affected. However, even with a limited picture, it seems clear that the regulator supply may be the key for these tissues and that the supply can be affected by dehydration.

For tissues that can be dehydrated to the air-dry state where water contents

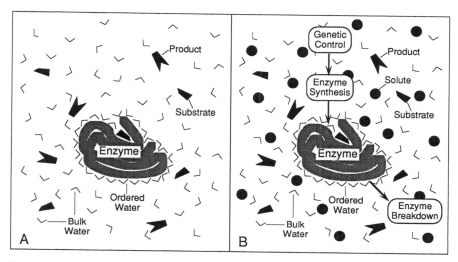

Figure 9.13   Schematic diagram of the difference between dehydration effects on isolated enzymes (A) and the same enzyme inside a cell (B). In (A), dehydration can act only on the water surrounding the enzyme and on the enzyme, substrate, or product as for urease in Fig. 9.12. In (B), dehydration has the additional possibility of altering genetic control, enzyme synthesis, enzyme breakdown, and the solute environment, as for nitrate reductase and nitrogenase in Fig. 9.12. These additional factors act as regulators of the enzyme response at dehydration levels that are less extreme than those needed for the changes in (A).

become very low, the response is more like that of urease (Fig. 9.12). As discussed in Chapter 10, photosynthesis in the marine alga *Fucus vesiculosus* does not decrease until dehydration is more severe than in sunflower and is completely lost only when water contents are less than 20% of the hydrated state (Fig. 10.2). Many seeds withstand similar dehydration and their metabolic activity becomes negligible only when water contents are extremely low (Vertucci, 1989; Vertucci and Leopold, 1987b). Desiccation tolerant *Selaginella* preserves membrane intactness after exposure to the air-dry state (Fig. 3.3). Therefore, it seems that some enzymes and cell structures can withstand very severe dehydration *in vivo* where they likely encounter water binding approaching monolayer thickness or less (Bruni and Leopold, 1991; Vertucci and Leopold, 1987a). However, regulatory molecules probably play a role in these cases as well. Chapter 12 describes some of the possibilities for protection of enzymes and membranes involving a role for sugars and specialized proteins as protective agents and it points out that gene regulation may be involved. Taken together, it seems that the regulatory control of enzymes and membranes may account for most of the behavior of metabolism during an episode of dehydration. Dehydration severe enough to remove all but monolayers of water probably includes these regulatory effects in addition to the direct effects of water on the internal motions of the peptides.

From these regulatory systems, the plant gains the possibility of modifying its response to dehydration. Thus, regulator levels can be varied before dehydration becomes severe, leading to acclimation. The ability of plants to acclimate to dehydration is a well-known phenomenon (Ackerson and Hebert, 1981; Ashton, 1956; Brown *et al.*, 1976; Matthews and Boyer, 1984; McCree, 1974; Shaner and Boyer, 1976b; Todd and Webster, 1965), and moderate episodes of dehydration appear to call into play systems that change some feature of the cell environment. This change at the molecular level may involve a diversity of regulators and suggests that acclimation might affect certain enzymes but not others and thus direct the metabolic response to dehydration.

SUMMARY

Salts of various inorganic ions are essential for plant growth and must be accumulated from the surrounding solution. They generally dissociate to ions that are freely mobile in solution, and in aquatic and marine environments the solutions generally are stirred and the nutrient ions reach the plant readily. In soil, the stirring does not occur and ions move to the root by a combination of bulk flow and diffusion. Bulk flow occurs as ions are carried along in water traveling to the root surface of transpiring plants. Diffusion occurs as gradients in ion concentrations form when the ions accumulate or are depleted next to the root surface according to their rate of uptake compared to water. Inside the plant, the rate of ion uptake is usually independent of the rate of water uptake, but very slow water entry can decrease ion entry apparently because the ion concentration in the root xylem becomes too high. Dehydration of the soil also decreases the rate of ion uptake but the rate more or less remains in balance with the decrease in dry mass occurring at the same time, hence elemental concentrations do not change much in the tissue with change in water supply.

Nitrogen is a special case because it undergoes extensive metabolism after being acquired by the cells. Nitrogen uptake by N_2 fixation involves the diffusion of N_2 and O_2 through the soil to nitrogen-fixing bacterial cells that are free-living or associated with host plants in symbiotic relationships such as with legumes. The N_2 is reduced to NH_3 by nitrogenase using energy from O_2-supported metabolism. Water-filled soil pores can restrict the diffusion of N_2 and O_2, and fixation often improves in the early phases of soil dehydration, but with more severe dehydration, fixation decreases. In legumes, the decrease results from a decreased flux of photosynthate to the nodules as photosynthesis decreases in the shoot. The decreased supply of photosynthate represents a molecular signal from the shoots to the roots that affects nitrogen fixation. There also is decreased oxygen diffusion into the nodules, which shrink and develop a strong barrier to O_2 entry, and their ability to use existing photosynthate becomes restricted.

In nonlegumes, nitrogen is acquired mostly by uptake of NO_3^- which is reduced first to NO_2^- and then to NH_3. The first reduction is catalyzed by nitrate reductase, which is synthesized in the presence of NO_3^-, and the second reduction is catalyzed by nitrite reductase, which is a constitutive enzyme in plastids. Nitrate reduction decreases during dehydration mostly because NO_3^- is transported to the sites of nitrate reductase synthesis more slowly in the transpiration stream because of the decreased uptake of NO_3^- by the roots. The decreased flux of NO_3^- decreases the synthesis of the enzyme, and the natural degradation of the enzyme in the cell depletes the cell of reductase activity. Thus, the decreased NO_3^- flux is a signal from the roots to the shoots that controls this aspect of shoot metabolism.

Protein synthesis also is generally inhibited by dehydration. The inhibition is not the same for each protein. The inhibition is not caused by general losses in messenger RNA or increases in the plant growth regulator abscisic acid. Decreased enzyme activities in the cell appear to result in part from this inhibition of protein synthesis followed by a decline in activity determined by the half-life of the enzyme in the cell.

The regulation of nitrogenase and nitrate reductase activities during dehydration indicates that the activities respond to regulator pools that often are supplied by other parts of the plant (or even the soil). The supply of regulatory molecules (photosynthate for nitrogenase, NO_3^- for nitrate reductase) represents a signal for the level of dehydration that changes the enzyme activity. The changes thus depend on root/shoot signals of a specific molecular nature for each enzyme system. The control of these enzymes contains the basic components of any feedback control system and indicates that dehydration effects on biochemistry will be understood only in the whole plant context.

A hypothesis is suggested that accounts for the effects of dehydration on enzymes and involves molecules other than water that act as regulators of enzyme action. The regulators change in concentration either because of changes in cell water content or altered transport or biosynthesis of the regulatory molecule. Enzymes subjected to dehydration outside of cells lack these regulatory systems but are directly affected by dehydration nonetheless. However, the dehydration must be considerably more severe than in most cells, which indicates that regulatory processes probably take precedence over direct dehydration effects inside the cells. The concept that regulatory molecules control the biochemical response to dehydration suggests that plant acclimation might be explained by alterations in the cell regulator environment that would predispose the plant for a particular biochemical response when dehydration occurs.

SUPPLEMENTARY READING

Beevers, L., and Hageman, R. H. (1983). Uptake and reduction of nitrate: Bacteria and higher plants. *In* "Encyclopedia of Plant Physiology" (A. Läuchli and R. L. Bieleski, eds.), Vol. 15A, pp. 351–375. Springer-Verlag, Berlin/Heidelberg.

Bothe, H., Yates, M. G., and Cannon, F. C. (1983). Physiology, biochemistry and genetics of di-nitrogen fixation. *In* "Encyclopedia of Plant Physiology" (A. Läuchli and R. L. Bieleski, eds.), Vol. 15A, pp. 241–285. Springer-Verlag, Berlin/Heidelberg.

Epstein, E. E. (1972). "Mineral Nutrition of Plants: Principles and Perspectives." Wiley, New York.

Gerdemann, J. W. (1968). Vesicular-arbuscular mycorrhiza and plant growth. *Annu. Rev. Phytopathol.* **6,** 397–418.

Harley, J. L., and Smith, S. E. (1983). "Mycorrhizal Symbiosis." Academic Press, New York.

Marschner, H. (1986). "Mineral Nutrition of Higher Plants." Academic Press, New York.

10

Photosynthesis and Respiration

INTRODUCTION

For most organisms, photosynthesis makes life possible. Plants use its products for their respiration and for building all the macromolecules and organic cofactors necessary for their metabolism and structure. Animals harvest plants and use the cell contents for their own metabolism and structure. As a consequence, animals have a simpler biochemistry than plants. They cannot synthesize many necessary metabolites and obtain them instead from plants. Their existence depends on the existence of plants.

Life began in the oceans, and photosynthesis appeared a little later. Water surrounded the cells and thus was abundant at the beginning. The earliest photosynthesis used light simply to move protons and electrons and to generate ATP. However, at a later time, light began to be used to oxidize water which generated a reductant for metabolic reactions, and water became a substrate. Oxygen was the byproduct and its release turned the atmosphere from a reducing one to an oxidizing one. Eventually, the oxygen reached the atmospheric levels of the present day (20.9%).

One of the main roles of the hydrogen obtained in the oxidation of water was the reduction of carbon in CO_2, which became an additional substrate and was abundant in the ocean mostly in the form of dissolved bicarbonate. The accumulated mass of hydrogen, carbon, and oxygen was thus a measure of the amount of photosynthesis. The cells used the photosynthetic products in part for respiration, which consumed the biomass and returned water and CO_2 to the environment, and in part to build new plant structure.

Table 10.1   Global Net Productivity by Photosynthesis in Terrestrial and Marine Plants[a]

Community	Production (10^9 tons dry mass · year^{-1})	Chlorophyll (10^7 tons)	Standing biomass (10^9 tons dry mass)	Area (% of surface)
Marine				70.78
Open Ocean	42	1.0	1.0	
Coastal[b]	13	0.8	2.9	
Terrestrial	118	23	1840	29.22

Note. The quantity of chlorophyll and standing plant biomass are also shown.
[a]After Whittaker and Likens (1975).
[b]Continental shelves, estuaries, seaweed beds, and reefs.

Rather late in this chain of evolutionary events, plants began to inhabit the land (Chapter 12). The availability of radiation appears to have been the primary force for this move but the plants risked dehydration, extreme temperatures, and wide variations in the availability of inorganic nutrients. Table 10.1 shows that the global production of biomass on land is now almost three times that in the ocean even though the area of the land is less than half that of the ocean. Thus, photosynthetic activity is greater on land than in the ocean mostly because of the higher radiation levels. Radiation is so abundant on land that stems, branches, and roots can be built even though they often do not carry on photosynthesis and instead consume photosynthetic products. The standing biomass of these structures is large compared to that in the ocean (Table 10.1) and serves to support the large photosynthetic surface. The amount of chlorophyll is higher (Table 10.1), and thus there is more radiation harvesting capability on the land than in the ocean.

The large amount of photosynthesis probably compensates somewhat for the extremes in environment that plants face on land. The extremes limit productivity more than in the ocean, and water can be particularly limiting, with large global effects on agriculture (see Chapter 12). The causes are complex because water is not only a substrate for photosynthesis, but also a solvent for the other substrate, CO_2, and is the medium in which all the reactions take place. Leaves are covered with an epidermis containing stomata that admit CO_2 from the atmosphere according to the availability of water. Therefore, the way in which water affects photosynthesis is not immediately apparent and has been the subject of several reviews (Boyer in Kozlowski, 1976; Boyer, 1990; Kaiser, 1987; Kriedemann and Downton, 1981; Ort and Boyer, 1985). This chapter gives an overview of the area with special emphasis on the role played by water. The reader is referred to texts by Foyer (1984) and Lawlor (1993) for detailed accounts of photosynthesis.

PHOTOSYNTHESIS AND WATER AVAILABILITY

For most plants, the overall reaction for photosynthesis can be written as

$$6CO_2 + 12H_2O = C_6H_{12}O_6 + 6O_2 + 6H_2O, \qquad (10.1)$$

which shows that carbon dioxide is the preferred substrate and that water also is required. All of the water is split into hydrogen and oxygen and the O_2 product is lost as a gas. The hydrogen reduces CO_2 to form the carbohydrate product ($C_6H_{12}O_6$). The C and O in the carbohydrate thus come entirely from CO_2. The remainder of the O in CO_2 is reduced to water in later carbon metabolism ($6H_2O$). The C and O are relatively heavy atoms compared to H, and most of the mass of the $C_6H_{12}O_6$ comes from the CO_2 molecule and thus most of the land biomass comes from CO_2. In the ocean, the supply of CO_2 and bicarbonate is quite stable but on land, the supply can be variable. The concentration of CO_2 is low in the atmosphere (0.035%) or, in terms of partial pressure, 35 Pa $\cdot$ $(10^5$ Pa$)^{-1}$. Inside the leaf the CO_2 partial pressure is regulated by the stomata and the rate of consumption in photosynthesis. Because the stomatal pores change diameter according to a number of factors, including water availability (see Chapter 8), the supply of CO_2 inside the leaves varies through the day.

Similarly for O_2, aquatic environments are generally stirred and the O_2 released by photosynthesis is mixed with surrounding water. The O_2 provides a substrate for respiratory activity during the night. On land, photosynthesis cannot occur below the soil surface because of the absence of light, and the soil solution is unstirred. The O_2 consumed in respiration of roots and microorganisms must be restored by diffusion from the atmosphere. The high O_2 content of the atmosphere readily supplies shoot tissues but diffusion into the soil depends on the amount of water in the soil pores (see Chapters 4 and 5).

Flooding and Dehydration of Soil

It was pointed out in Chapter 5 that flooding can decrease the diffusion of O_2 to roots which respond by losing some of their ability to conduct water to the shoot, leading to dehydration of the shoot. Flooding forces roots to obtain most of their O_2 from the shoot by diffusion through intervening tissues. Air spaces exist between the cells (intercellular spaces) and allow O_2 to diffuse (for example, see D. A. Barber *et al.*, 1962), but the movement is restricted by the amount and tortuosity of the spaces. In most land species, the intercellular spaces are small and so tortuous that flooding tends to decrease the O_2 available to the roots but in aquatic species there may be large tissue channels (aerenchyma) that facilitate O_2 diffusion to the roots.

Dehydrating the soil results first in improved gas diffusion as water is replaced by air in the soil pores. The effect can be seen in a study of soybean

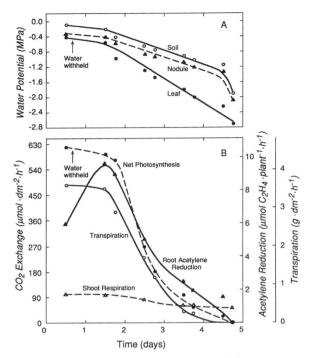

Figure 10.1 Water potential of leaves, nodules, and soil (A), and shoot photosynthesis, transpiration and respiration, and root acetylene reduction (B) in soybean after water was withheld from the soil for various times. All activities were measured simultaneously in the same intact plants in soil. Shoot activities are for the whole shoot and root activities are for the whole root system. Acetylene was supplied to the atmosphere around the soil–root system and the reduced product ethylene was measured in the same atmosphere after it diffused out of the soil. Acetylene reduction was a measure of nitrogenase activity in the roots *in situ* and also the ability of gas to diffuse through the soil and into the root nodules. After Huang *et al.* (1975a).

conducted by Huang *et al.* (1975a) and shown in Fig. 10.1. Soybean is a legume that fixes N_2 gas in the roots. As discussed in Chapter 9, the activity depends on a supply of N_2 and can be measured by supplying acetylene gas that diffuses through the soil to the roots where it is reduced to ethylene that diffuses out of the soil and is measured. The ethylene production measures not only the activity of nitrogenase but also how readily gas diffuses through the soil. Figure 10.1 shows that acetylene reducing activity was depressed in overwet soil and, as the soil drained, the activity increased. Photosynthesis and respiration were unaffected in the shoots. This suggests that nitrogen fixation had been curtailed by limited gas diffusion in the wet soil.

However, with further soil dehydration, acetylene reducing activity decreased. The dehydrated soil allowed very rapid gas diffusion, and the decrease thus resulted from unfavorable factors in the plant. The plants had only enough stored photosynthetic products to support acetylene reducing activity for a few hours and the dehydration required several days. Photosynthesis decreased as acetylene reducing activity declined (Fig. 10.1), suggesting that the supply of photosynthetic products may have become limiting (also see Chapter 9).

When severely dehydrated, net photosynthesis (the amount by which photosynthesis exceeds respiration) fell to zero in soybean (Fig. 10.1). Cell death was not an important factor because leaf respiration was only moderately inhibited. This indicates that photosynthesis was more labile than respiration and that the plant was deprived of its normal source of high energy compounds while the demand remained high. Note that the stomata closed as shown by the inhibition of transpiration (Fig. 10.1). The closure restricted the loss of water vapor and thus delayed the dehydration of the shoot. Dehydration was delayed for several days and this delaying effect is observed in nearly all leaves when they dehydrate.

The decrease of water loss reflects a general restriction of gas diffusion into and out of the leaf caused by stomatal closure. Respiration can continue under these circumstances because O_2 is so abundant in the atmosphere that internal consumption by respiration can generate a very large gradient in partial pressure in the inward direction, causing O_2 to enter fast enough to compensate for stomatal closure. Similarly, CO_2 can diffuse out during respiration. Leaves undergoing respiration in the dark build up concentrations of CO_2 inside the leaf large enough to cause outward diffusion of CO_2 even though stomata may be closed. This is not true for photosynthesis. CO_2 needs to be supplied by diffusion from outside the leaf, and the partial pressure of CO_2 is so low in the atmosphere that only a small gradient can develop as CO_2 is depleted inside the leaf. Therefore, the delaying effect of stomatal closure on leaf water loss has important implications for photosynthesis but not for respiration. This will be discussed later in more detail but it suffices to state that the more the stomata restrict water loss the more they restrict CO_2 entry for photosynthesis.

Plants differ in their photosynthetic response to dehydration (Boyer in Kozlowski, 1976; Kriedemann and Downton, 1981), even when photosynthesis occurs at similar rates in the hydrated plants. For example, leaves of vigorous sunflower plants display maximum rates of photosynthesis of about 60 μmol $\cdot$ m$^{-2} \cdot$ sec^{-1} when they are fully hydrated. *Fucus vesiculous*, an intertidal alga, displays a rate of about 40 μmol $\cdot$ m$^{-2} \cdot$ sec^{-1}. In Fig. 10.2A, sunflower lost most of its activity when its water content fell to 40 to 65% of full hydration while *Fucus* displayed only slight losses at those contents. *Fucus* had to be dehydrated to relative water contents below 20% before most of its activity was lost.

Figure 10.2B shows even more extreme differences when the responses are

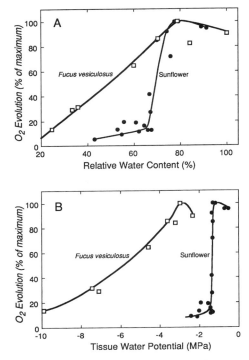

Figure 10.2 Photosynthesis during dehydration in *Fucus vesiculosus,* a marine plant, and sunflower, a land plant. (A) Oxygen evolution was measured at various tissue relative water contents (water content relative to fully turgid tissue). Full turgidity is achieved in seawater for *Fucus* but in wet soil for sunflower. (B) Oxygen evolution was measured at various tissue water potentials (note that, for *Fucus,* full hydration occurred in seawater having an osmotic potential of -2.4 MPa). Maximum photosynthesis was measured at 1% CO_2 in air and saturating radiation, and the rate at 100% of maximum was 40 and 60 $\mu mol \cdot m^{-2} \cdot sec^{-1}$ for *Fucus* and sunflower, respectively. Y. Kawamitsu and J. S. Boyer, unpublished data.

expressed at various tissue water potentials. Sunflower photosynthesis is markedly inhibited at water potentials supporting maximum photosynthesis in *Fucus.* Of course, *Fucus* is fully hydrated in seawater, which has a low water potential (-2.4 MPa) because of its large salt content. Hydrated sunflower encounters much higher water potentials because the soil is wetted with rainwater with low salt content. This illustrates the problem of expressing plant water status in terms of relative water content, as discussed in Chapter 2. Water that fully rehydrates *Fucus* is damaging to sunflower, and the significance of the water content becomes unclear whereas the water potential has no such ambiguity. Nevertheless, as pointed out later in this chapter, decreases in water content may be the cause of losses in activity of photosynthetic metabolism. The large

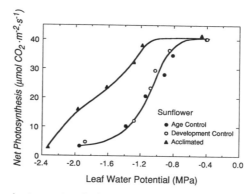

Figure 10.3 Photosynthesis at various leaf water potentials in sunflower acclimated for 2 weeks to partially dehydrated soil. The age control was given abundant water until the plants were the same age as the acclimated plants. The development control was given abundant water until the plants had developed to the same extent as the acclimated plants. After Matthews and Boyer (1984).

response of sunflower photosynthesis is curious when one considers that this land plant is probably dehydrated as often and occasionally as severely as the intertidal alga. Bewley (1979) reviews extreme desiccation tolerance in plants and points out that it is most common in primitive species such as *Fucus*.

Differences in photosynthetic response also can be found in a single species. Figure 10.3 shows that photosynthesis in sunflower decreased less than controls when the plants were exposed to moderately dehydrating conditions for long times before making the test dehydration (Matthews and Boyer, 1984). The pretreatment acclimated the plants, shifting the photosynthetic response. Growth conditions thus determine a significant part of the response of photosynthesis to dehydration.

MECHANISMS OF THE PHOTOSYNTHESIS RESPONSE

Respiration Changes

Under normal conditions, photosynthesis substantially exceeds respiration in plants. As a result, the surplus products of photosynthesis are available at times when photosynthesis cannot occur, as at night. It also provides substrate for building nonphotosynthetic parts such as roots. Photosynthesis produces products that can be moved and stored in various cells. These can be consumed at any time, thus preserving the viability of the cells. When photosynthesis is inhibited but respiration continues, photosynthetic reserves are depleted.

An important distinction between dehydration tolerant plants like *Fucus* and sensitive land plants like sunflower is the behavior of respiration. Boyer (1970, 1971a) showed that respiration continues in sunflower at water potentials low enough to completely prevent net photosynthesis (photosynthesis equaled respiration). Because respiration occurs, the tissue consumes stored photosynthetic products, and respiration has progressively less substrate. This can decrease growth and even cause some plant parts to die, as pointed out for some reproductive structures of crops in Chapter 12. By contrast, Quadir *et al.* (1979) showed that as photosynthesis approaches zero in *Fucus* during dehydration, respiration approaches zero much as in desiccation tolerant seeds. Thus, when photosynthesis is unable to occur because of dehydration, *Fucus* does not consume stored photosynthetic products.

Eventually, respiration becomes inhibited in all plants as dehydration progresses. In land plants, part of the inhibition undoubtedly comes from the scarcity of photosynthetic products to be used. However, there also are other effects. For example, Fig. 10.4 shows that isolated maize mitochondria lose activity even though malate and pyruvate are supplied at the levels expected if photosynthesis is occurring and ADP is supplied for phosphorylation (Bell *et al.*, 1971). This inhibition is thus a loss in the intrinsic activity for respiration rather than a lack of substrate, and all activity disappears at water potentials around − 3.5 MPa (Bell *et al.*, 1971) and is generally lethal. Typically, photosynthesis of land plants would have approached zero much earlier (at water potentials around − 2.0 MPa, Boyer, 1970; McPherson and Boyer, 1977; Westgate and Boyer, 1985a).

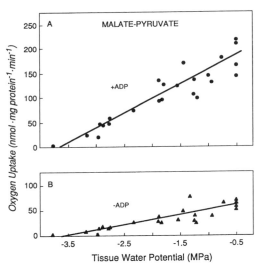

Figure 10.4 Mitochondrial respiration after isolation from maize mesocotyls having various tissue water potentials. Malate and pyruvate substrates were supplied at physiological concentrations. Rates were measured both with (A) and without (B) ADP in the medium. After Bell *et al.* (1971).

Although this inhibition of respiration occurs later than the loss of photosynthetic activity in land plants, the response is somewhat variable. Tomato, maize, soybean, and sunflower exhibit a simple decrease in respiration as the plants dehydrate (Boyer, 1970; Brix, 1962; Flowers and Hanson, 1969; Koeppe *et al.,* 1973) but young pine trees show an increase, then a decrease (Brix, 1962). The reason for this variation is unknown unless there is a temporary increase in substrate because of the action of increased amylase activity on starch as discussed later in this chapter.

Substrate Starvation

Water.   Just as respiration can be limited by the availability of products of photosynthesis that can be used as substrates, photosynthesis can be blocked by a lack of substrates. Water is a substrate whose availability often is altered by the water supply. As water is removed from the cell, the water potential decreases, the water content decreases, and the cell shrinks (Chapter 3). Enzyme-mediated reactions requiring water as substrate encounter less water in the cell. As pointed out in Chapter 9, enzymes require water for catalytic activity and sometimes also require water as substrate, as in the example urease. However, isolated urease required water potentials below -14 MPa before significant activity was lost, indicating that substrate water could not have been limiting at water potentials above -14 MPa. However, inside plant cells, metabolic activity is often affected at water potentials of -1.0 to -2.5 MPa (see Fig. 10.1 for photosynthesis). Moreover, respiration often occurs when photosynthesis is inhibited (see Fig. 10.1). Respiration requires considerable substrate water for some of the associated hydrolytic reactions, and water is a product of respiration. It would seem that if there is enough substrate water for respiration and its associated reactions, there must be enough for photosynthesis. Therefore, the loss in photosynthesis is unlikely to be caused by a lack of substrate water.

Carbon Dioxide. CO_2 is the other substrate of photosynthesis that could become limiting during dehydration. In seawater, bicarbonate is present at a concentration of about 2.5 mM and marine plants can use bicarbonate as a source of CO_2 according to the reaction

$$H_2O + CO_2 = H_2CO_3 = H^+ + HCO_3^-   \qquad (10.2)$$

which can occur in reverse. H_2CO_3 forms rapidly from HCO_3^- and H^+ but it dissociates slowly into H_2O and CO_2. Because photosynthesis uses CO_2 as substrate, the slow dissociation could be a problem for marine plants, but virtually all photosynthetic cells possess an enzyme, carbonic anhydrase, that increases the rate of H_2CO_3 dissociation (Graham and Smillie, 1976; Hatch and Burnell, 1990; Spalding *et al.,* 1983; Thielmann *et al.,* 1990). As a consequence, marine plants probably are able to obtain CO_2 readily under most conditions. For land plants, CO_2 is supplied by the atmosphere and diffuses through the epidermal barrier mostly through the stomatal pores. Once inside, it dissolves in the water in the walls of the cells of the leaf interior. The CO_2 probably is maintained inside the cell by carbonic anhydrase much as in marine plants (Hatch and Burnell, 1990).

It has long been clear that the stomata open and close according to the amount of light and the extent of plant dehydration, as discussed in Chapter 8. During dehydration, stomatal closure can decrease transpiration to less than 10% of the rate in hydrated plants (see Fig. 10.1), and transpiration may become almost undetectable in some desert species. While this slowdown is a major way of preserving the hydration of plant tissues, it also restricts CO_2 diffusion into leaves. The result is that if photosynthetic metabolism continues at substantial rates, CO_2 partial pressures will decrease in the intercellular space system of the leaf and photosynthetic metabolism will become limited by the low CO_2. On the other hand, if metabolism is also decreased, the demand for CO_2 may become less and CO_2 depletion may not occur.

Figure 10.5A shows the condition inside a hydrated leaf during active photosynthesis when CO_2 is diffusing in and being absorbed by the mesophyll cells. If stomata close but absorption continues, the CO_2 partial pressure will decrease as in Fig. 10.5B because the CO_2 is used faster than it enters. However, if photosynthetic metabolism also is blocked, CO_2 use decreases inside the leaf and CO_2 may accumulate, as shown in Fig. 10.5C. In this case, the CO_2 partial pressure can build inside the leaf until it equals the external partial pressure whereupon CO_2 entry ceases. The effect of stomatal closure thus depends on the response of photosynthetic metabolism to dehydration.

This important principle often was overlooked by early investigators. Pfeffer (1900) and Schneider and Childers (1941) understood that stomatal closure was correlated with losses in photosynthesis as leaves become dehydrated and they concluded that closure could *cause* the losses, as in Fig. 10.5B. A similar argument can occasionally be seen in recent papers (for example, Quick *et al.,* 1992). However, without knowing the response of photosynthetic metabolism or more particularly the partial pressure of CO_2 inside the leaf, such a conclusion cannot be made. It is likely that this problem accounts for much of the diversity in the literature on this subject. At present, it seems best for one to ignore those conclusions based solely on correlations between stomatal closure and photosynthesis and focus instead on results that take photosynthetic metabolism into account.

Various approaches have been used to explore the metabolic contribution. Correlations were noted between photosynthesis and decreased activities of isolated chloroplasts (see reviews by Boyer in Kozlowski, 1976; Farquhar and Sharkey, 1982; Ort and Boyer, 1985) and were considered evidence that the rate

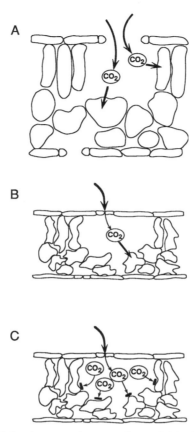

Figure 10.5 Hydrated leaf showing normal entry and use of CO_2 during active photosynthesis while stomata are open (A), dehydrated leaf showing depletion of CO_2 because of active photosynthetic metabolism while stomata are closed (B), and dehydrated leaf showing accumulation of CO_2 because of inhibited photosynthetic metabolism while stomata are closed (C). In (C), the use of CO_2 is limited more than the entry of CO_2 through the closed stomata and CO_2 builds up inside the leaf, indicating that photosynthesis is more affected by metabolism than by stomatal closure.

can be limited by metabolism and not stomatal closure. Calculations of the partial pressure of CO_2 inside leaves also indicated that CO_2 depletion was not occurring (Ehleringer and Cook, 1984; Forseth and Ehleringer, 1983; Matthews and Boyer, 1984; Radin and Ackerson, 1981). However, Terashima *et al.* (1988) and Downton *et al.* (1988a,b) found that the rate of photosynthesis may not be uniform throughout leaves and proposed that if patches of stomata close while others remain open, photosynthesis might appear to be inhibited by losses in metabolic activity whereas in reality the patchy closure was responsible

(Mansfield *et al.*, 1990; Terashima, 1992). However, Cheeseman (1991) argued against the importance of patchiness.

Patchy stomatal closure especially placed in doubt the calculation of the CO_2 partial pressures inside leaves. The calculations assume that H_2O molecules diffuse over the same path as CO_2 but in the opposite direction (Moss and Rawlins, 1963) and also that there is uniform stomatal opening across the leaf and uniform leaf temperatures. If temperatures are not uniform because of patchy closure, the partial pressure of CO_2 could be inaccurately calculated (Mansfield *et al.*, 1990; Terashima, 1992) and metabolism might be wrongly identified as the rate limitation.

The issue is further complicated by the hormonal control of stomatal aperture. As pointed out in Chapter 8, ABA is normally present at low levels in leaves; work with ABA mutants shows that stomatal opening is regulated by ABA (Imber and Tal, 1970; Neill and Horgan, 1985). In dehydrated plants, the leaf ABA normally rises (Beardsell and Cohen, 1975; Wright, 1969; Wright and Hiron, 1969). Figure 10.6 shows that leaf ABA content increased as leaf water potentials decreased, returning to normal after the plants were rewatered. The ABA rise was correlated with stomatal closure. Upon rewatering, the stomata reopened but somewhat more slowly than ABA disappeared so there was an

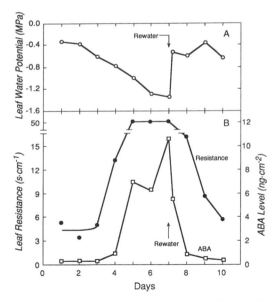

Figure 10.6   Leaf water potential (A) and abscisic acid level and leaf diffusive resistance (B) in maize from which water was withheld. The soil was rewatered on Day 7. After Beardsell and Cohen (1975).

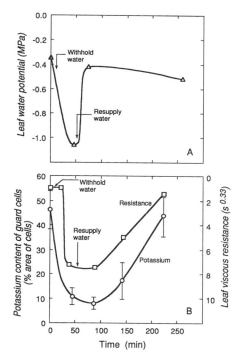

Figure 10.7   Leaf water potential (A) and guard cell K$^+$ content and stomatal resistance to the viscous flow of air (B) in a detached sunflower leaf from which water was withheld and resupplied. A numerically large resistance indicates stomatal closing. The authors also showed that high abscisic acid concentrations caused losses in K$^+$ and closure of stomata similar to those shown here for low water potentials. After Ehret and Boyer (1979).

after effect of the low water potentials not accounted for by ABA. However, high ABA can cause the loss of guard cell K$^+$ which is an essential osmoticum for stomatal opening (Mansfield and Jones, 1971), and guard cells of dehydrated leaves lose their K$^+$ (Ehret and Boyer, 1979). Figure 10.7 shows that K$^+$ was lost within a few minutes after leaf water potentials decreased. It reaccumulated slowly after rewatering. The stomatal opening was closely correlated with the amount of K$^+$ in the guard cells. Thus, it appears that high ABA causes stomatal closure because of a loss in K$^+$ and that the slow recovery after dehydration is caused by an inability of the guard cells to immediately recover K$^+$.

These findings show that ABA is an important regulator of stomata when the water supply varies. Leaves can be fed ABA through their petioles and will close their stomata, and Robinson *et al.* (1988), Downton *et al.* (1988a), and Terashima *et al.* (1988) used this approach to simulate the effects of low water potentials. Hydrated leaves fed ABA showed a patchy inhibition of photosynthesis that could be reversed at very high CO_2. Robinson *et al.* (1988) concluded that dehydration affects photosynthesis similarly by causing patchy stomatal closure. In support of this idea, Terashima *et al.* (1988) peeled the epidermis from ABA-fed leaves and observed a recovery of photosynthesis at high CO_2. Thus, the ABA results confirmed that patchy closure might occur.

On the other hand, Graan and Boyer (1990) showed that dehydration was not fully simulated by feeding ABA. Stomatal closure occurred in both cases but the dehydrated leaves did not fully recover photosynthesis at high CO_2 whereas the ABA-fed leaves did (Graan and Boyer, 1990). Lauer and Boyer (1992) directly measured the CO_2 partial pressure inside dehydrated leaves and found no decrease, but there was a decrease in ABA-fed leaves. Figure 10.8A shows the typical day/night response of the internal CO_2 (p_i) and external CO_2 (p_e) in well-watered sunflower and indicates that p_i was lower than p_e during the day, establishing an inwardly directed CO_2 gradient, and higher during the night, establishing an outward gradient (some time had to elapse at night before the outward gradient could be seen because of the measurement apparatus). In Fig. 10.8B, this diurnal pattern was disturbed by dehydration and p_i initially remained stable, then rose as the stomata closed. The inwardly directed gradient did not form on the second day. Without the gradient, CO_2 uptake could not occur. Thus, there was no evidence of CO_2 depletion inside dehydrated leaves.

In contrast, stomatal closure in ABA-fed leaves caused p_i to decrease and the inwardly directed gradient was steepened, indicating that CO_2 depletion had occurred (shown in Lauer and Boyer, 1992). This experiment indicated that stomatal closure had opposing effects on CO_2 inside the leaf depending on whether dehydration or ABA feeding caused the closure. With ABA, closure was the only response and CO_2 levels became lower inside the leaf as in Fig. 10.5B. With dehydration, there were effects in addition to those caused by closure, and the CO_2 did not become lower inside the leaf, as in Fig. 10.5C. Eventually, the CO_2 rose in the dehydrated leaves, indicating that CO_2 became more available, not less available. The additional effects were thus attributable to decreased photosynthetic metabolism that decreased the demand for CO_2 even though the stomata closed.

The problems associated with stomatal patchiness are caused mainly by nonuniform leaf temperatures that are difficult to measure accurately and are needed for calculations of p_i. In the Lauer and Boyer (1992) experiments, measuring p_i avoided the problem because leaf temperature was not involved, and the p_i was an average for the measured area of the leaf. A number of methods have been used to assess stomatal patchiness and have been reviewed (Terashima, 1992). Starch accumulation (Terashima *et al.*, 1988), autoradiography of fixed $^{14}CO_2$ (Downton *et al.*, 1988a,b; Gunasekera and Berkowitz, 1992; Sharkey and Seemann, 1989; Wise *et al.*, 1992), and fluorescence transients (Cornic *et al.*, 1989) and imaging (Daley *et al.*, 1989) have shown patchy photosynthesis under some

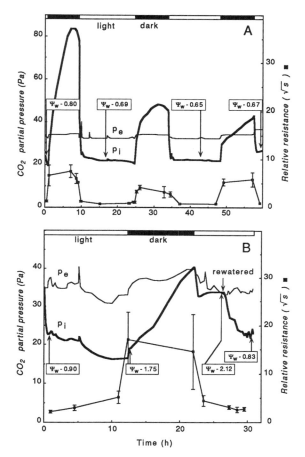

Figure 10.8 CO_2 partial pressure inside a sunflower leaf (p_i) and outside the same leaf (p_e), and stomatal resistance to the viscous flow of air through the same leaf in a hydrated control plant (A) and dehydrated plant (B). In (A), note the regular stomatal closure at night and the rise in p_i above p_e. During the day, p_i decreases below p_e because photosynthesis uses the CO_2 inside the leaf until an inward gradient forms that supplies CO_2 from the atmosphere as fast as it is used. In (B), water was withheld and p_i rises until it equals p_e. Water was resupplied as shown. Leaf water potentials are shown in boxes. The p_i was measured by equilibrating leaf CO_2 with the CO_2 in a cup attached to the underside of the leaf. After Lauer and Boyer (1992).

conditions. Although the results often have been attributed to patchy stomatal closure, the methods depend on photosynthetic metabolism and could as well reflect nonuniform metabolism (Lauer and Boyer, 1992; Wise *et al.*, 1992). Other methods are more specifically determined by stomatal aperture and include the infiltration of liquids (Alvim and Havis, 1954; Beyschlag and Pfanz,

1990; Beyschlag *et al.*, 1990, 1992; Molisch, 1912), stomatal impressions (Smith *et al.*, 1989; Weyers and Johansen, 1985), thermal imaging of leaves (Hashimoto *et al.*, 1984), and direct observations of stomata (Laisk *et al.*, 1980; Van Gardingen *et al.*, 1989). Of these, the direct observation of stomata gives the most unambiguous measure (Terashima, 1992). The infiltration of liquids is affected by patchy wetting of the stomatal pores, and stomatal impression materials may not uniformly enter the stomata. Thermal imaging usually does not have the required spatial resolution. The direct observation of stomata can indicate exactly which leaf areas might be deprived of CO_2 (Terashima *et al.*, 1988) but this has not been attempted on the scale necessary because closure must be demonstrated on a large number of stomata and on corresponding areas of the upper and lower leaf surfaces simultaneously if stomata occur on both surfaces. Several aspects of stomatal heterogeneity also are discussed in Chapter 8.

Metabolic Inhibition

Chloroplasts isolated from leaves having low water potentials display lower photosynthetic activity than chloroplasts isolated from the same leaves at high water potentials (Boyer and Bowen, 1970; Keck and Boyer, 1974; Potter and Boyer, 1973). The change reflects either a lowered photosynthetic activity *in vivo* or an increased susceptibility to chloroplast damage from isolation, but in either case fundamental metabolic change has occurred. Thylakoid membranes show less photosystem II, photosystem I, and photophosphorylating activity (Keck and Boyer, 1974; Mayoral *et al.*, 1981), and extracts of stromal enzymes display lower activities (Antolin and Sanchez-Diaz, 1993; Gunasekera and Berkowitz, 1993; Huffaker *et al.*, 1970; Johnson *et al.*, 1974; Mayoral *et al.*, 1981; O'Toole *et al.*, 1976). Crystalline structures sometimes are seen in the stroma and appear to be precipitated ribulose bisphosphate carboxylase (Fellows and Boyer, 1976; Freeman and Duysen, 1975; Gunning *et al.*, 1968; Shumway *et al.*, 1967).

Figure 10.9 shows an example in water-deficient wheat which lost activity for electron transport, photophosphorylation, and carboxylation of ribulose bisphosphate and phosphoenolpyruvate (Mayoral *et al.*, 1981). Of these activities, photophosphorylation was most inhibited. Keck and Boyer (1974) similarly found a large inhibition of photophosphorylation. An important aspect was that the extracts and assays were standard ones that would rehydrate membranes and enzymes. Thus, the inhibitions were not reversible after isolating the enzymes. This persistence of the inhibition after isolation is a central feature of losses in chloroplast activity caused by dehydration *in vivo*.

Electron transport and photophosphorylating activities are important for the photochemical activity of the chloroplasts and one would expect that if so much

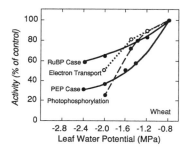

Figure 10.9   Activities for electron transport, photophosphorylation, ribulose bisphosphate carboxylase (RuBP Case), and phosphoenolpyruvate carboxylase (PEP Case) in extracts from wheat leaves having various water potentials. Electron transport, photophosphorylation, and RuBP Case are chloroplast activities. PEP Case is a cytoplasmic enzyme in this species. Activities were measured in standard media. After Mayoral *et al.* (1981).

inhibition was present in isolated chloroplasts, the overall photochemical activity also should be less in the intact leaves. Photochemical activity can be determined in leaves by measuring the number of CO_2 molecules fixed per quantum of radiation, termed the quantum yield of photosynthesis. In sunflower and soybean, quantum yields were reduced as the plants dehydrated (Matthews and Boyer, 1984; Mohanty and Boyer, 1976; Sharp and Boyer, 1986). Losses in photochemical activity also decreased the maximum rate of photosynthesis measured by saturating the leaf with CO_2 and radiation. In sunflower, maximum rates became only a small fraction of the control rates as plants dehydrated (Graan and Boyer, 1990; Matthews and Boyer, 1984; Sharp and Boyer, 1986). Measurements of photophosphorylation in the intact leaf (Ortiz-Lopez *et al.,* 1991) also confirmed the losses in photochemical activity but were not so severe when the leaves were pretreated in light. Light pretreatment activates ATP synthetase, the terminal phosphorylating enzyme sometimes called coupling factor (Hangarter *et al.*, 1987) which suggests that leaf dehydration may act in part on the activation of certain photosynthetic enzymes.

The electron microscope shows some of these dehydration-induced changes (see Chapter 3) when the dehydration is preserved during fixation (see Appendix 3.1). Fellows and Boyer (1976) showed that osmoticum can be added to the fixative to give the same water potential as the leaf tissue and thus preserve low water potentials during fixation. With this treatment, the chloroplast thylakoids did not show damage but instead displayed changes in conformation in response to the leaf water potential (Fellows and Boyer, 1976). Figure 10.10B shows that the illuminated thylakoid lamellae were 150 Å thick in the controls but only 120 to 130 Å thick in the dehydrated leaves. Because thinning normally occurs as part of electron transport and photophosphorylation when thylakoid

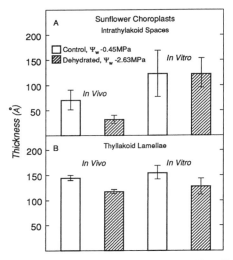

Figure 10.10   Thickness of intrathylakoid spaces (A) and thylakoid lamellae (B) measured with the electron microscope in chloroplasts of sunflower leaves (*in vivo*) or isolated from sunflower leaves (*in vitro*) having high or low water potentials. Open bars are from the control leaves and shaded bars are from the dehydrated leaves. The isolated membranes were suspended in identical medium and were exposed to light before and during fixation. The fixative had the same water potential as the leaves or isolation medium. Note that spaces (A) and lamellae (B) were thinner in dehydrated cells, but after isolation the spaces (A) swelled whereas the lamellae (B) remained thinner. The thickness of the spaces between lamellae responds to the osmotic potential significantly but the thickness of the lamellae themselves does not. After Fellows and Boyer (1976).

membranes are energized by light, the excessive thinning may be unfavorable. The excessive thinning persisted after the membranes were isolated which is consistent with the persistence in losses of chloroplast activities described previously.

The intrathylakoid spaces between the membranes also were thinner *in vivo*, but the differences did not persist after isolation (Fig. 10.10A). The thylakoid spaces respond to changes in osmotic potential across the membranes and the thinning *in vivo* indicates that dehydration had increased the potential difference. In effect, the intrathylakoid spaces acted as ultrastructural osmometers (Fellows and Boyer, 1976). In the cell, they showed how much water was removed. In the rehydrating assay medium, they swelled and showed how much rehydration occurred after isolation.

The persistence of dehydration effects in the thylakoid membranes could be seen in certain protein components of the membranes. Figure 10.11A shows that the membrane enzyme, coupling factor, changed its conformation in dehydrated leaves (Younis *et al.*, 1979), and the altered conformation was associated with a decreased binding affinity for the substrate ADP at the active site of

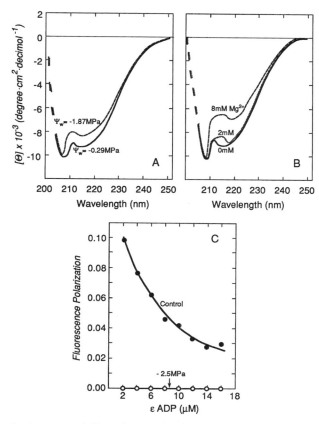

Figure 10.11 Conformation of chloroplast coupling factor (ATP synthetase) isolated from spinach leaves having high or low water potentials (A), or isolated from solutions having various Mg^{2+} concentrations (B). Isolation from Mg^{2+} solution simulated isolation from leaf. The 2 mM Mg^{2+} represents the concentration in control chloroplasts and 8 mM the possible concentration in dehydrated chloroplasts. Conformation was determined from differences in circular dichroism spectrum shown (θ). Note the similar conformation change caused by a low water potential and high Mg^{2+} concentration. (C) Binding of ADP analog (ϵADP) to coupling factor isolated from spinach leaves having high or low water potentials. The ϵADP binds to the active site of the enzyme. High binding is shown by high polarization at left that decreases with saturation of sites in control. Binding was undetectable when the coupling factor was isolated from leaves having a water potential of -2.5 MPa. After Younis *et al.* (1979) and Younis *et al.* (1983).

the enzyme (Fig. 10.11C). Thus, it seems that the altered conformation of the membranes seen in the electron microscope was traceable to an altered conformation of specific molecular components of the membranes. The changed conformation of the components might physically block access to the active site of some membrane enzymes such as coupling factor where the lack of ADP bind-

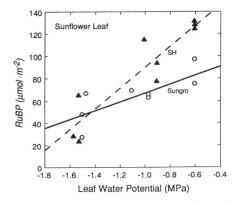

Figure 10.12   Ribulose bisphosphate (RuBP) concentrations in leaves of two sunflower cultivars (SH and Sungro) from which water was withheld. RuBP is regenerated in the carbon reduction cycle (Calvin cycle) with an input of products of the photochemical reactions. Decreased RuBP levels indicate a block may be present in the photochemical reactions or carbon reduction cycle. From Gimenez *et al.* (1992).

ing was observed (Fig. 10.11C). Such a block will also decrease the regeneration of ribulose bisphosphate (RuBP), which is the CO_2 acceptor in the carbon reduction cycle (Calvin cycle) of photosynthesis (Farquhar and Sharkey, 1982). RuBP regeneration was observed to be diminished in dehydrated sunflower leaves (Fig. 10.12) and was sufficient to account for the loss in activity for CO_2 fixation (Gimenez *et al.*, 1992). This supports the idea that the chloroplasts could lose important photochemical activities in dehydrated cells although other changes could alter RuBP regeneration as well (Boyer in Kozlowski, 1976; Farquhar and Sharkey, 1982; Ort and Boyer, 1985).

Plant Signals That Trigger the Metabolic Response

These changes in photosynthetic metabolism indicate that certain features of the cell are inhibitory for photosynthesis during dehydration but, since photosynthesis responds differently between species and within the same species, the factors must be variable and somehow under cell control. One possibility is that the removal of photosynthetic products is disrupted, and photosynthesis becomes inhibited by their accumulation. However, this seems unlikely in view of the lack of an immediate effect on photosynthesis when the phloem is disrupted (Huang *et al.*, 1975b) or when the leaves are detached. Photoinhibition also has been proposed as another cause of inhibited photosynthetic metabolism during dehydration (Björkman, 1981; Kriedemann and Downton, 1981; Osmond *et al.* in Turner and Kramer, 1980) and according to this hypothesis, stomatal closure would deprive the chloroplasts of the substrate CO_2 that normally accepts most of the hydrogen atoms from the photolysis of water. Continued photolysis would result in damage to the thylakoid membranes. There is evidence that photoinhibition can be detected in some species at low water potentials (Björkman and Powles, 1984; Ludlow and Björkman, 1984; Ögren and Öquist, 1985) but the effects on gas exchange appear to be negligibly small (Sharp and Boyer, 1986) and thus unimportant for CO_2 fixation.

Another hypothesis is that the low chemical potential of water may alter enzyme structure and thus inhibit metabolic reactions. Enzyme structure and activity rely on an ordered shell of water around proteins (Klotz, 1958; Rupley *et al.*, 1983) and it seems possible that decreasing the potential of water might alter the order of that water. However, as pointed out earlier and in Chapter 9, dehydration must be extreme before an effect is seen. Indeed, Boyer and Potter (1973) and Potter and Boyer (1973) showed that varying the chemical potential around the chloroplast thylakoid membranes by altering the turgor or osmotic potential had only small effects on their photochemical activity. However, the kind of solute causing the osmotic potential has a large effect on photochemical activity. Santarius (1969) and Santarius and Giersch (1984) convincingly showed that increasing concentrations of sucrose and glucose had little effect on electron transport or cyclic photophosphorylation whereas increasing NaCl was markedly inhibitory. Thus, direct ion effects are more important than the osmotic potential they create.

For a time it was thought that high ABA might alter photosynthetic metabolism (Raschke and Hedrich, 1985) but applying ABA to isolated chloroplasts and intact cells did not cause altered photophosphorylating activity (Boyer, 1973) and removing the epidermis removed the effects of ABA on photosynthesis (Terashima *et al.*, 1988). Thus, although leaf ABA content increases during dehydration and contributes to stomatal closure, no important alterations in photosynthetic metabolism have been substantiated.

As dehydration occurs, the solute environment around enzymes changes because water is removed but the solutes remain in the cell (Boyer, 1983). In sunflower leaves, photosynthesis is markedly inhibited when half the cell water is lost (Fig. 10.2). Solute concentrations would double in this situation, particularly for solutes that cannot be metabolized. Figure 10.13A shows that the photophosphorylating activity was lost in thylakoid membranes from dehydrated spinach leaves (Younis *et al.*, 1979) and that doubling the Mg^{2+} concentration around normal membranes similarly inhibited photophosphorylation (Fig. 10.13B; Younis *et al.*, 1983). Mg^{2+} is a known regulator of photophosphorylating activity (Anthon and Jagendorf, 1983; Pick and Bassilian, 1982; Racker, 1977; Tiefert *et al.*, 1977) and is markedly inhibitory when concentrations increase.

It is noteworthy that this effect persisted after isolating the membranes from the high Mg^{2+} concentrations just as the dehydration effect persists after isolat-

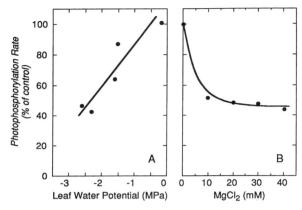

Figure 10.13   Photophosphorylation in thylakoid membranes isolated from spinach leaves having various water potentials (A) or isolated from solutions having various Mg^{2+} concentrations (B). Isolation from Mg^{2+} solutions simulated isolation from the leaf. Higher Mg^{2+} concentrations are expected at lower leaf water potentials. Concentrations of Mg^{2+} are normally about 3 mM around the thylakoid membranes in the light and could be significantly higher during dehydration. After Younis *et al.* (1979) and Younis *et al.* (1983).

ing the membranes from dehydrated cells (Fig. 10.13). Moreover, the chloroplast stromal concentration of Mg^{2+} is normally around 3mM in the light (Rao *et al.*, 1987) and could readily increase to inhibitory levels of 6 mM or higher as the cell loses half its water. Perhaps most significant, there is a change in conformation of coupling factor in high Mg^{2+} that is similar to the one caused by dehydration in the cell (Fig. 10.11B, cf. Fig. 10.11A).

Figure 10.14 shows the likely ionic environment around the chloroplast thylakoids where these effects occur. The change in enzyme conformation for the coupling factor may make the membrane appear thinner in the electron microscope and be associated with the lack of access of ADP to the active site for photophosphorylation. The Mg^{2+} may increase in concentration as dehydration occurs and bind increasingly to the coupling factor protein (Younis *et al.*, 1983). If Mg^{2+} remains bound during isolation of the enzyme, the large amount of bound Mg^{2+} could result in decreased activity *in vitro*. Thus, this conception of the membrane would account for both the *in vivo* and *in vitro* inhibition of photophosphorylation and the persistence of the effects *in vitro*.

In support of this concept of inhibitory solute concentrations, the K^+ in cells is an essential regulator for metabolism (Evans and Sorger, 1966; Evans and Wildes, 1971; Wilson and Evans, 1968), and Berkowitz and Whalen (1985) and Pier and Berkowitz (1987) observed an altered photosynthetic response when leaf tissue of differing K^+ status was exposed to low water potentials. Leaf tissue that was becoming depleted of K^+ appeared to be more inhibited than high K^+

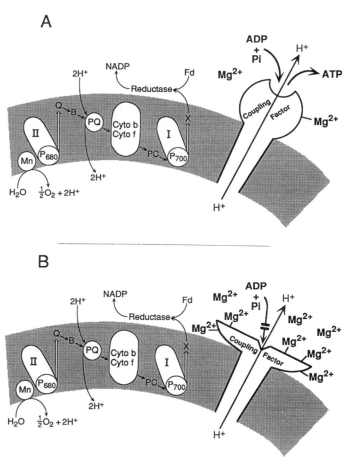

Figure 10.14 Ionic environment around chloroplast thylakoid membrane in a hydrated cell (A, high water potential) or a dehydrated cell (B, low water potential). Note the higher concentration of solutes in (B) than in (A) leading to the binding of more solute (Mg^{2+}) to the coupling factor protein required for photophosphorylation and the changed conformation of the coupling protein associated with a blocked access of ADP to the active site in (B). The binding of Mg^{2+} may account for the persistence of chloroplast inhibition after the membranes are isolated from the cell and for the changes in conformation of the membrane observed *in vivo* and *in vitro*.

tissue when low water potentials occurred. Berkowitz and Kroll (1988) and Gupta and Berkowitz (1987) showed that there may be a differential regulation of chloroplast water content compared to the surrounding cytoplasm. This could alter the ion concentrations in the chloroplast compartment. The solutes might be distributed differently between organelles as water is removed from the cell and might change the susceptibility to dehydration. Kaiser *et al.* (1986) found evidence that ionic constituents affect the photosynthetic response to low water potentials in an artificial stroma. The activity of ribulose bisphosphate carboxylase was decreased by slightly elevated concentrations of sulfate, phosphate, and occasionally Mg^{2+} in the medium. Such elevations might occur during dehydration. Santarius (1969) and Santarius and Giersch (1984) found that high concentrations of NaCl could inhibit electron transport and cyclic photophosphorylation, but similar concentrations of sucrose or glucose were not inhibitory and partially prevented the effect of NaCl.

These results indicate that the most likely cause of inhibited photosynthetic metabolism in dehydrated plants is the changed solute environment around the enzymes, particularly the ionic environment. Many of these solutes can change concentrations passively as a simple result of the removal of water. Some cannot be metabolized and others may be moved to different compartments. Changes in enzyme activity probably occur because the ions play regulatory roles, and changing the concentration increases or decreases the activity, as also discussed in Chapter 9.

ACCLIMATION

The fact that photosynthesis can acclimate to dehydrating conditions implies that some characteristic of the cell can be altered to put photosynthesis at less risk. Because altered stomatal behavior or cell ion status or compartmentation could be important contributors, the nutritional history of the plants could play a role. In agreement with this concept, Radin and his co-workers (Radin and Ackerson, 1981; Radin and Parker, 1979a,b) showed that plants grown in differing nitrogen regimes had altered stomatal closure at low water potentials. Plants having a low nitrogen status closed their stomata earlier (Radin and Parker, 1979a,b) and had higher ABA levels (Radin and Ackerson, 1981) than plants having a high nitrogen status. This delayed the onset of severe dehydration and represented a form of acclimation for the plant.

Matthews and Boyer (1984) showed that acclimation did not depend solely on stomatal closure but could be observed in photosynthetic metabolism of sunflower leaves. Rao *et al.* (1987) grew sunflower plants at differing Mg^{2+} levels and found that their leaves differed by fourfold in Mg^{2+} content. When the plants were dehydrated, the maximum rate of photosynthetic metabolism was more severely reduced in the high Mg^{2+} leaves than in the low Mg^{2+} leaves. Thus, it was possible to acclimate photosynthetic metabolism to dehydration by altering the Mg^{2+} concentration of the leaves.

Various enzymes were shown to be much less inhibited by photosynthetic products and metabolites such as sugars and amino acids than by inorganic ions like Mg^{2+}, K^+, or Na^+ (Hanson and Hitz, 1982; Yancey *et al.*, 1982; Wyn

Jones, 1980). Cells packed with sugars and amino acids are commonly seen during dehydration (Crowe and Crowe, 1986; Crowe *et al.*, 1987; Meyer and Boyer, 1981; Munns *et al.*, 1979; Yancey *et al.*, 1982). These compounds can accumulate apparently without inhibiting photosynthetic metabolism, and thus their contribution to acclimation is mostly through osmotic adjustment (see Chapter 3) in a way that allows photosynthesis to continue.

These results indicate that there are two features of photosynthetic acclimation that are important. One is a dehydration avoidance that results from delaying the onset of dehydration as seen with nitrogen or osmotic adjustment. Another is a dehydration tolerance that affects photosynthesis less at a particular dehydration as seen with magnesium. The two factors contribute by affecting stomatal closure and enzyme responsiveness to dehydration. With delays in dehydration and less enzyme inhibition, photosynthetic products can continue to be produced and contribute to osmotic adjustment, which maintains hydration levels and further delays the onset of metabolic inhibition. As a consequence, acclimation develops most fully when plants are dehydrated slowly enough to allow the adjustment of regulator pools and the production of noninhibitory solute from photosynthesis.

RECOVERY

The ability to recover photosynthetic capacity is important for the resumption of plant growth when water is resupplied. Plants differ genetically in this respect, and the extent and duration of the preceding dehydration have marked effects. An example is apparent when *Fucus* recovery is compared to sunflower (Fig. 10.15). *Fucus* readily recovers photosynthetic activity after dehydration to 5% relative water content and sunflower recovers readily from 40% water content. However, dehydrating sunflower to a 5% content is lethal.

In *Fucus,* water enters through the surface of the plant but in sunflower, water enters through the roots. The roots and vascular system must function in sunflower in order for rehydration to occur. Boyer (1971b) showed that part of the ability to rehydrate depended on the extent of vascular blockage. Dehydration puts the xylem water under tension and breaks often occur due to cavitations, forming gas embolisms that block water movement (Tyree and Sperry, 1989; also see Chapter 7). The more severe the dehydration, the larger the tension and more frequent the blockage. Leaf water potentials of only -2 MPa were sufficient to cause a major blockage in sunflower stems (Boyer, 1971b). Excising the leaf with its petiole under water removed the blocked part of the vascular system in the lower part of the plant and allowed rehydration and resumption of photosynthesis to occur (Boyer, 1971b).

Even if full rehydration occurred, the ability of photosynthesis to recover depended on the severity of dehydration. Fellows and Boyer (1978) found that

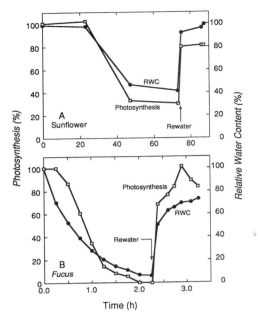

Figure 10.15   Photosynthesis and relative water content of sunflower (A) or *Fucus vesiculosus* (B) at various times after withholding and resupplying water. Sunflower was rehydrated by watering the soil whereas *Fucus* was rehydrated by submerging in seawater. Recovery was not observed in sunflower if relative water contents became as low as in *Fucus*. After Boyer (1971b) and T. Driscoll and J. S. Boyer (unpublished data).

sunflower leaf cells increasingly had breaks in the plasmalemma and/or tonoplast membranes as dehydration became more severe. Figure 10.16 shows that more breaks appeared as the water potentials became lower. The breaks resulted in lysis of the cells which limited the number of living cells that could recover when water was resupplied. Thus, even though the leaf recovered in water potential and appeared normal, the death of some of the cells limited the final extent of photosynthesis recovery. Interestingly, cell death appeared random but under severe conditions, clusters of cells were affected. Patchy photosynthesis, when it occurs during dehydration, may be attributable in part to this patchy cell death. Such patchy death would also cause patchy recovery.

Leopold *et al.* (1981) observed that leaf cells leaked internal solutes when dehydrated (see Fig. 3.3). The leakiness was a measure of the loss of plasmalemma function and was inversely correlated with the ability to recover from severe desiccation. If there was little leakage, recovery was complete but with significant leakage recovery was prevented. The leakage reported by these in-

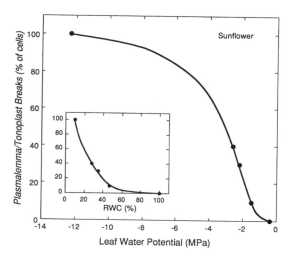

Figure 10.16 Plasmalemma and tonoplast breaks in cells of sunflower leaves having various water potentials (inset: various relative water contents). Breaks were observed under an electron microscope in tissue fixed at the same water potential as in the leaf. After Fellows and Boyer (1978).

vestigators could have resulted from the membrane breakage observed by Fellows and Boyer (1978).

It is clear from ultrastructural studies (Bewley and Pacey, 1978; Fellows and Boyer, 1978; Hallam and Gaff, 1978a,b) that dehydrated cells become severely deformed by their decrease in volume (see Chapter 3). The deformation of the plasmalemma is extreme and is forced on the cell by wall folding as the cell shrinks. The folding may become so severe that the plasmalemma and/or tonoplast become pressed inside the folds (Fellows and Boyer, 1978) which undoubtedly creates large local stresses on the membranes. Perhaps these stresses lead to membrane disruption. However, the chloroplasts generally are less deformed than the plasmalemma because they can move inside the cells.

It has been proposed that much of the ability of cells to survive severe dehydration depends on the ability to maintain the membranes intact (Crowe and Crowe, 1986; Crowe *et al.*, 1987; Leopold *et al.*, 1981). In addition to evidence of physical deformation of membranes in dehydrating cells, evidence has been presented that certain sugars such as trehalose or sucrose can replace water in maintaining membrane structure (Crowe and Crowe, 1986; Crowe *et al.*, 1987; also see Chapter 12). If so, the removal of water would disrupt the plasmalemma and/or tonoplast less if the sugars were abundant. It would be interesting to determine whether the ability of photosynthesis to recover from severe dehydration is related to the tissue content of sugars.

Another common effect of dehydration is an accelerated leaf senescence. Maize leaves senesce prematurely when the plants have insufficient water (Boyer and McPherson, 1975), and leaf senescence appears to be under genetic control through hydrolytic enzymes that increase in activity after the onset of dehydration (Todd in Kozlowski, 1972). Soon afterwards, leaf yellowing and death are observed. Jacobsen *et al.* (1986) found that the activity of α-amylase increased as leaf water potentials became low in seedlings of barley from which water was withheld. Because this activity appeared to be located in the cytoplasm, it was thought to be one of the general class of hydrolytic enzymes involved in the early phases of leaf senescence. The mRNA increased for the enzyme in the wilted leaves which indicated that the activity increase was regulated by gene action rather than by activation of pre-existing enzyme. The new activity appeared to result from the synthesis of new enzyme.

Thus, the genetic basis of accelerated leaf senescence can be altered at least in principle. In plant breeding, it is possible to select for genotypes whose leaves remain green during a water deficiency. The "stay green" character allows photosynthesis a better chance to quickly recover after rain than in genotypes whose leaves senesce. Aparicio-Tejo and Boyer (1983) concluded that premature leaf senescence is undesirable and should be minimized by selecting for genotypes whose leaves remain viable.

TRANSLOCATION

Leaf senescence is associated with the hydrolysis of leaf constituents and the transport of the hydrolytic products from the senescing leaves to viable plant parts. This has the effect of conserving dry mass and mineral nutrients for the plant although the effect is small. However, the ability to transport hydrolyzed products illustrates that the phloem remains viable. Fellows and Boyer (1978) observed that the ultrastructure of leaf phloem remained normal in appearance while that of the surrounding cells showed major disruptions when the plants were exposed to water deficits that caused leaf senescence. The phloem is thus one of the last parts of the leaf to senesce and it is often observed that veins remain green while interveinal tissues show signs of senescence. In general, translocation is one of the most stable plant activities (Munns and Pearson, 1974).

Several studies showed that, when $^{14}CO_2$ was supplied to dehydrated leaves and fixed photosynthetically, the photosynthetic products were translocated more slowly than in controls (Brevedan and Hodges, 1973; Johnson and Moss, 1976; Munns and Pearson, 1974; Sung and Krieg, 1979; Wardlaw, 1967; Watson and Wardlaw, 1981). Comparisons of losses in photosynthetic activity and translocating activity showed inhibitions for both processes (Brevedan and Hodges, 1973; Johnson and Moss, 1976; Munns and Pearson, 1974; Sung and

Krieg, 1979; Wardlaw, 1967). On the other hand, measurements of dry matter transport in dehydrated plants indicated that translocation occurs readily (Jurgens *et al.*, 1978; McPherson and Boyer, 1977; Westgate and Boyer, 1985a); Chapter 12 describes experiments in which the products of photosynthesis are fed to stems and maintain embryo viability in distant reproductive tissues, indicating that translocation is active. Jurgens *et al.* (1978) showed that for the first 24 hr after the onset of dehydration severe enough to inhibit photosynthesis, the translocation of fixed ^{14}C was markedly inhibited but recovered moderately several days later. Thus, much of the severe inhibition of ^{14}C translocation appeared to be transient. Many $^{14}CO_2$ measurements have been conducted for only a few hours after plant dehydration (Brevedan and Hodges, 1973; Johnson and Moss, 1976; Wardlaw, 1967) but the translocation of dry matter generally requires several days to measure (Jurgens *et al.*, 1978; McPherson and Boyer, 1977; Westgate and Boyer, 1985a). Therefore, many differences between results probably can be attributed to differences in times employed for the measurements, and it may be concluded that translocation generally remains active in water-deficient plants except for a few hours after the onset of the deficiency when severe inhibition may be observed.

SUMMARY

Plant productivity is determined mostly by photosynthetic CO_2 fixation, and water is a major limiting factor for photosynthesis in many environments. In land plants, photosynthesis is much more susceptible than respiration to dehydration effects, which implies that there are reactions or features specific to photosynthesis that are the cause. Large differences exist in susceptibility among species, and growth conditions can change the susceptibility within a species, leading to acclimation.

While respiration is active, stored reserves of photosynthetic products are drawn upon whenever photosynthesis is sufficiently inhibited, with some deleterious effects. The continued respiration indicates that water is adequately available as a substrate for metabolism, and photosynthesis probably fails for other reasons. With severe dehydration, respiration also can decrease and eventually will cease.

CO_2 is another substrate for photosynthesis and enters the leaf primarily through stomata, and much work centers on the role of stomatal closure. The stomata lose osmotic quantities of K^+ when they close, probably because abscisic acid levels increase as the leaf dehydrates, and the loss appears to cause the closure. Closure may or may not inhibit photosynthesis depending on the demand for CO_2 by photosynthetic metabolism. If metabolism remains active, CO_2 partial pressures decrease in the leaf as the stomata close, and the decrease can limit photosynthesis. If metabolism is inhibited, the demand for CO_2 diminishes and the partial pressure of CO_2 can rise even though stomata close. The inhibition is then attributed to the inhibited metabolism because the CO_2 has actually become more available through disuse. This indicates that the stomatal limitation of photosynthesis cannot be decided simply from the stomatal closure that accompanies losses in photosynthesis. Recent methods of measuring CO_2 levels inside the leaf promise to provide a way to test effects of stomatal closure.

Most metabolic stages in the photosynthetic process are inhibited by dehydration, and the inhibition can be seen in the intact leaf. Photophosphorylation appears to be heavily involved, and there are molecular changes in enzyme conformation that correlate with the decreased access of substrates to the photophosphorylating enzymes. Certain causes can be ruled out such as reduced translocation of products, losses in ultrastructural integrity, damage by photoinhibition, and effects of the chemical potential of water directly on the photosynthetic enzymes. Instead, altered concentrations of regulatory solutes, particularly inorganic ions, appear most likely to be responsible. Concentrations of these ions change because of the passive concentrating effects of dehydration and perhaps because of altered compartmentation of the solutes inside the cells. Sugars and amino acids accumulate in dehydrated leaves but do not have the inhibitory action seen with high concentrations of inorganic ions. The accumulation of sugars and amino acids depends on photosynthesis to supply the appropriate substrates and in turn protects photosynthetic metabolism.

This probably explains why photosynthetic metabolism is affected by the water content of cells. Factors that delay the loss of water such as early stomatal closure also delay the concentrating of the cell solution and thus diminish the inhibition of metabolism. Delays in dehydration and changes in cell composition appear to be major mechanisms of leaf acclimation to dehydrating conditions.

During dehydration, leaf cells undergo lysis that becomes pervasive as dehydration becomes more severe. The extent of lysis determines in part how much recovery of photosynthesis will occur when water is resupplied to the leaves. Vascular emboli form during dehydration and can hinder the recovery from dehydration because of inhibited water transport, and photosynthesis may be suppressed for several days after a dehydration/recovery cycle because of this effect or because of incomplete stomatal opening due to inability of the guard cells to accumulate K^+. Leaves also undergo accelerated senescence during dehydration, at least some of which appears to be under genetic control. Translocation continues to be active in dehydrated plants, although there is evidence of transient inhibition for a few hours after dehydration occurs.

SUPPLEMENTARY READING

Boyer, J. S. (1976). Water deficits and photosynthesis. *In* "Water Deficits and Plant Growth" (T. T. Kozlowski, ed.), Vol. 4, pp. 153–190. Academic Press, New York.

Foyer, C. H. (1984). "Photosynthesis." Wiley, New York.

Lawlor, D. W. (1993). "Photosynthesis: Molecular, Physiological and Environmental Processes," 2nd Ed. Longman Scientific & Technical, Essex.

Ort, D. R., and Boyer, J. S. (1985). Plant productivity, photosynthesis and environmental stress. *In* "Changes in Eukaryotic Gene Expression in Response to Environmental Stress' (B. G. Atkinson and D. B. Walden, eds.), pp. 279–313. Academic Press, New York.

Yancey, P. H., Clark, M. E., Hand, S. C., Bowlus, R. D., and Somero, G. N. (1982). Living with water stress: Evolution of osmolyte systems. *Science* **217**, 1214–1222.

Growth

11

INTRODUCTION

Growth is the most obvious feature of plant activity because it causes plants to become larger or cells to become more numerous in a population. All the work of metabolism is directed toward it, and the ability to reproduce depends on it. Water plays a central part as can be seen in a maize plant that by the time it flowers has gained 800 g of weight of which 85% is water. Only 100 g make up the organic part of the plant together with some mineral salts. By this time, the plant has grown to about 300 cm and Table 11.1 shows that the various organs of maize grow rapidly when conditions are favorable; the leaves become visibly longer almost minute-by-minute. To grow this rapidly, the plant must gain an average of 14 ml of water content every day until flowering, and the water must be incorporated into each cell according to its rate of enlargement. Similarly in a population of single-celled plants, growth of the population occurs by cell division, and the daughter cells enlarge before they can reproduce again by dividing but much of the enlargement comes from an increase in water content of the cells.

This indicates that most of what we call plant growth is actually increased cellular water content. However, as discussed in Chapter 3, cells can change in size not only by growing but also by swelling and shrinking reversibly as they hydrate and dehydrate. The reversible swelling and shrinking are caused by the elastic nature of the cell walls as they are stretched, causing the cells to become turgid when water enters but flaccid as water departs. Gains in water content for growth need to be carefully distinguished from these reversible elastic effects

Table 11.1   Growth Rates in Various Parts
of a Maize Plant[a]

Part	Growth rate $(mm \cdot hr^{-1})$
Leaf	4.0
Style (silk)	2.6
Stem	1.3
Root (nodal)	1.1

Note. The rates were measured for 24 hr in plants supplied with adequate soil water and growing in day temperatures of 30°C and night temperatures of 20°C. Root and leaf rates are from the same plants and style and stem rates are from older plants.
[a]From Westgate and Boyer (1985b).

and we will define plant growth to be an irreversible increase in size to separate it from the reversible changes. To determine whether an increased water content is growth or simply an elastic change, the tissue can be dehydrated to its original water content and the size remeasured. An increased size at the original water content indicates that enlargement was irreversible and growth occurred. Also, continuously measuring the size of plant parts allows one to distinguish between the reasonably steady long-term increases characteristic of growth and the short term but reversible changes caused by elastic enlargement. In mature tissues, elastic changes are the only ones that occur and they are easily detected.

Growth is a prerequisite of reproduction in unicellular plants because the cell usually cannot divide unless it enlarges to about the size of the cell from which it came. Similarly, in multicellular plants the dividing cells of the meristematic tissues must double in size before they can divide again. Cell division generally is restricted to specific regions in multicellular plants, but cells outside of those regions can enlarge often by much more than doubling. It is common for cells in tissues to enlarge 10- to 15-fold after division and they divide no further but instead contribute to the increased size and water content of the tissue. In the process, the cells expand in directions that develop the form of the plant part. They also differentiate to form specialized tissues. Thus, division, enlargement, and differentiation together give rise to plant organs such as leaves, roots, and flowers.

Because water plays such a large role in these processes, its availability often becomes limiting. In the ocean, marine plants are surrounded by water but high concentrations of salt create an environment that can be osmotically unfavorable for some of them. On land, the roots are in contact with soil water and supply the rest of the plant but there is often an inadequate supply. Understanding how growth is limited under these conditions has many applications in agriculture and in the management of natural plant communities.

GROWTH OF SINGLE CELLS

Much of our understanding of cell enlargement comes from studies of single cells of plants such as the large-celled algae *Nitella flexilis* and *Chara corallina* (Fig. 11.1). Among the algae, these species are probably some of the closest

Figure 11.1 *Chara corallina* showing a giant internode cell and branches at the nodes. The internode cell is about 0.7 μm in diameter and frequently reaches a length of 10–15 cm.

relatives of land plants (Chapter 12). The internodal cells are large enough to observe easily, are in direct contact with the surrounding medium, and are photosynthetic and self-sufficient for substrate which simplifies growth experiments. Immature cells of these plants elongate steadily (Green *et al.,* 1971; Zhu and Boyer, 1992) and additionally undergo reversible changes in length as they hydrate and dehydrate (Kamiya *et al.,* 1963; Zhu and Boyer, 1992). They quickly develop turgor pressure when water enters (Green *et al.,* 1971; Zhu and Boyer, 1992), and in dilute solution the pressure rises to around 0.6 MPa where it balances the osmotic potential inside the cell (Zhu and Boyer, 1992). Bisson and Bartholomew (1984) found that there can be osmotic adjustment in some species as the external solution becomes more concentrated.

The structure which limits the expansion of the cell is the wall, which must be strong enough to prevent rupture under the force of internal pressure but also able to enlarge in a controlled pattern. Because the turgor pressure is multidirectional and would be expected to generate a spherical cell if the wall was isotropic, it follows that a cylindrical cell like an internode cell of *Nitella* or *Chara* (Fig. 11.1) must have an anisotropic wall that is reinforced to prevent lateral expansion. Probine and Preston (1962) found that isolated walls of *Nitella* stretched more longitudinally than laterally when subjected to a unidirectional pull, confirming the anisotropic nature of the wall. Cellulose microfibrils are embedded in the wall matrix and in young walls they are oriented in a transverse direction that strengthens the wall against lateral extension but allows longitudinal growth by spreading apart. The longitudinal growth is thus determined by the spreading of the matrix in which the strengthening microfibrils are embedded (Taiz *et al.,* 1981). The composition of the walls resembles that of dicotyledonous plants (Morrison *et al.,* 1993) and is about 32% cellulose, 26% hemicellulose, 34% pectin, and 5% protein with the hemicellulose/pectin/protein forming the matrix for the cellulose microfibrils (Taiz *et al.,* 1981).

As the cell grows, the wall does not get thinner but tends to thicken instead (Green, 1958; Métraux, 1982), indicating that wall polymers are continually synthesized. There are no instances of sustained wall growth without synthesis (Taiz *et al.,* 1981), which is strong evidence that synthesis is part of the growth process, but growth and synthesis often correlate only loosely. Growth often occurs only in localized bands of *Nitella* walls while deposition of new polymers occurs similarly in growing and nongrowing bands (Taiz *et al.,* 1981). Richmond *et al.* (1980) present evidence that the inner layers of the wall next to the plasmalemma control the directionality of growth. Green (1958) found that most synthesis occurred in the inner layers in *Nitella* but Ray (1967) also observed deposition throughout the wall in oat coleoptile cells. Synthesis continues after *Nitella* cells mature (Morrison *et al.,* 1993), and the new layers are richer in cellulose with a distinctly layered appearance. The new layers make up the secondary wall and it is possible that they play a part in preventing further growth of the cells. It seems that a detailed knowledge of polymer deposition will aid in explaining the relationship between growth and wall synthesis.

Because turgor can increase cell size elastically, there has been much interest in the role of turgor in irreversible growth. Pfeffer (1900) considered turgor to be important and Probine and Preston (1962) showed that isolated walls of *Nitella* become irreversibly longer when they are stretched for long times. The lengthening occurs steadily and resembles the plastic deformation termed "creep" in polymers. Creep was faster in walls isolated from rapidly growing cells than in walls from slowly growing cells and was seen only when the force was above a threshold and in a range expected from the turgor of an intact cell. Green *et al.* (1971) measured turgor and growth simultaneously in intact *Nitella* cells and similarly observed a threshold turgor, although they thought the threshold was changeable.

Based mostly on the Probine and Preston (1962) finding that creep was faster when more force was applied above a threshold, Lockhart (1965a,b) formalized the concepts in an equation of the form

$$G = M(\Psi_p - Y), \tag{11.1}$$

where G is the relative growth rate (rate of change in volume per unit of cell volume in units of $m \cdot m^{-3} \cdot sec^{-1}$ or sec^{-1}), Ψ_p is the turgor pressure (MPa), Y is the yield threshold turgor (MPa), and M is the wall extensibility ($m \cdot m^{-3} \cdot sec^{-1} \cdot MPa^{-1}$ or $sec^{-1} \cdot MPa^{-1}$). This equation describes a straight line of slope M as shown in Fig. 11.2A. It considers the wall to act as a deformable polymer having ideal linear properties. Turgor is the driving force, and the amount above the yield threshold is the growth-active turgor ($\Psi_p - Y$) for the deformation of the wall. According to this equation, growth occurs more rapidly when Ψ_p increases.

However, when Green *et al.* (1971) decreased Ψ_p in *Nitella*, elongation ceased and then resumed at the original rate. Ortega *et al.* (1989) observed a similar behavior when Ψ_p was increased in single cells of a fungus. Although this behavior disagrees with Eq. (11.1), Green *et al.* (1971) thought that Y could change, thus maintaining growth when turgor changed. However, Green *et al.* (1971) altered the turgor with external osmotica that necessarily changed the composition of the wall solution. All elongation was considered to be irreversible growth and no elastic responses were reported, but Kamiya *et al.* (1963) observed elastic responses in mature *Nitella* cells when turgor was changed.

These differences between theory and observation caused Zhu and Boyer (1992) to do additional experiments with *Chara*. They were able to repeat the results of Green *et al.* (1971) when external osmotica were supplied but by using a microcapillary filled with cell solution, they also could inject cell solution. Thus for the first time, turgor alone could be changed without using osmotica in the external medium. Enlargement was monitored in the same cell, and the

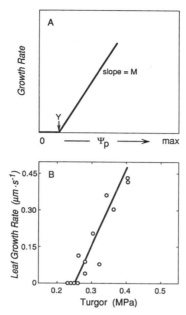

Figure 11.2   Relationship between turgor and growth. (A) Lockhart relationship showing yield threshold Y and wall extensibility M (slope of the line) obtained when the growth rate G is plotted as a function of the turgor Ψ_p. (B) Experiment showing growth of leaves at various turgor pressures in sunflower (from Matthews *et al.*, 1984). The experiment was done by withholding water from the soil and measuring growth for several days.

relationship among turgor, growth, and elastic responses could be investigated while avoiding the complications of previous experiments.

Figure 11.3 shows that when turgor was above a threshold of about 0.4 MPa, growth occurred in *Chara*. Below the threshold only elastic effects were present (Fig. 11.3, steps 1 and 2). Above the threshold, elastic changes were apparent initially but steady growth was seen after the elastic effects were completed (Fig. 11.3, steps 0, 3, and 4). Except for the threshold, turgor had no other effect on growth rate (Fig. 11.4). Since only the turgor was varied in these cells, this interpretation seems very strong. Zhu and Boyer (1992) also used inhibitors of energy metabolism and observed that growth was immediately inhibited without affecting the turgor, suggesting that growth was highly dependent on metabolism. A plastic-like wall deformation was observed only when turgor was forced to a higher level than would be possible by osmosis in the cells. In this situation, increased growth occurred for a short time but was soon irretrievably lost, indicating wounding. Therefore at normal turgors, Zhu and Boyer (1992) proposed that the growth rate may be controlled by enzymes involved in alter-

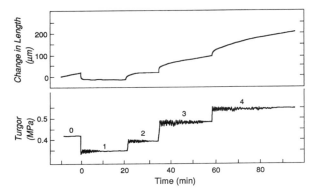

Figure 11.3   Relationship between enlargement and turgor pressure in *Chara*. Note that growth was zero at turgors 1 and 2 but growth occurred at turgors above a threshold (turgors 0, 3, 4). However, above the threshold, variations in turgor did not affect the growth rate. The initial shrinkage or swelling at each turgor step was caused by rapid elastic effects of the pressure change not related to growth. The turgor was changed by injecting the cell solution into the cell or removing the solution from the cell. After Zhu and Boyer (1992).

ing wall structure and synthesizing a new wall (Fry, 1989a,b; Hayashi, 1989) but above normal turgors, there may be a transient extensibility response like that observed by Probine and Preston (1962). The transient response would have little role to play because turgor could not reach the levels required. Why

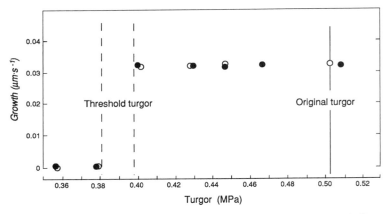

Figure 11.4   Growth at various turgor pressures in a *Chara* cell in an experiment similar to that in Fig. 11.3. The turgor pressure extends from zero to the original pressure in the cell, which was the maximum that could develop because of the cell osmotic potential. The turgor pressure was varied by injecting or withdrawing the cell solution as in Fig. 11.3. Solid points, injections; open points, withdrawals. Note that there is no growth at low pressures and that growth begins abruptly at a threshold pressure of 0.38 to 0.40 MPa. The growth rate remains constant at any pressure between this threshold and the original (maximum) pressure (adapted from Zhu and Boyer, 1992).

a threshold turgor is required in the normal turgor range is a mystery unless turgor is needed for the incorporation of intermediates into the wall as suggested by Robinson and Cummins (1976) or for straining wall polymers to a suitable conformation for enzyme action as proposed for bacterial growth (Koch, 1990).

These observations taken together indicate that the wall is complex and functions more dynamically than might at first be thought, especially when growth occurs. In order to explore the process, elastic responses and growth need to be carefully distinguished, and the role of new wall synthesis needs investigation. It seems that detailed studies of wall deposition may be helpful, and mutants of wall composition might be used such as those recently described by Kokubo *et al.* (1989, 1991). It is particularly curious that a threshold turgor is involved in growth, a fact about which all experiments are in unanimous agreement. It is likely that any theory of growth will need to include not only the role of synthesis but also how the threshold is involved. From work with these cells, it is possible to build a concept of growth from basic principles but the extent to which they apply to more complex plants needs to be explored.

GROWTH IN COMPLEX TISSUES

There is a crucial difference between single cells bathed in water and cells in a tissue competing with each other for water. In a complex tissue, the cells closest to the supply must conduct water fast enough to provide for all the other cells. This requirement adds new constraints to the growth process and is treated later.

In other respects, many of the principles observed with single cells carry over to complex plants, and Cleland (1971) and Taiz (1984) provide reviews of this area. The structure limiting the expansion of the cell is the wall, but epidermal tissues also appear to influence the rate and direction of growth (Kutschera, 1989; Kutschera and Briggs, 1988) and sometimes vascular tissue can have an effect. Cells packed inside a plant organ can create a tension in the epidermis much as pressure in a cell creates turgor in the wall. The enlargement of the epidermis then contributes to the shape and size of the organ.

As in large-celled algae, cellulose microfibrils are oriented to strengthen the cell walls in certain directions in plant organs but growth can occur in other directions by spreading the microfibrils apart, and the embedding wall matrix appears to determine growth rates (Frey-Wyssling, 1976; Preston, 1974; Taiz, 1984). When turgor is high enough to allow growth to occur (Cleland, 1959), the cell walls and cytoplasm maintain their thickness despite the 10-fold enlargement that typical cells will undergo (Bonner, 1934; Bret-Harte *et al.*, 1991) and thus a new wall and cytoplasm are synthesized at about the rate the cells enlarge (Ray, 1967). Sugars supplied to growing cells and tissues are rapidly incorporated into the walls (Gibeaut and Carpita, 1991; Kutschera and Briggs,

1987) and some wall polymers show evidence of turnover (Gibeaut and Carpita, 1991; Labavitch and Ray, 1974). If sugars are not available, the walls become thinner (Bret-Harte *et al.*, 1991) and the content of some wall polymers decreases (Loescher and Nevins, 1972, 1973). There is a rapid decrease in enlargement if inhibitors of biosynthetic metabolism are supplied to the cells (Brummell and Hall, 1985; Masuda, 1985; Robinson and Ray, 1977). If the cells are prevented from enlarging by exposure to osmotica, the walls become thicker for at least a few hours as synthesis continues in the absence of enlargement (Bret-Harte *et al.*, 1991; Loescher and Nevins, 1973).

In both large-celled algae and more complex multicellular plants, the ionic environment of the wall can influence growth rates. A low pH of the wall solution enhances rates and high Ca^{2+} diminishes them (Cleland, 1973, 1975, 1977, 1983, 1986; Cleland and Rayle, 1978; Taiz *et al.*, 1981). There is evidence that the plasmalemma secretes protons (H^+) as a result of ion accumulation by the cell and it was suggested that the wall solution could become acidified to pH's as low as 4 to 5 (Cleland, 1973, 1975, 1983). Vanderhoef *et al.* (1976, 1977b) point out that sustained growth associated with these low pH's would still require the synthesis of new wall polymers and recent studies have been conflicting, some claiming that the role for protons was an artifact of the experimental technique (Schopfer, 1989; Vanderhoef *et al.*, 1977a) and others that it was not (Lüthen *et al.*, 1990). There is a large buffering capacity of cell walls in the pH range between 5 and 6 (Vanderhoef *et al.*, 1977a) and the deposition of new wall material adds new buffering capacity. Thus, while it is likely that proton extrusion plays a role, its contribution to growth is still debated.

The role of turgor also is uncertain. Equation (11.1) has been applied extensively to plants of many kinds, and Fig. 11.2B shows a typical result with a land plant growing in gradually dehydrating soil. The growth fits the linear relationship predicted by Eq. (11.1) and exhibits a threshold turgor at about 0.25 MPa. Nevertheless, Nonami and Boyer (1989) observed decreased growth without a turgor decrease under similar conditions in soybean stems, and Shackel *et al.* (1987) found little relationship between turgor and growth in grape leaves. Soybean seedlings whose stem growth was slowed by low temperature had increased rather than decreased turgor and the ($\Psi_p - Y$) did not change (Boyer, 1993). Seedlings also showed increased turgor despite slower stem growth when deprived of auxin (Maruyama and Boyer, 1994) and little change in turgor as the tissue matured (Cavalieri and Boyer, 1982), which are counter to the theory of turgor-driven growth by plastic extension or "creep." It may be possible to reconcile these results by postulating a change in the wall extensibility M or the yield threshold Y in Eq. (11.1), as suggested by Serpe and Matthews (1992), but it is also possible that some other turgor relationship applies such as that of Zhu and Boyer (1992) observed in *Chara*. Thus far, it has not been possible to do experiments in multicellular tissues like those in *Chara* because of technical difficulties.

Table 11.2   Potentials Measured in the Cell Interior and in the Cell Exterior (Apoplast)

		Interior (MPa)		Apoplast (MPa)	
		Ψ_s	Ψ_P	Ψ_s	Ψ_P
Chara[a]	Growing	-0.60 ± 0.05	0.58 ± 0.06	-0.01 ± 0.00	0.0
	Mature	-0.62 ± 0.05	0.61 ± 0.05	-0.01 ± 0.00	0.0
Soybean[b,c]	Growing	-0.68 ± 0.05	0.42 ± 0.02	-0.04 ± 0.01	-0.24 ± 0.04
	Mature	-0.56 ± 0.02	0.50 ± 0.02	-0.04 ± 0.01	0.00 ± 0.00

Note. Measurements are in individual cells of intact plants: internode cells in *Chara* and stem cells in soybean. *Chara* cells were bathed in growth medium whereas soybean seedlings had roots in wet vermiculite and shoots in saturated air that prevented transpiration. For *Chara*, the cell apoplast was in growth medium under atmospheric pressure which is defined to have Ψ_P of zero. For soybean, apoplast Ψ_P was measured and was lower than zero (Nonami and Boyer, 1987).
[a] Zhu and Boyer (1992).
[b] Nonami and Boyer (1989).
[c] Nonami and Boyer (1987).

Growth-Induced Water Potentials

The competition for water between growing cells in a tissue can be demonstrated by raising the humidity sufficiently to prevent transpiration, thus bringing the tissue to the same condition as *Chara* surrounded by water. In this situation, the only water transport is for growth and metabolism of the cells. Table 11.2 compares the osmotic potential and turgor pressure in *Chara* with the same potentials in stems of soybean seedlings treated this way. In *Chara* growing in a medium whose water potential is near zero, the turgor pressure essentially balances the osmotic potential in the protoplasts (Zhu and Boyer, 1992). In soybean, the turgor pressure of the growing region does not balance the osmotic potential even though it nearly does in the mature part of the same stems (Nonami and Boyer, 1987, 1989). This lack of balance indicates that the cells in the growing tissue have a water potential associated with the growth process, termed a growth-induced water potential, that is scarcely detectable in cells surrounded by water or in mature cells.

It was proposed that growth-induced water potentials arise from the enlargement of the cell walls which prevents turgor pressure from becoming as high as it otherwise would if the walls were rigid (Boyer, 1968). The low pressure would result in a low water potential that would favor water uptake by the cells because the potential would be transmitted to the cell wall solution as a tension (negative pressure) that would move water out of the xylem and into the apoplast from which it would enter the cells. Tensions of an appropriate magnitude can be detected in the apoplast of growing tissues using a pressure chamber and are near zero in the adjacent mature tissues (Nonami and Boyer, 1987). The tensions are able to mobilize water from mature tissues to supply the growing cells when water uptake by the roots is prevented (Matyssek *et al.*, 1991a,b).

Another theory is that growth-induced water potentials are generated by high concentrations of solute in the apoplast of growing tissues, and Cosgrove and Cleland (1983a) detected high concentrations of solute in extracts of the apoplast. Meshcheryakov *et al.* (1992) similarly proposed that large and rapid changes in solute concentrations could occur in the apoplast but did not measure them. However, repeating the Cosgrove and Cleland (1983a) experiments, Nonami and Boyer (1987) found solutes released by phloem and cell membranes disturbed by the experimental procedures. When these problems were avoided, concentrations were too small to account for growth-induced water potentials (Nonami and Boyer, 1987). Others also detected only low solute concentrations in the apoplast of complex tissues (Boyer, 1967a; Jachetta *et al.*, 1986; Klepper and Kaufmann, 1966; Scholander *et al.*, 1965), with the only exception being in the placental tissues of developing seeds where concentrations can be significant (Bradford, 1994; Maness and McBee, 1986).

It was also proposed that growth-induced water potentials are artifacts of excision (Cosgrove *et al.*, 1984) that might cause the cell walls to relax because growth would continue without water entry and turgor would decrease (Boyer *et al.*, 1985; Cosgrove *et al.*, 1984). Cosgrove *et al.* (1984) and Cosgrove (1985, 1987) found large decreases in water potential after excision and considered them to account for the growth-induced potential. However, the Cosgrove experiments (Cosgrove, 1985, 1987; Cosgrove *et al.*, 1984) were done with excised tissues having some mature or slowly growing tissue attached. Matyssek *et al.* (1988) showed that water was moved to the growing cells in this situation and the water potential of the whole tissue decreased substantially so that relaxation was delayed and appeared much larger than it actually was. When water movement from mature tissue was prevented by excising only the rapidly growing tissue, relaxation was much faster and smaller than 0.1 MPa. Also, growth-induced water potentials were detected in plants that were completely intact where no excision artifacts were possible (Boyer, 1968, 1974; Boyer *et al.*, 1985; Cavalieri and Boyer, 1982; Nonami and Boyer, 1987, 1989, 1990a).

Thus, growth-induced potentials appear to arise from the growth activity of the cells in complex tissues where there is competition for water from a distant vascular system. The enlargement of walls keeps turgor from rising to a level that would be achieved if the walls were rigid, and a growth-induced water potential results. Water is pulled into the growing cells by this potential, indicating that the growth of the wall is the cause and the underlying mechanism is thus tied to the metabolism of the wall. Metabolism determines the wall composition and, as pointed out earlier for *Chara*, is involved in the growth of the wall when cells enlarge irreversibly.

Gradients in Water Potential during Growth

The mechanism for generating growth-induced water potentials does not involve any particular specialization of the growing cells. As pointed out in Chapter 3, each cell is nearly in water potential equilibrium with its own wall (Boyer, 1985, 1988; Molz and Ferrier, 1982) because the hydraulic conductivity of the cell membranes is large (see Table 3.1 in Chapter 3) and water for growth flows so slowly that there are minimal potential gradients just as in *Chara*. Over distances of several cells, however, significant gradients begin to appear (Boyer, 1985, 1988; Molz and Boyer, 1978; Molz and Ferrier, 1982; Silk and Wagner, 1980) because water must move from cell to cell across many wall-cytoplasm barriers and through the apoplast. Additionally, water must move through the cells next to the xylem to reach all the outlying cells. The gradients in water potential are thus steepest in cells close to the xylem where the flux of water is the greatest (Molz and Boyer, 1978).

Figure 11.5A shows the gradient predicted from the hydraulic characteristics of stem tissue of soybean seedlings (Molz and Boyer, 1978) and Fig. 11.5B gives

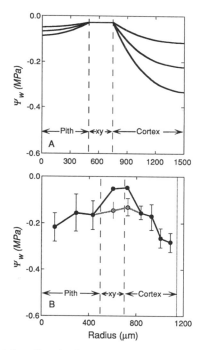

Figure 11.5 Water potential gradient in the growing region of soybean stems. The plants were grown in the dark in a water-saturated atmosphere so that the water potentials reflected only those for the growth process. The stem has a central pith in the center of a vascular cylinder and has cortical tissue covered by an epidermis on the outside, as shown. (A) Water potentials calculated at various positions along the radius of the stems assuming transport and anatomical properties characteristic of plant cells (adapted from Molz and Boyer, 1978). (B) Water potentials measured in cells at various positions along the radius of the stems by determining cell turgor and cell osmotic potential and summing the two. Values were corrected for mixing of solutions in the microcapillary used to sample for osmotic potential. Shaded values were uncorrected. From Nonami and Boyer (1993).

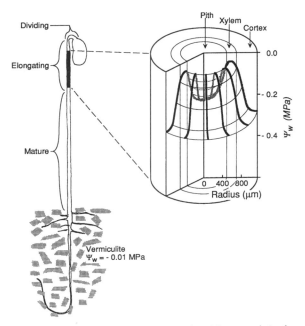

Figure 11.6   Three-dimensional water potential field induced by growth in the stem of soybean. The field is shown next to the stem in which it was measured, as in Fig. 11.5. The water potential is highest in the xylem and decreases in the surrounding pith and cortex tissues. Compare with the two-dimensional view of Fig. 11.5. After Nonami and Boyer (1993).

the gradient observed by direct measurements of water potentials in cells of the tissue (Nonami and Boyer, 1993). Water potentials were highest in the xylem (about -0.05 MPa, Fig. 11.5B) and lowest at the outermost layer of growing cells (-0.28 MPa, Fig. 11.5B). The three-dimensional shape of this gradient is shown in Fig. 11.6 next to the soybean plants in which it was measured. The three-dimensional nature of the gradient creates a potential field around the xylem that moves water radially inward and outward to the growing cells.

The field can be detected from measurements of the average water potential of whole tissues that are not transpiring. The average water potential is about -0.15 to -0.4 MPa (Barlow, 1986; Boyer, 1968; Cavalieri and Boyer, 1982; Molz and Boyer, 1978; Nonami and Boyer, 1989; Sharp and Davies, 1979; Westgate and Boyer, 1984) and can vary as conditions change around the plant. As temperatures decrease, for example, plant growth decreases and the field becomes less steep as expected from the smaller demand for water movement through the tissue (Boyer, 1993). Similarly, when growth decreases because the auxin supply is removed, the field becomes less steep (Maruyama and Boyer, 1994).

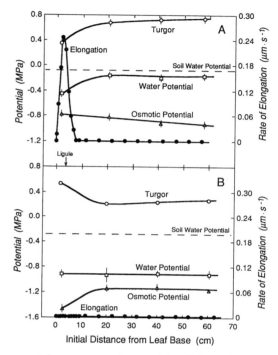

Figure 11.7 Water potential, turgor, osmotic potential, and elongation at various positions along a maize leaf at night. (A) High soil water potential. (B) Low soil water potential. Note in (A) that elongation occurred in basal tissue (left part of figure) but not in mature tissue far from the base (to the right in the figure). There was a low water potential in the basal tissue but not in the mature tissue, and a similar pattern was observed for the turgor pressure. The water potential of the mature tissue was similar to that of the soil (dashed line), indicating that the intervening xylem had a similar water potential. Thus, the low Ψ_w of the elongating tissue at the leaf base occurred in the tissues outside of the xylem. In (B), the decreased soil water potential prevented leaf elongation. Wilting was observed in the mature blade beyond 30 cm from the leaf base, and turgor pressure was low. There was a large decrease in the water potential of all leaf tissues compared to (A), and there was no difference in water potential between the elongating and mature tissue, indicating that the favorable water potential gradient for growth shown in (A) had collapsed. However, turgor pressure remained high in the elongating tissues. The collapse of the gradient thus appeared to inhibit leaf growth, and the turgor could not account for the effect. Adapted from Westgate and Boyer (1985b).

Westgate and Boyer (1985b) showed that these water potential gradients exist in all rapidly growing parts of maize plants. Roots, stems, leaves, and stigmas of developing florets exhibit substantial growth-induced water potentials compared with nearby mature tissues. For example, Fig. 11.7A shows that in a maize leaf, growth occurs at the leaf base and not in the exposed blade. The water potential of the blade is close to that of the soil at night when transpiration is negligible. The xylem connects the roots to the blade hydraulically, and the similarity of the blade and soil water potentials shows that the connecting xylem has a similar water potential as well. The water potential in the growing basal part of the leaf is substantially below that of the soil and blade. Since the connecting xylem passes through the growing zone, the lower water potential in this zone is outside of the xylem in the growing cells. In this situation, the surrounding tissue forms a three-dimensional gradient with the xylem, and the presence of these gradients in each growing part of the plant (Westgate and Boyer, 1985b) indicates that the gradients probably are universal in growing tissues.

It is worth noting that the turgor pressure of the basal growing zone also is below that of the mature blade in Fig. 11.7A. The lower turgor is consistent with the concept that the growth-induced water potential originates from the growth of the walls that prevents turgor pressure from developing as much as it would if the walls were rigid. This pattern was observed in each of the organs of the maize plant (Westgate and Boyer, 1985b).

TRANSPIRATION AND GROWTH

Land plants differ from single cells of algae not only in their multicellular tissues but also in the large amounts of water that flow through the tissues for transpiration. A sunflower plant can transpire enough to completely replace all of the water in the leaves every 20 min (Boyer, 1974, 1977), but in the same time rapid leaf growth will import only about 0.2 or 0.3% of that amount. What happens to water uptake for growth when xylem water potentials decrease because of transpiration?

As discussed in Chapter 7, the water potential of the xylem decreases as transpiration becomes rapid mostly because dehydration of the leaf creates a tension in the apoplast water that is transmitted to the xylem and ultimately through the roots into the soil. The tension develops until water enters the leaves at the rate it is lost, preventing further dehydration. Since water for shoot growth must be extracted from the same xylem, the increased xylem tension affects the growth of surrounding tissues. In effect, water obtained for growth is obtained in competition with transpiration.

The competition is decided by the water potentials that can be exerted on water in the xylem. One theory (Boyer, 1974, 1985) is that water to be used for transpiration evaporates close to the xylem vessels and thus bypasses many of the cells outside of the xylem. There is increasing evidence that this occurs (Boyer, 1985; Nonami and Schulze, 1989; Nonami et al., 1991) and that water for growth follows a much longer path since it must enter all the cells of a growing tissue. As a consequence, water lost by transpiration would encounter a low resistance path (Boyer, 1974, 1977; Raney and Vaadia, 1965a,b; Rayan and Matsuda, 1988) and water for growth would encounter a high resistance path. The low resistance for transpiration would allow large flows to occur without developing xylem tensions that would inhibit growth (Boyer, 1974, 1977, 1985).

Another theory is that water evaporates from surfaces close to the stomatal pores. Meidner (1975, 1976a,b) showed convincingly that water will evaporate from surfaces close to a pore if all the surfaces are uniformly wet, and he proposed that water moves from the veins along bundle sheath extensions to the epidermis where it can pass to the guard cells and evaporate, which is also supported by Maercker (1965) and Maier-Maercker (1979a,b, 1983). However, the internal surfaces of leaves probably are not uniformly wet. There is a waxy cuticle covering cell walls inside leaves and it is particularly thick close to the stomatal pores (Boyer, 1985; Leon and Bukovac, 1978; Nonami *et al.*, 1991; Norris and Bukovac, 1968; Scott, 1964, 1966). Because the inner cuticle undoubtedly decreases evaporation from guard cells and the cells near them, most evaporation probably occurs from deeper surfaces close to the xylem within the leaf (Boyer, 1985; Nonami and Schulze, 1989; Nonami *et al.*, 1991). In this situation, evaporation occurs before water moves far from the xylem and most of the leaf cells are bypassed by water for transpiration. Therefore, the accumulating evidence argues against near-stomatal transpiration. This is also discussed in Chapter 7.

Despite the low resistance of the transpiration path, there is an impact on growth-induced water potentials as transpiration-induced tensions develop. For example, leaf growth occurs both during the day and night in maize, and Fig. 11.8A shows that the water potential remains lower in the growing basal leaf tissue than in the exposed mature blade at both times (Westgate and Boyer, 1984), indicating that the water potential of the growing tissue is below that in the nearby xylem. Figure 11.8B shows that the turgor pressure also is much lower in the growing tissue than in the mature tissue, but the reverse tends to be true for the osmotic potential, especially during the day, as shown in Fig. 11.8C. Nevertheless, the osmotic potential during the day is somewhat lower than at night in the growing tissue and together the lower turgor pressure and osmotic potential in the daytime decrease the water potential in the growing tissues enough to compensate for the decreased water potential in the xylem. The water potential in the growing cells thus is favorable for water uptake from the xylem regardless of transpiration.

Growth can occur only if transpiration develops slowly enough to allow sufficient change in osmotic potential and turgor to generate the growth-induced water potentials needed to maintain water movement into the cells. If xylem water potentials change rapidly, as during sudden illumination or shading of the plant, it is unlikely that the osmotic potential of the enlarging cells will keep pace (McNeil, 1976; Meyer and Boyer, 1981; Schmalstig and Cosgrove, 1990), and gradients in water potential might rapidly become unfavorable. This may ex-

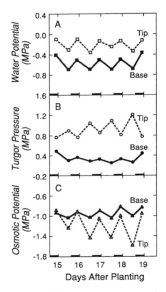

Figure 11.8   Night and day water potential, turgor pressure, and osmotic potential in the elongating tissue (leaf base, closed symbols) and mature tissue (leaf tip, open symbols) of a maize leaf. Night is shown by black bars on the X axis, day by open intervals between black bars. In (A), there was a lower water potential in the growing base than in the mature tip at night when transpiration was negligible and during the day when transpiration occurred. Assuming the mature tip indicated the xylem water potential (see text), the gradient between the xylem and growing base favored water uptake from the xylem during night and day. In (B), there was less turgor pressure in the growing base than in the mature tip during night and day. In (C), the osmotic potential tended to be lower (more negative) in the tip than in the base during the day, presumably because solute was produced by photosynthesis in the tip but not the base. However, the osmotic potential also was lower during the day than at night in the base, reflecting osmotic adjustment. Adapted from Westgate and Boyer (1984).

plain why growth can show rapid fluctuations under these conditions (McIntyre and Boyer, 1984; Milligan and Dale, 1988a,b; Smith and Dale, 1988).

GROWTH AT LOW WATER POTENTIALS

The variations in xylem water potential brought about by transpiration are similar to those caused by the depletion of soil water, but the latter often is of long duration and can cause very low water potentials. Several approaches have been used to study the phenomenon. One shown in Fig. 11.9 has been to grow seedlings in vermiculite in the dark at constant temperature and saturating humidities (Meyer and Boyer, 1972). Transpiration is negligible and soil water potentials remain stable for days while the seedlings consume photosynthetic products stored in the cotyledons. Water limitation is imposed by carefully

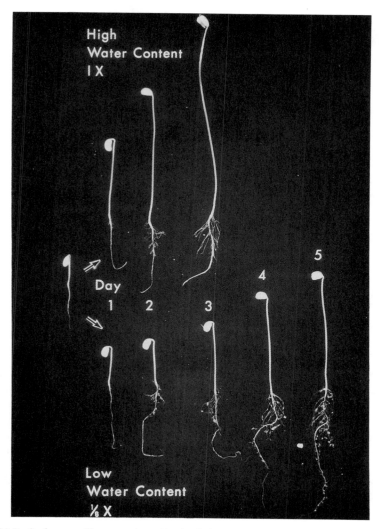

Figure 11.9   Soybean seedlings germinated in the dark and transplanted either to vermiculite containing adequate water (1×) or limited water (1/8×). The 1/8× vermiculite contained one-eighth of the water in the 1× vermiculite and had a water potential of −0.3 MPa compared to −0.01 MPa in the 1× control. Note the marked inhibition of stem (hypocotyl) growth for the first 2 days in 1/8× vermiculite followed by a modest resumption of growth. Roots continued to develop as fast at 1/8× as at 1×.

transplanting the young seedlings to vermiculite having a low water content (Fig. 11.9). Methods also have proven useful that slowly dehydrate the root medium (Jones and Turner, 1978) or expose only a few roots to water by split-

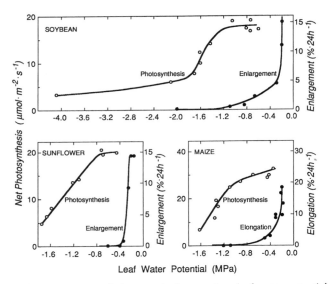

Figure 11.10   Leaf enlargement and photosynthesis at various leaf water potentials in soybean, sunflower, and maize plants. After Boyer (1970).

ting the root system into two or more volumes of soil (Blackman and Davies, 1985; Gowing *et al.,* 1990). Also, water can be supplied at a rate less than the plant requires (Matthews *et al.,* 1984; McPherson and Boyer, 1977). Each approach is a compromise that depends on the objectives of the experiments. Osmotica have sometimes been used to create water deficits but the solutes can directly inhibit growth (e.g., Zhu and Boyer, 1992) and this approach should be avoided. This problem is discussed in Chapter 4.

When soil water is gradually depleted, a number of plant functions are inhibited. Leaf growth is one of the first to diminish and Fig. 11.10 shows that it is nearly abolished before photosynthesis is affected except in maize where photosynthesis also is slightly inhibited (Boyer, 1970). Respiration is even less affected (Boyer, 1970), which illustrates that growth is inhibited by some factor other than the availability of photosynthetic products or the ability to use them in respiration. Figure 11.11 shows that growth is affected to varying degrees depending on the organ (Westgate and Boyer, 1985b). Leaves ceased growing whereas roots continued to grow rapidly on the same plants at the same tissue water potential, which probably served to reduce the development of a new transpirational surface and increase the access to soil water. The large inhibition of stigma (silk) and stem growth disrupts floral development and pollination (Herrero and Johnson, 1981; Westgate and Boyer, 1985b), a common problem in agriculture that is treated more fully in Chapter 12. Westgate and Boyer

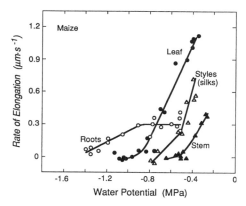

Figure 11.11   Elongation of various maize organs at various water potentials of the elongating tissues. At low water potentials, elongation often became negative, indicating that the tissue shrank. Adapted from Westgate and Boyer (1985b).

(1985b) conclude that there must be an internal control of growth that could depend on different metabolic and/or physical mechanisms in different tissues.

Primary Signals

During development, the amount of water passing through plants vastly exceeds the amount needed for growth. Fully developed maize plants normally consume about $1500 \, cm^3$ of water every day and during a drought might absorb as little as $150 \, cm^3 \cdot day^{-1}$ but only $14 \, cm^3 \cdot day^{-1}$ are required for rapid growth (Aparicio-Tejo and Boyer, 1983; Boyer, 1970). Some factor other than water quantity must be responsible and for many years it was thought that loss in turgor was the problem (e.g., Hsiao, 1973; Pfeffer, 1900). Wilted leaves indicate an obvious loss of turgor, and growth generally is not observed during wilting (Boyer, 1970; Davies and Van Volkenburgh, 1983; Matthews *et al.*, 1984; Radin and Boyer, 1982). However, in leaves of grasses growth occurs at the leaf base which is hidden from view, and Michelena and Boyer (1982) and Westgate and Boyer (1985b) found high turgor in the basal growth zone when growth was completely prevented and the exposed blade was markedly wilted as shown in Fig. 11.7B. Barlow (1986), Cutler *et al.* (1980b), and Passioura (1988a) also noted that turgor loss often could not explain growth losses.

Another theory is that changes in the growth-induced water potential prevent growth. Figure 11.12C shows that stem growth immediately slowed when soybean seedlings were transplanted to vermiculite having one-eighth the usual amount of water (as shown in Fig. 11.9), which has also been seen by others in leaves (Hsiao *et al.*, 1970). Because of the transplanting, the water potential of

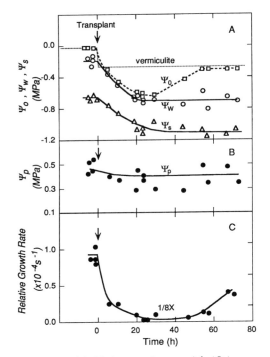

Figure 11.12   Tissue water potentials (Ψ_w), osmotic potentials (Ψ_s), turgor pressures (Ψ_p), and relative growth rates in the elongating region of soybean stems after transplanting to $1/8 \times$ vermiculite as in Fig. 11.9. Also shown are water potentials of the vermiculite (dotted line) and the basal nongrowing stem tissue (Ψ_o, dashed line). In (A), note that the basal tissue had Ψ_o close to that of the vermiculite before transplanting and at the end of the experiment, indicating that the Ψ_o monitored the xylem water potential. Before transplanting, the Ψ_w was below Ψ_o, indicating that a favorable water potential gradient existed for water uptake for growth. After transplanting, Ψ_o became similar to Ψ_w and the favorable water potential gradient collapsed. Growth was inhibited (C). Later, Ψ_o increased, reestablishing a favorable gradient and allowing growth to resume at a moderate rate. However, the lack of full recovery indicates that the growth limitation had shifted from the collapsed water potential gradient to a blockage of metabolism. (B) The tissue turgor pressure was virtually unchanged throughout the experiment. After Nonami and Boyer (1990a).

the vermiculite rapidly went from -0.01 to -0.3 MPa (Fig. 11.12A, dotted line) and the xylem water potential Ψ_o decreased faster than the water potential of the growing region Ψ_w. The potential difference ($\Psi_o - \Psi_w$) representing the growth-induced water potential collapsed as a result, indicating that the gradients in water potential became unfavorable for water extraction from the xylem, depriving the enlarging cells of water (Nonami and Boyer, 1990a). Growth ceased at the same time (Fig. 11.12C). After a few hours, osmotic adjustment occurred (Ψ_s decreased, while Ψ_p remained essentially stable as discussed in

Chapter 3) and the ($\Psi_o - \Psi_w$) recovered (seen as a rise in Ψ_o above Ψ_w after about 40 hr in Fig. 11.12A). Growth resumed slowly (Fig. 11.12C). The average tissue turgor was virtually unaffected throughout this time (Fig. 11.12B). This collapse and recovery of the growth-induced water potential indicates that there can be rapidly transmitted hydraulic signals between roots and shoots that can affect growth. Others (Boyer, 1969, 1971c; Cutler *et al.*, 1980b; Raschke, 1970) also have seen very rapid transmission of these signals, and simply opening the xylem by cutting at the apical end inhibits growth within a few seconds (Matyssek *et al.*, 1991a).

Nonami and Boyer (1989, 1990a) found that hydraulic signals could be rapidly transmitted to cells next to the xylem, detected as a change in turgor pressure, and Fig. 11.13 diagrams this dynamic system in elongating stem tissues. Decreasing the water potential of the xylem inverts the water potential gradient next to the xylem, rapidly decreasing the turgor pressure Ψ_p of the nearby cells. Growth is inhibited because water can only move down a potential gradient and it is blocked by the inverted gradient, depriving the outlying cells of the water they need for growth. This type of rapid inhibition occurred even though the

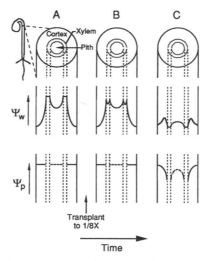

Figure 11.13 Diagrammatic representation of dynamics of water potential (Ψ_w) and turgor pressure (Ψ_p) at various radial positions in the elongating region of soybean stems before and after transplanting to dehydrated (1/8×) vermiculite as in Figs. 11.9 and 11.12. (A) Water potential gradient before transplanting when growth is rapid. In this case, turgor is uniform in the tissue. (B) Decreased xylem water potential immediately after transplanting to dehydrated vermiculite. Growth is inhibited because the gradient is inverted next to xylem and prevents water from being extracted. (C) Decreased water potentials after transplanting for several hours. Turgor is decreased next to xylem. Subsequent to (C), the gradient reestablishes and growth resumes slowly. Adapted from Nonami and Boyer (1989).

outlying cells did not change in water potential or turgor pressure, and it probably is common when transpiration rates increase or soil water is depleted.

Another factor that possibly is important is the production of plant growth regulators as dehydration signals (Gowing *et al.*, 1990; Saab and Sharp, 1989; Saab *et al.*, 1990; Zhang and Davies, 1989a,b). As discussed in Chapters 5 and 9, some molecular signals are transmitted from roots to shoots, and some may involve plant growth regulators but evidence for the latter is difficult because they may be produced in many plant tissues and their mode of detection and regulation in target tissues is uncertain. As discussed next, responses to plant growth regulators sometimes resemble those to water limitation but for different molecular reasons.

Metabolic Changes

It is difficult to identify which changes among the many in the cell may be important for growth when the water supply is limited. Obvious factors such as changes in turgor pressure or supply of photosynthetic products often fail to account for growth losses, and molecular changes such as osmotic adjustment, polyribosome alteration, and hormone accumulation tend to occur simultaneously as water is slowly depleted in the soil. The most fruitful approach has been to subject plants to realistic water limitation rapidly and observe the sequence of molecular events that follows. The idea is to decrease water availability just enough to inhibit shoot growth in a plant system like that of Fig. 11.9. Because the roots continue to grow in this condition, they can be used as a control on the same plant.

From these types of studies, it was found that cell division and cell enlargement decreased simultaneously and to about the same extent when soybean stem growth was inhibited in this way (Meyer and Boyer, 1972). Others found that cell division and enlargement were both reduced (Kirkham *et al.*, 1972) or division was reduced more than enlargement in leaves (Terry *et al.*, 1971) or the elongating region became shorter and narrower during a severe water limitation in roots (Sharp *et al.*, 1988) and the cells became smaller (Fraser *et al.*, 1990). One report of changes in cell size in stems (Paolillo, 1989) appears to have been confounded by transplant shock that itself decreased growth, making the results difficult to interpret.

Cell division and enlargement consume sugars and amino acids for biosynthesis and respiration, and new molecules of low molecular weight must be acquired to maintain both processes (McNeil, 1976; Schmalstig and Cosgrove, 1990; Sharp *et al.*, 1990). Under favorable growth conditions, there is evidence (Meyer and Boyer, 1981; McNeil, 1976) that import generally balances use and that the concentration of solutes remains practically constant, making the cell osmotic potential fairly stable. As growth slows with water limitation, however, biosynthesis slows but import remains high (Meyer and Boyer, 1981; Sharp *et al.,* 1988, 1990). This causes import to exceed use, and the imported solute accumulates. The osmotic potential becomes more negative (osmotic adjustment, see Chapter 3 and also Fig. 11.12), but after a few hours it stabilizes as import decreases and comes back into balance with use (Meyer and Boyer, 1981). The osmotic adjustment gives an increased ability of the cells to extract water from the xylem and ultimately the soil (Barlow, 1986; Meyer and Boyer, 1972; Morgan, 1984). Wilson and Ludlow (1984) observed maximum osmotic adjustment of 0.8 to 1 MPa in grass leaves.

Osmotic adjustment allows growth to continue at a faster rate than it would without adjustment, albeit still at a slow rate (Meyer and Boyer, 1972; Michelena and Boyer, 1982). Figure 11.14 shows that maize leaves that adjusted osmotically (Control) grew faster at the same water potential than leaves whose adjustment was diminished (Dark 48 hr). In other instances, osmotic adjustment was not as extensive and the effect on growth was less apparent (Acevedo *et al.,* 1971; Cutler and Rains, 1977a,b; Cutler *et al.,* 1977, 1980a; Hsiao *et al.,* 1976; Jones and Turner, 1978; Turner *et al.,* 1978). Because osmotic adjustment involves a decreased use of imported solutes like sugars and amino acids (Meyer and Boyer, 1981; Morgan, 1984; Munns and Weir, 1981) but their import often continues at high rates, it follows that there must be a metabolic block in their use for biosynthesis of new cells. Barlow *et al.* (1976) present

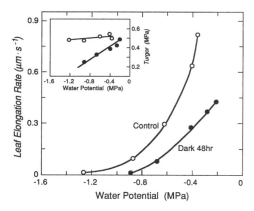

Figure 11.14 Rates of maize leaf elongation at various water potentials measured in the elongating tissues. (Inset) Turgor at various water potentials in the same elongating tissues. Darkening the plants for 48 hr (darkened points) prevented photosynthesis and decreased the transport of photosynthetic products to the elongating tissues. Osmotic adjustment was diminished and turgor was not maintained as much as in the undarkened controls (open points). After the dark treatment, leaf growth was slower and ceased at higher water potentials than in the controls. Adapted from Michelena and Boyer (1982).

evidence that ATP concentrations decrease in leaf growing regions about this time, which suggests that metabolic energy may be less available for biosynthesis. Thus, the cells appear to lose some biosynthesis but gain increased osmoticum and water that permit slow growth where otherwise none would occur.

Additional evidence for decreased biosynthesis comes from observations of decreased rates of deposition of dry matter (Meyer and Boyer, 1981), reduced DNA synthesis (Meyer and Boyer, 1972), and fewer polyribosomes in the enlarging stem tissues of water-deficient plants (Hsiao, 1970; Morilla *et al.*, 1973; Mason and Matsuda, 1985; Mason *et al.*, 1988a). The decreased biosynthesis is correlated with changes in physical characteristics of the cell walls. Using a psychrometer method to monitor the relaxation of cell walls (Nonami and Boyer, 1990a) and an extensiometer method to stretch the walls (Nonami and Boyer, 1990b), wall deformability was monitored in stems of soybean seedlings grown as in Fig. 11.9. Figure 11.15 shows that there was a gradual decrease in deformability that reached a minimum at 24 hr and showed a slow recovery thereafter. Also, there was a decreased rate of proton extrusion into the apoplast

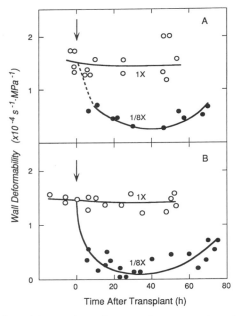

Figure 11.15 Wall deformability in elongating stem tissues at various times after transplanting soybean seedlings to vermiculite of low water content. (A) Wall deformability measured from wall relaxation after removing the water supply by excising the rapidly elongating tissues. (B) Wall deformability measured by pulling the stems in the direction of growth. Deformation was measured after subtracting the elastic component. Adapted from Nonami and Boyer (1990a,b).

of growing maize tissues (Van Volkenburgh and Boyer, 1985). Since the walls control the size of the cell, the decrease in deformability and proton extrusion may have been related to the inhibition of growth.

When the cells of the soybean stems were disrupted and the wall fraction was isolated (Bozarth *et al.*, 1987), a 28-kDa protein normally present in small amounts was found in large quantities in the walls from the elongating tissues (Fig. 11.16). It also was present in the cytoplasm and its mRNA was present among the polyribosomes, indicating that it was being synthesized in the enlarging tissues (Mason *et al.*, 1988b). Further work indicated that the protein was one of the acid phosphatases (DeWald *et al.*, 1992; Williamson and Colwell, 1991) which are common in plant tissues and release phosphate from a range of phosphorylated intermediates involved in biosynthesis and respiratory me-

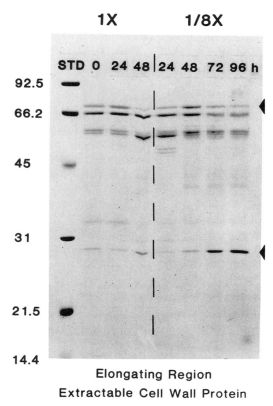

Elongating Region
Extractable Cell Wall Protein

Figure 11.16 Accumulation of protein in cell walls of elongating stem tissues of soybean seedlings at various times after transplanting to vermiculite of low water content. Note the prevalence of a 28-kDa protein after several days in the drier vermiculite. From Bozarth *et al.* (1987).

tabolism. Their specific role is not well understood, however, and their presence in the cell walls raises many as yet unanswered questions.

Nevertheless, certain features of the accumulation may be significant. The 28-kDa protein and its mRNA accumulated in the enlarging tissues of the stem at low water potentials but not in the root (Surowy and Boyer, 1991). The antibodies to the 28-kDa protein also identified an antigenically related protein of about 31-kDa that did not accumulate in cell walls of shoot tissues exposed to low water potentials. The mRNA for the 31-kDa protein accumulated in the root tissues but not in the stem (Surowy and Boyer, 1991). The growth of the stem was inhibited but not the roots (Meyer and Boyer, 1981; Surowy and Boyer, 1991). The appearance of the mRNAs was thus specific for parts of the plant showing different growth responses and different protein accumulation in the cell wall fraction during water limitation. Moreover, the accumulation of the mRNA for the 28-kDa protein was particularly apparent in the epidermal cells of the stem (Mason and Mullet, 1990). Epidermal tissues are thought to contribute to the rate and form of development of plant organs (Kutschera, 1989; Kutschera and Briggs, 1988). The 31-kDa protein appears to be another acid phosphatase or part of a dimer between the 28- and 31-kDa forms (DeWald *et al.*, 1992; Mason *et al.*, 1988b).

These molecular changes were so specific that they were likely to be closely regulated. As pointed out in Chapters 8 and 9, abscisic acid is a plant growth regulator that accumulates in plants at low water potentials (Beardsell and Cohen, 1975; Neill and Horgan, 1985; Wright and Hiron, 1969; Zabadal, 1974) and it accumulates as well in the enlarging tissues of seedlings exposed to low water potentials as described earlier (Bensen *et al.*, 1988). Supplying ABA to the roots or shoots raised internal concentrations to those occurring at low water potentials and caused similar growth effects (Creelman *et al.*, 1990) but the mRNA for the 28-kDa protein showed only modest accumulation (Mason and Mullet, 1990) and, importantly, the polyribosome content of the tissue did not decrease in the ABA-treated tissue (Creelman *et al.*, 1990). The mRNA population from the polyribosomes responded differently to ABA than to low water potentials both in the shoot and in the root growing tissues (Creelman *et al.*, 1990), although Bray (1988) found a similar change in the mRNA population for the two treatments, perhaps because she used relatively mature tomato leaves that may not have been comparable to growing tissues. Others (Davies and Zhang, 1991; Davies *et al.*, 1990; Saab *et al.*, 1990) also noted similarities between growth effects of high ABA and low water potentials but failed to measure mRNA changes. Therefore, despite the similarity in growth responses, the molecular responses to ABA may be different from those occurring at low water potentials.

Another plant growth regulator, jasmonic acid, was much more effective than ABA in enhancing the mRNA content of the tissue for the 28-kDa protein

(Mason and Mullet, 1990). Jasmonic acid is released when epidermal cells are wounded. The 28-kDa protein was observed to accumulate in the vacuoles of soybean leaves when pods were removed from the reproductive plants (Staswick, 1988, 1989a,b,c) and was termed a vegetative storage protein. The jasmonic acid results make it likely that the protein was accumulating in response to the wound created by the pod removal.

Figure 11.17 shows the sequence of molecular events that have been observed in stems of soybean seedlings after transplanting to water-deficient vermiculite. The earliest hydraulic inhibition sets in motion a biosynthetic block that leads to osmotic adjustment and other metabolic changes eventually causing alterations in the cell walls. Upon rewatering, the growth inhibition is rapidly reversed, which indicates that the metabolic changes are reversible.

It is intriguing that growth regulators should give growth responses in many ways similar to those at low water potential but often not involving the same

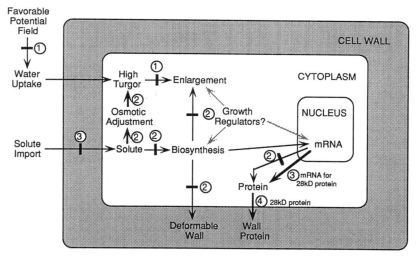

Figure 11.17   Sequence of events in elongating stem tissues after transplanting soybean seedlings to vermiculite of low water content. The order of events is shown by numbers 1–4. A block is shown by a heavy bar and enhancement is shown by increased arrow thickness. (1) Inversion of the potential field occurs first because of a decrease in water potential of the xylem which deprives the elongating cells of water and slows elongation. (2) Biosynthesis is blocked and osmotic adjustment occurs, maintaining turgor. Cell walls become less deformable. Less protein synthesis occurs. (3) Despite a decreased protein synthesis, an increased mRNA content occurs for a 28-kDa protein, a phosphatase, in soybean stems. Solute import slows, bringing import back into balance with solute use. (4) Phosphatase accumulates in cell walls. In roots, 1–3 occur rapidly but growth recovers to the control rate. The increased mRNA is for a 31-kDa protein and not the 28-kDa protein of the stem.

molecular responses. Also, it is curious that two proteins (28 and 31 kDa) should be so closely tied to the growth controlling tissues and growth responses of the plant, but have an enzymatic activity that so far has not been related to growth. Perhaps further investigation will show other aspects of regulation or wall biosynthesis that will make the mechanism clearer.

ECOLOGICAL AND AGRICULTURAL SIGNIFICANCE

Plants respond to dehydrating soils by continuing to grow roots but decreasing the growth of shoots. This causes the quantity of roots to increase in comparison to the shoots (Bennett and Doss, 1960; El Nadi *et al.*, 1969; Gales, 1979; Malik *et al.*, 1979). In these circumstances, roots tend to extend to deeper layers of soil and proliferate there (Klepper in Stewart and Nielsen, 1990; Klepper *et al.*, 1973). Figure 11.18 shows that cotton root systems in soils with and without irrigation are markedly different in root distribution. Lower soil layers had more roots in the nonirrigated treatment than in the irrigated treatment. The proliferation occurred between July 8 and July 29 while some roots disappeared from the upper layers, indicating that roots grew rapidly deep in the soil but died in shallower layers. Figure 5.13 also illustrates the rapid proliferation and death that can occur in root systems.

There is evidence that soil becomes more difficult to penetrate as it dries and that root growth is favored in wetter regions (Greacen and Oh, 1972). Thus, as the upper soil layers dehydrate, root growth slows because of increased fric-

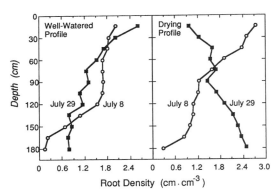

Figure 11.18   Change in root density profile upon withholding water from cotton plants. Root densities showed a typical declining density with increased depth when water was supplied (left graph) but showed an inverted profile after water was withheld (right graph). Note that there was a proliferation of roots in the deep layers but a loss of roots in the shallow layers, which dehydrated first. The changes occurred in only 3 weeks, indicating that root growth and death were occurring rapidly at this stage of development. Adapted from Klepper (in Stewart and Nielsen, 1990).

tional resistance and shifts to those roots in the wetter deeper soil at the periphery of the root system.

However, although this effect exists, there also appears to be an internal regulation of root growth. Because of the presence of growth-induced water potentials, one might expect that growing regions could obtain water from nongrowing tissues. Such a situation exists when stored onions or potatoes grow stems or cut tree stems sprout leaves even though there is no external supply of water. Figure 11.19 shows that when the roots of soybean seedlings were removed from the medium and stored in humid air, stem growth was immediately inhibited but continued slowly for many hours (Matyssek *et al.*, 1991a). The roots also grew relatively rapidly (Matyssek *et al.*, 1991a). The plants did not gain water from the humid air (Matyssek *et al.*, 1991a) and all the water for growth came from nongrowing parts of the plant (Matyssek *et al.*, 1991a).

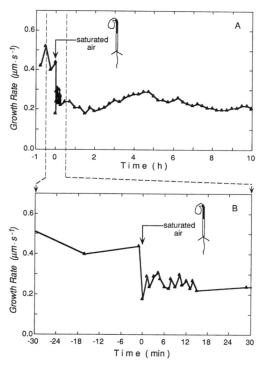

Figure 11.19 Stem growth of soybean seedling removed from the water supply and exposed to saturated air (arrow), preventing virtually all water exchange with the seedling. Continued stem growth relied on mobilization of internal water from nongrowing tissues mostly in lower parts of the stem. Expanded scale (B) shows that growth diminished immediately when the external water was removed. Adapted from Matyssek *et al.* (1991a).

Growth on internal water appears to depend not only on the ability of the enlarging cells to generate the growth-induced water potential but also on the volume of water stored in the surrounding tissue (Matyssek *et al.*, 1991a). This has practical application during transplanting where the external supply of water is disrupted and reestablishment depends on root growth to reconnect the roots to the external supply. The roots grow with water extracted from the mature tissues and reconnection is favored. It follows that plants having larger volumes of mature tissue will have greater success reestablishing after transplanting.

A similar situation applies to plants from which soil water is merely withheld. Internal water can provide a reservoir for continued growth activity. Tree stems and fleshy fruits often shrink during times of water shortage as water is withdrawn by other parts of the plant (Kozlowski in Kozlowski, 1972), and growing tissues sometimes appear to be the beneficiaries.

In this respect, it is important to note that roots can transport water not only to the shoot but also to other roots located in dry layers of soil. Several investigators observed the hydration of dry soil with water transported from roots in wet soil (Baker and Van Bavel, 1986; Caldwell and Richards, 1989; Mooney *et al.*, 1980). The net effect is to allow roots to grow where otherwise little growth would occur, and the wetting of the dry soil creates a small reservoir for the next day. This is discussed in Chapter 5 as hydraulic lift.

There are genetic differences in the ability of roots to penetrate deep layers of soil (Boyer *et al.*, 1980; Jordan *et al.*, 1979; O'Toole and Bland, 1987; Taylor *et al.*, 1978). In some instances, the differences are inherited simply. Ekanayake *et al.* (1985) and Armenta-Soto *et al.* (1983) showed that in rice the difference in depth of rooting was controlled by only a few genes. Boyer *et al.* (1980) found evidence that the high yields of modern soybean cultivars appeared traceable to less midday dehydration of the leaves resulting from deeper rooting than in older cultivars. Although this was contradicted by Frederick *et al.* (1990), there seem to be questions about the technology used by these authors, as described in Chapter 5. These experiments suggest that root systems can be genetically modified for improved performance in dehydrating soils. Because it is becoming increasingly possible to map genes, it eventually may be possible to determine the characteristics of root systems by looking at the genetic map without looking at the roots, which could enhance progress in this area.

SUMMARY

Growth is an irreversible increase in size and is the most central feature of plant activity because the plant must grow sufficiently to reproduce itself and thus ensure its representation in the next generation. Most of the increased size and weight that we consider growth consists of increased water content. However, the increased water content must be carefully distinguished from the reversible effects of hydration and dehydration of the tissues, which can affect the size of the cells but is not growth.

The bulk of the increased water content associated with growth is caused by the enlargement of individual cells. Water uptake is fundamentally driven by the high concentration of cellular solutes, and the turgor that develops plays a role. There is general agreement that a certain amount of turgor must be present before enlargement begins but there is debate about whether turgor also determines the growth rate. The evidence seems to favor a control of the rate by energy metabolism rather than turgor.

In multicellular tissues, water uptake must occur in competition with other cells and significant gradients in water potential can build up that extract water from the xylem fast enough to meet the demand. The faster the growth rate, the larger the gradient. When the xylem water potential falls, the gradient is reversed locally and inhibits growth, sometimes very rapidly. The evidence indicates that the gradient builds up because the enlargement of the cell compartment prevents turgor from becoming as high as it otherwise would and the water potential of the cell interior is lowered. This is transmitted mostly as a tension to the water in the cell walls and eventually the xylem. Provided the tension is greater than in the xylem, water moves out and into the growing cells. The competition for xylem water also involves transpiration which can change xylem tensions. When transpiration is rapid, it appears that the tension must increase in the cell walls of the growing cells if growth is to continue.

When water becomes in short supply around the roots, growth responds in a few seconds probably because of changes in these water potential gradients. Shoot tissues decrease in rate whereas root tissues tend to maintain their rate with obvious advantages for the plant. Subsequently, metabolic changes take place that inhibit the incorporation of small substrate molecules into the polymers needed to grow new cells, and a metabolic inhibition prevails after a few hours. The substrates accumulate, leading to osmotic adjustment, and turgor is maintained. Growth occurs more rapidly than if the accumulation had not occurred but the growth is slow.

At about this time, the molecular nature of the cell walls changes and the walls become less deformable. In particular, a phosphatase present in the cytoplasm appears to accumulate in the cell walls. A related but different protein appears to accumulate in the roots. These changes are accompanied by changes in growth regulators such as abscisic acid but, although the growth response can be mimicked by abscisic acid, the molecular changes appear different and the role of growth regulators remains unclear.

SUPPLEMENTARY READING

Boyer, J. S. (1985). Water transport. *Annu. Rev. Plant Physiol.* **36**, 473–516.
Cleland, R. E. (1971). Cell wall extension. *Annu. Rev. Plant Physiol.* **22**, 197–222.

Frey-Wyssling, A. (1976). The plant cell wall. *In* "Handbuch der Pflanzenanatomie" (M. Zimmermann, S. Carlquist, and H. D. Wulff, eds.), Vol. 4. Gebrüder Borntraeger, Berlin.

Fry, S. C. (1989). The structure and functions of xyloglucan. *J. Exp. Bot.* **40**, 1–11.

Fry, S. C. (1989). Cellulases, hemicelluloses and auxin-stimulated growth: A possible relationship. *Physiol. Plant.* **75**, 532–536.

Preston, R. D. (1974). "The Physical Biology of Plant Cell Walls." Chapman and Hall, London.

Taiz, L. (1984). Plant cell expansion: Regulation of cell wall mechanical properties. *Annu. Rev. Plant Physiol.* **35**, 585–657.

12

Evolution
and Agricultural
Water Use

INTRODUCTION

The earth is about 4.6 billion years old and photosynthetic organisms appeared around 3 billion years ago, but land plants have existed for only about 0.5 billion years (Chapman, 1985). The invasion of the land probably was initiated by special conditions existing 0.5 billion years ago, including a favorable location of the continents (Bambach *et al.*, 1980; Fischer, 1984; Ziegler *et al.*, 1981), tidal flooding of vast areas of the continents (Fischer, 1982; Hallam, 1984; Parrish, 1983; Ziegler *et al.*, 1981), and moderate temperatures with high humidity and rainfall (Fischer, 1984; Holland, 1978). The early land plants appear to have been relatives of present-day Charaphytes such as *Nitella* and *Chara* that were anchored to bottom sediments with rhizoids and needed only a thin cuticle to resist moderate dehydration (Chapman, 1985; Graham, 1985).

After that time, the land climate became more severe as the continents moved and larger land masses became exposed (Fischer, 1982; Hallam, 1984; Ziegler *et al.*, 1981), and structural and functional adaptations developed rapidly among the plants. These included roots that permitted absorption of water and minerals from large soil volumes, a vascular system that facilitated rapid transport of water and photosynthetic products, and a well-developed cuticle with stomata that permitted CO_2 to enter but controlled water loss from the moist tissues beneath. A major force encouraging the invasion of the land was the availability of light that could support vigorous photosynthesis, but atmospheric CO_2 was consumed until it became only a trace gas in the atmosphere (Holland, 1984), and the photosynthetic apparatus developed special adaptations for operating at a low CO_2 concentration. C_4 photosynthesis appeared in certain species and involved temporary CO_2 fixation in four-carbon compounds followed by its release at high concentrations near the site of more permanent CO_2 fixation in the leaf cells. Crassulacean acid metabolism (CAM) developed in other species and allowed the temporary fixation of CO_2 to occur at night and release the following day. Both adaptations permitted more photosynthesis at declining atmospheric CO_2 levels for the same water loss.

MEASURING EVOLUTIONARY PRESSURES

The expansion of the land area and the increased exposure of plants to environmental extremes created evolutionary pressures for further adaptation, and the fossil record indicates that these pressures have changed as climates changed because many land species have disappeared and been replaced by species better adapted to the new climatic conditions. It therefore seems probable that significant evolutionary pressures are still present. One way to measure evolutionary pressure is to determine the ability to reproduce. The more intense the pressure, the less the reproductive success and the more rapidly unadapted species will disappear from the population. In ideal environments, little evolutionary pressure is present and the maximum potential for reproduction can be approached ($R_{potential}$). On the other hand, in average environments that represent the usual conditions, evolutionary pressure may be present and reproduction may be suppressed to give ($R_{average}$). The fraction of the potential that is achieved is the reproductive success $\dfrac{R_{average}}{R_{potential}}$ and the fraction that is lost is $(1 - \dfrac{R_{average}}{R_{potential}})$. The evolutionary pressure P_{ev} is the fraction that is lost

$$P_{ev} = 1 - \frac{R_{average}}{R_{potential}} \qquad (12.1)$$

and when $R_{average}$ is as high as $R_{potential}$, P_{ev} approaches 0, indicating there is little evolutionary pressure, and growth conditions allow the full expression of the reproductive potential of the plants. When $P_{ev} = 1$, evolutionary pressures are so large that the plants cannot reproduce and will disappear from the next generation. Values of P_{ev} between these extremes indicate evolutionary pressures of various intensities, resulting in varying degrees of adaptation to the environment.

Data for natural communities of plants are sparse but there are numerous data for agricultural communities having economically valuable reproductive structures. Table 12.1 shows that maximum (record) yields were high in eight major crops in 1975 in the United States. Six of these had valuable reproductive structures (maize, wheat, soybean, sorghum, oat, and barley). Because the rec-

Table 12.1 Record Yields, Average Yields, and Yield Losses Due to Diseases, Insects, Weeds, and Unfavorable Physicochemical Environments for Major U.S. Crops[a]

Crop	Record[b] yield	Average[b] yield	Average losses[c]			
			Diseases	Insects	Weeds	Physicochemical[d]
Maize	19,300	4,600	836	836	697	12,300
Wheat	14,500	1,880	387	166	332	11,700
Soybean	7,390	1,610	342	73	415	4,950
Sorghum	20,100	2,830	369	369	533	16,000
Oat	10,600	1,720	623	119	504	7,630
Barley	11,400	2,050	416	149	356	8,430
Potato	94,100	28,200	8,370	6,170	1,322	50,000
Sugar beet	121,000	42,600	10,650	7,990	5,330	54,400
Mean percentage of record yield	100	21.5	5.1	3.0	3.5	66.9

Note. Values are kilograms per hectare. Record and average yields are as of 1975.

[a] In the original work (Boyer, 1982), weed losses were considered to be physicochemical because the losses were attributable to competition for light, nutrients, and so on. On the other hand, weeds are of biological origin and it may be argued that the losses should be included with insects and diseases. For simplicity, the latter approach is taken here, which slightly alters the values calculated for each loss in comparison with Boyer (1982).

[b] From Wittwer (1975).

[c] Calculated according to U.S. Department of Agriculture (1965).

[d] Physicochemical losses calculated as record yield—(average yield + disease loss + insect loss + weed loss).

ord yield was measured under conditions that virtually eliminated pests and competing weeds, and nutrients and water were supplied in copious amounts, the record yield should have been an approximate measure of the maximum potential for reproduction ($R_{potential}$). In contrast, the average yield was obtained on farms under average agricultural conditions and should have been an estimate of the degree of suppression of the potential yield by the environment ($R_{average}$). All crops showed average yields that were much less than record yields. From Eq. (12.1), the resulting calculated value of P_{ev} was $(1 - 0.215/1.00) = 0.785$ for all the crops and 0.822 for the reproductive crops. Thus, environmental conditions are exerting a marked evolutionary pressure on crops. It seems likely that P_{ev} is substantial in natural communities as well.

From these principles we see that plants probably are evolving rapidly at present and what we perceive as stable species in natural populations appear that way only because the time scale of our observation is so brief. The changes brought about by agriculture are mostly accelerating and modifying change that is occurring naturally, which raises the possibility of modifying plants to allow them to reproduce at higher levels in average environments. Agriculture has approached the problem mostly by raising the genetic potential (e.g., many hybrids) or by changing the environment (e.g., irrigation and fertilization), and to a lesser extent by adapting plants to the existing environment. The latter approach probably will receive increased attention as the magnitude of loss in potential yield becomes better perceived (Boyer, 1982).

ENVIRONMENTAL LIMITATIONS ON YIELD

The causes of losses in potential yield include biotic factors (diseases, insects, and weeds in Table 12.1) that account for about 12% of the genetic potential. This represents the residual pest losses after intense measures for pest control, and a much larger loss would occur if pest control was not practiced. As a consequence, in natural communities where biotic factors are left only to natural defenses, the evolutionary pressure from pests is likely to be larger than in agricultural systems.

After accounting for biotic losses, Table 12.1 shows that nearly 70% of the genetic potential is lost because of physicochemical causes which include water and nutrient availability, temperature, daylength, soil pH, aeration, and excessive salt concentrations in the soil. Thus, these abiotic causes are partly attributable to soil problems and Table 12.2 shows that permanently dehydrated soils and shallow soils subject to frequent dehydration occupy about 45% of the U.S. land area, cold soils are present on nearly 17%, and wet soils cover almost 16% of the surface. Alkaline and saline soils occupy about 7% of the surface and only 12% of the soils are classified as being free of physicochemical problems (Boyer, 1982). The soils of the world (Dudal, 1976) have a similar classifica-

Table 12.2 Area of the U.S. Land Surface
Subject to Environmental
Limitations of Various Types[a]

Environmental limitation	Area of U.S. affected (%)
Drought	25.3
Shallowness	19.6
Cold	16.5
Wet	15.7
Alkaline salts	2.9
Saline or no soil	4.5
Other	3.4
None	12.1

[a] From U.S. Department of Agriculture (1975).

tion (Table 9.1). Table 12.3 shows that insurance payments to U.S. farmers for crop losses mirror these soil classifications and the biotic and abiotic effects (cf. Tables 12.1, 12.2, and 12.3).

It is clear from these data that the physicochemical environment, especially the effect of dehydration, is the dominant factor suppressing the productivity of land plants in the present world. As a consequence, desiccation continues as a major force in plant adaptation. Because of the intensive evolutionary pressure, it is likely that plants are still evolving better systems for coping with the land environment. Rice is a likely example because its genotypes extend from deep-water rices that are semiaquatic to upland rices that require well-drained soils

Table 12.3 Distribution of Insurance
Indemnities for Crop Losses
in the United States from 1939
to 1978[a]

Cause of crop loss	Proportion of payments (%)
Drought	40.8
Excess water	16.4
Cold	13.8
Hail	11.3
Wind	7.0
Insect	4.5
Disease	2.7
Flood	2.1
Other	1.5

[a] From U.S. Department of Agriculture (1979).

and exhibit many characters that are necessary for colonizing the land. The upland rices have thicker cuticle than the deep-water and paddy rices (O'Toole, 1982), and dry atmospheric conditions can desiccate reproductive structures and cause abortion (O'Toole *et al.*, 1984). The upland rices have deep roots that are often more extensive than those of deep-water and paddy rices (Chang *et al.*, 1974). Interestingly, only a few genes appear to control the differing root morphologies (Armenta-Soto *et al.*, 1983; Ekanayake *et al.*, 1985; O'Toole and Bland, 1987). The stems of upland rices have slow stem growth whereas rapid elongation occurs in deep-water rices and keeps part of the shoot and the flowers above rising flood waters in delta areas fed by monsoons (Raskin and Kende, 1985). Thus, it seems clear that in a crop that spans aquatic to upland habitats, extensive genetic adaptation to the land environment is occurring and there is a high degree of variability that could be used to change crop performance.

The traditional solution to water shortages has been irrigation, as pointed out in Chapter 4. Irrigation has made agriculture possible in many otherwise nonproductive areas and has the advantage that water can be made available as needed, and production is more predictable so that investment in other favorable cropping practices also can be undertaken. Large supplies of water are necessary because most of the irrigation water is evaporated by the crop. As a result, the water is consumed and not returned for other uses, and in the United States more water is consumed by irrigation than by all other uses combined (U.S. Department of the Interior, 1977). Salt-laden water cannot be used for irrigation because evaporation removes the water but leaves most of the salt, degrading the soil. New supplies of low salt water have diminished and municipalities compete for the same water, so new irrigation is becoming less possible than in the past, even if farmers can justify the large capital costs of the equipment and expense of moving the water (Boyer, 1982). As this trend grows there is increasing interest in improving the efficiency of water use in irrigation and determining whether plants can yield well in water-deficient conditions. If genetic manipulation or altered cultural practices can contribute to this goal, there could be less degradation of soils and water supplies and more cost effective irrigation.

A number of methods exist for improving the efficiency of water use and have been summarized by Taylor *et al.* (1983) and by Stewart and Nielsen (1990). The methods can be classified in three broad categories: (1) increasing the efficiency of water delivery and the timing of water application, (2) increasing the efficiency of water use by the plants, and (3) increasing the drought tolerance of the plants. The first method is practiced most because it depends mostly on engineering and minimally on the crop. Transporting water with minimal evaporation, preventing runoff, storing water in catchments, delivering water only to the root zone, and timing irrigation to the needs of the plant have been successful in improving productivity per unit of water delivered to the farm.

There are estimates that just by improving irrigation timing, the amount of water can be decreased by half in some crops while maintaining high levels of production (e.g., Bordovsky *et al.*, 1974). The second and third methods depend on understanding the biology of the crop and whether it can be manipulated to achieve the same productivity with less water. The state of knowledge in this area is the focus of the remainder of this chapter.

WATER USE EFFICIENCY

Water use efficiency is defined as the total dry matter produced by plants per unit of water used

$$\text{WUE} = \frac{D}{W}, \tag{12.2}$$

where WUE is the water use efficiency, D is the mass of dry matter produced, and W is the mass of water used. For a field experiment, D and W would be expressed on the basis of land area. For a single plant experiment, D and W would be measured in the same plant and expressed on the basis of the whole plant. One may also consider the water use efficiency for a single leaf and so on. The higher the dry matter production per unit of water use, the higher the efficiency.

There is extensive evidence that WUE of single plants varies among species in the same environment (see Table 7.3) and among climates for the same crop (Briggs and Shantz, 1914; de Wit, 1958; Hanks in Taylor *et al.*, 1983; Tanner and Sinclair in Taylor *et al.*, 1983). Taking advantage of the species and climate effects can help manage limited water supplies in agriculture. For example, alfalfa has a lower water use efficiency than maize when grown in nearby sites in the same year (Hanks in Taylor *et al.*, 1983; Table 7.3). Thus, changing crops can significantly reduce water consumption with little sacrifice in dry matter production. Relocating production to a new climate with lower evapotranspiration is another possible approach. For economic reasons, however, these options are not often employed and probably will not be until the cost of water rises to a level that forces change. What then are the prospects for improving water use efficiency within a species, or protecting against yield loss in a particular climate when irrigation is not possible?

It is first necessary to consider some principles of plant productivity and transpiration. Because land plants fix CO_2 from the air and the C and O atoms of this molecule account for most of the dry mass of the plant (see Chapter 10), D of Eq. (12.2) represents mostly photosynthetic activity. The CO_2 must diffuse into the leaf and dissolve in the wet surface of the cells before it becomes available for photosynthesis. The wet surfaces are exposed to the atmosphere inside the leaf and transpiration is inevitable. The photosynthesizing cells dehydrate to varying degrees depending on how readily the lost water can be replaced. Land plants generally absorb water from the soil and have a shoot covered by a waxy layer containing stomata that regulate water loss as CO_2 is fixed. Land plants have a much greater control of evaporation and water acquisition than their aquatic counterparts, and depending on the leaf anatomy and physiology, the dry matter produced per unit of water used can vary widely. Nevertheless, water use is affected by physical factors in addition to those imposed by the plant. CO_2 enters the leaf by diffusing down a concentration gradient to the leaf interior, and the water vapor in the intercellular spaces inside the leaf likewise diffuses in the opposite direction. The lower the external humidity, the faster transpiration will be when all other factors are constant. Leaf temperature plays an important role by affecting the vapor pressure of water in the leaf (see Chapters 2 and 7). The higher the leaf temperature, the higher the vapor pressure (Fig. 7.3) and the more rapid the transpiration. Water use will differ among sites and seasons for these reasons and the water use efficiency in Eq. (12.2) thus reflects a complex of plant and environmental factors.

Briggs and Shantz (1914) conducted an extensive survey of the water use efficiency of crops, and they expressed it as the water requirement, that is, the amount of water used per unit of dry matter produced which is the reciprocal of the water use efficiency. They grew the plants in large containers of soil and measured plant dry weight and the water used at the end of the entire growing season. This had the advantage that a large number of crops could be compared in a uniform climate during a single season. In their experiments, the transpiration ratio of maize, sorghum, and millet was less than for the other crops and, although Briggs and his co-workers could not have known at the time, the three crops are C_4 species possessing a special anatomy and biochemistry that allows CO_2 to be concentrated around the site of fixation. This resulted in more photosynthesis per unit of water transpired and accounted for the lower transpiration ratio.

After the experiments of Briggs and his co-workers, various investigators measured water use efficiency under field conditions where all the adaptations of the crop could express themselves (de Wit, 1958; Hanks in Taylor *et al.,* 1983). Figure 12.1 gives examples for several crops near Logan, Utah, that were grown with varying amounts of irrigation. It is striking that there is a linear relationship between water use and dry matter production (de Wit, 1958; Hanks in Taylor *et al.,* 1983). The linearity is mostly caused by the diffusion link between photosynthesis and transpiration because the visible radiation input is almost completely absorbed by all crops after the canopy closes and, in a given climate, the input is partitioned in a constant proportion between energy for transpiration and energy for photosynthesis. The slope of the linear relationship is the water use efficiency [Eq. (12.2)] and the straight line indicates that the water use efficiency does not change as the availability of water varies. However, it differs among species, especially between maize and the other crops

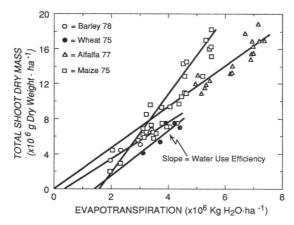

Figure 12.1 Production of aboveground shoot dry matter at various levels of water use in several crops near Logan, Utah. The years in which the crops were grown are shown in the symbol key. Water use was controlled by varying the amount of irrigation and is shown as combined evaporation from the soil and transpiration from the plants. A positive evapotranspiration intercept indicates the amount of water obtained from soil stores. The slope of the linear relation is the water use efficiency which was 2.11 g of dry weight per kg of H_2O for barley, 2.50 for wheat, 2.36 for alfalfa, and 4.49 for maize. Note that maize is a C_4 plant and the others are C_3. Maize and wheat were grown in the same year. Adapted from Hanks (in Taylor *et al.,* 1983).

(Fig. 12.1). These experiments confirm the differences noted by Briggs and his co-workers and further indicate that water use efficiency does not differ under varying availabilities of soil water. However, it differs among species, climates, and from year to year (Briggs and Shantz, 1914; Brown and Simmons, 1979; Garrity *et al.,* 1982; Hanks in Taylor *et al.,* 1983; Kawamitsu *et al.,* 1987; Pandey *et al.,* 1984a,b; Robichaux and Pearcy, 1984; Tanner and Sinclair in Taylor *et al.,* 1983). More work may be needed on the constancy of water use efficiency with different mineral availabilities, plant spacing, and other cropping practices.

In this respect, it is important to note that while differences between C_4 and C_3 species are apparent, tests have been made only under limited conditions in species exhibiting CAM. Species such as pineapple exhibit CAM and concentrate CO_2 by temporarily fixing it at night in organic acids from which it is released the next day for photosynthesis. During release, the stomata are closed and water is conserved (see Fig. 8.3). This allows CAM plants to achieve even higher water saving than C_4 plants, and estimates of water use efficiency for pineapple are about 20 g dry mass $\cdot$ kg^{-1} water, for C_4 plants about 3 to 5 g dry mass $\cdot$ kg^{-1} water, and for C_3 plants only about 2 to 3 g dry mass $\cdot$ kg^{-1} water with variations depending on the evaporative environment (Briggs and Shantz, 1914; Hanks in Taylor *et al.,* 1983; Joshi *et al.,* 1965; Neales *et al.,* 1968).

It has been argued that water use efficiencies should not be expressed as absolute dry mass gained per unit of water mass used but should be normalized for evaporative demand (de Wit, 1958; Tanner and Sinclair in Taylor *et al.*, 1983) and often for the potential productivity of the crop (Hanks in Taylor *et al.*, 1983). Thus, modified expressions of WUE have been used such as

$$\frac{D}{D_{max}} = \frac{W}{W_{max}}, \tag{12.3}$$

where the fractional dry mass is D/D_{max} and is expressed relative to the maximum dry mass produced with optimum water D_{max}. The fractional water use W/W_{max} is likewise expressed relative to the maximum transpiration that would occur at the site with optimum water. The approach has the advantage that for a water use of, say, half the potential transpiration, half the maximum dry mass would be predicted. This simplifies the job of predicting the impact of water shortages. However, it requires a knowledge of the maximum dry matter yield and transpiration of the crop for the year, which will vary.

It must be kept in mind that farm income is generally based on the absolute dry mass or economic yield, and expense is based on the absolute amounts of water used. A farmer contemplating whether to irrigate semiarid land needs to have high absolute production of dry mass to justify pumping larger amounts of water than a farmer in a humid region. Even better, he should know the absolute production of marketable yield, which may be only a part of the total dry mass. Figure 12.2 shows that the water use efficiency for producing grain in

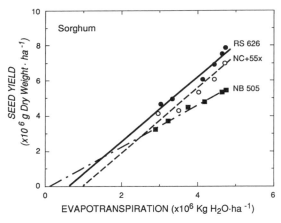

Figure 12.2 Seed yield at various levels of water use by three sorghum genotypes in west central Nebraska. The water use efficiency for seed yield is the slope of the line and was 1.8 g of dry weight per kg of H_2O in RS626, 1.9 in NC+55X, and 1.2 in NB505. Note that the absolute seed yield of RS626 was superior to that in NC+55X even though the water use efficiency was essentially the same. The water use efficiency for total shoot dry matter production was 3.3 in RS626, 3.2 in NC+55X, and 2.0 in NB505. Adapted from Garrity *et al.* (1982).

sorghum was greater in the genotypes RS626 and NC+55X than in NB505. Reducing water use by half in each genotype would give half of the maximum yield but would not distinguish which genotype would give the highest grain production for a particular amount of water used. The farmer would profit more by planting RS626 because it not only had a better absolute water use efficiency than NB505 but also the highest absolute yield (Garrity *et al.*, 1982).

Measuring Water Use Efficiency

Measuring water use efficiency in the field is the most accurate means of determining how dry matter production will be affected by water availability but it is labor intensive and costly. Less expensive methods have been sought, and one has been to measure directly the CO_2 and H_2O exchange of individual leaves (Bierhuizen and Slatyer, 1965; Brown and Simmons, 1979; Robichaux and Pearcy, 1984). Because the carbon dioxide molecule contributes most of the dry mass, the gas exchange efficiency can be defined as the ratio of the mass of CO_2 gained to the mass of H_2O lost. Figure 12.3 shows an example of water use efficiency measured as gas exchange efficiency in comparison with the actual water use efficiency for the whole growing season measured in the usual way in tomato (*Lycopersicon esculentum* Mill.) and its wild relative *L. pennellii* (Cor.) D'Arcy. The relationship is poor because of the additional factors affecting dry mass accumulation but not gas exchange (Martin and Thorstenson, 1988). The mass of the plant is determined by long-term net dry mass accumulation which is affected by respiratory losses at night and partitioning to nonphotosynthetic organs as well as CO_2 uptake. It is altered by temperature and the molecular composition of the dry mass. Gas exchange for short times during the day does not detect these additional factors. Therefore, while the gas exchange efficiency gives valuable insight into the physiologic and metabolic controls that might operate during photosynthesis and transpiration, the method is being used less frequently than in the past.

Another method involves measuring the relative abundance of natural isotopes in plant tissue. Although most of the CO_2 in the atmosphere is $^{12}CO_2$, a small amount is $^{13}CO_2$. Because the $^{12}CO_2$ is lighter, it diffuses into the leaf faster than $^{13}CO_2$. Also, ribulose bisphosphate carboxylase fixes the lighter isotope faster. The cells accumulate relatively less ^{13}C than ^{12}C, and the unused ^{13}C in the intercellular spaces of the leaf diffuses out according to the extent of stomatal opening. This outward diffusion is correlated with transpiration. Because the inward diffusion and use of $^{12}CO_2$ correlates with photosynthesis and dry mass but the outward diffusion of $^{13}CO_2$ correlates with transpiration, the relative uptake of ^{12}C and ^{13}C correlates with the water use efficiency. Farquhar and his colleagues demonstrated differences in water use efficiency between genotypes of wheat (*Triticum aestivum* L.), peanut (*Arachis hypogaea* L.), barley (*Hordeum vulgare* L.), and other crops by measuring the ratio of ^{12}C to ^{13}C

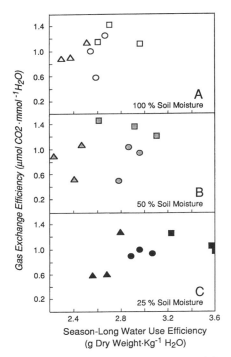

Figure 12.3   Relation between the water use efficiency determined from measurements of CO_2 fixed : H_2O transpired (gas exchange efficiency) and the season-long water use efficiency determined from the total plant dry mass accumulated per water mass used in tomato. (A–C) Plants grown at 100% (open symbols), 50% (half closed symbols), and 25% (closed symbols) of soil field capacity, respectively. Triangles are for *Lycopersicon esculentum*, squares are for *L. pennellii*, and circles are for the F_1 hybrids. Note that the variability is so high that none of the relationships are significant at the $P<0.05$ level. However, the water use efficiency measured for the whole season is generally higher for *L. pennellii* than for *L. esculentum*. Adapted from Martin and Thorstenson (1988).

isotope content of plant tissue relative to that in a standard (Bowman *et al.,* 1989; Brugnoli *et al.,* 1988; Condon *et al.,* 1987, 1990; Hubick and Farquhar, 1990; Hubick *et al.,* 1986).

This ratio technique makes it possible to survey a large number of plants with a simple analysis of the leaf tissue. Differences integrate the conditions over which the plants are grown. Analyzing the entire shoot indicates the water use efficiency for the time required to grow the shoot whereas analyzing only leaf starch indicates the water use efficiency during the time necessary to accumulate the starch. One may integrate over long or short times with the method and this avoids one of the problems of the gas exchange technique.

Martin and Thorstenson (1988) used this technique to show that differences

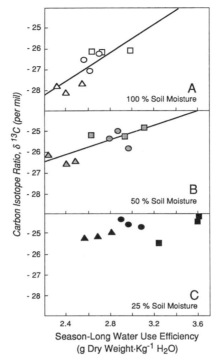

Figure 12.4   Relation between the carbon isotope ratio of leaf tissue and the water use efficiency for the whole growing season in tomato. The tomato species are the same as in Fig. 12.3 and the symbols are the same. Note that the correlations between isotope ratio and water use efficiency are better than for gas exchange in Fig. 12.3. The correlations in (A) and (B) are significant at the $P<0.01$ and $P<0.05$ levels, respectively. Adapted from Martin and Thorstenson (1988).

in water use efficiency were present between the domestic tomato species and *L. pennellii* and their hybrids. Figure 12.4 shows that the differences in water use efficiency were detectable in isotope ratio data between the parents and the hybrids particularly when water was optimally available. The domesticated parent had the lowest efficiency, the wild parent had the highest efficiency, and the hybrids showed intermediate behavior. Because the species could be crossed, it was possible to correlate the differences in water use efficiency with restriction fragment maps of the tomato DNA (Martin *et al.*, 1989). Three loci were found to be predictors of the variation in water use efficiency in field-grown tomato. This landmark effort indicates that water use efficiencies are determined by relatively few genetic loci and implies not only that agriculturally relevant differences exist but that they might be genetically manipulated in a simple fashion.

The success of the method suggests that differences in water use efficiency not only exist within individual species but might be incorporated into breeding programs (Bowman *et al.*, 1989; Brugnoli *et al.*, 1988; Condon *et al.*, 1987, 1990; Hubick and Farquhar, 1990; Hubick *et al.*, 1986), although this is still in its infancy. A significant amount of variability is sometimes present in the data, but it is becoming clearer that selecting for extremes in carbon isotope ratios will select for extremes in water use efficiency. It is surprising that a complex trait like water use efficiency should be controlled by only a few genetic loci. Thus far it has not been determined whether each locus corresponds to more than one gene. It remains possible that the trait is in fact complex. Despite this situation, further studies of the genetic basis for differences in water use efficiency seem warranted, and it is likely that the differences will be heritable.

DROUGHT TOLERANCE

Plants showing improved growth with limited water are considered to tolerate drought regardless of how the improvement occurs. Some species can avoid drought by maturing rapidly before the onset of dry conditions or reproducing only after rain. Examples of these drought avoiders are ephemerals such as California poppy (*Eschscholtzia californica*) that can complete their life cycle in a few weeks or tree crops such as coffee, cacao, and mango that flower and fruit after moderate drought followed by rain (Alvim, 1960, 1985, cover photograph). Others can postpone dehydration by growing deep roots or sealing themselves tightly against transpiration or accumulating large stores of water in fleshy tissues. Examples of dehydration postponers are upland rice with deep roots compared to paddy rice (Chang *et al.*, 1974) or agave or saguaro cactus with thick cuticle or fleshy tissue adaptations. Still other species allow dehydration of the tissues and simply tolerate it by continuing to grow when dehydrated or surviving severe dehydration. *Fucus vesiculosus* is an example of a dehydration tolerator, and the acclimation of sunflower illustrates that dehydration tolerance can be increased as discussed in Chapter 10.

These effects are generally distinct from the factors controlling water use efficiency. Drought avoiders often reproduce themselves after a minimal accumulation of dry weight and their success ensures that they are represented in the next generation. Their adaptation centers on timing development and thus is under internal control. Dehydration postponers having deep roots may have a water use efficiency identical to that of other species but will accumulate more dry weight because of their ability to gain access to a larger amount of water than shallow rooted types. In effect the slope of the water use efficiency relation in Fig. 12.5 may be the same but the deep rooted species work farther out on the curve. Their adaptations are mostly structural and take time to build, requiring the expenditure of photosynthetic products. Finally, dehydration toler-

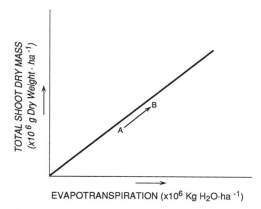

TOTAL SHOOT DRY MASS
(x10^6 g Dry Weight · ha^{-1})

EVAPOTRANSPIRATION (x10^6 Kg H$_2$O·ha^{-1})

Figure 12.5   Effect of increasing the amount of water available to a crop without changing the water use efficiency. Production moves from A to B. An example might be increasing rooting depth.

ators may have the same water use efficiency as dehydration sensitive species when water is available but the tolerators can grow at tissue hydration levels at which the other species cannot.

Of the three forms of drought tolerance, dehydration tolerance is most intriguing because it often requires only slight repartitioning of dry mass. An example is osmotic adjustment (Morgan, 1984; Munns, 1988) which occurs because dry mass normally used to synthesize new cells instead accumulates in the cells as solute (Meyer and Boyer, 1972, 1981) or is deposited in fewer or smaller cells (Fraser *et al.*, 1990; Sharp *et al.*, 1990). Only a brief decrease in biosynthesis of tissue is necessary to accomplish this (Meyer and Boyer, 1981), but the increased concentration of solutes can markedly increase the ability of the cells to extract water from the soil. The increased solute is present only under dry conditions. In other words, there is little cost to the plant when water is scarce and no cost when water is plentiful.

Improvement of Drought Tolerance

From these examples it can be seen that crop improvement under conditions of limited water involves more than water use efficiency. Often, breeding programs for drought tolerance are based on the productivity of plants when water is plentiful. The idea is that, for a given climate, water use efficiency will be highest when dry matter production is highest. Because productivity is linearly proportional to water use (Figs. 12.1 and 12.2), the high productivity of dry matter should carry over to drought conditions. However, it is clear that many opportunities will be missed if superior selections are based only on this concept of water use efficiency. Characters such as osmotic adjustment are called into play only during a water deficit. Roots may penetrate deeper soil layers or leaves may persist better during a water deficit in some genotypes than in others, and so on. Without plant selection under water-deficient conditions, these beneficial traits will be missed.

The design of a crop improvement program for drought tolerance seems difficult at first because water is so ubiquitously involved in growth and metabolism that identifying targets seems impossible. Moreover, the multitude of possible targets implies that effects might involve enormous numbers of genes and that improvements might be only incremental or, worse still, may cause problems at other genetic loci. However, there are examples of successful approaches that have resulted in significant improvements in the drought tolerance of plants. Jensen and Cavalieri (1983) described the improvement in drought tolerance they achieved by breeding maize after field testing at a large number of locations varying in water availability. Genotypes were identified that had all the combinations of yield performance under optimum and water-deficient conditions: a high yield in both conditions (Hybrids 3377 and 3358 in Fig. 12.6), a high yield in optimum conditions but a low yield under deficit conditions (Hybrid 3323 in Fig. 12.6), and a low yield in optimum conditions but a high yield under deficit conditions (Hybrid 3388 in Fig. 12.6). The first kind of response is the preferred one but the last response seems worthy of some consideration.

The study by Jensen and Cavalieri (1983) is particularly important because it tested whether improved yield under water limited conditions sacrificed yield under optimum conditions. Because grain yield from about 500 field replications was used to evaluate the germplasm and whole season yield performance, characters associated with particular environments or parts of the life cycle were included. Their experiments give the strongest possible evidence that improvement under water limited conditions need not sacrifice yield under favorable conditions. The number of replications, genotypes, and field sites was so large that this principle now seems beyond doubt (Fig. 12.6).

The principle was confirmed by Morgan (1983) in a completely different experiment. He selected wheat for superior osmotic adjustment under dehydrating conditions and observed improved yields that were at no cost to yield in optimum conditions. An important feature of this study was that the test genotypes had essentially the same genetic background. Therefore, osmotic adjustment was the main difference between the standard commercial genotype and the genotype with superior drought performance. Quisenberry *et al.* (1984) also tested the effectiveness of osmotic adjustment in cotton but concluded that it had little benefit. However, they failed to compare plants of similar genetic backgrounds and the results could have been caused by features other than osmotic adjustment. Therefore, the work of Morgan (1983) provides the better

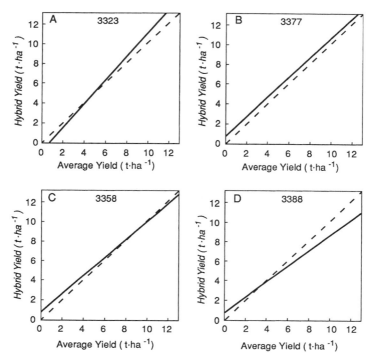

Figure 12.6 Regressions of seed yield of four maize hybrids grown at various locations in the United States over 3 years. The dashed line indicates the average yield for all hybrids at each location (1:1) and the solid line shows the yield of an individual hybrid for comparison. A solid line above the dashed line indicates a better than average yield for the hybrid. (A) Hybrid 3323, (B) hybrid 3377, (C) hybrid 3358, and (D) hybrid 3388. Except for (B), the slopes of each hybrid regression differed significantly from the slope of the dashed line ($P<0.01$). The r^2 values were between 0.67 and 0.82 for the regressions of the four hybrids. Regressions were formed for 399 genotypes and, in most instances, for over 500 sites. Adapted from Jensen and Cavalieri (1983).

test and indicates that there can be a benefit of osmotic adjustment without sacrificing yield under optimum conditions. This probably is explained by the low metabolic cost of osmotic adjustment together with the lack of osmotic adjustment under optimum conditions.

Another example of genetic improvement of drought tolerance is the selection for improved seedling establishment in native range grasses in the western United States. Wright and Jordan (1970) showed rapid improvement in the establishment of boer lovegrass (*Eragrostis curvula* Nees) selected for seedling growth in dehydrated soil. The character that appeared most improved was the thickness of the cuticle covering the shoot tissues of the young seedlings (Hull *et al.*, 1978). These selections allowed the establishment of grasses to become more reliable when rooting was shallow, rainfall was sporadic, and germination had to occur with limited water.

Burton *et al.* (1954, 1957) showed that deep rooted Bermuda grass [*Cynodon dactylon* (L.) Pers.] exhibited increased pasture productivity compared to more shallow rooted types in humid regions subjected to sporadic drought. Deep rooted rice is another example of this approach to improving drought tolerance (Chang *et al.*, 1974), which works best in deep soils that allow deep rooting.

Another approach has been to control the life cycle so that growth only occurs when water is most available. Hall and Grantz (1981) selected early flowering cowpeas [*Vigna unguiculata* L. (Walp.)] that escaped late season drought. Because the reproductive tissues were the valuable structures, genetic selection for earliness restricted growth to the part of the season when water was available. Similarly, Passioura (1972) demonstrated that wheat using only water stored in the soil produced grain if the roots were pruned to reduce early season water use by the shoot but not if the roots were unpruned and the plants consumed most of the soil water by the time of grain fill. This experiment suggests that genetic means of controlling root morphology might be sought in wheat that matures grain on water stored in the soil. These kinds of genetic manipulation of development are valuable for climates where late season drought is predictable.

Although each of these approaches is unique, there are some common concepts among them. First, each investigator made selections under conditions of realistic water limitation in soils. Approaching the problem this way ensured that drought-adaptive factors were called into play and had an opportunity to express themselves. Important traits for drought performance were identified because they were present. This avoided the problem of selecting only crops yielding well in favorable environments and hoping they will not "crash" in water-limited environments. Second, in most cases, there was an intimate knowledge of the soil, climate, and physiology and biology of the crop. This knowledge increased the rate at which superior genotypes could be found. Morgan (1983) was aware that osmotic adjustment could benefit the crop. Without that knowledge, he would have been restricted to selections for yield alone. Third, selections often were for single traits. Reducing the problem to a few specific traits simplifies the selection effort. The number of selections and range of conditions are fewer and the program more easily fits into the available resources.

The complexity of drought tolerance has tempted many to take shortcuts such as selecting seedlings for growth in osmotica or using single biochemical tests for performance. In general, the results do not carry over into field situations. For example, Sammons *et al.* (1978, 1979) showed that physiological tests on plants grown in controlled environment chambers gave tolerance rankings that differed from those for yield in the greenhouse. These differed again when ranked for field performance (Mederski and Jeffers, 1973). Therefore, for drought tolerance there seems to be no substitute for growing the plant under realistic field conditions or carefully simulated field conditions.

WATER DEFICITS AND REPRODUCTION

The principles discussed in the preceding section can be applied to any aspect of plant development and have the potential to improve individual features of performance. However, reproductive development holds particular interest because a large part of agricultural production is devoted to the reproductive parts of plants. In the United States, reproductive crops (grain, fruit, nut, vegetable) account for about 78% of the harvested area of land. Moreover, early stages of reproduction are more susceptible to losses from limited water than any other stage of development in reproductive crops (Claassen and Shaw, 1970; Salter and Goode, 1967). A good example is maize where part of the problem is caused by a high susceptibility of floral parts to inhibition of cell enlargement (Herrero and Johnson, 1981; Westgate and Boyer, 1985b). This susceptibility exists in part because the cells enlarge dramatically in the floral tissues during normal development, and water deficit can prevent the enlargement (see Chapter 11). However, more than cell enlargement is involved because Damptey *et al.* (1978) observed losses in inflorescence development in maize treated with abscisic acid during floral initiation before most enlargement of reproductive structures had begun. Losses in reproductive activity also were reported because of megagametophyte sterility (Moss and Downey, 1971), asynchronous floral development (Herrero and Johnson, 1981), and nonreceptive silks (Lonnquist and Jugenheimer, 1943), depending on when dehydration occurred. When gamete and floral development are normal and plants are hand-pollinated, reproductive failure still occurs and can be induced by only a few days of dehydration (Westgate and Boyer, 1985b, 1986b). The loss is caused by irreversibly arrested embryo development (Westgate and Boyer, 1986b). This indicates that, provided there is good floral development, pollination and fertilization can be successful. Nevertheless, a complete block in embryo growth can remain even when everything has been normal up to the time of cell division in the newly formed zygotes.

In other crops such as wheat and barley, drought during microsporogenesis caused pollen sterility (Morgan, 1980; Saini and Aspinall, 1981, 1982; Saini *et al.*, 1984). Well-watered plants whose stems were fed ABA (Saini *et al.*, 1984) or whose shoots were sprayed with ABA (Morgan, 1980) showed a similar pollen abortion, thus implicating high ABA levels during dehydrating conditions. However, the high ABA may have acted by closing stomata and inhibiting photosynthesis. Increasing CO_2 pressures around wheat plants overcame some of the reproductive losses (Gifford, 1979), which supports an involvement of photosynthesis. In rice (*Oryza sativa* L.), dehydration of the soil caused especially severe dehydration of reproductive tissues, and death and bleaching of florets followed (O'Toole *et al.*, 1984). The cuticle is only poorly developed on the floral tissues of rice and may be inadequate to prevent excessive dehydration (O'Toole *et al.*, 1984). Therefore, in various crops, there is increasing evidence for metabolic and growth regulator effects and some direct dehydration effects that might account for the susceptibility of early reproduction to water limitation. CO_2 and ABA are involved, and photosynthesis also may play a role but each might act in concert or separately, depending on the crop.

More insight may be possible from studies of embryo development in maize. Westgate and Boyer (1985a) found that the block in embryo development was correlated with low photosynthetic reserves in the maternal plant. Because photosynthesis was inhibited during the treatment, the lack of reserves could have caused embryo starvation. Westgate and Thomson Grant (1989) observed that the sugar content of maize embryos was not significantly different in hydrated and dehydrated plants but concluded that the flux of sugar might differ. Schussler and Westgate (1991a,b) found that the uptake of sugars was less in maize ovules isolated from dehydrated plants, even though the sugar content was high which further confirms that the flux of sugars was more important than the sugar content of the developing grain. Myers *et al.* (1990) showed an inhibition of endosperm cell division in maize when high ABA levels were present 5 to 10 days after fertilization.

Boyle *et al.* (1991a,b) took advantage of the finding of Westgate and Boyer (1985a, 1986a) that a few days of low water potentials can prevent embryo growth and developed a system to feed stems a complete medium for embryo growth during this time. This allowed photosynthetic products and other salts and metabolites to be supplied to the plants at normal levels without rehydrating the plants. The controls yielded well (Fig. 12.7A), but withholding water for a few days virtually eliminated grain production (Fig. 12.7B). Nevertheless, production was almost fully restored when the plants were infused with the complete medium as low Ψ_w developed (Fig. 12.7C). Infusing the same amount of water alone showed no restorative activity (Fig. 12.7D). Therefore, sufficient water was available to the embryos so that water itself was not the limiting factor. Reproduction was maintained by feeding other substances normally supplied by the parent plant during embryo development that the parent plant failed to supply. Thus, reproductive loss is a biochemical problem.

This type of experiment offers the promise of eventually identifying components that are required for reproductive development in plants and which may be lacking when dehydration occurs. Supplying these nutrients might be possible in superior genotypes or with cultural conditions that would allow large

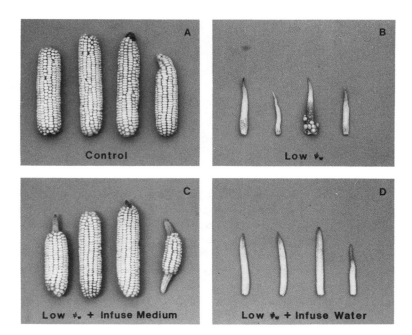

Figure 12.7   Grain yield at maturity for soil-grown maize plants subjected to a water deficit during pollination. Treatments were adequately watered controls (A), water-deficient plants from which water was withheld for 6 days (B), plants treated as in (B) but with stem infusion of a complete medium for embryo culture (C), and plants treated as in (B) but with stem infusion of water (D) in the same amount as in (C). All plants were rewatered on the sixth day. Plants were hand-pollinated. The grain weight of the plants infused with medium at low water potentials was about 80% of the weight of the controls. The infusion did not change the water potential or photosynthesis of the plants. From Boyle *et al.* (1991b).

amounts of the missing constituents to be present at the right time. This form of dehydration tolerance might protect against losses in reproduction, at least for short periods of water deficiency.

It also demonstrates that the reproductive fraction of the plant can vary from zero to nearly normal during a drought. This implies that successful protection of reproductive development may be possible under otherwise inhibiting drought conditions. It will be important to determine the active ingredients in the medium and whether any other aspects of reproduction also can be protected in the same manner.

From this work, it seems that through genetic manipulation the impact of dehydration could be minimized during the early stages of reproduction. For example, Edmeades and his co-workers found that the time between pollen shed and silking can be changed by genetic means in maize (Bolaños and Edmeades, 1993a; Bolaños and Edmeades, 1993b; Bolaños *et al.*, 1993; Edmeades *et al.*, 1992, 1993), and early silking during drought may indicate vigorous development of the ear perhaps because the plant supplies more of the biochemical requirements for ear growth. Genetic selections in this direction might then increase the ability of the plants to maintain grain during a drought.

DESICCATION

When seeds mature, it is common for them to dehydrate as part of the maturation process. Barlow *et al.* (1980) found water potentials as low as -5 MPa in maturing wheat grain. Westgate and Boyer (1986c) observed water potentials of -7 to -8 MPa or lower late in the growing season in maize grain. These seeds are exposed somewhat to the atmosphere and are known to desiccate to a large extent by evaporation to the air. Seeds surrounded by a fleshy fruit show a similar but less severe desiccation. Welbaum and Bradford (1988) observed that water potentials of melon seeds decreased to about -2 MPa during maturation, and the surrounding fleshy fruit decreased similarly in water potential. Bradford (1994) considers high solute concentrations to be present in the apoplast surrounding embryos and proposes that structures may exist to keep the solutes localized there. The low osmotic potential of the apoplast solution may explain how the seeds are dehydrated inside fleshy fruits. Regardless of whether the seeds air dry or are dehydrated osmotically inside a fruit, it is clear that embryos become exceptionally tolerant of desiccation late in maturation despite their susceptibility to the effects of water limitation when they are young.

Plants lower in the evolutionary scale than seed plants sometimes show a similar tolerance to desiccation. Some fungi, algae from the intertidal zone, and a few mosses and lycopods can be desiccated to the air-dry state without losing viability (Bewley, 1979). Some specialized seed plants (*Craterostigma* species, *Myrothamnus flabellifolia*, *Xerophyta* species) that can tolerate desiccation also exist (Gaff, 1971, 1977; Gaff and Churchill, 1976). However, desiccation tolerance is virtually nonexistent in most agricultural species except for the seeds. It is curious that most seed plants, which are descendants of plants that crossed the intertidal zone, should have lost the ability to tolerate the desiccation that is so prevalent in that zone. In land plants, desiccation tolerance often evolved as part of the seed habit because an aqueous medium generally was absent and the embryo was exposed to drying conditions during dispersal. In agriculture, this property makes it possible to store seeds and allows uniform planting times.

An important aspect of severe desiccation is that water contents can become so low in the cells that enzyme activities can be directly inhibited by the lack of water, as described by Vertucci and Leopold (1987a,b). As discussed in Chapter 9, enzymes equilibrated in humidities around 60% or below are affected directly because the hydration shells next to the protein are necessary for catalytic activity and may become modified. Substrates probably are unable to reach the active site of the enzyme because the aqueous medium is no longer continuous (Skujins and McLaren, 1967). Seeds desiccated to the air-dry state are likely to be affected by these phenomena. Most can return to activity when they are rehydrated, provided water contents have not become so low that the tightly bound water required for viability is lost (Vertucci and Leopold, 1987a,b).

On the other hand, leaves generally are susceptible to desiccation damage and in most crop species show a breakdown of compartmentation that releases cell constituents to the apoplast (Leopold *et al.*, 1981), and the plasmalemma and tonoplast show breakage followed by a loss of organelle structure (Fellows and Boyer, 1978). In leaves of tolerant species, the membranes and organelles remain intact although they often are distorted (Hallam and Gaff, 1978a,b). Therefore, an important distinction between tolerance and sensitivity to severe desiccation appears to be the maintenance of membrane structure and an ability of enzyme activity to return upon rehydration.

It has been proposed from work with desiccation tolerant animals that a possible mechanism to account for preservation of enzymes and cell structure might be an accumulation of specific sugars such as trehalose or sucrose whose structure resembles water in certain respects (Crowe and Crowe, 1986). Sugars having the appropriate stereostructure might form hydrogen bonds with cell membranes where water would ordinarily bind. Because the sugars would remain as water is removed, the bonding would be stable and membrane structure might be maintained where otherwise it would become disorganized.

Evidence that the sugar replacement hypothesis may have merit is the accumulation of sugars such as sucrose and raffinose in developing seeds (Caffrey *et al.*, 1988; Koster and Leopold, 1988). Species such as maize have seeds that can tolerate desiccation to the air-dry state, and their sugar concentration, while not high for the seed as a whole, becomes high in the remaining residual water and could have a stabilizing influence at local sites. As germination proceeds, the stabilizing sugars are metabolized to nonstabilizing ones such as glucose and fructose, and desiccation tolerance is lost (Koster and Leopold, 1988). A related hypothesis is that certain sugars may be converted to the glassy state during dehydration (Williams and Leopold, 1989). The glassy state is common in sugars such as sucrose used to make candy, and evidence for the existence of glassy sugars is accumulating for embryos of dehydrated seeds (Williams and Leopold, 1989).

A similar role has been proposed for certain proteins in seeds (Crowe and Crowe, 1986; Dure *et al.*, 1989). The developing seeds of a range of crops accumulate hydrophilic proteins in the embryo as normal desiccation begins (Dure *et al.*, 1989). The proteins have been variously called dehydrins, embryo

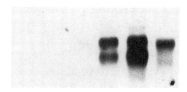

Figure 12.8   mRNAs for wheat dehydrins. A barley cDNA was used to detect the level of dehydrin mRNA in a total mRNA extract from severely dehydrated seedlings and in hydrated controls. al, aleurone tissue; sh, shoots; and root, root system. From Close and Chandler (1990).

maturation (Em) proteins, or late embryogenesis abundant (LEA) proteins (Dure *et al.*, 1989). Common to all of them is a high content of hydrophilic amino acids so that the proteins as a whole are highly water soluble. In some of them, an α-helix is present that could remain structurally stable during desiccation and it has been proposed that this portion of the protein could act like a membrane-stabilizing sugar (Crowe and Crowe, 1986).

The mRNAs for these proteins are not readily detected in leaves or roots of hydrated plants but can be induced by severe desiccation in very young rape (Harada *et al.*, 1989) and maize and barley seedlings (Close *et al.*, 1989; Close and Chandler, 1990). Figure 12.8 shows the marked increase in dehydrin mRNAs when young wheat seedlings were dehydrated soon after germination (Close and Chandler, 1990). The mRNA expression was especially increased in shoots, which are most exposed to dehydration under natural conditions. Also, the mRNAs can be induced by treating hydrated seedlings or immature embryos with high ABA concentrations (Galau *et al.*, 1986; Hong *et al.*, 1988; Mundy and Chua, 1988). ABA levels normally increase in plants subjected to dehydration and they become high in maturing dehydrating seeds (Ihle and Dure, 1972). The induction of these mRNAs suggests that the new dehydrin-Em-LEA proteins may be involved in the development of desiccation tolerance of young seedlings and embryos.

ANTITRANSPIRANTS

The possibility of reducing evaporation from plants and plant parts is attractive, and covering stem ends with protective coatings to prevent splitting or covering grafts with waxes to preserve viability has been practiced for many years. It has been possible to decrease the evaporation from large bodies of water by covering them with suitable films such as cetyl alcohol and, in principle, similar effects on transpiration could save considerable water and might reduce the inhibitory effects of dehydration on leaf metabolism. However, the plant carries on many activities besides transpiration and the effectiveness of an antitranspirant is determined not only by its water-saving capability but also by the way in which it alters other aspects of plant performance.

Most antitranspirants act on the stomata by closing them or by covering them with a substance that decreases diffusion. Other approaches such as increasing the reflectiveness of leaves for infrared radiation or increasing the CO_2 partial pressure around leaves have been tried, but these have not proven practical because the infrared reflectance is already high for most leaves and increasing the CO_2 is too expensive. Therefore, most antitranspirants center on stomatal function, which was discussed in Chapter 8, and Gale and Hagan (1966) reviewed those types.

Stomata control not only the loss of water vapor but also the diffusion of CO_2 into the leaf. When an antitranspirant is applied that decreases water loss there is also the probability that the diffusion of CO_2 into leaves will be reduced, and most interest in side effects of antitranspirants has centered on photosynthesis and growth (which reflects photosynthesis to some degree as described in Chapters 10 and 11). At times when photosynthesis is not limited by the availability of CO_2 in the leaf such as in low light intensities, it should be possible to decrease transpiration without inhibiting photosynthesis. Similarly, under other conditions in which the photosynthesis rate depends on the internal partial pressure of CO_2, it was argued that decreasing stomatal apertures should inhibit transpiration more than photosynthesis (Gaastra, 1959; Gale and Hagan, 1966; Slatyer and Bierhuizen, 1964b). As explained in Chapter 7, transpiration is determined mostly by the resistances to water vapor diffusion from the evaporating surface to the bulk air outside the leaf [$r_{leaf} + r_{air}$ in Eq. (7.2)], but CO_2 encounters an additional resistance to diffusion in the liquid of the mesophyll cells. The argument pointed out that because the total resistance to diffusion is thus smaller for transpiration than for photosynthesis, increasing the resistance of the leaf with an antitranspirant should have less effect on photosynthesis than on transpiration (Gaastra, 1959; Gale and Hagan, 1966; Slatyer and Bierhuizen, 1964b). Recognition of this fact gave considerable impetus to antitranspirant work (Gale and Hagan, 1966).

Nevertheless, it has been difficult to identify a chemical that reduces transpiration by closing the stomata without inhibiting photosynthesis or growth at least as much. Slatyer and Bierhuizen (1964b) tested several chemicals that decreased transpiration but all except phenylmercuric acetate caused photosynthesis to decrease as much as transpiration. Other workers found variable effects on photosynthesis and growth with this compound (Shimshi, 1963a,b; Zelitch and Waggoner, 1962a,b).

Film-forming chemicals such as hexadecanol can inhibit transpiration when coated on leaf surfaces containing the stomata. An ideal film would inhibit water vapor diffusion more than CO_2 diffusion or at least have equal effects on both. However, Woolley (1967) measured the diffusive characteristics of a number of film-forming polymers and found that all inhibited the diffusion of CO_2 more than water vapor, some by large amounts. CO_2 is a larger molecule than H_2O and this result may reflect that fact. Woolley (1967) concluded that favorable polymers may not exist. Therefore, film-forming polymers have had their largest use in protecting overwintering ornamentals from desiccation damage or transplanting stock from water loss during storage. In short, the polymers work mostly in cases where photosynthesis is not important.

Most tests of the effectiveness of antitranspirants have been made on single leaves or plants, but crop canopies have much larger overall dimensions than the leaves making them up because the diffusive resistance of the air in the unstirred boundary layer and the turbulent boundary layer above is much larger in a canopy than when the individual plants and leaves are measured separately. As a consequence, the crop boundary layer is generally more limiting than the diffusive resistance of the leaves, and the stomata need to close much more to increase the diffusive resistance of the crop than in a single leaf of the same crop. An additional factor is the energy budget for the crop canopy (see Chapter 7) which absorbs radiation and partitions it between latent heat loss (energy lost in evapotranspiration) and sensible heat loss (energy lost because of a temperature difference between the leaf and air). According to the energy budget, decreasing stomatal apertures by large amounts to decrease transpiration will force more of the radiation to be partitioned to sensible heat loss which results in increased leaf temperature. The increased temperature increases the vapor pressure of water in the leaf, reversing somewhat the effects of stomatal closure. Therefore, canopy transpiration can be inhibited only by marked stomatal closure under many conditions, and this causes marked inhibition of CO_2 uptake and photosynthesis. With the demonstration of this principle by investigators such as Van Bavel (1966, 1967), Van Bavel and Ehrler (1968), Brown and Rosenberg (1973), and Johns *et al.* (1983), research with antitranspirants was largely abandoned.

SUMMARY

The development of the land habit was a dramatic phase in evolution and caused major changes in plant form and function. These appear to be continuing amid strong evolutionary pressures imposed by the inherently dehydrating conditions presently existing on land. As a consequence, land plants are not yet optimally adapted to meet dehydration and we would likely see large improvements in dehydration performance if this chapter could be written a few hundred million years hence. However, modern methods of plant breeding and genetic modification can speed the transition to increased tolerance of dehydration so that it can occur within a few years.

Indeed, considerable success has already been achieved. Water acquisition can be improved by deep rooting and strong osmotic adjustment, earliness in reproduction can be used to avoid late season droughts, and cuticular characters can be modified to conserve water in some cases, although antitranspirants have not proven generally practical. It also appears increasingly possible to improve water use efficiency by genetic means using new techniques for screening for this trait. Central to these efforts is the use of realistic drought conditions, and selection for a small number of traits that might be valuable for crop performance with limited water. It is now clear that successful improvement of drought performance can come at no sacrifice to performance under favorable conditions.

Water is required for biological activity, and studies of plant water use efficiency show that total dry matter production is linearly proportional to the amount of water used. Total water acquisition can be enhanced by means such as larger deeper root systems with concomitant increases in dry matter production. However, it also is possible to partition dry matter production differently to valuable plant parts. This is relatively independent of the overall water use efficiency and provides a means for maintaining a fraction of plant development that may be important without having to improve the productive capacity of the plant as a whole. This approach could lead to large benefits.

Metabolic changes have developed during the course of evolution that have improved the ability of plants to withstand limited water supplies, particularly in photosynthesis. The recent evolutionary development of C_4 photosynthesis and Crassulacean acid metabolism are clear examples of this type of metabolic adaptation, leading to increased water use efficiency in those species possessing it. The demonstration that reproductive losses usually associated with drought often are a biochemical problem instead of a direct water deficiency problem further supports the notion that metabolic modification may be important for improving plant performance with limited supplies of water.

The molecular mechanisms of reproductive losses and desiccation tolerance are being elucidated, and the work with desiccation tolerance shows that changes in expression of specific genes are correlated with a decreased lethality of severe desiccation at least during late seed development. Plant growth regulators such as abscisic acid can trigger this protective mode and it will be interesting to determine whether the impact of desiccation can be ameliorated by manipulating this regulation chemically or genetically.

SUPPLEMENTARY READING

Bambach, R. K., Scotese, C. R., and Ziegler, A. M. (1980). Before Pangea: The geographies of the paleozoic world. *Am. Sci.* **68**, 26–38.

Boyer, J. S. (1982). Plant productivity and environment. *Science* **218**, 443–448.

Briggs, L. J., and Shantz, H. L. (1914). Relative water requirement of plants. *J. Agric. Res.* **3**, 1–63.

Close, T. J., and Bray, E. A., eds. (1993). "Plant Responses to Cellular Dehydration during Environmental Stress." Current Topics in Plant Physiology, Vol. 10, American Society of Plant Physiologists, Rockville, MD.

Dure, L., III, Crouch, M., Harada, J., Ho, T.-H. D., Mundy, J., Quatrano, R., Thomas, T., and Sung, Z. R. (1989). Common amino acid sequence domains among the LEA proteins of higher plants. *Plant Mol. Biol.* **12,** 475–486.

O'Toole, J. C. (1982). Adaptation of rice to drought-prone environments. *In* "Drought Resistance in Crops with Emphasis on Rice," pp. 195–213. International Rice Research Institute, Manila, Philippines.

Salter, P. J., and Goode, J. E. (1967). "Crop Responses to Water at Different Stages of Growth." Commonwealth Agricultural Bureau, Farnham Royal, Bucks, England.

Taylor, H. M., Jordan, W. R., and Sinclair, T. R. (1983). "Limitations to Efficient Water Use in Crop Production." American Society of Agronomy, Madison, WI.

References

Abell, C. A., and Hursh, C. R. (1931). Positive gas and water pressures in oaks. *Science* 73, 449.

Acevedo, E., Hsiao, T. C., and Henderson, D. W. (1971). Immediate and subsequent growth responses of maize leaves to changes in water status. *Plant Physiol.* 48, 631–636.

Ackerson, R. C. (1980). Stomatal response of cotton to water stress and abscisic acid as affected by water stress history. *Plant Physiol.* 65, 455–459.

Ackerson, R. C., and Hebert, R. R. (1981). Osmoregulation in cotton in response to water stress. I. Alterations in photosynthesis, leaf conductance, translocation, and ultrastructure. *Plant Physiol.* 67, 484–488.

Ackerson, R. C., and Krieg, D. R. (1977). Stomatal and nonstomatal regulation of water use in cotton, corn and sorghum. *Plant Physiol.* 60, 850–853.

Adams, S., Strain, B. R., and Adams, M. S. (1970). Water-repellent soils, fire, and annual plant cover in a desert scrub community of southeastern California. *Ecology* 51, 696–700.

Adamsen, F. J. (1992). Irrigation method and water quality effects on corn yield in the mid-Atlantic coastal plain. *Agron. J.* 84, 837–843.

Addoms, R. M. (1946). Entrance of water into suberized roots of trees. *Plant Physiol.* 21, 109–111.

Aesbacher, R. A., Schiefelbein, J. W., and Benfey, P. N. (1994). The genetic and molecular basis of root development. *Annu. Rev. Plant Physiol. Plant Mol. Biol.* 45, 25–45.

Aldrich, W. W., Work, R. A., and Lewis, M. R. (1935). Pear root concentration in relation to soil-moisture extraction in heavy clay soil. *J. Agric. Res.* (Washington, D.C.) 50, 975–988.

Alieva, S. A., Tageeva, S. V., Tairbekov, M. G., Kasatkina, V. S., and Vagabova, M. E. (1971). Structural and functional condition of the chloroplasts as a function of the water regime. *Sov. Plant Physiol.* 18, 416–422.

Altus, D. P., and Canny, M. J. (1985). Water pathways in wheat leaves. I. The division of fluxes between different vein types. *Aust. J. Plant Physiol.* 12, 173–181.

Alvim, P. deT. (1960). Moisture stress as a requirement for flowering of coffee. *Science* 132, 354.

Alvim, P. deT. (1965). A new type of porometer for measuring stomatal opening and its use in irrigation studies. *UNESCO Arid Zone Res.* 25, 325–329.

Alvim, P. deT. (1985). Theobroma cacao. *In* "Handbook of Flowering" (A. H. Halevy, ed.), Vol. V, pp. 357–365. CRC Press, Boca Raton, FL.

Alvim, P. deT., and Havis, J. R. (1954). An improved infiltration series for studying stomatal opening as illustrated with coffee. *Plant Physiol.* **29**, 97–98.

Amato, I. (1992). A new blueprint for water's architecture. *Science* **256**, 1764.

Amthor, J. S. (1989). "Respiration and Crop Productivity." Springer-Verlag, New York.

Anderson, C. P., Markhart, A. H., III, Dixon, R. K., and Sucoff, E. L. (1988). Root hydraulic conductivity of vesicular-arbuscular mycorrhizal green ash seedlings. *New Phytol.* **109**, 465–471.

Anderson, L. E., and Bourdeau, P. F. (1955). Water relations in two species of terrestrial mosses. *Ecology* **36**, 206–212.

Anderson, M. C. (1981). The geometry of leaf distribution in some south-eastern Australian forests. *Agric. Meteorol.* **25**, 195–205.

Andrews, D. L., Cobb, B. G., Johnson, J. R., and Drew, M. C. (1993). Hypoxic and anoxic induction of alcohol dehydrogenase in roots and shoots of seedlings of *Zea mays. Plant Physiol.* **101**, 407–414.

Andrews, D. L., Drew, M. C., Johnson, J. R., and Cobb, B. G. (1994). The response of maize seedlings of different ages to hypoxic and anoxic stress. *Plant Physiol.* **105**, 53–60.

Andrews, F. C. (1976). Colligative properties of simple solutions. *Science* **194**, 567–571.

Andrews, R. E., and Newman, E. I. (1968). The influence of root pruning on the growth and transpiration of wheat under different soil moisture conditions. *New Phytol.* **67**, 617–630.

Anthon, G. E., and Jagendorf, A. T. (1983). Effect of methanol on spinach thylakoid ATPase. *Biochim. Biophys. Acta* **723**, 358–365.

Antolín, M. C., and Sánchez-Díaz, M. (1993). Effects of temporary droughts on photosynthesis of alfalfa plants. *J. Exp. Bot.* **44**, 1341–1349.

Aparicio-Tejo, P. M., and Boyer, J. S. (1983). Significance of accelerated leaf senescence at low water potentials for water loss and grain yield in maize. *Crop Sci.* **23**, 1198–1202.

Aphalo, P. J., and Jarvis, P. G. (1991). Do stomata respond to relative humidity? *Plant Cell Environ.* **14**, 127–132.

Armenta-Soto, J., Chang, T. T., Loresto, G. C., and O'Toole, J. C. (1983). Genetic analysis of root characters in rice. *Sabrao J.* **15**, 103–116.

Armstrong, J., and Armstrong, W. (1991). A convective through-flow of gases in *Phragmites australis* (Cav.) Trin. ex Steud. *Aquatic Bot.* **39**, 75–88.

Armstrong, J., Armstrong, W., and Beckett, P. M. (1992). *Phragmites australis:* Venturi and humidity-induced pressure flows enhance rhizome aeration and rhizosphere oxidation. *New Phytol.* **120**, 197–207.

Armstrong, W. (1968). Oxygen diffusion from the roots of woody species. *Physiol. Plant.* **21**, 539–543.

Armstrong, W., Justin, S. H. F. W., Beckett, P. M., and Lythe, S. (1991). Root adaptation to soil waterlogging. *Aquatic Bot.* **39**, 57–73.

Arndt, C. H. (1945). Temperature-growth relations of the roots and hypocotyls of cotton seedlings. *Plant Physiol.* **20**, 200–220.

Arntzen, C. J., Haugh, M. F., and Bobick, S. (1973). Induction of stomatal closure by *Helminthosporium maydis* pathotoxin. *Plant Physiol.* **52**, 569–574.

Ashton, F. M. (1956). Effects of a series of cycles of alternating low and high soil water contents on the rate of apparent photosynthesis in sugar cane. *Plant Physiol.* **31**, 266–274.

Askenasy, E. (1895). Ueber das saftsteigen. *Bot. Centrbl.* **62**, 237–238.

Assmann, S. M., and Grantz, D. A. (1990). Stomatal response to humidity in sugarcane and soybean: Effect of vapour pressure difference on the kinetics of the blue light response. *Plant Cell Environ.* **13**, 163–169.

Atkins, W. R. G. (1916). "Some Recent Researches in Plant Physiology." Whitaker and Co., London.

Auchter, E. C. (1923). Is there normally a cross transfer of foods, water, and mineral nutrients in woody plants? Md. Agr. Exp. Sta. Bull. 257.

Augé, R. M., and Duan, X. (1991). Mycorrhizal fungi and nonhydraulic root signals of soil drying. *Plant Physiol.* 97, 821–824.

Avery, G. S., Jr. (1933). Structure and development of the tobacco leaf. *Am. J. Bot.* 20, 565–592.

Axelsson, E., and Axelsson, B. (1986). Changes in carbon allocation patterns in spruce and pine trees following irrigation and fertilization. *Tree Physiol.* 2, 189–204.

Aylor, D. E., Parlange, J.-Y., and Krikorian, A. D. (1973). Stomatal mechanics. *Am. J. Bot.* 60, 163–171.

Ayres, P. G., and Boddy, L., eds. (1986). "Water, Fungi, and Plants." Cambridge University Press, Cambridge, England.

Azaizeh, H., Gunse, B., and Steudle, E. (1992). Effects of NaCl and $CaCl_2$ on water transport across root cells of maize (*Zea mays* L.) seedlings. *Plant Physiol.* 99, 886–894.

Azevedo, J., and Morgan, D. L. (1974). Fog precipitation in coastal California forests. *Ecology* 55, 1135–1141.

Baarstad, L. L., Rickman, R. W., Wilkins, D., and Morita, S. (1993). A hydraulic soil sample providing minimum field plot disruption. *Agron. J.* 85, 178–181.

Baker, J. M., and Van Bavel, C. H. M. (1986). Resistance of plants roots to water loss. *Agron. J.* 78, 641–644.

Baldocchi, D. D., Luxmoore, R. J., and Hatfield, J. L. (1991). Discerning the forest from the trees: An essay on scaling canopy stomatal conductance. *Agric. For. Meteorol.* 54, 197–226.

Ball, J. T., and Berry, J. A. (1982). The C_i-C_s ratio: A basis for predicting stomatal control of photosynthesis. *Carnegie Inst. Washington Yearbook* 81, 88–92.

Ball, J. T., Woodrow, I. E., and Berry, J. A. (1987). A model predicting stomatal conductance and its contribution to the control of photosynthesis under different environmental conditions. *In* "Progress in Photosynthesis research" (J. Biggins, ed.), Vol. 4, pp. 221–224. Nijhoff, Dordrecht.

Balling, A., and Zimmermann, U. (1990). Comparative measurements of the xylem pressure of *Nicotiana* plants by means of the pressure bomb and pressure probe. *Planta* 182, 325–338.

Bambach, R. K., Scotese, C. R., and Ziegler, A. M. (1980). Before Pangea: The geographies of the paleozoic world. *Am. Sci.* 68, 26–38.

Bange, G. G. J. (1953). On the quantitative explanation of stomatal transpiration. *Acta Bot. Neerl.* 2, 255–297.

Barber, D. A., Ebert, M., and Evans, N. T. S. (1962). The movement of [15]O through barley and rice plants. *J. Exp. Bot.* 13, 397–403.

Barber, D. A., and Martin, J. K. (1976). The release of organic substances by cereal roots. *New Phytol.* 76, 69–80.

Barber, S. A. (1962). A diffusion and mass-flow concept of soil nutrient availability. *Soil Sci.* 93, 39–49.

Barber, S. A., Walker, J. M., and Vasey, E. H. (1962a). Principles of ion movement through the soil to the plant root. *In* "Transactions of the International Society of Soil Science, Commissions IV and V," pp. 121–124. International Soil Conference, Soil Bureau, P. B. Lower Hutt, New Zealand, 1963.

Barber, S. A., Walker, J. M., and Vasey, E. H. (1962b). Mechanisms for the movement of plant nutrients from the soil and fertilizer to the plant roots. *J. Agric. Food Chem.* 11, 204–207.

Bardzik, J. M., Marsh, H. V., Jr., and Havis, J. R. (1971). Effects of water stress on the activities of three enzymes in maize seedlings. *Plant Physiol.* 47, 828–831.

Barley, K. P. (1962). The effect of mechanical stress on the growth of roots. *J. Exp. Bot.* 13, 95–110.

Barley, K. P. (1970). The configuration of the root system in relation to nutrient uptake. *Adv. Agron.* 22, 159–201.

Barlow, E. W. R. (1986). Water relations of expanding leaves. *Aust. J. Plant Physiol.* **13**, 45–58.

Barlow, E. W. R., Ching, T. M., and Boersma, L. (1976). Leaf growth in relation to ATP levels in water stressed corn plants. *Crop Sci.* **16**, 405–407.

Barlow, E. W. R., Lee, J. W., Munns, R., and Smart, M. G. (1980). Water relations of the developing wheat grain. *Aust. J. Plant Physiol.* **7**, 519–525.

Barney, C. W. (1951). Effects of soil temperature and light intensity on root growth of loblolly pine seedlings. *Plant Physiol.* **26**, 146–163.

Baron-Epel, O., Gharyal, P. K., and Schindler, M. (1988). Pectins as mediators of wall porosity in soybean cells. *Planta* **175**, 389–395.

Barrs, H. D. (1968). Effect of cyclic variations in gas exchange under constant environmental conditions on the ratio of transpiration to net photosynthesis. *Physiol. Plant.* **21**, 918–929.

Barrs, H. D. (1971). Cyclic variations in stomatal aperture, transpiration, and leaf water potential under constant environmental conditions. *Annu. Rev. Plant Physiol.* **22**, 223–236.

Barrs, H. D., and Klepper, B. (1968). Cyclic variations in plant properties under constant environmental conditions. *Physiol. Plant.* **21**, 711–730.

Barta, A. L. (1984). Ethanol synthesis and loss from flooded roots of *Medicago sativa* L. and *Lotus corniculatus* L. *Plant Cell Environ.* **7**, 187–191.

BassiriRad, H., and Radin, J. W. (1992). Temperature-dependent water and ion transport properties of barley and sorghum roots. II. Effects of abscisic acid. *Plant Physiol.* **99**, 34–37.

BassiriRad, H., Radin, J. W., and Matsuda, K. (1991). Temperature-dependent water and ion transport properties of barley and sorghum roots. I. Relationship to leaf growth. *Plant Physiol.* **97**, 426–432.

Bates, L. M., and Hall, A. E. (1981). Stomatal closure with soil water depletion not associated with changes in bulk leaf water status. *Oecologia* **50**, 62–65.

Baver, L. D. (1948). "Soil Physics," 2nd Ed. Wiley, New York.

Baxter, P., and West, D. (1977). The flow of water into fruit trees. II. Water intake through a cut limb. *Ann. Appl. Biol.* **87**, 103–112.

Bayliss, W. M. (1924). "Principles of General Physiology," 4th Ed., Chap. 8. Longmans, Green and Co., London.

Beardsell, M. F., and Cohen, D. (1975). Relationships between leaf water status, abscisic acid levels, and stomatal resistance in maize and sorghum. *Plant Physiol.* **56**, 207–212.

Beardsell, M. F., Jarvis, P. G., and Davidson, B. (1972). A null-balance diffusion porometer suitable for use with leaves of many shapes. *J. Appl. Ecol.* **9**, 677–690.

Beasley, E. W. (1942). Effects of some chemically inert dusts upon the transpiration rate of yellow coleus plants. *Plant Physiol.* **17**, 101–108.

Beasley, R. S. (1976). Contribution of subsurface flow from the upper slopes of forested watersheds to channel flow. *Soil Sci. Soc. Am. Proc.* **40**, 955–957.

Beevers, L., and Hageman, R. H. (1969). Nitrate reduction in higher plants. *Annu. Rev. Plant Physiol.* **20**, 495–522.

Beevers, L., and Hageman, R. H. (1983). Uptake and reduction of nitrate: bacteria and higher plants. *In* "Encyclopedia of Plant Physiology" (A. Läuchli and R. L. Bieleski, eds.), Vol. 15A, pp. 351–375. Springer-Verlag, Berlin.

Beevers, L., Schrader, L. E., Flesher, D., and Hageman, R. H. (1965). The role of light and nitrate in the induction of nitrate reductase in radish cotyledons and maize seedlings. *Plant Physiol.* **40**, 691–698.

Bell, D. T., Koeppe, D. E., and Miller, R. J. (1971). The effects of drought stress on respiration of isolated corn mitochondria. *Plant Physiol.* **48**, 413–415.

Benfey, P. N., and Schiefelbein, J. W. (1994). Insights into root development from *Arabidopsis* root mutants. *Plant Cell Environ.* **17**, 675–680.

Bengough, A. G., and Mullins, C. E. (1991). The resistance experienced by roots growing in a pressurized cell. *Plant Soil* **123**, 73–82.

Bengtson, C., Larsson, S., and Liljenberg, C. (1978). Effects of water stress on cuticular transpiration rate and amount and composition of epicuticular wax in seedlings of six oat varieties. *Physiol. Plant.* 44, 319–324.

Bennett, O. L., and Doss, B. D. (1960). Effect of soil moisture level on root distribution of cool-season forage species. *Agron. J.* 52, 204–207.

Bensen, R. J., Boyer, J. S., and Mullet, J. E. (1988). Water deficit-induced changes in abscisic acid content, growth, polysomes, and translatable RNA in soybean hypocotyls. *Plant Physiol.* 88, 289–294.

Benson, S. W., and Siebert, E. D. (1992). A simple two-structure model for liquid water. *J. Am. Chem. Soc.* 114, 4269–4276.

Bental, M., Degani, H., and Avron, M. (1988a). ^{23}Na-NMR studies of the intracellular sodium ion concentration in the halotolerant alga *Dunaliella salina*. *Plant Physiol.* 87, 813–817.

Bental, M., Oren-Shamir, M., Avron, M., and Degani, H. (1988b). ^{31}P and ^{13}C-NMR studies of the phosphorus and carbon metabolites in the halotolerant alga, *Dunaliella salina*. *Plant Physiol.* 87, 320–324.

Berkowitz, G. A., and Kroll, K. S. (1988). Acclimation of photosynthesis in *Zea mays* to low water potentials involves alterations in protoplast volume reduction. *Planta* 175, 374–379.

Berkowitz, G. A., and Whalen, C. (1985). Leaf K^+ interaction with water stress inhibition of nonstomatal-controlled photosynthesis. *Plant Physiol.* 79, 189–193.

Bernal, J. D. (1965). The structure of water and its biological implications. *Symp. Soc. Exp. Biol.* 19, 17–32.

Bernstein, L. (1961). Osmotic adjustment of plants to saline media. I. Steady state. *Am. J. Bot.* 48, 909–918.

Bernstein, L., Gardner, W. R., and Richards, L. A. (1959). Is there a vapor gap around roots? *Science* 129, 1750–1753.

Berry, L. J. (1949). The influence of oxygen on the respiratory rate in different segments of onion roots. *J. Cell Comp. Physiol.* 33, 41–66.

Bethlenfalvay, G. J., Brown, M. S., Mihara, K. L., and Stafford, A. E. (1987). *Glycine-Glomus-Rhizobium* symbiosis. V. Effects of mycorrhiza on nodule activity and transpiration in soybeans under drought stress. *Plant Physiol.* 85, 115–119.

Bewley, J. D. (1979). Physiological aspects of desiccation tolerance. *Annu. Rev. Plant Physiol.* 30, 195–238.

Bewley, J. D., and Larsen, K. M. (1982). Differences in the responses to water stress of growing and non-growing regions of maize mesocotyls: Protein synthesis on total, free and membrane-bound polyribosome fractions. *J. Exp. Bot.* 33, 406–415.

Bewley, J. D., and Pacey, J. (1978). Desiccation-induced ultrastructural changes in drought-sensitive and drought-tolerant plants. *In* "Dry Biological Systems" (J. H. Crowe and J. S. Clegg, eds.), pp. 53–73. Academic Press, New York.

Beyschlag, W., and Pfanz, H. (1990). A fast method to detect the occurrence of nonhomogeneous distribution of stomatal aperture in heterobaric plant leaves: Experiments with *Arbutus unedo* L. during the diurnal course. *Oecologia* 82, 52–55.

Beyschlag, W., Pfanz, H., and Ryel, R. J. (1992). Stomatal patchiness in Mediterranean evergreen sclerophylls. *Planta* 187, 546–553.

Beyschlag, W., Phibbs, A., and Pfanz, H. (1990). The role of temperature and humidity in controlling the diurnal stomatal behavior of *Arbutus unedo* L. during the dry season. *Biochem. Physiol. Pflanzen* 186, 265–271.

Bhat, K. K. S., and Nye, P. H. (1974). Diffusion of phosphate to plant roots in soil. II. Uptake along the roots at different times and the effect of different levels of phosphorus. *Plant Soil* 41, 365–382.

Bialzyk, J., and Lechowski, L. (1992). Absorption of HCO_3^- by roots and its effect on carbon metabolism of tomato. *J. Plant Nutr.* 15, 293–312.

Bible, B. B., Cuthbert, R. L., and Corolus, R. L. (1968). Responses of some vegetable crops to atmospheric modifications under field conditions. *J. Am. Soc. Hortic. Sci.* **92**, 590–594.

Bierhuizen, J. F., and Slatyer, R. O. (1965). Effect of atmospheric concentration of water vapor and CO_2 in determining transpiration-photosynthesis relationships of cotton leaves. *Agric. Meteorol.* **2**, 259–270.

Biles, C. L., and Abeles, F. B. (1991). Xylem sap proteins. *Plant Physiol.* **96**, 597–601.

Billings, W. D. (1978). "Plants and the Ecosystem." Wadsworth, Belmont, CA.

Billings, W. D., and Godfrey, P. J. (1967). Photosynthetic utilization of internal carbon dioxide by hollow-stemmed plants. *Science* **158**, 121–123.

Bingham, E. C., and Jackson, R. F. (1918). Standard substances for the calibration of viscometers. *Bull. Bureau Standards* **14**, 59–86.

Bingham, L. J., and Stevenson, E. A. (1993). Control of root growth: Effects of carbohydrates on the extension, branching and rate of respiration of different fractions of wheat roots. *Physiol. Plant.* **88**, 149–158.

Bisson, M. A., and Bartholomew, D. (1984). Osmoregulation or turgor regulation in *Chara?* *Plant Physiol.* **74**, 252–255.

Biswell, H. H. (1935). Effect of the environment upon the root habits of certain deciduous forest trees. *Bot. Gaz.* (Chicago) **96**, 676–708.

Björkman, E. (1942). Über die Bedingungen der Mykorrhizabildung bei Kiefer und Fichts. *Symb. Bot. Ups.* **6**, 1–190.

Björkman, O. (1981). Responses to different quantum flux densities. *In* "Encyclopedia of Plant Physiology" (O. L. Lange, P. S. Nobel, C. B. Osmond, and H. Ziegler, eds.), Vol. 12A, p. 57. Springer-Verlag, Berlin.

Björkman, O., and Powles, S. B. (1984). Inhibition of photosynthetic reactions under water stress: Interaction with light level. *Planta* **161**, 490–504.

Black, C. R. (1979a). The relationship between transpiration rate, water potential, and resistances to water movement in sunflower (*Helianthus annuus* L.). *J. Exp. Bot.* **30**, 235–243.

Black, C. R. (1979b). The relative magnitude of the partial resistances to transpirational water movement in sunflower (*Helianthus annuus* L.). *J. Exp. Bot.* **30**, 245–253.

Blackman, P. G., and Davies, W. J. (1985). Root to shoot communications in maize plants of the effects of soil drying. *J. Exp. Bot.* **36**, 39–48.

Blake, G. R., Allred, E. R., Van Bavel, C. H. M., and Whisler, F. D. (1960). Agricultural drought and moisture excesses in Minnesota. Univ. Minn. Tech. Bull. 235.

Blizzard, W. E., and Boyer, J. S. (1980). Comparative resistance of the soil and the plant to water transport. *Plant Physiol.* **66**, 809–814.

Bloodworth, J. E., Page, J. B., and Cowley, W. R. (1956). Some applications of the thermoelectric method for measuring water flow in plants. *Agron. J.* **48**, 222–228.

Blum, A. (1979). Genetic improvement of drought resistance in crop plants: A case for sorghum. *In* "Stress Physiology in Crop Plants" (H. Mussell and R. C. Staples, eds.), pp. 429–445. Wiley, New York.

Blum, A., Johnson, J. W., Ramseur, E. L., and Tollner, E. W. (1991). The effect of a drying top soil and a possible non-hydraulic root signal on wheat growth and yield. *J. Exp. Bot.* **42**, 1225–1231.

Blum, A., Schertz, K. F., Toler, R. W., Welch, R. I., Rosenow, D. T., Johnson, J. W., and Clark, L. E. (1978). Selection for drought avoidance in sorghum using aerial infra-red photography. *Agron. J.* **70**, 472–477.

Böhm, J. (1893). Capillarität und Saftsteigen. *Ber. Dtsch. Bot. Ges.* **11**, 203–212.

Böhm, W. (1979). "Methods of Studying Root Systems." Springer-Verlag, Berlin.

Bolaños, J., and Edmeades, G. O. (1993a). Eight cycles of selection for drought tolerance in lowland tropical maize. I. Responses in grain yield, biomass, and radiation utilization. *Field Crops Res.* **31**, 233–252.

Bolaños, J., and Edmeades, G. O. (1993b). Eight cycles of selection for drought tolerance in lowland tropical maize. II. Responses in reproductive behavior. *Field Crops Res.* **31**, 253–268.

Bolaños, J., Edmeades, G. O., and Martinez, L. (1993). Eight cycles of selection for drought tolerance in lowland tropical maize. III. Responses in drought-adaptive physiological and morphological traits. *Field Crops Res.* **31**, 269–286.

Bole, J. B. (1973). Influence of root hairs in supplying soil phosphorus to wheat. *Can J. Soil Sci.* **53**, 169–175.

Bollard, E. G. (1960). Transport in the xylem. *Annu. Rev. Plant Physiol.* **11**, 141–166.

Bonner, J. (1934). Studies on the growth hormone of plants. V. The relation of cell elongation to cell wall formation. *Proc. Natl. Acad. Sci. USA* **20**, 393–397.

Bonner, J. (1959). Water transport. *Science* **129**, 447–450.

Booker, F. L., Blum, U., and Fiscus, E. L. (1992). Short-term effects of ferulic acid on ion uptake and water relations in cucumber seedlings. *J. Exp. Bot.* **43**, 649–655.

Borchert, R. (1973). Simulation of rhythmic tree growth under constant conditions. *Physiol. Plant.* **29**, 173–180.

Bordovsky, D. G., Jordan, W. R., Hiler, E. A., and Howell, T. A. (1974). Choice of irrigation timing indicator for narrow row cotton. *Agron. J.* **66**, 88–91.

Borghetti, M., Edwards, W. R. N., Grace, J., Jarvis, P. G., and Raschi, A. (1991). The refilling of embolized xylem in *Pinus sylvestris* L. *Plant Cell Environ.* **14**, 357–369.

Bormann, F. H. (1957). Moisture transfer between plants through intertwined root systems. *Plant Physiol.* **32**, 48–55.

Bormann, F. H., and Graham, B. F., Jr. (1959). The occurrence of natural root grafting in eastern white pine, *Pinus strobus* L., and its ecological implications. *Ecology* **40**, 677–691.

Bormann, F. H., and Graham, B. F. (1960). Translocation of silvicides through root grafts. *J. For.* **58**, 402–403.

Bosch, J. M., and Hewlett, J. D. (1982). A review of catchment experiments to determine the effect of vegetation changes on water yield and evaporation. *J. Hydrol.* (Amsterdam) **55**, 3–23.

Bose, J. C. (1923). "Physiology of the Ascent of Sap." Longmans Green and Co., London.

Bothe, H., Yates, M. G., and Cannon, F. C. (1983). Physiology, biochemistry and genetics of dinitrogen fixation. *In* "Encyclopedia of Plant Physiology" (A. Läuchli and R. L. Bieleski, eds.), Vol. 15A, pp. 241–285. Springer-Verlag, Berlin/Heidelberg.

Bottomley, P. A., Rogers, H. H., and Prior, S. A. (1993). NMR imaging of root water distribution in intact *Vicia faba* L. plants in elevated atmospheric CO_2. *Plant Cell Environ.* **16**, 335–338.

Boucherie, A. (1840). Rapport sur une mémoire de M. le docteur Boucherie relatif á la conservation des bois. *Compt. Rendus* **10**, 685–689.

Bouma, J. (1991). Influence of soil macroporosity on environmental quality. *Adv. Agron.* **46**, 1–37.

Bouma, J., Belmans, C. F. M., and Dekker, L. W. (1982). Water infiltration and redistribution in a silt loam subsoil with vertical worm channels. *Soil Sci. Soc. Am. J.* **46**, 917–921.

Bouyoucos, G. J. (1954). New type electrode for plaster of paris moisture blocks. *Soil Sci.* **78**, 339–342.

Bowen, G. D. (1984). Tree roots and the use of soil nutrients. *In* "Nutrition of Plantation Forests" (G. D. Bowen and E. K. S. Nambier, eds.), pp. 147–179. Academic Press, London.

Bowes, G. (1993). Facing the inevitable: Plants and increasing atmospheric CO_2. *Annu. Rev. Plant Physiol. Plant Mol. Biol.* **44**, 309–332.

Bowling, D. J. F. (1973). Measurement of a gradient of oxygen partial pressure across the intact root. *Planta* **111**, 323–328.

Bowman, W. D., Hubick, K. T., von Caemmerer, S., and Farquhar, G. (1989). Short-term changes in leaf carbon isotope discrimination in salt- and water-stressed C_4 grasses. *Plant Physiol.* **90**, 162–166.

Box, J. E., Jr., and Hammond, L. C., eds. (1990). "Rhizosphere Dynamics." Westview Press, Boulder, CO.

Box, J. E., Jr., and Ramseur, E. L. (1993). Minirhizotron wheat root data: Comparisons to soil core root data. *Agron. J.* 85, 1058–1060.

Boyce, S. G. (1954). The salt spray community. *Ecol. Monogr.* 24, 29–67.

Boyer, J. S. (1965). Effects of osmotic water stress on metabolic rates of cotton plants with open stomata. *Plant Physiol.* 40, 229–234.

Boyer, J. S. (1967a). Leaf water potentials measured with a pressure chamber. *Plant Physiol.* 42, 133–137.

Boyer, J. S. (1967b). Matric potentials of leaves. *Plant Physiol.* 42, 213–217.

Boyer, J. S. (1968). Relationship of water potential to growth of leaves. *Plant Physiol.* 43, 1056–1062.

Boyer, J. S. (1969). Free-energy transfer in plants. *Science* 163, 1219–1220.

Boyer, J. S. (1970). Leaf enlargement and metabolic rates in corn, soybean, and sunflower at various leaf water potentials. *Plant Physiol.* 46, 233–235.

Boyer, J. S. (1971a). Nonstomatal inhibition of photosynthesis in sunflower at low leaf water potentials and high light intensities. *Plant Physiol.* 48, 532–536.

Boyer, J. S. (1971b). Recovery of photosynthesis in sunflower after a period of low leaf water potential. *Plant Physiol.* 47, 816–820.

Boyer, J. S. (1971c). Resistances to water transport in soybean, bean and sunflower. *Crop Sci.* 11, 403–407.

Boyer, J. S. (1973). Response of metabolism to low water potentials in plants. *Phytopathology* 63, 466–472.

Boyer, J. S. (1974). Water transport in plants: mechanism of apparent changes in resistance during absorption. *Planta* 117, 187–207.

Boyer, J. S. (1977). Regulation of water movement in whole plants. *Soc. Expt. Biol. Symp.* 31, 455–470.

Boyer, J. S. (1982). Plant productivity and environment. *Science* 218, 443–448.

Boyer, J. S. (1983). Subcellular mechanisms of plant response to low water potential. *Agricultural Water Management* 7, 239–248.

Boyer, J. S. (1985). Water transport. *Annu. Rev. Plant Physiol.* 36, 473–516.

Boyer, J. S. (1988). Cell enlargement and growth-induced water potentials. *Physiol. Plant.* 73, 311–316.

Boyer, J. S. (1989). Water potential and plant metabolism: Comments on Dr. P. J. Kramer's article "Changing concepts regarding plant water relations," Vol. 11, Number 7, pp. 565–568, and Dr. J. B. Passioura's response, pp. 569–571. *Plant Cell Environ.* 12, 213–216.

Boyer, J. S. (1990). Photosynthesis in dehydrating plants. *Bot. Mag. Tokyo Special Issue* 2, 73–85.

Boyer, J. S. (1993). Temperature and growth-induced water potential. *Plant Cell Environ.* 16, 1099–1106.

Boyer, J. S. (1995). "Measuring the Water Status of Plants and Soil." Academic Press, San Diego.

Boyer, J. S., and Bowen, B. L. (1970). Inhibition of oxygen evolution in chloroplasts isolated from leaves with low water potentials. *Plant Physiol.* 45, 612–615.

Boyer, J. S., Cavalieri, A. J., and Schulze, E.-D. (1985). Control of cell enlargement: Effects of excision, wall relaxation, and growth-induced water potentials. *Planta* 163, 527–543.

Boyer, J. S., Johnson, R. R., and Saupe, S. G. (1980). Afternoon water deficits and grain yields in old and new soybean cultivars. *Agron. J.* 72, 981–986.

Boyer, J. S., and Knipling, E. B. (1965). Isopiestic technique for measuring leaf water potentials with a thermocouple psychrometer. *Proc. Natl. Acad. Sci. USA* 54, 1044–1051.

Boyer, J. S., and McPherson, H. G. (1975). Physiology of water deficits in cereal crops. *Adv. Agron.* 27, 1–23.

Boyer, J. S., and Potter, J. R. (1973). Chloroplast response to low leaf water potentials. I. Role of turgor. *Plant Physiol.* 51, 989–992.

Boyle, M. G., Boyer, J. S., and Morgan, P. W. (1991a). Stem infusion of maize plants. *Crop Sci.* 31, 1241–1245.

Boyle, M. G., Boyer, J. S., and Morgan, P. W. (1991b). Stem infusion of liquid culture medium prevents reproductive failure of maize at low water potentials. *Crop Sci.* 31, 1246–1252.

Bozarth, C. S., Mullet, J. E., and Boyer, J. S. (1987). Cell wall proteins at low water potentials. *Plant Physiol.* 85, 261–267.

Bradford, K. J. (1994). Water stress and the water relations of seed development: A critical review. *Crop Sci.* 34, 1–11.

Bradford, K. J., and Yang, S. F. (1980). Xylem transport of 1-aminocyclopropane-1-carboxylic acid, an ethylene precursor, in waterlogged tomato plants. *Plant Physiol.* 65, 322–326.

Bradshaw, A. D. (1965). Evolutionary significance of phenotypic plasticity in plants. *Adv. Genetics* 13, 115–155.

Bray, E. A. (1988). Drought- and ABA-induced changes in polypeptide and mRNA accumulation in tomato leaves. *Plant Physiol.* 88, 1210–1214.

Bray, J. R. (1963). Root production and the estimation of net productivity. *Can. J. Bot.* 41, 65–72.

Breazeale, E. L., and McGeorge, W. T. (1953). Exudation pressure in roots of tomato plants under humid conditions. *Soil Sci.* 75, 293–298.

Bremner, P. M., Preston, G. K., and Fazekas de St. Groth, C. (1986). A field comparison of sunflower (*Helianthus annuus*) and sorghum (*Sorghum bicolor*) in a long drying cycle. I. Water extraction. *Aust. J. Agr. Res.* 37, 483–493.

Bresler, E. (1977). Trickle-drip irrigation: Principles and application to soil-water management. *Adv. Agron.* 29, 344–393.

Bret-Harte, M. S., Baskin, T. I., and Green, P. B. (1991). Auxin stimulates both deposition and breakdown of material in the pea outer epidermal cell wall, as measured interferometrically. *Planta* 185, 462–471.

Bret-Harte, M. S., and Silk, W. K. (1994). Nonvascular, symplasmic diffusion of sucrose cannot satisfy the carbon demands of growth in the primary root tip of *Zea mays* L. *Plant Physiol.* 105, 19–33.

Brevedan, E. R., and Hodges, H. F. (1973). Effects of moisture deficits on ^{14}C translocation in corn (*Zea mays* L.). *Plant Physiol.* 52, 436–439.

Briggs, L. J., and Shantz, H. L. (1911). A wax seal method for determining the lower limit of available soil moisture. *Bot. Gaz.* (Chicago) 51, 210–219.

Briggs, L. J., and Shantz, H. L. (1912). The relative wilting coefficients for different plants. *Bot. Gaz.* 53, 229–235.

Briggs, L. J., and Shantz, H. L. (1914). Relative water requirement of plants. *J. Agric. Res.* (Washington, DC) 3, 1–63.

Brix, H. (1962). The effect of water stress on the rates of photosynthesis and respiration in tomato plants and loblolly pine seedlings. *Physiol. Plant.* 15, 10–20.

Brix, H. (1990). Uptake and photosynthetic utilization of sediment-derived carbon by *Phragmites australis* (Cav.) Trin. ex Steudel. *Aquatic Bot.* 38, 377–389.

Brouwer, R. (1965). Ion absorption and transport in plants. *Annu. Rev. Plant Physiol.* 16, 241–266.

Brown, D. P., Pratum, T. K., Bledsoe, C., Ford, E. D., Cothren, J. S., and Perry, D. (1991). Noninvasive studies of conifer roots: Nuclear magnetic resonance (NMR) imaging of Douglas fir seedlings. *Can. J. For. Res.* 21, 1559–1566.

Brown, E. M. (1939). Some effects of temperature on the growth and chemical composition of certain pasture grasses. Missouri Agr. Exp. Sta. Res. Bull. 299.

Brown, H. T., and Escombe, F. (1900). Static diffusion of gases and liquids in relation to the assimilation of carbon and translocation of plants. *Phil. Trans. Roy. Soc. London Ser. B* 193, 223–291.

Brown, J. M., Johnson, G. A., and Kramer, P. J. (1986). *In vivo* magnetic resonance microscopy of changing water content in *Pelargonium hortorum* roots. *Plant Physiol.* **82**, 1158–1160.

Brown, J. M., Kramer, P. J., Cofer, G. P., and Johnson, G. A. (1990). Use of nuclear magnetic resonance microscopy for noninvasive observations of root-soil water relations. *Theor. Appl. Climatol.* **42**, 229–236.

Brown, K. W., Jordan, W. R., and Thomas, J. C. (1976). Water stress induced alteration in the stomatal response to leaf water potential. *Plant Physiol.* **37**, 1–5.

Brown, K. W., and Rosenberg, N. J. (1973). A resistance model to predict evapotranspiration and its application to a sugar beet field. *Agron. J.* **65**, 341–347.

Brown, R. H., and Simmons, R. E. (1979). Photosynthesis of grass species differing in CO_2 fixation pathways. I. Water-use efficiency. *Crop Sci.* **19**, 375–379.

Brown, R. W., and Oosterhuis, D. M. (1992). Measuring plant and soil water potentials with thermocouple psychrometers: Some concerns. *Agron. J.* **84**, 78–86.

Brown, V. K., and Gange, A. C. (1991). Effects of root herbivory on vegetation dynamics. *In* "Plant Root Growth" (D. Atkinson, ed.), pp. 453–470. Blackwell, London.

Brown, W. V., and Pratt, G. A. (1965). Stomatal inactivity in grasses. *Southwestern Naturalist* **10**, 48–56.

Brugnoli, E., Hubick, K. T., von Caemmerer, S., Wong, S. C., and Farquhar, G. D. (1988). Correlation between the carbon isotope discrimination in leaf starch and sugars of C_4 plants and the ratio of intercellular and atmospheric partial pressures of carbon dioxide. *Plant Physiol.* **88**, 1418–1424.

Brummell, D. A., and Hall, J. L. (1985). The role of cell wall synthesis in sustained auxin-induced growth. *Physiol. Plant.* **63**, 406–412.

Bruni, F., and Leopold, A. C. (1991). Hydration, protons and onset of physiological activities in maize seeds. *Physiol. Plant.* **81**, 359–366.

Buckland, S. T., Campbell, C. D., Mackie-Dawson, L. A., Horgan, G. W., and Duff, E. I. (1993). A method for counting roots observed in minirhizotrons and their theoretical conversion to root length density. *Plant Soil* **153**, 1–9.

Bunce, J. A. (1978). Effects of shoot environment on apparent root resistance to water flow in whole soybean and cotton plants. *J. Exp. Bot.* **29**, 595–601.

Bunce, J. A. (1981). Relationship between maximum photosynthetic rates and photosynthetic tolerance of low leaf water potentials. *Can. J. Bot.* **59**, 769–774.

Bunce, J. A., Miller, L. N., and Chabot, B. F. (1977). Competitive exploitation of soil water by five eastern North American tree species. *Bot. Gaz.* (Chicago) **138**, 168–173.

Bunger, M. T., and Thomson, H. J. (1938). Root development as a factor in the success or failure of windbreak trees in the southern high plains. *J. For.* **36**, 790–803.

Burger, D. W., Hartin, J. S., Hodel, D. R., Gukazewski, T. A., Tjosvold, S. A., and Wagner, S. A. (1987). Water use in California's ornamental nurseries. *Calif. Agric.* **41**, 7–8.

Burke, M. J., Bryant, R. G., and Weiser, C. J. (1974). Nuclear magnetic resonance of water in cold acclimating red osier dogwood stem. *Plant Physiol.* **54**, 392–398.

Burke, M. J., Gusta, L. V., Quamme, H. A., Weiser, C. J., and Li, P. H. (1976). Freezing and injury in plants. *Annu. Rev. Plant Physiol.* **27**, 507–528.

Burrows, W. J., and Carr, D. J. (1969). Effects of flooding the root system of sunflower plants on the cytokinin content in the xylem sap. *Physiol. Plant.* **22**, 1105–1112.

Burström, H. (1959). Growth and intercellularies in root meristems. *Physiol. Plant.* **12**, 371–385.

Burton, G. W., DeVane, E. H., and Carter, R. L. (1954). Root penetration, distribution and activity in southern grasses measured by yields, drought symptoms and P^{32} uptake. *Agron. J.* **46**, 229–233.

Burton, G. W., Prine, G. M., and Jackson, J. E. (1957). Studies of drought tolerance and water use of several southern grasses. *Agron. J.* **49**, 498–503.

Buswell, A. M., and Rodebush, W. H. (1956). Water. *Sci. Am.* **194**, 77–89.

Byott, G. S., and Sheriff, D. W. (1976). Water movement into and through *Tradescantia virginiana* (L.) leaves. II. Liquid flow pathways and evaporative sites. *J. Exp. Bot.* **27**, 634–639.

Byrne, G. F., Begg, J. E., and Hansen, G. K. (1977). Cavitation and resistance to water flow in plant roots. *Agric. Meteorol.* **18**, 21–25.

Caffrey, M., Fonseca, V., and Leopold, A. C. (1988). Lipid-sugar interactions: Relevance to anhydrous biology. *Plant Physiol.* **86**, 754–758.

Caldwell, M. M., and Richards, J. H. (1989). Hydraulic lift: Water efflux from upper roots improves effectiveness of water uptake by deep roots. *Oecologia* **79**, 1–5.

Campbell, W. A., and Copeland, O. L. (1954). Little leaf disease of shortleaf and loblolly pines. *U.S. Dept. Agric. Circ.* 940.

Cannell, R. Q., Belford, R. K., Gales, K., Thomson, R. J., and Webster, C. P. (1984). Effects of waterlogging and drought on winter wheat and winter barley grown on a clay and a sandy loam soil. *Plant Soil* **80**, 53–66.

Canny, M. J. (1990). What becomes of the transpiration stream? *New Phytol.* **114**, 341–368.

Canny, M. J., and Huang, C. X. (1993). What is in the intercellular spaces of roots? Evidence from the cryo-analytical-scanning microscope. *Physiol. Plant.* **87**, 561–568.

Canny, M. J., and McCully, M. E. (1988). The xylem sap of maize roots: Its collection, composition, and formation. *Aust. J. Plant Physiol.* **15**, 557–566.

Cardon, Z. G., and Berry, J. (1992). Effects of O_2 and CO_2 concentration on the steady-state fluorescence yield of single guard cell pairs in intact leaf discs of *Tradescantia albiflora*. *Plant Physiol.* **99**, 1238–1244.

Cardon, Z. G., Berry, J. A., and Woodrow, I. E. (1994). Dependence on the extent and direction of average stomatal response in *Zea mays* L. and *Phaseolus vulgaris* L. on the frequency of fluctuations in environmental stimuli. *Plant Physiol.* **105**, 1007–1013.

Carlson, T. N., and Lynn, B. (1991). The effects of plant water storage on transpiration and radiometric surface temperatures. *Agric. For. Meteorol.* **57**, 171–186.

Carlson, W. C., Harrington, C. A., Farnum, P., and Hollgren, S. W. (1988). Effects of root severing treatments on loblolly pine. *Can. J. For. Res.* **18**, 1376–1385.

Carmi, A. (1993). Effects of root zone restriction on amino acid status and bean plant growth. *J. Exp. Bot.* **44**, 1161–1166.

Carmi, A., Hesketh, J. D., Enos, W. T., and Peters, D. B. (1983). Interrelationships between shoot growth and photosynthesis as affected by root growth restriction. *Photosynthesis* **17**, 240–245.

Carpita, N. C. (1982). Limiting diameters of pores and the surface structure of plant cell walls. *Science* **218**, 813–814.

Carpita, N. C., Sabularse, D., Montezinos, D., and Delmer, D. P. (1979). Determination of the pore size of cell walls of living plant cells. *Science* **205**, 1144–1147.

Carter, J. C. (1945). Wetwood of elms. *Illinois Natural History Survey* **23**, 401–448.

Cassel, D. K., and Klute, A. (1986). Water potential: Tensiometry. *In* "Methods of Soil Analysis" (A. Klute, ed.), Vol. 9, pp. 563–596. American Society of Agronomy, Madison, WI.

Catska, V., Vancura, V., Hudska, G., and Prikryl, Z. (1982). Rhizosphere microorganisms in relation to the apple replant problem. *Plant Soil* **69**, 187–197.

Caughey, M. F. (1945). Water relations of pocosins or bog shrubs. *Plant Physiol.* **20**, 671–689.

Cavalieri, A. J., and Boyer, J. S. (1982). Water potentials induced by growth in soybean hypocotyls. *Plant Physiol.* **69**, 492–496.

Cermák, J., Cienciala, E., Kučera, J., Lindroth, A., and Hallgren, J. E. (1992). Radial velocity profiles of water flow in stems of spruce and oak and response of spruce tree to severing. *Tree Physiol.* **10**, 367–380.

Cermák, J., and Kučera, J. (1981). The compensation of natural temperature gradient at the measuring point during the sap flow rate determination in trees. *Biol. Plant.* **23**, 409–477.

Chamel, A., Pineri, M., and Escoubes, M. (1991). Quantitative determination of water sorption by plant cuticles. *Plant Cell Environ.* **14**, 87–95.

Chaney, W. R. (1981). Sources of water. *In* "Water Deficits and Plant Growth" (T. T. Kozlowski, ed.), Vol. 6, pp. 1–47. Academic Press, New York.

Chang, T. T., Loresto, G. C., and Tagumpay, O. (1974). Screening rice germ plasm for drought resistance. *SABRAO J.* **6**, 9–16.

Chapman, A. G. (1935). The effects of black locust on associated species with special reference to forest trees. *Ecol. Monographs* **5**, 37–60.

Chapman, D. J. (1985). Geological factors and biochemical aspects of the origin of land plants. *In* "Geological Factors and the Evolution of Plants" (B. H. Tiffney, ed.), pp. 23–44. Yale University Press, New Haven, CT.

Chason, J. W., Baldocchi, D. D., and Huston, M. A. (1991). A comparison of direct and indirect methods for estimating forest canopy leaf area. *Agric. For. Meteorol.* **57**, 107–128.

Chatfield, C., and Adams, G. (1940). Proximate composition of American food materials. U.S. Dept. Agric. Circ. 549.

Cheeseman, J. M. (1991). PATCHY: Simulating and visualizing the effects of stomatal patchiness on photosynthetic CO_2 exchange studies. *Plant Cell Environ.* **14**, 593–599.

Chrispeels, M. J., and Maurel, C. (1994). Aquaporins: The molecular basis of facilitated water movement through living plant cells? *Plant Physiol.* **105**, 9–13.

Chung, H.-H., and Kramer, P. J. (1975). Absorption of water and ^{32}P through suberized and unsuberized roots of loblolly pine. *Can. J. For. Res.* **5**, 229–235.

Claassen, M. M., and Shaw, R. H. (1970). Water deficit effects on corn. II. Grain components. *Agron. J.* **62**, 652–655.

Clark, J., and Gibbs, R. D. (1957). Studies in tree physiology. IV. Further investigations of seasonal changes in moisture content of certain Canadian forest trees. *Can. J. Bot.* **35**, 219–253.

Clark, W. S. (1874). The circulation of sap in plants. *Mass. State Board Agr. Annu. Rep.* **21**, 159–204.

Clark, W. S. (1875). Observations upon the phenomena of plant life. *Mass. State Board Agr. Annu. Rep.* **22**, 204–312.

Clarke, J. M. (1986). Effect of leaf rolling on leaf water loss in *Triticum* spp. *Can. J. Plant Sci.* **66**, 885–891.

Clarkson, D. T. (1976). The influence of temperature on the exudation of xylem sap from detached root systems of Orye (*Secale cereale*) and barley (*Hordeum vulgare*). *Planta* **132**, 297–304.

Clarkson, D. T. (1985). Factors affecting mineral nutrient acquisition by plants. *Annu. Rev. Plant Physiol.* **36**, 77–115.

Clarkson, D. T. (1993). Roots and the delivery of solutes and water to the xylem. *Phil. Trans. Roy. Soc. London Ser. B* **341**, 5–17.

Clarkson, D. T., Mercer, E. R., Johnson, M. G., and Mattam, D. (1975). The uptake of nitrogen (ammonium and nitrate) by different segments of the roots of intact barley plants. *Annu. Rep. Agric. Res. Counc., Letcombe Lab*, pp. 10–13.

Cleland, R. (1959). Effect of osmotic concentration on auxin-action and on irreversible and reversible expansion of the *Avena* coleoptile. *Physiol. Plant.* **12**, 809–825.

Cleland, R. E. (1971). Cell wall extension. *Annu. Rev. Plant Physiol.* **22**, 197–222.

Cleland, R. E. (1973). Auxin-induced hydrogen ion excretion from *Avena* coleoptiles. *Proc. Natl. Acad. Sci. USA* **70**, 3092–3093.

Cleland, R. E. (1975). Auxin-induced hydrogen ion excretion: correlation with growth, and control by external pH and water stress. *Planta* **127**, 233–242.

Cleland, R. E. (1977). The control of cell enlargement. *Symp. Soc. Exp. Biol.* **31**, 101–115.

Cleland, R. E. (1983). The capacity for acid-induced wall loosening as a factor in the control of *Avena* coleoptile cell elongation. *J. Exp. Bot.* **34**, 676–680.

Cleland, R. E. (1986). The role of hormones in wall loosening and plant growth. *Aust. J. Plant Physiol.* **13**, 93–103.

Cleland, R. E., and Rayle, D. L. (1978). Auxin, H$^+$-excretion and cell elongation. *Bot. Mag. Tokyo Spec. Issue* **1**, 125–139.

Clements, F. E., and Long, F. L. (1934). The method of collection films for stomata. *Am. J. Bot.* **21**, 7–17.

Clements, H. F. (1934). Significance of transpiration. *Plant Physiol.* **9**, 165–171.

Close, T. J., and Bray, E. A., eds. (1993). "Plant Responses to Cellular Dehydration During Environmental Stress." American Society of Plant Physiologists Series, Rockville, Maryland.

Close, T. J., and Chandler, P. M. (1990). Cereal dehydrins: Serology, gene mapping and potential functional roles. *Aust. J. Plant Physiol.* **17**, 333–334.

Close, T. J., Kortt, A. A., and Chandler, P. M. (1989). A cDNA-based comparison of dehydration-induced proteins (dehydrins) in barley and corn. *Plant Mol. Biol.* **13**, 95–108.

Cochard, H. (1992). Vulnerability of several conifers to air embolism. *Tree Physiol.* **11**, 73–83.

Cohen, Y., Fuchs, M., Falkenflug, V., and Moreshet, S. (1988). Calibrated heat pulse method for determining water uptake in cotton. *Agron. J.* **80**, 398–402.

Cohen, Y., Takeuchi, S., Nozaka, J., and Yano, T. (1993). Accuracy of sap flow measurement using heat balance and heat pulse methods. *Agron. J.* **85**, 1080–1086.

Coile, T. S. (1937). Distribution of forest tree roots in North Carolina Piedmont soils. *J. For.* **35**, 247–257.

Coile, T. S. (1940). Soil changes associated with loblolly pine succession on abandoned agricultural land of the Piedmont Plateau. Duke Univ. School of For. Bull. **5**.

Cole, F. D., and Decker, J. P. (1973). Relation of transpiration to atmospheric vapor pressure. *J. Arizona Acad. Sci.* **8**, 74–75.

Colire, C., LeRumeur, E., Gallier, J., deCertaines, J., and Larker, F. (1988). An assessment of proton nuclear magnetic resonance as an alternative method to describe water status of leaf tissue in wilted plants. *Plant Physiol. Biochem.* **26**, 767–776.

Collatz, G. J., Ball, J. T., Grivet, C., and Berry, J. A. (1991). Physiological and environmental regulation of stomatal conductance, photosynthesis and transpiration: A model that includes a laminar boundary layer. *Agric. For. Meteorol.* **54**, 107–136.

Collatz, G. J., Ribas-Carbo, M., and Berry, J. A. (1992). Coupled photosynthesis-stomatal conductance model for leaves of C$_4$ plants. *Aust. J. Plant Physiol.* **19**, 519–538.

Colombo, M. F., Rau, D. C., and Parsegian, V. A. (1992). Protein solvation in allosteric regulation: A water effect on hemoglobin. *Science* **256**, 655–659.

Colton, C. E., and Einhellig, F. E. (1980). Allelopathic mechanisms of velvet leaf (*Abutilon theophrasti*, Medic., Malvaceae) on soybean. *Am. J. Bot.* **67**, 1407–1413.

Condon, A. G., Farquhar, G. D., and Richards, R. A. (1990). Genotypic variation in carbon isotope discrimination and transpiration efficiency in wheat: Leaf gas exchange and whole plant studies. *Aust. J. Plant Physiol.* **17**, 9–22.

Condon, A. G., Richards, R. A., and Farquhar, G. D. (1987). Carbon isotope discrimination is positively correlated with grain yield and dry matter production in field-grown wheat. *Crop Sci.* **27**, 996–1001.

Constable, J. V. H., Grace, J. B., and Longstreth, D. J. (1992). High carbon dioxide concentrations in aerenchyma of *Typha latifolia*. *Am. J. Bot.* **79**, 415–418.

Cook, G. D., Dixon, J. R., and Leopold, A. C. (1964). Transpiration: Its effects on plant leaf temperature. *Science* **144**, 546–547.

Cope, F. W. (1967). NMR evidence for complexing of Na$^+$ in muscle, kidney, and brain, and by actomyosin: The relation of cellular complexing of Na$^+$ to water structure and to transport kinetics. *J. Gen. Physiol.* **50**, 1353–1375.

Corak, S. J., Blevins, D. G., and Pallardy, S. G. (1987). Water transfer in an alfalfa-maize association: Survival of maize during drought. *Plant Physiol.* **84**, 582–586.

Corey, A. T., and Klute, A. (1985). Application of the potential concept to soil water equilibrium and transport. *Soil Sci. Soc. Am. J.* **49**, 3–11.

Cornic, G., Le Gouallec, J.-L., Briantais, J. M., and Hodges, M. (1989). Effect of dehydration and high light on photosynthesis of two C_3 plants (*Phaseolus vulgaris* L. and *Elatostema repens* (Lour.) Hall f.). *Planta* **177**, 84–90.

Cortes, P. M. (1992). Analysis of the electrical coupling of root cells: Implications for ion transport and the existence of an osmotic pump. *Plant Cell Environ.* **15**, 351–363.

Cosgrove, D. J. (1985). Cell wall yield properties of growing tissue: Evaluation by *in vivo* stress relaxation. *Plant Physiol.* **78**, 347–356.

Cosgrove, D. J. (1987). Wall relaxation in growing stems: Comparison of four species and assessment of measurement techniques. *Planta* **171**, 266–278.

Cosgrove, D. J., and Cleland, R. E. (1983a). Solutes in the free space of growing stem tissues. *Plant Physiol.* **72**, 326–331.

Cosgrove, D. J., and Cleland, R. E. (1983b). Osmotic properties of pea internodes in relation to growth and auxin action. *Plant Physiol.* **72**, 332–338.

Cosgrove, D. J., and Steudle, E. (1981). Water relations of growing pea epicotyl segments. *Planta* **153**, 343–350.

Cosgrove, D. J., Van Volkenburgh, E., and Cleland, R. E. (1984). Stress relaxation of cell walls and the yield threshold for growth: Demonstration and measurement by micropressure probe and psychrometer techniques. *Planta* **162**, 46–54.

Couchat, Ph., and Lasceve, G. (1980). Tritiated water vapour exchange method for the evaluation of whole plant diffusion resistance. *J. Exp. Bot.* **31**, 1217–1222.

Couchat, Ph., Moutonnet, P., Honelle, M., and Picard, D. (1980). *In situ* study of corn seedling root and shoot growth by neutron radiography. *Agron. J.* **72**, 321–324.

Coutts, M. P. (1983). Root architecture and tree stability. *Plant Soil* **71**, 171–188.

Cowan, I. R. (1965). Transport of water in the soil-plant-atmosphere system. *J. Appl. Ecol.* **2**, 221–239.

Cowan, I. R. (1972). Oscillations in stomatal conductance and plant functioning associated with stomatal conductance. I. Observations and a model. *Planta* (Berl.) **106**, 185–219.

Cowan, I. R. (1977). Stomatal behavior and environment. *In* "Advances in Botanical Research" (R. D. Preston and H. W. Woolhouse, eds.), Vol. 4, pp. 117–228. Academic Press, New York.

Cowan, I. R. (1982). Regulation of water use in relation to carbon gain in higher plants. *In* "Encyclopedia of Plant Physiology" (O. L. Lange and J. D. Bewley, eds.), Vol. 12B, pp. 535–562. Springer-Verlag, Berlin.

Crafts, A. S. (1936). Further studies on exudation in cucurbits. *Plant Physiol.* **11**, 63–79.

Crafts, A. S., and Broyer, T. C. (1938). Migration of salts and water into xylem of the roots of higher plants. *Am. J. Bot.* **25**, 529–535.

Crafts, A. S., Currier, H. B., and Stocking, C. R. (1949). "Water in the Physiology of the Plant." Chronica Botanica Co., Waltham, MA.

Cramer, G. R., and Bowman, D. C. (1991). Short-term leaf elongation kinetics in response to salinity are independent of the root. *Plant Physiol.* **95**, 965–967.

Crawford, N. M., and Campbell, W. H. (1990). Fertile fields. *Plant Cell* **2**, 829–835.

Crawford, R. M. M. (1976). Tolerance of anoxia and the regulation of glycolysis in tree roots. *In* "Tree Physiology and Yield Improvement" (M. G. R. Cannell and F. T. Last, eds.), pp. 387–401. Academic Press, New York.

Creelman, R. A., Mason, H. S., Bensen, R. J., Boyer, J. S., and Mullet, J. E. (1990). Water deficit and abscisic acid cause differential inhibition of shoot *versus* root growth in soybean seedlings. *Plant Physiol.* **92**, 205–214.

Criswell, J. G., Havelka, U. D., Quebedeaux, B., and Hardy, R. W. F. (1976). Adaptation of nitrogen fixation by intact soybean nodules to altered rhizosphere pO_2. *Plant Physiol.* **16**, 131–140.

Crook, M. J., and Ennos, A. R. (1993). The mechanics of root lodging in winter wheat, *Triticum aestivum* L. *J. Exp. Bot.* **44**, 1219–1224.

Crowe, J. H., and Crowe, L. M. (1986). Stabilization of membranes in anhydrobiotic organisms. *In* "Membranes, Metabolism and Dry Organisms" (A. C. Leopold, ed.), pp. 188–209. Comstock

Publishing Association, Ithaca, NY.

Crowe, J. H., Crowe, L. M., Carpenter, J. F., Rudolph, A. S., Wistrom, C. A., Spargo, B. J., and Anchordoguy, T. J. (1988). Interactions of sugars with membranes. *Biochim. Biophys. Acta* **947**, 367–384.

Crowe, J. H., Spargo, B. J., and Crowe, L. M. (1987). Preservation of dry liposomes does not require retention of residual water. *Proc. Natl. Acad. Sci. USA* **84**, 1537–1540.

Crowe, L. M., Mouradian, R., Crowe, J. H., Jackson, S. A., and Womersley, C. (1984). Effects of carbohydrates on membrane stability at low water activities. *Biochim. Biophys. Acta* **769**, 141–150.

Crowe, L. M., Womersley, C., Crowe, J. H., Reid, D., Appel, L., and Rudolph, A. (1986). Prevention of fusion and leakage in freeze-dried liposomes by carbohydrates. *Biochim. Biophys. Acta* **861**, 131–140.

Cruz, R. T., Jordan, W. R., and Drew, M. C. (1992). Structural changes and associated reduction of hydraulic conductance in roots of *Sorghum bicolor* L. following exposure to water deficits. *Plant Physiol.* **99**, 203–212.

Cumming, J. R., and Weinstein, L. H. (1990). Aluminum-mycorrhizal interactions in the physiology of pitch pine seedlings. *Plant Soil* **125**, 7–18.

Curl, E. A., and Truelove, B. (1986). "The Rhizosphere." Springer-Verlag, Berlin.

Currie, D. J., and Paquin, V. (1987). Large-scale biogeographical patterns of species richness of trees. *Nature (London)* **329**, 326–327.

Curtis, L. C. (1944). The exudation of glutamine from lawn grass. *Plant Physiol.* **19**, 1–5.

Curtis, O. F. (1936). Leaf temperatures and the cooling of leaves by radiation. *Plant Physiol.* **11**, 343–364.

Cutler, J. M., and Rains, D. W. (1977a). Effects of water stress and hardening on the internal water relations and osmotic constituents of cotton leaves. *Physiol. Plant.* **42**, 261–268.

Cutler, J. M., and Rains, D. W. (1977b). Effect of irrigation history on responses of cotton to subsequent water stress. *Crop Sci.* **17**, 329–335.

Cutler, J. M., Rains, D. W., and Loomis, R. S. (1977). Role of changes in solute concentration in maintaining favorable water balance in field-grown cotton. *Agron. J.* **69**, 773–779.

Cutler, J. M., Shannon, K. W., and Steponkus, P. L. (1980a). Influence of water deficits and osmotic adjustment on leaf elongation in rice. *Crop Sci.* **20**, 314–318.

Cutler, J. M., Steponkus, P. L., Wach, M. J., and Shahan, K. W. (1980b). Dynamic aspects and enhancement of leaf elongation in rice. *Plant Physiol.* **66**, 147–152.

Dacey, J. W. H. (1981). Pressurized ventilation in the yellow water lily. *Ecology* **62**, 1137–1147.

Dacey, J. W. H. (1987). Krudsen-transitional flow and gas pressurization in leaves of *Nelumbo*. *Plant Physiol.* **85**, 199–203.

Dainty, J. (1963). Water relations of plant cells. *Adv. Bot. Res.* **1**, 279–326.

Dainty, J., and Ginzburg, B. Z. (1964). The reflection coefficient of plant cell membranes for certain solutes. *Biochim. Biophys. Acta* **79**, 129–137.

Dakora, F. D., and Atkins, C. A. (1989). Diffusion of oxygen in relation to structure and function in legume root nodules. *Aust. J. Plant Physiol.* **16**, 131–140.

Daley, P. F., Raschke, K., Ball, J. T., and Berry, J. A. (1989). Topography of photosynthetic activity of leaves obtained from video images of chlorophyll fluorescence. *Plant Physiol.* **90**, 1233–1238.

Dalton, F. N. (1988). Plant root water extraction studies using stable isotopes. *Plant Soil* **111**, 217–221.

Dalton, F. N., Raats, P. A. C., and Gardner, W. R. (1975). Simultaneous uptake of water and solutes by plant roots. *Agron. J.* **67**, 334–339.

Damptey, H. B., Coombe, B. G., and Aspinall, D. (1978). Apical dominance, water deficit, and axillary inflorescence growth in *Zea Mays:* The role of abscisic acid. *Ann. Bot.* (London) **42**, 1447–1458.

Daniels, B. A., Hetrick, D., Gerschefske, K., and Wilson, G. T. (1987). Effects of drought stress on growth response in corn, sudan grass, and big bluestem to *Glomus etunicatum. New Phytol.* **105**, 403–410.

Darwin, C. (1880). "The Power of Movement in Plants." Murray, London.

Darwin, C. (1881). "The Formation of Vegetable Mould through the Action of Worms." Murray, London.

Darwin, F., and Pertz, D. F. M. (1911). On a new method of estimating the aperture of stomata. *Proc. Roy. Soc.* (London) **B84**, 136–154.

Dasberg, S., and Dalton, F. N. (1985). Time domain reflectometry field measurements of soil water content and electrical conductivity. *Soil Sci. Soc. Am. J.* **49**, 293–297.

Dasgupta, J., and Bewley, J. D. (1984). Variations in protein synthesis in different regions of greening leaves of barley seedlings and effects of imposed water stress. *J. Exp. Bot.* **35**, 1450–1459.

Daughters, M. R., and Glenn, D. S. (1946). The role of water in freezing foods. *Refrig. Engin.* **52**, 137–148.

Daum, C. R. (1967). A method for determining water transport in trees. *Ecology* **48**, 425–431.

Davey, A. G., and Simpson, R. J. (1990). Nitrogen fixation by subterranean clover at varying stages of nodule dehydration. I. Carbohydrate status and short-term recovery of nodulated root respiration. *J. Exp. Bot.* **41**, 1175–1187.

Davidson, O. W. (1945). Salts in old greenhouse soils stunt flowers and vegetables. *Florists Rev.* **95**, 17–19.

Davies, W. J. (1977). Stomatal responses to water stress and light in plants grown in controlled environments and in the field. *Crop Sci.* **17**, 735–740.

Davies, W. J. (1986). Transpiration and the water balance of plants. *In* "Plant Physiology" (F. C. Stewart, J. F. Sutcliffe and J. E. Dale, eds.), Vol. IX, pp. 49–154. Academic Press, Orlando, FL.

Davies, W. J., Kozlowski, T. T., and Pereira, J. (1974). Effects of wind on transpiration and stomatal aperture of woody plants. *Bull. Roy. Soc. N.Z.* **12**, 433–438.

Davies, W. J., Mansfield, T. A., and Hetherington, A. M. (1990). Sensing of soil water status and the regulation of plant growth and development. *Plant Cell Environ.* **13**, 709–719.

Davies, W. J., Metcalfe, J., Lodge, T. A., and DaCosta, A. R. (1986). Plant growth substances and the regulation of growth under drought. *Aust. J. Plant Physiol.* **13**, 105–125.

Davies, W. J., Tardieu, F., and Trejo, C. L. (1994). How do chemical signals work in plants that grow in drying soil? *Plant Physiol.* **104**, 309–314.

Davies, W. J., and Van Volkenburgh, E. (1983). The influence of water deficit on the factors controlling the daily pattern of growth of *Phaseolus* trifoliates. *J. Exp. Bot.* **34**, 987–999.

Davies, W. J., and Zhang, J. (1991). Root signals and the regulation of growth and development of plants in drying soil. *Annu. Rev. Plant Physiol. Plant Mol. Biol.* **42**, 55–76.

Davis, R. E., Rosseau, D. L., and Board, R. D. (1971). "Polywater": Evidence from electron spectroscopy for chemical analysis (ESCA) of a complex salt matrix. *Science* **171**, 167–171.

Dawson, R. F. (1942). Accumulation of nicotine in reciprocal grafts of tomato and tobacco. *Am. J. Bot.* **29**, 66–71.

Dawson, T. E. (1993). Hydraulic lift and water use by plants: Implications for water balance, performance and plant-plant interactions. *Oecologia* **95**, 565–574.

Day, T. A., Heckathorn, S. A., and DeLucia, E. H. (1991). Limitations of photosynthesis in *Pinus taeda* L. (loblolly pine) at low soil temperatures. *Plant Physiol.* **96**, 1246–1254.

de Candolle, A. P. (1832). "Physiologie Végétale." Bechet Jeune, Paris.

de Saussure, N. T. (1804). "Recherches Chimiques sur la Végétation." Madame Huzard, Paris.

de Vries, H. (1877). "Untersuchungen über die mechanischen Ursachen der Zellstreckung." W. Engelmann, Leipzig.

de Wit, C. T. (1958). Transpiration and crop yields. *In* "Institute of Biological and Chemical Research on Field Crops and Herbage," Wageningen, The Netherlands, Verslagen Landbouwkundige Onderzoekingen 64.6, 1–88.

Decker, J. P., Gaylor, W. G., and Cole, F. D. (1962). Measuring transpiration of undisturbed tamarisk shrubs. *Plant Physiol.* **37**, 393–397.

Decker, J. P., and Skau, C. M. (1964). Simultaneous studies of transpiration rate and sap velocity in trees. *Plant Physiol.* **39**, 213–215.

Delhaize, E., Ryan, P. R., and Randall, P. J. (1993). Aluminum tolerance in wheat (*Triticum aestivum* L.). II. Aluminum-stimulated excretion of malic acid from root apices. *Plant Physiol.* **103**, 695–702.

DeLucia, E. H. (1986). Effect of low-root temperature on net photosynthesis, stomatal conductance and carbohydrate concentration in Engelmann spruce (*Picea engelmannii* Parry ex Engelm.) seedlings. *Tree Physiol.* **2**, 143–154.

Delves, A. C., Higgins, A. V., and Greenough, P. M. (1987). Shoot control of supernodulation in a number of mutant soybeans, *Glycine max* (L.) Merr. *Aust. J. Plant Physiol.* **14**, 689–694.

Denison, R. F., Hunt, S., and Layzell, D. B. (1992). Nitrogenase activity, nodule respiration and O_2 permeability following detopping of alfalfa and birdsfoot trefoil. *Plant Physiol.* **98**, 894–900.

Denison, R. F., Smith, D. L., Legros, T., and Layzell, D. B. (1991). Noninvasive measurement of internal oxygen concentration of field-grown soybean nodules. *Agron. J.* **83**, 166–169.

Denmead, O. T., and Shaw, R. H. (1962). Availability of soil water to plants as affected by soil moisture content and meteorological conditions. *Agron. J.* **54**, 385–390.

Desai, M. C. (1937). Effect of certain nutrient deficiencies on stomatal behavior. *Plant Physiol.* **12**, 253–283.

DeWald, D. B., Mason, H. S., and Mullet, J. E. (1992). The soybean vegetative storage proteins VSPα and VSPβ are acid phosphatases active on polyphosphates. *J. Biol. Chem.* **267**, 15958–15964.

Dexter, A. R. (1987). Mechanics of root growth. *Plant Soil* **98**, 303–312.

Dhindsa, R. S., and Bewley, J. D. (1976). Plant desiccation: Polysome loss not due to ribonuclease. *Science* **191**, 181–182.

Dhindsa, R. S., and Bewley, J. D. (1978). Messenger RNA is conserved during drying of the drought-tolerant moss *Tortula ruralis*. *Proc. Natl. Acad. Sci. USA* **75**, 842–846.

Dimond, A. E. (1966). Pressure and flow relations in vascular bundles of the tomato plant. *Plant Physiol.* **41**, 119–131.

Dittmer, H. J. (1937). A quantitative study of the roots and root hairs of a winter rye plant (*Secale cereale*). *Am. J. Bot.* **24**, 417–420.

Dixon, H. H., and Joly, J. (1895). The path of the transpiration current. *Ann. Bot.* (London) **9**, 416–419.

Dixon, M. A., Grace, J., and Tyree, M. T. (1984). Concurrent measurements of stem density, leaf and stem water potential, stomatal conductance, and cavitation on a sapling of *Thuja occidentalis* L. *Plant Cell Environ.* **7**, 615–618.

Dixon, R. K., Pallardy, S. G., Garrett, H. E., and Cox, G. S. (1983). Comparative water relations of container-grown and bare root-grown ectomycorrhizal and nonmycorrhizal *Quercus velutina* seedlings. *Can. J. Bot.* **61**, 1559–1565.

Döbereiner, J. (1983). Dinitrogen fixation in rhizosphere and phyllosphere associations. *In* "Encyclopedia of Plant Physiology" (A. Läuchli and R. L. Bieleski, eds.), Vol. 15A, pp. 330–350. Springer-Verlag, Berlin/Heidelberg.

Dong, Z., Canny, M. J., McCully, M. E., Roboredo, M. R., Cabadilla, C. F., Ortega, E., and Rodés, R. (1994). A nitrogen-fixing endophyte of sugarcane stems. *Plant Physiol.* **105**, 1139–1147.

Downton, W. J. S., Loveys, B. R., and Grant, W. J. R. (1988a). Stomatal closure fully accounts for the inhibition of photosynthesis by abscisic acid. *New Phytol.* **108**, 263–266.

Downton, W. J. S., Loveys, B. R., and Grant, W. J. R. (1988b). Non-uniform stomatal closure induced by water stress causes putative non-stomatal inhibition of photosynthesis. *New Phytol.* **110**, 503–509.

Drew, M. C., Chamel, A., Garrec, J.-P., and Fourcy, A. (1980). Cortical air spaces (aerenchyma) in roots of corn subjected to oxygen stress: Structure and influence on uptake and translocation of [86]rubidium ions. *Plant Physiol.* **65**, 506–511.

Drew, M. C., Jackson, M. B., and Giffard, S. (1979a). Ethylene-promoted adventitious rooting and development of cortical air spaces (aerenchyma) in roots may be adaptive responses to flooding in *Zea mays* L. *Planta* **147**, 83–88.

Drew, M. C., Siswaro, E. J., and Saker, L. R. (1979b). Alleviation of waterlogging damage to young barley plants by application of nitrate and a synthetic cytokinin, and comparison between the effects of waterlogging, nitrogen deficiency and root excision. *New Phytol.* **82**, 315–329.

Drost-Hansen, W. (1965). Forms of water in biologic systems. *Ann. N.Y. Acad. Sci.* **125**, 249–272.

Drost-Hansen, W., and Clegg, J. S., eds. (1979). Cell-Associated Water." Academic Press, New York.

Dudal, R. (1976). Inventory of the major soils of the world with special reference to mineral stress hazards. *In* "Plant Adaptation to Mineral Stress in Problem Soils" (M. J. Wright, ed.), pp. 3–14. Cornell University Agricultural Experiment Station, Ithaca, NY.

Duddridge, J. A., Malibari, A., and Read, D. J. (1980). Structure and function of mycorrhizal rhizomorphs with special reference to their role in water transport. *Nature* **287**, 834–836.

Duell, R. W., and Markus, D. K. (1977). Guttation deposits on turfgrass. *Agron. J.* **69**, 891–894.

Dugas, W. A., Wallace, J. S., Allen, S. J., and Roberts, J. M. (1993). Heat balance porometer, and deuterium estimates of transpiration from potted trees. *Agric. For. Meteorol.* **64**, 47–62.

Duke, S. H., Schrader, L. E., Henson, C. A., Servaites, J. C., Vogelzang, R. D., and Pendleton, J. W. (1979). Low root temperature effects on soybean nitrogen metabolism and photosynthesis. *Plant Physiol.* **63**, 956–962.

Dumbroff, E. B., and Peirson, D. R. (1971). Probable sites for passive movement of ions across the endodermis. *Can. J. Bot.* **49**, 35–38.

Dunham, R. J., and Nye, P. H. (1973). I. Soil water content gradients near a plane of onion roots. *J. Appl. Ecol.* **10**, 585–598.

Duniway, J. M. (1977). Changes in resistance to water transport in safflower during the development of *Phytophthora* root rot. *Phytopath.* **67**, 331–337.

Durbin, R. D. (1967). Obligate parasites: Effect on the movement of solutes and water. *In* "The Dynamic Role of Molecular Constituents in Plant–Parasite Reactions" (C. J. Mirocha and I. Uritani, eds.), pp. 80–99. Am. Phytopath. Soc., St. Paul, MN.

Dure, L., III, Crouch, M., Harada, J., Ho, T.-H. D., Mundy, J., Quatrano, R., Thomas, T., and Sung, Z. R. (1989). Common amino acid sequence domains among the LEA proteins of higher plants. *Plant Mol. Biol.* **12**, 475–486.

Dutrochet, H. J. (1837). "Memoires Pour Servir a l'Histoire Anatomique et Physiologie des Végétaux et des Animaux." J. B. Baillière et Fils, Paris.

Duvdevani, S. (1953). Dew gradients in relation to climate, soil and topography. *Desert Res. Proc. Int. Symp. 1952 Spec. Publ.* **2**, 136–152.

Dye, P. J., and Olbrich, B. W. (1993). Estimating transpiration from 6-year-old *Eucalyptus grandis*: Development of a canopy conductance model and comparison with independent sap flow measurements. *Plant Cell Environ.* **16**, 45–53.

Dye, P. J., Olbrich, B. W., and Calder, I. R. (1992). A comparison of the heat pulse method and deuterium tracing method for measuring transpiration from *Eucalyptus grandis* trees. *J. Exp. Bot.* **43**, 337–343.

Eames, A. J., and MacDaniels, L. H. (1947). "An Introduction to Plant Anatomy," 2nd Ed. McGraw-Hill, New York.

Eaton, F. M. (1927). The water requirement and cell-sap concentration of Australian saltbush and wheat as related to the salinity of the soil. *Am. J. Bot.* **14**, 212–226.

Eaton, F. M. (1930). Cel-sap concentration and transpiration as related to age and development of cotton leaves. *J. Agr. Res.* **40**, 791–805.

Eaton, F. M. (1931). Root development as related to character of growth and fruitfulness of the cotton plant. *J. Agr. Res.* **43**, 875–883.

Eaton, F. M. (1942). Toxicity and accumulation of chloride and sulfate salts in plants. *J. Agr. Res.* **64**, 357–399.

Eaton, F. M., and Harding, R. B. (1959). Foliar uptake of salt constituents of water by citrus plants during intermittent sprinkling and immersion. *Plant Physiol.* **34**, 22–26.

Eavis, B. W., and Taylor, H. M. (1979). Transpiration of soybeans as related to leaf area, root length, and soil water content. *Agron. J.* **71**, 441–445.

Eck, H. V., and Musick, J. T. (1979). Plant water stress effects on irrigated grain sorghum. II. Effects on nutrients in plant tissues. *Crop Sci.* **19**, 592–598.

Edlefsen, N. E., and Bodman, G. B. (1941). Field measurements of water movement through a silt loam soil. *Am. Soc. Agron. J.* **33**, 713–731.

Edmeades, G. O., Bolaños, J., Hernández, M., and Bello, S. (1993). Causes for silk delay in a lowland tropical maize population. *Crop Sci.* **33**, 1029–1035.

Edmeades, G. O., Bolaños, J., and Lafitte, H. R. (1992). Progress in breeding for drought tolerance in maize. *In* "Proceedings of the Forty-Seventh Annual Corn and Sorghum Industry Research Conference," pp. 93–111.

Edsall, J. T., and McKenzie, H. A. (1978). Water and proteins. I. The significance and structure of water: Its interaction with electrolytes and non-electrolytes. *Adv. Biophys.* **10**, 137–207.

Ehleringer, J. R., and Cook, C. S. (1984). Photosynthesis in *Encelia farinosa* Gray in response to decreasing leaf water potential. *Plant Physiol.* **75**, 688–693.

Ehleringer, J. R., and Field, C. B. (1993). "Scaling Physiological Processes." Academic Press, San Diego.

Ehleringer, J. R., Hall, A. E., and Farquhar, G. D., eds. (1993). "Stable Isotopes and Plant Carbon-Water Relations." Academic Press, San Diego.

Ehlig, C. F. (1962). Measurement of energy status of water in plants with a thermocouple psychrometer. *Plant Physiol.* **37**, 288–290.

Ehret, D. L., and Boyer, J. S. (1979). Potassium loss from stomatal guard cells at low water potentials. *J. Exp. Bot.* **30**, 225–234.

Eisenberg, D., and Kauzmann, W. (1969). "The Structure and Properties of Water." Oxford University Press, New York.

Ekanayake, I. J., O'Toole, J. C., Garrity, D. P., and Masajo, T. M. (1985). Inheritance of root characters and their relations to drought resistance in rice. *Crop Sci.* **25**, 927–933.

Ekern, P. C. (1965). Evapotranspiration of pineapple in Hawaii. *Plant Physiol.* **40**, 736–739.

El Nadi, A. H., Brouwer, R., and Locker, J. Th. (1969). Some responses of the root and the shoot of *Vicia faba* plants to water stress. *Neth. J. Agric. Sci.* **17**, 133–142.

Elfving, F. (1882). Ueber die Wasserleitung im Holz. *Bot. Zeitung* **42**, 706–723.

Emerson, W. W., Bond, R. D., and Dexter, A. R., eds. (1978). "Modification of Soil Structure." Wiley, New York.

England, C. B., and Lesesne, E. H. (1962). Evapotranspiration research in western North Carolina. *Agr. Eng.* **43**. 526–528.

Ennos, A. R., Crook, M. J., and Grimshaw, C. (1993). The anchorage mechanics of maize, *Zea mays. J. Exp. Bot.* **44**, 147–153.

Epstein, A. H. (1978). Root graft transmission of tree pathogens. *Annu. Rev. Phytopathol.* **16**, 181–192.

Epstein, E. E. (1972). "Mineral Nutrition of Plants: Principles and Perspectives." Wiley, New York.

Errera, L. (1886). Ein Transpirationsversuch. *Ber. deut. bot. Ges.* **4**, 16–18.

Esau, K. (1965). "Plant Anatomy," 2nd Ed. Wiley, New York.

Evans, H. J., and Sorger, G. J. (1966). Role of mineral elements with emphasis on the univalent cations. *Annu. Rev. Plant Physiol.* **17**, 47–76.

Evans, H. J., and Wildes, R. A. (1971). Potassium and its role in enzyme activation. *In* "Potassium in Biochemistry and Physiology," pp. 13–38. International Potash Institute, Bern, Switzerland.

Evans, P. S., and Klett, J. E. (1985). The effects of dormant branch thinning on total leaf, shoot, and root production from bare-root *Prunus cerasifera* "Newportii." *J. Arboric.* **11**, 149–151.

Everard, J. D., and Drew, M. C. (1989). Mechanisms controlling changes in water movement through the roots of *Helianthus annuus* L. during continuous exposure to oxygen deficiency. *J. Exp. Bot.* **40,** 95–104.

Faiz, S. M. A., and Weatherley, P. E. (1978). Further investigations into the location and magnitude of the hydraulic resistances in the soil-plant system. *New Phytol.* **81,** 19–28.

Faiz, S. M. A., and Weatherley, P. E. (1982). Root contraction in transpiring plants. *New Phytol.* **92,** 333–343.

Falk, M., and Kell, G. S. (1966). Thermal properties of water: Discontinuities questioned. *Science* **154,** 1013–1015.

Faroud, N., Lynch, D. R., and Entz, T. (1993). Potato water content impact on soil moisture measurement by neutron meter. *Plant Soil* **148,** 101–106.

Farquhar, G. D., and Sharkey, T. D. (1982). Stomatal conductance and photosynthesis. *Annu. Rev. Plant Physiol.* **33,** 317–345.

Farquhar, G. D., Wetselaar, R., and Firth, P. M. (1979). Ammonia volatilization from senescing leaves of maize. *Science* **203,** 1257–1258.

Farris, F., and Strain, B. R. (1978). The effects of water stress on leaf $H_2^{18}O$ enrichment. *Rad. Environ. Biophys.* **15,** 167–202.

Fayle, D. C. F. (1978). Poor vertical root development may contribute to suppression in a red pine plantation. *For. Chron.* **54,** 99–103.

Feldman, L. J. (1984). Regulation of root development. *Annu. Rev. Plant Physiol.* **35,** 223–242.

Fellows, R. J., and Boyer, J. S. (1976). Structure and activity of chloroplasts of sunflower leaves having various water potentials. *Planta* **132,** 229–239.

Fellows, R. J., and Boyer, J. S. (1978). Altered ultrastructure of cells of sunflower leaves having low water potentials. *Protoplasma* **93,** 381–395.

Fellows, R. J., Patterson, R. P., Raper, C. D., and Harris, D. (1987). Nodule activity and allocation of photosynthate of soybean during recovery from water stress. *Plant Physiol.* **84,** 456–460.

Fensom, D. S. (1957). The bioelectric potentials of plants and their functional significance. *Can. J. Bot.* **35,** 573–582.

Fensom, D. S. (1958). The bioelectric potentials of plants and their functional significance. II. The patterns of bioelectric potential and exudation rate in excised sunflower roots and stems. *Can. J. Bot.* **36,** 367–383.

Fereres, E., Kiflas, P. M., Goldfein, R. E., Pruitt, W. O., and Hagan, R. M. (1981). Simplified but scientific irrigation scheduling. *Calif. Agric.* **36,** 19–21.

Ferguson, H., and Gardner, W. H. (1962). Water content measurement in soil columns by gamma ray absorption. *Soil Sci. Soc. Am. Proc.* **26,** 11–14.

Fick, A. (1855). Über Diffusion. *Poggendorffs Annalen* **94,** 59–86.

Fischer, A. G. (1982). Long-term climatic oscillations recorded in stratigraphy. *In* "Climate in Earth History" (W. H. Berger and J. C. Crowell, eds.), pp. 97–104. Studies in Geophysics, Natl. Acad. Press, Washington, DC.

Fischer, A. G. (1984). The two phanerozoic supercycles. *In* "Catastrophes and Earth History" (W. A. Berggren and J. A. van Couvering, eds.), pp. 129–150. Princeton University Press, Princeton, NJ.

Fischer, D. B., and Frame, J. M. (1984). A guide to the use of the exuding-stylet technique in phloem physiology. *Planta* **161,** 385–393.

Fischer, R. A. (1968a). Stomatal opening in isolated epidermal strips of *Vicia faba*. I. Response to light and to CO_2-free air. *Plant Physiol.* **43,** 1947–1952.

Fischer, R. A. (1968b). Stomatal opening: Role of potassium uptake by guard cells. *Science* **168,** 784–785.

Fischer, R. A. (1970). After-effect of water stress on stomatal opening potential. II. Possible causes. *J. Exp. Bot.* **21,** 386–404.

Fischer, R. A. (1971). Role of potassium in stomatal opening in the leaf of *Vicia faba*. *Plant Physiol.* **47,** 555–558.

Fischer, R. A., and Hsiao, T. C. (1968). Stomatal opening in isolated epidermal strips of *Vicia faba*. II. Responses to KCl concentration and the role of potassium absorption. *Plant Physiol.* **43**, 1953–1958.

Fiscus, E. L. (1975). The interaction between osmotic- and pressure-induced water flow in plant roots. *Plant Physiol.* **55**, 917–922.

Fiscus, E. L. (1977). Determination of hydraulic and osmotic properties of soybean root systems. *Plant Physiol.* **59**, 1013–1020.

Fiscus, E. L. (1981). Analysis of the components of area growth of bean root systems. *Crop Sci.* **21**, 909–913.

Fiscus, E. L. (1986). Diurnal changes in volume and solute transport coefficients of *Phaseolus* roots. *Plant Physiol.* **80**, 752–759.

Fiscus, E. L., and Kramer, P. J. (1970). Radial movement of oxygen in plant roots. *Plant Physiol.* **45**, 667–669.

Fiscus, E. L., and Kramer, P. J. (1975). General model for osmotic and pressure-induced flow in plant roots. *Proc. Natl. Acad. Sci. USA* **72**, 3114–3118.

Fiscus, E. L., Klute, A., and Kaufmann, M. R. (1983). An interpretation of some whole plant water transport phenomena. *Plant Physiol.* **71**, 810–817.

Fiscus, E. L., Mahbub-Ul Alam, A. N. M., and Hirasawa, T. (1991). Fractional integrated stomatal opening to control water stress in the field. *Crop Sci.* **31**, 1001–1008.

Fiscus, E. L., and Markhart, A. H., III (1979). Relationships between root system water transport properties and plant size in *Phaseolus*. *Plant Physiol.* **64**, 770–773.

Fiscus, E. L., Parsons, L. R., and Alberte, R. S. (1973). Phyllotaxy and water relations in tobacco. *Planta* **112**, 285–292.

Fiscus, E. L., Wullschleger, S. D., and Duke, H. R. (1984). Integrated stomatal opening as an indicator of water stress in *Zea*. *Crop Sci.* **24**, 245–249.

Fitter, A. H. (1987). An architectural approach to the comparative ecology of plant root systems. *New Phytol.* **106**(Suppl.), 61–77.

Flanagan, L. B., Comstock, J. P., and Ehleringer, J. R. (1991). Comparison of modeled and observed environmental influence on the stable oxygen and hydrogen isotope composition of leaf water in *Phaseolus vulgaris* L. *Plant Physiol.* **96**, 588–596.

Flanagan, L. B., Ehleringer, J. R., and Marshall, J. D. (1992). Differential uptake of summer precipitation among co-occurring trees and shrubs in a pinyon-juniper woodland. *Plant Cell Environ.* **15**, 831–836.

Flores, H. E., Dai, Y.-R., Cuello, J. L., Maldonado-Mendoza, I. E., and Loyola-Vargas, V. M. (1993). Green roots: Photosynthesis and photoautotrophy in an underground plant organ. *Plant Physiol.* **101**, 363–371.

Flowers, T. J., and Hanson, J. B. (1969). The effect of reduced leaf water potential on soybean mitochondria. *Plant Physiol.* **44**, 939–945.

Flowers, T. J., Troke, P. F., and Yeo, A. R. (1977). The mechanism of salt tolerance in halophytes. *Annu. Rev. Plant Physiol.* **28**, 89–121.

Fonteyn, P. J., Schlesinger, W. H., and Marion, G. M. (1987). Accuracy of soil thermocouple hygrometer measurements in desert ecosystems. *Ecology* **68**, 1121–1124.

Forseth, I. N., and Ehleringer, J. R. (1983). Ecophysiology of two solar tracking desert winter annuals. III. Gas exchange responses to light, CO_2 and VPD in relation to long-term drought. *Oecologia* **57**, 344–351.

Foster, R. G. (1981). The ultrastructure and histochemistry of the rhizosphere. *New Phytol.* **89**, 263–273.

Fowler, D., Cope, J. N., Deans, J. D., Leith, I. D., Murray, M. B., Smith, R. I., Sheppard, L. J., and Unsworth, M. H. (1989). Effects of acid mist on the frost hardiness of red spruce seedlings. *New Phytol.* **113**, 321–335.

Foyer, C. H. (1984). "Photosynthesis." Wiley, New York.

Franco, C. M., and Magalhaes, A. C. (1965). Techniques for the measurement of transpiration of individual plants. *Arid Zone Res.* **25**, 211–224, UNESCO, Paris.

Franks, F., ed. (1975). "Water: A Comprehensive Treatise," Vol. 5. Plenum Press, New York.

Franks, F. (1981). A scientific gold rush. *Science* **213**, 1104–1105.

Fraser, D. A., and Mawson, C. A. (1953). Movement of radioactive isotopes in yellow birch and white pine as detected with a portable scintillation counter. *Can. J. Bot.* **31**, 324–333.

Fraser, T. E., Silk, W. K., and Rost, T. L. (1990). Effects of low water potential on cortical cell length in growing regions of maize roots. *Plant Physiol.* **93**, 648–651.

Frederick, J. R., Woolley, J. T., Hesketh, J. D., and Peters, D. B. (1990). Water deficit development in old and new soybean cultivars. *Agron. J.* **82**, 76–81.

Freeman, T. P., and Duysen, M. E. (1975). The effect of imposed water stress on the development and ultrastructure of wheat chloroplasts. *Protoplasma* **83**, 131–145.

Frensch, J., and Hsiao, T. C. (1993). Hydraulic propagation of pressure along immature and mature xylem vessels of roots of *Zea mays* measured by pressure-probe techniques. *Planta* **190**, 263–270.

Frensch, J., and Steudle, E. (1989). Axial and radial hydraulic resistance to roots of maize (*Zea mays* L.). *Plant Physiol.* **91**, 719–726.

Frey-Wyssling, A. (1976). The plant cell wall. *In* "Handbuch der Pflanzenanatomie" (M. Zimmermann, S. Carlquist, and H. D. Wulff, eds.), Vol. 4. Gebrüder Borntraeger, Berlin.

Friesner, R. C. (1940). An observation on the effectiveness of root pressure in the ascent of sap. *Butler Univ. Bot. Studies* **4**, 226–227.

Fritschen, L. J., Cox, L., and Kinerson, R. (1973). A 28-meter Douglas fir in a weighing lysimeter. *For. Sci.* **19**, 256–261.

Fritts, H. C. (1976). "Tree Rings and Climate." Academic Press, New York/London.

Fromm, J., and Eschrich, W. (1993). Electric signals released from roots of willow (*Salix viminalis* L.) change transpiration and photosynthesis. *J. Plant Physiol.* **141**, 673–680.

Fromm, J., and Spanswick, R. (1993). Characteristics of action potentials in willow (*Salix viminalis* L.). *J. Exp. Bot.* **44**, 1119–1125.

Fry, K. E., and Walker, R. B. (1967). A pressure-infiltration method for estimating stomatal opening in conifers. *Ecology* **48**, 155–157.

Fry, S. C. (1989a). The structure and functions of xyloglucan. *J. Exp. Bot.* **40**, 1–11.

Fry, S. C. (1989b). Cellulases, hemicelluloses and auxin-stimulated growth: A possible relationship. *Physiol. Plant.* **75**, 532–536.

Furr, J. R., and Reeve, J. O. (1945). The range of soil-moisture percentages through which plants undergo permanent wilting in some soils from semiarid irrigated areas. *J. Agric. Res.* **71**, 149–170.

Furr, J. R., and Taylor, C. A. (1933). The cross-transfer of water in mature lemon trees. *Proc. Am. Soc. Hort. Sci.* **30**, 45–51.

Fusseder, A. (1983). A method for measuring length, spatial distribution and distance of living roots *in situ*. *Plant Soil* **73**, 441–445.

Gaastra, P. (1959). Photosynthesis of crop plants as influenced by light, carbon dioxide, temperature, and stomatal diffusion resistance. *Meded. Landbouwhogeschool, Wageningen* **59**, 1–68.

Gaff, D. F. (1971). Desiccation-tolerant flowering plants in Southern Africa. *Science* **174**, 1033–1034.

Gaff, D. F. (1977). Desiccation tolerant vascular plants of South Africa. *Oecologia* **31**, 95–109.

Gaff, D. F., and Churchill, D. M. (1976). *Borya nitida* Labill: An Australian species in the Liliaceae with desiccation-tolerant leaves. *Aust. J. Bot.* **24**, 209–224.

Gaiser, R. N. (1952). Root channels and roots in forest soils. *Soil Sci. Soc. Am. Proc.* **16**, 62–65.

Galau, G. A., Hughes, D. W., and Dure, L., III (1986). Abscisic acid induction of cloned cotton late embryogenesis-abundant (Lea) mRNAs. *Plant Mol. Biol.* **7**, 155–170.

Gale, J., and Hagan, R. M. (1966). Plant antitranspirants. *Annu. Rev. Plant Physiol.* **17**, 269–282.

Gales, K. (1979). Effects of water supply on partitioning of dry matter between roots and shoots in *Lolium perenne. J. Appl. Ecol.* **16**, 863–877.

Gambles, R. L., and Dengler, N. G. (1974). The leaf anatomy of hemlock, *Tsuga canadensis. Can. J. Bot.* **52**, 1049–1056.

Ganns, R. A., Zasada, J. C., and Phillips, C. (1981). Sap production of paper birch in the Tanana Valley, Alaska. *For. Chron.* **58**, 19–22.

Gardner, B. R., Blad, B. L., and Watts, D. G. (1981). Plant and air temperatures in differentially irrigated corn. *Agric. Meteorol.* **25**, 207–217.

Gardner, W. R. (1958). Some steady state solutions of the unsaturated moisture flow equation with application to evaporation from a water table. *Soil Sci.* **85**, 228–237.

Gardner, W. R. (1960). Dynamic aspects of water availability to plants. *Soil Sci.* **89**, 63–73.

Gardner, W. R., and Ehlig, C. F. (1962). Impedance to water movement in soil and plant. *Science* **138**, 522–523.

Gardner, W. R., and Fireman, M. (1958). Laboratory studies of evaporation from soil columns in the presence of a water table. *Soil Sci.* **85**, 244–249.

Gardner, W. R., and Nieman, R. H. (1964). Lower limit of water availability to plants. *Science* **143**, 1460–1462.

Garrity, D. P., Watts, D. G., Sullivan, C. Y., and Gilley, J. R. (1982). Moisture deficits and grain sorghum performance: Evapotranspiration-yield relationships. *Agron. J.* **74**, 815–820.

Gartner, B. L. (1991). Stem hydraulic properties of vines vs. shrubs of western poison oak, *Toxicodendron diversloburn. Oecologia* **87**, 180–189.

Gartner, B. L., Bullock, S. H., Mooney, H. A., Brown, V. B., and Whitbeck, J. L. (1990). Water transport properties of vine and tree stems in a tropical deciduous forest. *Am. J. Bot.* **77**, 742–749.

Gates, D. M. (1968). Transpiration and leaf temperature. *Annu. Rev. Plant Physiol.* **19**, 211–238.

Gates, D. M. (1980). "Biophysical Ecology." Springer-Verlag, New York.

Gerdemann, J. W. (1968). Vesicular-arbuscular mycorrhiza and plant growth. *Annu. Rev. Phytopathology* **6**, 397–418.

Geurten, I. (1950). Untersuchungen über den Gaswechsel von Baumrinden. *Forstwiss. Centrolbl.* **69**, 704–743.

Gibbs, J. W. (1875–1876). On the equilibrium of heterogeneous substances. *Trans. Conn. Acad. Sci.* **III**, 108–248.

Gibbs, J. W. (1931). "The Collected Works of J. Willard Gibbs," Vol. 1. Longmans, Green and Co., New York.

Gibeaut, D. M., and Carpita, N. C. (1991). Tracing cell wall biogenesis in intact cells and plants. *Plant Physiol.* **97**, 551–561.

Giddings, J. L. (1962). Development of tree-ring dating as an archeology aid. *In* "Tree Growth" (T. T. Kozlowski, ed.), pp. 119–132. Ronald Press, New York.

Gifford, R. M. (1979). Growth and yield of CO_2-enriched wheat under water-limited conditions. *Aust. J. Plant Physiol.* **6**, 367–378.

Gilbert, D. E., Myer, J. L., Kessler, J. J., La Vine, P. D., and Carlson, C. V. (1970). Evaporative cooling of vineyards. *Calif. Agric.* **24**, 12–14.

Giles, K. L., Beardsell, M. F., and Cohen, D. (1974). Cellular and ultrastructural changes in mesophyll and bundle sheath cells of maize in response to water stress. *Plant Physiol.* **54**, 208–212.

Giles, K. L., Cohen, D., and Beardsell, M. F. (1976). Effects of water stress on the ultrastructure of leaf cells of *Sorghum bicolor. Plant Physiol.* **57**, 11–14.

Gill, W. R., and Bolt, G. H. (1955). Pfeffer's studies of the root growth pressures exerted by plants. *Agron. J.* **47**, 166–168.

Gimenez, C., Mitchell, V. J., and Lawlor, D. W. (1992). Regulation of photosynthetic rate of two sunflower hybrids under water stress. *Plant Physiol.* **98**, 516–524.

Gimmler, H., and Möller, E.-M. (1981). Salinity-dependent regulation of starch and glycerol metabolism in *Dunaliella parva*. *Plant Cell Environ.* **4**, 367–375.

Gindel, I. (1973). "A New Ecophysiological Approach to Forest-Water Relationships in Arid Climates." Junk, The Hague.

Ginsburg, H., and Ginzburg, B. Z. (1970). Radial water and solute flows in the roots of *Zea mays*. *J. Exp. Bot.* **21**, 580–592.

Givnish, T. J. (1986). Optimal stomatal conductance, allocation of energy between leaves and roots, and the marginal cost of transpiration. *In* "On the Economy of Plant Form and Function" (T. J. Givnish, ed.), pp. 171–213. Cambridge University Press, Cambridge, UK.

Glinski, J., and Lipiec, J. (1990). "Soil Physical Conditions and Plant Roots." CRC Press, Boca Raton, FL.

Glinski, J., and Stepniewski, W. (1985). "Soil Aeration and its Role for Plants." CRC Press, Boca Raton, FL.

Gollan, T., Passioura, J. B., and Munns, R. (1986). Soil water status affects the stomatal conductance of fully turgid wheat and sunflower leaves. *Aust. J. Plant Physiol.* **13**, 459–464.

Gortner, R. A. (1938). "Outlines of Biochemistry." 2nd Ed. Wiley, New York.

Gowing, D. J. G., Davies, W. J., and Jones, H. G. (1990). A positive root-sourced signal as an indicator of soil drying in apple, *Malus x domestica* Borkh. *J. Exp. Bot.* **41**, 1535–1540.

Gowing, D. J. G., Davies, W. J., Trejo, C. L., and Jones, H. G. (1993). Xylem-transported chemical signals and the regulation of plant growth and physiology. *Phil. Trans. Roy. Soc. London Ser. B* **341**, 41–47.

Graan, T., and Boyer, J. S. (1990). Very high CO_2 partially restores photosynthesis in sunflower at low water potentials. *Planta* **181**, 378–384.

Gračanin, M. (1963). Über Unterscheide in der Transpiration von Blattspreite und Stamm. *Phyton* **10**, 216–224.

Grace, J. (1977). "Plant Response to Wind." Academic Press, New York.

Gradmann, H. (1928). Untersuchungen über die Wasserverhältnisse des Bodens als Grundlage des Pflanzenwachstums. *Jahrb. Wiss. Bot.* **69**, 1–100.

Graham, D., and Smillie, R. M. (1976). Carbonate dehydratase in marine organisms of the Great Barrier Reef. *Aust. J. Plant Physiol.* **3**, 113–119.

Graham, J. H., Syvertsen, J. P., and Smith, M. G., Jr. (1987). Water relations of mycorrhizal and phosphorus-fertilized nonmycorrhizal *Citrus* under drought stress. *New Phytol.* **105**, 441–419.

Graham, L. E. (1985). The origin of the life cycle of land plants. *Am Sci.* **73**, 178–186.

Graham, T. (1862). Liquid diffusion applied to analysis. *Phil. Trans. Roy. Soc. London* **151**, 183–224.

Granier, A., Bobay, V., Gash, J. H. C., Gelpe, J., Saugier, B., and Shuttleworth, W. J. (1990). Vapour flux density and transpiration rate comparisons in a stand of Maritime pine (*Pinus pinaster* Ait.) in Les Lander forest. *Agric. For. Meteorol.* **51**, 309–319.

Grant, M. C. (1993). The trembling giant. *Discover* **14**, 82–89.

Grantz, D. A., Graan, T., and Boyer, J. S. (1985a). Chloroplast function in guard cells of *Vicia faba* L.: Measurement of the electrochromic absorbance change at 518 nm. *Plant Physiol.* **77**, 956–962.

Grantz, D. A., Ho, T. D., Uknes, S. J., Cheeseman, J. M., and Boyer, J. S. (1985b). Metabolism of abscisic acid in guard cells of *Vicia faba* L. and *Commelina communis* L. *Plant Physiol.* **78**, 51–56.

Gray, A. (1868). "Lessons in Botany and Vegetable Physiology." Ivison, Blakeman and Taylor, New York.

Greacen, E. L., and Oh, J. S. (1972). Physics of root growth. *Nature New Biol.* **235**, 24–25.

Greaves, J. E., and Carter, E. G. (1923). The influence of irrigation water on the composition of grains and the relationship to nutrition. *J. Biol. Chem.* **58**, 531–541.

Green, J. R. (1914). "A History of Botany in the United Kingdom From the Earliest Times to the End of the 19th Century." J. M. Dent & Sons, Ltd., London.

Green, P. B. (1958). Concerning the site of the addition of new wall substances to the elongating *Nitella* cell wall. *Am. J. Bot.* **45**, 111–116.

Green, P. B., Erickson, R. O., and Buggy, J. (1971). Metabolic and physical control of cell elongation rate: *In vivo* studies in *Nitella*. *Plant Physiol.* **47**, 423–430.

Green, W. N., Ferrier, J. M., and Dainty, J. (1979). Direct measurement of water capacity of *Beta vulgaris* storage tissue sections using a displacement transducer, and resulting values for cell membrane hydraulic conductivity. *Can. J. Bot.* **57**, 921–985.

Greenidge, K. N. H. (1957). Ascent of sap. *Annu. Rev. Plant Physiol.* **8**, 237–256.

Greenidge, K. N. H. (1958). A note on the rates of upward travel of moisture in trees under differing experimental conditions. *Can. J. Bot.* **36**, 357–361.

Greenwood, D. J. (1967). Studies on the transport of oxygen through the stems and roots of vegetable seedlings. *New Phytol.* **66**, 337–347.

Gregoriou, C., and Economides, C. V. (1993). Tree growth, yield and fruit quality of Ortanique tangor in eleven rootstocks in Cyprus. *J. Am. Soc. Hort. Sci.* **118**, 335–338.

Gregory, F. G., and Pearse, H. L. (1934). The resistance porometer and its application to the study of stomatal movement. *Proc. Roy. Soc.* (London) **B114**, 477–493.

Gregory, P. J., Lake, J. V., and Rose, D. A., eds. (1987). "Root Development and Function." Cambridge University Press, Cambridge, England.

Grier, C. C., Vogt, K. A., Keyes, M. R., and Edmonds, R. L. (1981). Biomass distribution and above- and below-ground production in young and mature *Abies amabilis* zone ecosystems of the Washington Cascades. *Can. J. For. Res.* **11**, 155–167.

Gries, G. A. (1943). Juglone (5-hydroxy-1,4-naphthoquinone): A promising fungicide. *Phytopathology* **33**, 1112.

Grimmond, C. S. B., Isard, S. A., and Belding, M. J. (1992). Development and evaluation of continuously weighing mini-lysimeters. *Agric. For. Meteorol.* **62**, 205–218.

Groot, A., and King, K. M. (1992). Measurement of sap flow by the heat balance method: Numerical analysis and application to coniferous seedlings. *Agric. For. Meteorol.* **59**, 289–308.

Grose, M. J., and Hainsworth, J. M. (1992). Soil water extraction, measured by computer-assisted tomography, in seedling *Lupinus angustifolius* cv. Yandee when healthy and infected with *Phytophthora cinnamomi. J. Exp. Bot.* **43**, 121–127.

Grosse, W., Büchel, H. B., and Tiebel, H. (1991). Pressurized ventilation in wetland plants. *Aquatic Bot.* **39**, 89–98.

Grosse, W., Frye, J., and Latterman, S. (1992). Root aeration in wetland trees by pressurized gas transport. *Tree Physiol.* **10**, 285–295.

Grumet, R., and Hanson, A. D. (1986). Genetic evidence for an osmoregulatory function of glycinebetaine accumulation in barley. *Aust. J. Plant Physiol.* **13**, 353–364.

Guerin, V., Trinchant, J.-C., and Rigaud, J. (1990). Nitrogen fixation (C_2H_2 reduction) by broad bean (*Vicia faba* L.) nodules and bacteroids under water-restricted conditions. *Plant Physiol.* **92**, 595–601.

Guerrero, F. D., Jones, J. T., and Mullet, J. E. (1990). Turgor-responsive gene transcription and RNA levels increase rapidly when pea shoots are wilted: Sequence and expression of three inducible genes. *Plant Mol. Biol.* **15**, 11–26.

Gunasekera, D., and Berkowitz, G. A. (1992). Heterogeneous stomatal closure in response to leaf water deficits is not a universal phenomenon. *Plant Physiol.* **98**, 660–665.

Gunasekera, D., and Berkowitz, G. A. (1993). Use of transgenic plants with ribulose-1,5-bisphosphate carboxylase/oxygenase antisense DNA to evaluate the rate limitation of photosynthesis under water stress. *Plant Physiol.* **103**, 629–635.

Gunning, B. E. S., Steer, M. W., and Cochrane, M. P. (1968). Occurrence, molecular structure and induced formation of the "stroma centre" in plastids. *J. Cell Sci.* **3**, 445–456.

Gupta, A. S., and Berkowitz, G. A. (1987). Osmotic adjustment, symplast volume, and nonstomatally mediated water stress inhibition of photosynthesis in wheat. *Plant Physiol.* **85**, 1040–1047.

Gurr, C. G. (1962). Use of gamma rays in measuring water content and permeability in unsaturated columns of soil. *Soil Sci.* **94**, 224–229.

Gurr, C. G., Marshall, T. J., and Hutton, J. T. (1952). Movement of water in soil due to a temperature gradient. *Soil Sci.* **74**, 335–345.

Guttenberger, M., and Hampp, R. (1992). Ectomycorrhizins: Symbiosis-specific or artifactual polypeptides from ectomycorrhizas? *Planta* **188**, 129–142.

Haas, A. R. C. (1948). Effect of the rootstock on the composition of citrus trees and fruit. *Plant Physiol.* **23**, 309–330.

Hagan, R. M. (1949). Autonomic diurnal cycles in the water relations of nonexuding detopped root systems. *Plant Physiol.* **24**, 441–454.

Hagan, R. M., Haise, H. R., and Edminster, T. W., eds. (1967). "Irrigation of Agricultural Lands." Amer. Soc. Agron., Madison, WI.

Hagan, R. M., Vaadia, Y., and Russell, M. B. (1959). Interpretation of plant responses to soil moisture regimes. *Adv. Agron.* **11**, 77–98.

Hajibagheri, M. A., and Flowers, T. J. (1989). X-ray microanalysis of ion distribution within root cortical cells of the halophyte *Suaeda maritima* (L.) Dum. *Planta* **177**, 131–134.

Hajibagheri, M. A., Gilmour, D. J., Collins, J. C., and Flowers, T. J. (1986). X-ray microanalysis and ultrastructural studies of cell compartments of *Dunaliella parva. J. Exp. Bot.* **37**, 1725–1732.

Hales, S. (1727). "Vegetable Staticks." W. & J. Innys and T. Woodward, London. [Reprinted by Scientific Book Guild, London.]

Hall, A. E., Chandler, W. F., Van Bavel, C. H. M., Reid, P. H., and Anderson, J. H. (1953). A tracer technique to measure growth and activity of plant root systems. N.C. Agric. Exp. Sta. Tech. Bull. 101.

Hall, A. E., and Grantz, D. A. (1981). Drought resistance of cowpea improved by selecting for early appearance of mature pods. *Crop Sci.* **21**, 461–464.

Hall, A. E., and Kaufman, M. R. (1975). Stomatal response to environment with *Sesamum indicum* L. *Plant Physiol.* **55**, 455–459.

Hallam, A. (1984). Pre-quaternary sea-level changes. *Annu. Rev. Earth Planet. Sci.* **12**, 205–243.

Hallam, N. D., and Gaff, D. F. (1978a). Regeneration of chloroplast structure in *Talbotia elegans:* A desiccation tolerant plant. *New Phytol.* **81**, 657–662.

Hallam, N. D., and Gaff, D. F. (1978b). Re-organization of fine structure during rehydration of desiccated leaves of *Xerophyta villosa. New Phytol.* **81**, 349–355.

Ham, J. M., and Heilman, J. L. (1990). Dynamics of a heat balance stem flow gauge during high flow. *Agron. J.* **82**, 147–152.

Hamada, S., Ezaki, S., Hayashi, K., Toko, K., and Yamafuji, K. (1992). Electric current precedes emergence of a lateral root in higher plants. *Plant Physiol.* **100**, 614–619.

Hamblin, A. P. (1985). The influence of soil structure on water movement, crop root growth and water uptake. *Adv. Agron.* **38**, 95–158.

Hammel, H. T. (1967). Freezing of xylem sap without cavitation. *Plant Physiol.* **42**, 55–66.

Hammel, H. T. (1976). Colligative properties of a solution. *Science* **192**, 748–756.

Hammel, H. T., and Scholander, P. F. (1976). "Osmosis and Tensile Solvent." Springer-Verlag, Berlin.

Hangarter, R. P., Grandoni, P., and Ort, D. R. (1987). The effects of chloroplast coupling factor reduction on the energetics and efficiency of ATP formation. *J. Biol. Chem.* **262**, 13513–13519.

Hanson, A. D., and Hitz, W. D. (1982). Metabolic responses of mesophytes to plant water deficits. *Annu. Rev. Plant Physiol.* **33**, 163–203.

Hanson, B. R., and Dickey, G. L. (1993). Field practices affect neutron moisture meter accuracy. *Calif Agric.* **47**, 29–31.

Hanson, P. J., Sucoff, E. I., and Markhart, A. H., III (1985). Quantifying apoplastic flux through red pine root systems using trisodium, 3-hydroxy-5,8,10-pyrenetrisulfonate. *Plant Physiol.* **77**, 21–24.

Harada, J. J., DeLisle, A. J., Baden, C. S., and Crouch, M. L. (1989). Unusual sequence of an abscisic acid-inducible mRNA which accumulates late in *Brassica napus* seed development. *Plant Mol. Biol.* **12**, 395–401.

Harborne, J. B. (1988). "Introduction to Ecological Biochemistry," 3rd Ed. Academic Press, London.

Harding, R. B., Miller, M. P., and Fireman, M. (1958). Absorption of salts by citrus leaves during sprinkling with water suitable for surface irrigation. *Proc. Am. Soc. Hort. Sci.* **71**, 248–256.

Harley, J. L., and Smith, S. E. (1983). "Mycorrhizal Symbiosis." Academic Press, New York.

Harmond, U., Schaesberg, N., Graham, J., and Syvertsen, J. (1987). Salinity and flooding stress effects on VAM and non-VAM citrus rootstock seedlings. *Plant Soil* **104**, 37–43.

Harris, D. G. (1971). Simultaneous measurements of solution absorption, transpiration, relative plant water content and growth. *Agron. J.* **63**, 840–845.

Harris, J. A. (1934). "The Physico-chemical Properties of Plant Saps in Relation to Phytogeography." University of Minnesota Press, Minnesota.

Harris, M. J., and Outlaw, W. H., Jr. (1991). Rapid adjustment of guard-cell abscisic acid levels to current leaf-water status. *Plant Physiol.* **95**, 171–173.

Harris, W. F., Kinerson, R. S., Jr., and Edwards, N. T. (1977). Comparison of below-ground biomass of natural deciduous forests and loblolly pine plantations. Range Sci. Dep. Sci. Ser. (Colo. State Univ.), Fort Collins, CO.

Hartenstein, R. (1986). Earthworm biotechnology and global biogeochemistry. *Adv. Ecol. Res.* **15**, 379–409.

Hartwig, U., Boller, B. C., Baur-Hoch, B., and Nosberger, J. (1990). The influence of carbohydrate reserves in the response of nodulated white clover to defoliation. *Ann. Bot.* **65**, 97–105.

Hasegawa, S. (1986). Changes in soil water contents in the vicinity of soybean roots. *Trans. 13th Congr. Int. Soc. Soil Sci.* **2**, 73–74.

Hasegawa, S., and Sato, T. (1987). Water uptake by roots in cracks and water movement in clayey subsoil. *Soil Sci.* **143**, 381–386.

Hashimoto, Y., Ino, T., Kramer, P. J., Naylor, A. W., and Strain, B. R. (1984). Dynamic analysis of water stress of sunflower leaves by means of a thermal image processing system. *Plant Physiol.* **76**, 266–269.

Hashimoto, Y., Kramer, P. J., Nonami, H., and Strain, B. R., eds. (1990). "Measurement Techniques in Plant Science." Academic Press, San Diego.

Hatch, M. D., and Burnell, J. N. (1990). Carbonic anhydrase activity in leaves and its role in the first step of C_4 photosynthesis. *Plant Physiol.* **93**, 825–828.

Häussling, M., Jorns, C. A., Lehmbecker, G., Hecht-Buchholz, C., and Marschner, H. (1988). Ion and water uptake in relation to root development in Norway spruce (*Picea abies* (L.) Karst). *J. Plant Physiol.* **133**, 486–491.

Hayashi, T. (1989). Xyloglucans in the primary cell wall. *Annu. Rev. Plant Physiol. Plant Mol. Biol.* **40**, 139–168.

Hayward, H. E., and Long, E. M. (1942). The anatomy of the seedling and roots of the Valencia orange. U.S. Dep. Agric. Tech. Bull. 786.

Hayward, H. E., Long, E. M., and Uhvits, R. (1946). Effect of chloride and sulfate salts on the growth and development of the Elberta peach on Shalil and Lovell rootstocks. U.S. Dept. Agr. Tech. Bull. 922.

Hayward, H. E., and Spurr, W. B. (1943). Effects of osmotic concentration of substrate on the entry of water into corn roots. *Bot. Gaz.* **105**, 152–164.

Hayward, H. E., and Spurr, W. (1944). Effects of isosmotic concentrations of inorganic and organic substrates on entry of water into corn roots. *Bot. Gaz.* **106**, 131–139.

Head, G. C. (1964). A study of "exudation" from the root hairs of apple roots by time-lapse cinephotomicrography. *Ann. Bot.* **28**, 495–498.

Head, G. C. (1967). Effects of seasonal changes in shoot growth on the amount of unsuberized root on apple and plum trees. *J. Hort. Sci.* **42**, 169–180.

Hecht-Buchholz, Ch., and Foy, C. D. (1981). Effect of aluminum toxicity on root morphology of barley. *In* "Structure and Function of Plant Roots" (R. Brouwer, O. Gasparikova, J. Kolek, and B. G. Loughman, eds.), pp. 343–345. Nijhoff/Junk, The Hague.

Heeramon, D. A., and Juma, N. G. (1993). A comparison of minirhizotron, core, and monolith methods for quantifying barley (*Hordeum vulgare* L.) and faba bean (*Vicia faba* L.) root distribution. *Plant Soil* **148**, 29–41.

Hegde, R. S., and Miller, D. A. (1990). Allelopathy and autotoxicity in alfalfa: Characterization and effects of preceding crops and residue incorporation. *Crop Sci.* **30**, 1255–1259.

Hegde, R. S., and Miller, D. A. (1992). Scanning electron microscopy for studying root morphology and anatomy in alfalfa autotoxicity. *Agron. J.* **84**, 618–620.

Heimovaara, T. J. (1993). Design of triple-wire time domain reflectometry probes in practice and theory. *J. Soil Sci. Soc. Am.* **57**, 1410–1417.

Hellkvist, J., Richards, G. P., and Jarvis, P. G. (1974). Vertical gradients of water potential and tissue water relations in Sitka spruce trees measured with the pressure chamber. *J. Appl. Ecol.* **11**, 637–667.

Hellmers, H., Horton, J. S., Juhren, G., and O'Keefe, J. (1955). Root systems of some chaparral plants in southern California. *Ecology* **36**, 667–678.

Henderson, L. J. (1913). "The Fitness of the Environment." The Macmillan Co., New York.

Hepler, P. K., and Wayne, R. O. (1985). Calcium and plant development. *Annu. Rev. Plant Physiol.* **36**, 397–439.

Herkelrath, W. N., Miller, E. E., and Gardner, W. R. (1977). Water uptake by plants. II. The root contact model. *Soil Sci. Soc. Am. J.* **41**, 1039–1043.

Herrero, M. P., and Johnson, R. R. (1981). Drought stress and its effect on maize reproductive systems. *Crop Sci.* **21**, 105–110.

Hershey, D. R. (1990). Container-soil physics and plant growth. *BioScience* **40**, 685.

Hewlett, J. D. (1961). Soil moisture as a source of base flow from steep mountain watersheds. U.S. Dept. Agr. Forest Serv., Southeastern Forest Exp. Sta. Paper 132.

Hewlett, J. D., and Hibbert, A. R. (1961). Increases in water yield after several times of forest cutting. *Int. Assoc. Sci. Hydrol. Bull.* **6**, 5–17.

Heyl, J. G. (1933). Der Einfluss von Aussenfaktoren auf das Bluten der Pflanzen. *Planta* **20**, 294–353.

Hilbert, J.-L., Costa, G., and Martin, F. (1991). Ectomycorrhizin synthesis and polypeptide changes during the early stage of Eucalypt mycorrhiza development. *Plant Physiol.* **97**, 977–984.

Hiler, E. A., and Clark, R. N. (1971). Stress day index to characterize effects of water stress on crop yields. *Trans. A.S.A.E.* **14**, 757–761.

Hillel, D. (1980a). "Fundamentals of Soil Physics." Academic Press, New York.

Hillel, D. (1980b). "Applications of Soil Physics." Academic Press, New York.

Hirasawa, T., Gotou, T., and Ishihara, K. (1992). On resistance to water transport from roots to the leaves at the different positions on a stem in rice plants. *Japan. J. Crop Sci.* **61**, 153–158.

Hirasawa, T., Iida, Y., and Ishihara, K. (1988). Effect of leaf water potential and air humidity on photosynthetic rate and diffusive conductance in rice plants. *Japan. J. Crop Sci.* **57**, 112–118.

Hite, D. R. C., Outlaw, W. H., Jr., and Tarczynski, M. C. (1993). Elevated levels of both sucrose-phosphate synthase and sucrose synthase in *Vicia* guard cells indicate cell-specific carbohydrate interconversions. *Plant Physiol.* **101**, 1217–1221.

Ho, T. Y., and Mishkind, M. L. (1991). The influence of water deficits on mRNA levels in tomato. *Plant Cell Environ.* **14**, 67–75.

Hodgson, R. H. (1953). "A Study of the Physiology of Mycorrhizal Roots on *Pinus taeda* L." M.A. Thesis, Duke University, Durham, NC.

Hodnett, M. G. (1986). The neutron probe for soil moisture measurement. *In* "Advanced Agricultural Instrumentation" (W. G. Gensler, ed.), pp. 148–192. M. Nijhoff Publisher, Dordrecht.

Hoffman, G. T., Rawlins, S. L., Garber, M. J., and Cullen, E. M. (1971). Water relations and growth of cotton as influenced by salinity and relative humidity. *Agron. J.* **63**, 822–826.

Hofmeister, W. (1862). Ueber Spannung, Ausflusemenge und Ausflusegeschwindigheit von Säften lebender Pflanzen. *Flora* **45**, 97–108, 113–120, 138–144, 145–152, 170–175.

Holbrook, N. M., Burns, M. J., and Sinclair, T. R. (1992). Frequency and time-domain dielectric measurements of stem water content in the arborescent palm, *Sabal palmetto. J. Exp. Bot.* **43**, 111–119.

Holch, A. E. (1931). Development of roots and shoots of certain deciduous tree seedlings in different forest sites. *Ecology* **12**, 259–298.

Hole, D. J., Cobb, B. G., Hole, P. S., and Drew, M. C. (1992). Enhancement of anaerobic respiration in root tips of *Zea mays* following low-oxygen (hypoxic) acclimation. *Plant Physiol.* **99**, 213–218.

Hollaender, A. (ed.) (1956). "Radiation Biology." McGraw-Hill, New York.

Holland, H. D. (1978). "The Chemistry of Oceans and Atmospheres." Wiley, New York.

Holland, H. D. (1984). "The Chemical Evolution of the Atmosphere and Oceans." Princeton University Press, Princeton, NJ.

Holmgren, P., Jarvis, P. G., and Jarvis, M. S. (1965). Resistances to carbon dioxide and water vapour transfer in leaves of different plant species. *Physiol. Plant.* **18**, 557–573.

Honert, T. H. van den (1948). Water transport in plants as a catenary process. *Disc. Faraday Soc.* **3**, 146–153.

Hong, B., Uknes, S. J., and Ho, T.-H. D. (1988). Cloning and characterization of a cDNA encoding a mRNA rapidly-induced by ABA in barley aleurone layers. *Plant Mol. Biol.* **11**, 495–506.

Hook, D. D., Brown, C. L., and Kormanik, P. P. (1971). Inductive flood tolerance in swamp tupelo (*Nyssa sylvatica* var. *biflora* [Walt.] Sarg.). *J. Exp. Bot.* **22**, 78–89.

Hoover, M. D. (1944). Effect of removal of forest vegetation upon water yield. *Trans. Am. Geophys. Union* **25**, 969–977.

Hoover, M. D. (1949). Hydrologic characteristics of South Carolina Piedmont forest soils. *Proc. Soil Sci. Soc. Am.* **14**, 353–358.

Hooymans, J. J. M. (1969). The influence of the transpiration rate on uptake and transport of potassium in barley plants. *Planta* **88**, 369–371.

Horst, W. J., Wagner, A., and Marschner, H. (1982). Mucilage protects root meristems from aluminum injury. *Z. Pflanzenphysiol.* **105**, 435–444.

Hough, W. A., Woods, F. W., and McCormack, M. L. (1965). Root extension of individual trees in surface soils of a natural longleaf pine-turkey oak stand. *For. Sci.* **11**, 223–242.

Howard, A. (1925). The effect of grass on trees. *Roy. Soc. London Proc.* **B97**, 284–321.

Hsiao, T. C. (1970). Rapid changes in levels of polyribosomes in *Zea mays* in response to water stress. *Plant Physiol.* **46**, 281–285.

Hsiao, T. C. (1973). Plant responses to water stress. *Annu. Rev. Plant Physiol.* **24**, 519–570.

Hsiao, T. C., Acevedo, E., and Henderson, D. W. (1970). Maize leaf elongation: Continuous measurements and close dependence on plant water status. *Science* **168**, 590–591.

Hsiao, T. C., Acevedo, E., Fereres, E., and Henderson, D. W. (1976). Water stress, growth, and osmotic adjustment. *Phil. Trans. Roy. Soc. London Ser. B* **273**, 479–500.

Hsiao, T. C., Allaway, W. G., and Evans, L. T. (1973). Action spectra for guard cell Rb$^+$ uptake and stomatal opening in *Vicia faba. Plant Physiol.* **51**, 82–88.

Hsu, S.-T., and Goodman, R. N. (1978). Production of a host-specific, wilt-inducing toxin in apple cell suspension cultures inoculated with *Erwinia amylovora. Phytopathology* **68**, 351–354.

Huang, C. Y., Boyer, J. S., and Vanderhoef, L. N. (1975a). Acetylene reduction (nitrogen fixation) and metabolic activities of soybean having various leaf and nodule water potentials. *Plant Physiol.* **56**, 222–227.

Huang, C. Y., Boyer, J. S., and Vanderhoef, L. N. (1975b). Limitation of acetylene reduction (nitrogen fixation) by photosynthesis in soybean having low water potentials. *Plant Physiol.* **56**, 228–232.

Huang, R. S., Smith, W. K., and Yost, R. S. (1985). Influence of vesicular-arbuscular mycorrhiza on growth, water relations, and leaf orientation in *Leucaena leucocephala* (Lam.) DeWit. *New Phytol.* **99**, 229–243.

Huang, Y.-H., and Morris, J. T. (1991). Evidence for hygrometric pressurization in the internal gas space of *Spartina alterniflora*. *Plant Physiol.* **96**, 166–171.

Huber, B. (1924). Die Beurteilung des Wasserhaushaltes der Pflanze. *Jahrb. Wiss. Bot.* **64**, 1–120.

Huber, B. (1932). Beobachtung und Messung pflanzlichen Saftströme. *Ber. deut. bot. Ges.* **50**, 89–109.

Huber, B. (1935). Die physiologische Bedeutung der Ring- und Zerstreutporigkeit. *Ber. deut. bot. Ges.* **53**, 711–719.

Huber, B. (1956). Die Transpiration von Sprossachsen und anderen nicht foliosen Organen. *In* "Encycl. Plant Physiol.," Vol. 3, pp. 427–435. Springer-Verlag, Berlin.

Huber, B., and Schmidt, E. (1937). Eine Kompensations-methode zur thermoelektrischen Messung Langsamer Saftstrome. *Ber. deut. bot. Ges.* **50**, 514–529.

Hubick, K., and Farquhar, G. (1990). Carbon isotope discrimination and the ratio of carbon gained to water lost in barley cultivars. *Plant Cell Environ.* **12**, 795–804.

Hubick, K. T., Farquhar, G. D., and Shorter, R. (1986). Correlation between water-use efficiency and carbon isotope discrimination in diverse peanut (*Arachis*) germplasm. *Aust. J. Plant Physiol.* **13**, 803–816.

Huck, M. G. (1970). Variation in taproot elongation rate as influenced by composition of the soil air. *Agron. J.* **62**, 815–818.

Huck, M. G. (1983). Root distribution, growth, and activity with reference to agroforestry. *In* "Plant Research and Agroforestry" (P. A. Huxley, ed.), pp. 527–542. Int. Council for Res. in Agroforestry, Nairobi, Kenya.

Huck, M. G., Klepper, B., and Taylor, H. M. (1970). Diurnal variations in root diameter. *Plant Physiol.* **45**, 529–530.

Huckenpahler, B. J. (1936). Amount and distribution of moisture in a living shortleaf pine. *J. For.* **34**, 399–401.

Hudson, J. P. (1960). Relations between root and shoot growth in tomatoes. *Sci. Hort.* **14**, 49–54.

Huffaker, R. C., Radin, T., Kleinkopf, G. E., and Cox, E. L. (1970). Effects of mild water stress on enzymes of nitrate assimilation and of the carboxylative phase of photosynthesis in barley. *Crop Sci.* **10**, 471–474.

Hull, H. M., Wright, L. N., and Bleckmann, C. A. (1978). Epicuticular wax ultrastructure among lines of *Eragrostis lehmanniana* Nees developed for seedling drouth tolerance. *Crop Sci.* **18**, 699–704.

Hulugalle, N. R., and Willatt, S. T. (1983). The role of soil resistance in determining water uptake by plant root systems. *Aust. J. Soil Res.* **21**, 571–574.

Hunt, S., King, B. J., Canvin, D. T., and Layzell, D. B. (1987). Steady and nonsteady state gas exchange characteristics of soybean nodules in relation to the oxygen diffusion barrier. *Plant Physiol.* **84**, 164–172.

Hunt, S., and Layzell, D. B. (1993). Gas exchange of legume nodules and the regulation of nitrogenase activity. *Annu. Rev. Plant Physiol. Plant Mol. Biol.* **44**, 483–511.

Hurd, E. A. (1974). Phenotype and drought tolerance in wheat. *Agric. Meteorol.* **14**, 39–55.

Hüsken, D., Steudle, E., and Zimmermann, U. (1978). Pressure probe technique for measuring water relations of cells in higher plants. *Plant Physiol.* **61**, 158–163.

Hylmö, B. (1953). Transpiration and ion absorption. *Physiol. Plant.* **6**, 333–405.

Idso, S. B., and Baker, D. G. (1967). Relative importance of reradiation, convection, and transpiration in heat transfer from plants. *Plant Physiol.* **42**, 631–640.

Idso, S. B., Clawson, K. L., and Anderson, M. G. (1986). Foliage temperature: Effects on environmental factors with implications for plant water stress assessment and the CO_2 climate connection. *Water Resour. Res.* **22**, 1702–1716.

Ihle, J. N., and Dure, L., III (1972). Developmental biochemistry of cottonseed embryogenesis and germination. III. Regulation of the biosynthesis of enzymes utilized in germination. *J. Biol. Chem.* **247**, 5048–5055.

Imber, D., and Tal, M. (1970). Phenotypic reversion of flacca, a wilty mutant of tomato, by abscisic acid. *Science* **169**, 592–593.

Imsande, J., and Touraine, B. (1994). N demand and the regulation of nitrate uptake. *Plant Physiol.* **105**, 3–7.

Incoll, L. D., Long, S. P., and Ashmore, M. R. (1977). SI units in publications in plant science. *Curr. Adv. Plant Sci.* **9**, 331–343.

Ingle, J., Joy, K. W., and Hageman, R. H. (1966). The regulation of activity of the enzymes involved in the assimilation of nitrate by higher plants. *Biochem. J.* **100**, 577–588.

Inoue, Y., Kimball, B. A., Jackson, R. D., and Pinter, P. J., Jr. (1990). Remote estimation of leaf transpiration rate and stomatal resistance based on IR thermometry. *Agr. For. Meteorol.* **51**, 21–34.

Irigoyen, J. J., Emerich, D. W., and Sánchez-Díaz, M. (1992a). Phosphoenolpyruvate carboxylase, malate and alcohol dehydrogenase activities in alfalfa (*Medicago sativa*) nodules under water stress. *Physiol. Plant.* **84**, 61–66.

Irigoyen, J. J., Emerich, D. W., and Sánchez-Díaz, M. (1992b). Water stress induced changes in concentrations of proline and total soluble sugars in nodulated alfalfa (*Medicago sativa*) plants. *Physiol. Plant.* **84**, 55–60.

Irigoyen, J. J., Sánchez-Díaz, M., and Emerich, D. W. (1992c). Transient increase of anaerobically-induced enzymes during short-term drought of alfalfa root nodules. *J. Plant Physiol.* **139**, 397–402.

Isebrands, J. G., Promnitz, L. C., and Dawson, D. H. (1977). Leaf area development in short rotation intensive cultured *Populus* plots. TAPPI For. Biol. Wood Chem. Conf. (Conf. Pap.), pp. 201–209.

Ishihara, K., Iida, O., Hirasawa, T., and Ogura, T. (1978). Relationship between potassium content in leaf blades and stomatal aperture in rice plants. *Japan. J. Crop Sci.* **47**, 719–720.

Israel, D. W., Giddens, J. E., and Powell, W. W. (1973). The toxicity of peach tree roots. *Plant Soil* **39**, 103–112.

Itai, C., and Vaadia, Y. (1965). Kinetin-like activity in root exudate of water-stressed sunflower plants. *Physiol. Plant.* **18**, 941–944.

Itoh, S., and Barber, S. A. (1983). Phosphorus uptake by six plant species as related to root hairs. *Agron. J.* **75**, 457–461.

Ivanov, V. B. (1980). Specificity of spatial and time organization of root cell growth in connection with functions of the root. *Soviet Plant Physiol.* **26**(5, P+1), 720–728.

Iwata, S., Tabuchi, T., and Warkentin, B. P. (1988). "Soil-Water Interactions: Mechanisms and Applications." Dekker, New York.

Jaafar, M. N., Stone, L. R., and Goodrum, D. E. (1993). Rooting depth and dry matter development of sunflower. *Agron. J.* **85**, 281–286.

Jachetta, J. J., Appleby, A. P., and Boersma, L. (1986). Use of the pressure vessel to measure concentrations of solutes in apoplastic and membrane-filtered symplastic sap in sunflower leaves. *Plant Physiol.* **82**, 995–999.

Jackson, M. B. (1985). Ethylene and responses of plants to soil waterlogging and submergence. *Annu. Rev. Plant Physiol.* **36**, 145–174.

Jackson, M. B. (1991). Regulation of water relationships in flooded plants by ABA from leaves, roots and xylem sap. *In* "Abscisic Acid" (W. J. Davies and H. G. Jones, eds.), pp. 217–226. Bios Scientific Publications, Oxford, U.K.

Jackson, M. B., Davies, D. D., and Lambers, H. (eds.) (1991). "Plant Life under Oxygen Deprivation." SPB Publ. Co., The Hague.

Jackson, M. B., and Hall, K. C. (1987). Early stomatal closure in waterlogged pea plants is mediated by abscisic acid in the absence of foliar water deficits. *Plant Cell Environ.* **10**, 121–130.

Jackson, M. B., Herman, B., and Goodenough, A. (1982). An examination of the importance of ethanol in causing injury to flooded plants. *Plant Cell Environ.* **5**, 163–172.

Jackson, M. B., Young, S. F., and Hall, K. C. (1988). Are roots a source of abscisic acid for the shoots of flooded pea plants? *J. Exp. Bot.* **39**, 1631–1637.

Jackson, R. D. (1982). Canopy temperature and plant water stress. *Adv. Irrig.* **7**, 43–85.

Jackson, W. T. (1955). The role of adventitious roots in recovery of shoots following flooding of the original root systems. *Am. J. Bot.* **42**, 816–819.

Jacobsen, J. V., Hanson, A. D., and Chandler, P. C. (1986). Water stress enhances expression of an α-amylase gene in barley leaves. *Plant Physiol.* **80**, 350–359.

Jacoby, G. C., Jr., Sheppard, P. R., and Sieh, K. E. (1988). Irregular occurrence of large earthquakes along the San Andreas fault: Evidence from trees. *Science* **241**, 196–199.

Jamison, V. C. (1946). The penetration of irrigation and rain water into sandy soils of central Florida. *Soil Sci. Soc. Am. Proc.* **10**, 25–29.

Janes, B. E. (1948). The effect of varying amounts of irrigation on the composition of two varieties of snap beans. *Proc. Amer. Soc. Hort. Sci.* **51**, 457–462.

Janos, D. P. (1980). Vesicular-arbuscular mycorrhizae affect lowland tropical rain forest plant growth. *Ecology* **61**, 151–162.

Jäntti, A., and Kramer, P. J. (1957). Regrowth of pastures in relation to soil moisture and defoliation. *In* "Proc. Seventh Int. Grasslands Congr.," pp. 1–12.

Jarvis, P. G. (1985). Transpiration and assimilation of tree and agricultural crops: The "omega" factor. *In* "Trees as Crop Plants" (M. G. R. Cornell and J. E. Jackson, eds.), pp. 461–480. Inst. Terrestrial Ecol., Huntingdon, England.

Jarvis, P. G., and Mansfield, T. A., eds. (1981). "Stomatal Physiology." Cambridge University Press, Cambridge, UK.

Jarvis, P. G., and McNaughton, K. G. (1986). Stomatal control of transpiration: Scaling up from leaf to region. *Adv. Ecol. Res.* **15**, 1–49.

Jarvis, P. G., Rose, C. W., and Begg, J. E. (1967). An experimental and theoretical comparison of viscous and diffusive resistances to gas flow through amphistomatous leaves. *Agr. Meteorol.* **4**, 103–117.

Jarvis, P. G., and Slatyer, R. O. (1966). A controlled environment chamber for studies of gas exchange by each surface of a leaf. CSIRO Div. Land Res., Tech. Paper 29.

Jarvis, P. G., and Slatyer, R. O. (1970). The role of the mesophyll cell wall in leaf transpiration. *Planta* **90**, 303–322.

Jeffree, C. E., Johnson, R. P. C., and Jarvis, P. G. (1971). Epicuticular wax in the stomatal antechamber of Sitka spruce and its effects on the diffusion of water vapour and carbon dioxide. *Planta* **98**, 1–10.

Jemison, G. M. (1944). The effect of basal wounding by forest fires on the diameter growth of some southern Appalachian hardwoods. Duke Univer. School Forestry Bull. 9.

Jenks, M. A., Joly, R. J., Peters, P. J., Rich, P. J., Axtell, J. D., and Ashworth, E. N. (1994). Chemically induced cuticle mutation affecting epidermal conductance to water vapor and disease susceptibility in *Sorghum bicolor* (L.) Moench. *Plant Physiol.* **105**, 1239–1245.

Jenne, E. A., Rhoades, H. F., Yien, C. H., and Howe, O. W. (1958). Change in nutrient element accumulation by corn with depletion of soil moisture. *Agron. J.* **50**, 71–74.

Jenner, C. F., Xia, Y., Eccles, C. D., and Callaghan, P. T. (1988). Circulation of water within wheat grain revealed by nuclear magnetic resonance micro-imaging. *Nature* (London) **336**, 399–402.

Jenniskens, P., and Blake, D. F. (1994). Structural transitions in amorphous water ice and astrophysical implications. *Science* **265**, 753–756.

Jenny, H., and Grossenbacher, K. (1963). Root-soil boundary zones as seen in the electron micro-scope. *Soil Sci. Soc. Am. Proc.* **27**, 273–277.

Jensen, S. D., and Cavalieri, A. J. (1983). Drought tolerance in U.S. maize. *In* "Plant Production and Management under Drought Conditions" (J. F. Stone and W. O. Willis, eds.), pp. 223–236. Developments in Agricultural and Managed-Forest Ecology 12, Elsevier Science Publishers, New York.

Johansen, A., Jacobsen, I., and Jensen, E. S. (1993). External hyphae of vesicular-arbuscular mycor-rhizal fungi associated with *Trifolium subterraneum* L. 3. Hyphal transport of ^{32}P and ^{15}N. *New Phytol.* **124**, 61–68.

Johns, D., Beard, J. B., and Van Bavel, C. H. M. (1983). Resistances to evapotranspiration from a St. Augustinegrass turf canopy. *Agron. J.* **75**, 419–422.

Johnson, G. A., Brown, J., and Kramer, P. J. (1987). Magnetic resonance microscopy of changes in water content in stems of transpiring plants. *Proc. Natl. Acad. Sci. USA* **84**, 2752–2755.

Johnson, H. B. (1975). Plant pubescence: An ecological perspective. *Bot. Rev.* **41**, 233–258.

Johnson, I. R., Melkonian, J. J., Thornley, J. H. M., and Riha, S. J. (1991). A model of water flow through plants incorporating shoot/root "message" control of stomatal conductance. *Plant Cell Environ.* **14**, 531–544.

Johnson, I. R., and Thornley, J. H. M. (1987). A model of shoot:root partitioning with optimal growth. *Ann. Bot.* **60**, 133–142.

Johnson, J. F., Allan, D. L., and Vance, C. P. (1994). Phosphorus stress-induced proteoid roots show altered metabolism in *Lupinus albus*. *Plant Physiol.* **104**, 657–665.

Johnson, J. O. (1984). A rapid technique for estimating total surface area of pine needles. *For. Sci.* **30**, 913–921.

Johnson, L. P. V. (1945). Physiological studies on sap flow in the sugar maple, *Acer saccharum* Marsh. *Can. J. Res. Sec. C.* **23**, 192–197.

Johnson, N. C., Copeland, P. J., Crookston, R. K., and Pfleger, F. L. (1992). Mycorrhizae: Possible explanation for yield decline with continuous corn and soybean. *Agron. J.* **84**, 387–390.

Johnson, R. R., Frey, N. M., and Moss, D. N. (1974). Effect of water stress on photosynthesis and transpiration of flag leaves and spikes of barley and wheat. *Crop Sci.* **14**, 728–731.

Johnson, R. R., and Moss, D. N. (1976). Effect of water stress on $^{14}CO_2$ fixation and translocation in wheat during grain filling. *Crop Sci.* **16**, 697–701.

Johnson, R. W., Tyree, M. T., and Dixon, M. A. (1987). A requirement for sucrose in xylem sap flow from dormant maple trees. *Plant Physiol.* **84**, 495–500.

Jones, C. G., Edson, A. W., and Morse, W. J. (1903). The maple sap flow. Bull. 103, Vermont Agric. Exp. Sta.

Jones, H. G., Hamer, P. J. C., and Higgs, K. H. (1988a). Evaluation of various heat-pulse methods for estimation of sap flow in orchard trees: Comparison with micro-meteorological estimates of evaporation. *Trees* **2**, 250–260.

Jones, H. G., and Higgs, K. H. (1980). Resistance to water loss from the mesophyll cell surface in plant leaves. *J. Exp. Bot.* **31**, 545–553.

Jones, H. G., Leigh, R. A., Wyn Jones, R. G., and Tomos, A. D. (1988b). The integration of whole-root and cellular hydraulic conductivities in cereal roots. *Planta* **174**, 1–7.

Jones, H. G., Tomos, A. D., Leigh, R. A., and Wyn Jones, R. G. (1983). Water-relation parame-ters of epidermal and cortical cells in the primary root of *Triticum aestivum* L. *Planta* **158**, 230–236.

Jones, M. M., and Turner, N. C. (1978). Osmotic adjustment in leaves of sorghum in response to water deficits. *Plant Physiol.* **61**, 122–126.

Jordan, W. R., Miller, F. R., and Morris, D. E. (1979). Genetic variation in root and shoot growth of sorghum in hydroponics. *Crop Sci.* **19**, 468–472.

Jordan, W. R., and Ritchie, J. T. (1971). Influence of soil water stress on evaporation, root absorp-tion, and internal water status of cotton. *Plant Physiol.* **48**, 783–788.

Joshi, M. C., Boyer, J. S., and Kramer, P. J. (1965). Growth, carbon dioxide exchange, transpiration, and transpiration ratio of pineapple. *Bot. Gaz.* **126**, 174–179.

Jupp, A. P., and Newman, E. I. (1987). Morphological and anatomical effects of severe drought on the roots of *Lolium perenne* L. *New Phytol.* **105**, 393–402.

Jurgens, S. K., Johnson, R. R., and Boyer, J. S. (1978). Dry matter production and translocation in maize subjected to drought during grain fill. *Agron. J.* **70**, 678–682.

Kaiser, W. M. (1987). Effects of water deficit on photosynthetic capacity. *Physiol. Plant.* **71**, 142–149.

Kaiser, W. M., Schröppel-Meier, G., and Wirth, E. (1986). Enzyme activities in an artificial stroma medium: An experimental model for studying effects of dehydration on photosynthesis. *Planta* **167**, 292–299.

Kalela, A. (1954). Mantysiemenpuiden japuustojen juuroisuhteista (On root relations of pine seed trees). *Acta For. Fenn.* **61**, 1–17.

Kamiya, N., and Tazawa, M. (1956). Studies on water permeability of a single plant cell by means of transcellular osmosis. *Protoplasma* **46**, 394–422.

Kamiya, N., Tazawa, M., and Takata, T. (1963). The relation of turgor pressure to cell volume in *Nitella* with special reference to mechanical properties of the cell wall. *Protoplasma* **57**, 501–521.

Kanemasu, E. T., Thurtell, G. W., and Tanner, C. B. (1969). Design, calibration and field use of a stomatal diffusion porometer. *Plant Physiol.* **44**, 881–885.

Kargol, M. (1992). The graviosmotic hypothesis of xylem transport of water in plants. *Gen. Physiol. Biophys.* **11**, 469–487.

Karsten, K. S. (1939). Root activity and the oxygen requirement in relation to soil fertility. *Am. J. Bot.* **26**, 855–860.

Kasuga, M. C. M., Muchovej, R. M. C., and Muchovej, J. J. (1990). Influence of aluminum on *in vitro* formation of *Pinus caribaea* mycorrhizae. *Plant Soil* **124**, 73–77.

Katz, C., Oren, R., Schulze, E.-D., and Milburn, J. A. (1989). Uptake of water and solutes through twigs of *Picea abies* (L.) Karst. *Trees* **3**, 33–37.

Kaufmann, M. R. (1968). Water relations of pine seedlings in relation to root and shoot growth. *Plant Physiol.* **43**, 281–288.

Kaufmann, M. R. (1976). Stomatal response of Engelmann spruce to humidity, light, and water stress. *Plant Physiol.* **57**, 899–901.

Kaufmann, M. R. (1981). Automatic determination of conductance, transpiration, and environmental conditions in forest trees. *For. Sci.* **27**, 817–827.

Kaufmann, M. R. (1982). Evaluation of season, temperature, and water stress effects on stomata using a leaf conductance model. *Plant Physiol.* **69**, 1023–1026.

Kaufmann, M. R. (1984). A canopy model (RM-CWU) for determining transpiration of subalpine forests. I. Model development. *Can. J. For. Res.* **14**, 218–226.

Kaufmann, M. R. (1985). Annual transpiration in subalpine forests: Large differences among four tree species. *For. Ecol. & Mngt.* **13**, 235–246.

Kaufmann, M. R., and Fiscus, E. L. (1985). Water transport through plants: Internal integration of edaphic and atmospheric effects. *Acta Hort.* **171**, 83–93.

Kaufmann, M. R., and Landsberg, J. J., eds. (1991). "Advancing Toward Closed Models of Forest Ecosystems." Heron Publishing, Victoria, B.C., Canada.

Kaufmann, M. R., and Troendle, C. A. (1981). The relationship of leaf area and foliage biomass to sapwood conducting area in four subalpine forest tree species. *For. Sci.* **27**, 477–482.

Kauss, H. (1983). Volume regulation in *Poterioochromonas*. *Plant Physiol.* **71**, 169–172.

Kauss, H., and Thomson, K.-S. (1982). Biochemistry of volume control in *Poterioochromonas*. *In* "Plasmalemma and Tonoplast: Their Functions in the Plant Cell" (D. Marmé, E. Marrè, and R. Hertel, eds.), pp. 255–261. Elsevier Biomedical Press, B.V.

Kavanau, J. L. (1964). "Water and Solute-Water Interactions." Holden-Day, Inc., San Francisco.

Kawamitsu, Y., Agata, W., and Miura, S. (1987). Effects of vapour pressure difference on CO_2 assimilation rate, leaf conductance and water use efficiency in grass species. *J. Fac. Agr. Kyushu Univ.* **31**, 1–10.

Kawase, M. (1979). Role of cellulase in aerenchyma development in sunflower. *Am. J. Bot.* **66**, 183–190.

Kearns, E. V., and Assman, S. M. (1993). The guard cell-environment connection. *Plant Physiol.* **102**, 711–715.

Keck, R. W., and Boyer, J. S. (1974). Chloroplast response to low leaf water potentials. III. Differing inhibition of electron transport and photophosphorylation. *Plant Physiol.* **53**, 474–479.

Kedem, O., and Katchalsky, A. (1958). Thermodynamic analysis of the permeability of biological membranes to non-electrolytes. *Biochim. Biophys. Acta* **27**, 229–246.

Keever, C. (1950). Causes of succession on old fields of the Piedmont, North Carolina. *Ecol. Monogr.* **20**, 229–250.

Keller, R. (1930). Der elektrische Faktor des Wassertransporte in Luhte der Vitalfarbung. *Ergeb. Physiol.* **30**, 294–407.

Kende, H. (1965). Kinetinlike factors in the root exudate of sunflowers. *Proc. Natl. Acad. Sci. USA* **53**, 1302–1307.

Ketring, D. L., and Reid, J. L. (1993). Growth of peanut roots under field conditions. *Agron. J.* **85**, 80–85.

Kevekordes, K. G., McCully, M. E., and Canny, M. J. (1988). Later maturation of large metaxylem vessels in soybean roots: Significance for water and nutrient supply to the shoot. *Ann. Bot.* **62**, 105–117.

Khalil, A. A. M., and Grace, J. (1993). Does xylem sap ABA control the stomatal behaviour of water-stressed sycamore (*Acer pseudoplatanus* L.) seedlings? *J. Exp. Bot.* **44**, 1127–1134.

Kiesselbach, T. A., Russell, J. C., and Anderson, A. (1929). The significance of subsoil moisture in alfalfa production. *J. Am. Soc. Agron.* **21**, 241–268.

Killian, Ch., and Lemée, G. (1956). Les xerophytes: Leur économie d'eau. *Encyl. Plant Physiol.* **3**, 787–824.

Kimmerer, T. W., and MacDonald, R. C. (1987). Acetaldehyde and ethanol biosynthesis in leaves of plants. *Plant Physiol.* **84**, 1204–1209.

Kimmerer, T. W., and Stringer, M. A. (1988). Alcohol dehydrogenase and ethanol in stems of trees. *Plant Physiol.* **87**, 693–697.

King, B. J., Hunt, S., Weagle, G. E., Walsh, K. B., Pottier, R. H., Canvin, D. T., and Layzell, D. B. (1988). Regulation of O_2 concentration in soybean nodules observed by *in situ* spectroscopic measurement of leghemoglobin oxygenation. *Plant Physiol.* **87**, 296–299.

King. D. A. (1993). A model analysis of the influence of root and foliage allocation on forest production and competition between trees. *Tree Physiol.* **12**, 119–135.

Kinman, C. F. (1932). A preliminary report on root growth studies with some orchard trees. *Proc. Am. Soc. Hort. Sci.* **29**, 220–224.

Kirkham, M. B., Gardner, W. R., and Gerloff, G. C. (1972). Regulation of cell division and cell enlargement by turgor pressure. *Plant Physiol.* **49**, 961–962.

Kitano, M., and Eguchi, H. (1992a). Dynamics of whole-plant water balance and leaf growth in response to evaporative demand. I. Effect of change in irradiance. *Biotronics* **21**, 39–50.

Kitano, M., and Eguchi, H. (1992b). Dynamics of whole-plant water balance and leaf growth in response to evaporative demand. II. Effect of change in wind velocity. *Biotronics* **21**, 51–60.

Kitano, M., and Eguchi, H. (1993). Dynamic analyses of water relations and leaf growth in cucumber plants under midday water deficit. *Biotronics* **22**, 73–85.

Klebs, G. (1910). Alterations in the development and forms of plants as a result of environment. *Proc. Roy. Soc. London* **82B**, 547–558.

Klebs, G. (1913). Über das Verhaltnisse der Aussenwelt zur Entwicklung der Pflanzen. *Sitzber. Heidelberg Akad. Wiss. Abt.* **B5**, 1–47.

Klepper, B., and Kaufmann, M. R. (1966). Removal of salt from xylem sap by leaves and stems of guttating plants. *Plant Physiol.* **41**, 1743–1747.

Klepper, B., Taylor, H. M., Huck, M. G., and Fiscus, E. L. (1973). Water relations and growth of cotton in drying soil. *Agron. J.* **65**, 307–310.

Klotz, I. M. (1958). Protein hydration and behavior. *Science* **128**, 815–822.

Kluge, M., and Ting, I. P. (1978). "Crassulacean Acid Metabolism." Springer-Verlag, Berlin.

Knight, D. H., Fahey, T. J., Running, S. W., Harrison, A. T., and Wallace, L. L. (1981). Transpiration from 100-year old lodgepole pine forests estimated with whole tree potometers. *Ecology* **62**, 717–726.

Knoerr, K. R. (1967). Contrasts in energy balances between individual leaves and vegetated surfaces. *In* "International Symposium on Forest Hydrology" (W. E. Sopper and H. W. Lull, eds.), pp. 391–401. Pergamon Press, New York.

Koch, A. L. (1990). Growth and form of the bacterial cell wall. *Amer. Sci.* **78**, 327–341.

Koeppe, D. E., Miller, R. J., and Bell, D. T. (1973). Drought-affected mitochondrial processes as related to tissue and whole plant responses. *Agron. J.* **65**, 566–569.

Koide, R. T., and Schreiner, R. P. (1992). Regulation of vesicular-arbuscular mycorrhizal symbiosis. *Annu. Rev. Plant Physiol. Plant Mol. Biol.* **43**, 557–591.

Kokubo, A., Kuraishi, S., and Sakurai, N. (1989). Culm strength of barley: Correlation among maximum bending stress, cell wall dimensions, and cellulose content. *Plant Physiol.* **91**, 876–882.

Kokubo, A., Sakurai, N., Kuraishi, S., and Takeda, K. (1991). Culm brittleness of barley (*Hordeum vulgare* L.) mutants is caused by smaller number of cellulose molecules in cell wall. *Plant Physiol.* **97**, 509–514.

Kolata, G. B. (1979). Water structure and ion binding: a role in cell physiology? *Science* **192**, 1220–1222.

Kolattukudy, P. E. (1981). Structure, biosynthesis, and biodegradation of cutin and suberin. *Annu. Rev. Plant Physiol.* **32**, 539–567.

Korstian, C. F. (1924). Density of cell sap in relation to environmental conditions in the Wasatch Mountains of Utah. *J. Agr. Res.* **28**, 845–909.

Koster, K. L., and Leopold, A. C. (1988). Sugars and desiccation tolerance in seeds. *Plant Physiol.* **88**, 829–832.

Kozlowski, T. T., ed. (1968). "Water Deficits and Plant Growth," Vol. 2. Academic Press, New York.

Kozlowski, T. T., ed. (1972). "Water Deficits and Plant Growth," Vol. 3. Academic Press, New York.

Kozlowski, T. T., ed. (1973). "Shedding of Plant Parts." Academic Press, New York.

Kozlowski, T. T., ed. (1976). "Water Deficits and Plant Growth," Vol. 4. Academic Press, New York.

Kozlowski, T. T., ed. (1978). "Water Deficits and Plant Growth," Vol. 5. Academic Press, New York.

Kozlowski, T. T., ed. (1984). "Flooding and Plant Growth." Academic Press, Orlando, FL.

Kozlowski, T. T., and Cooley, J. C. (1961). Root grafting in northern Wisconsin. *J. For.* **59**, 105–107.

Kozlowski, T. T., Kramer, P. J., and Pallardy, S. G. (1991). "The Physiological Ecology of Woody Plants." Academic Press, San Diego.

Kozlowski, T. T., and Winget, C. H. (1963). Patterns of water movement in forest trees. *Bot. Gaz.* **124**, 301–311.

Kramer, P. J. (1932). The absorption of water by root systems of plants. *Am. J. Bot.* **19**, 148–164.

Kramer, P. J. (1933). The intake of water through dead root systems and its relation to the problem of absorption by transpiring plants. *Am. J. Bot.* **20**, 481–492.

Kramer, P. J. (1937). The relation between rate of transpiration and rate of absorption of water in plants. *Am. J. Bot.* **24**, 10–15.

Kramer, P. J. (1938). Root resistance as a cause of the absorption lag. *Am. J. Bot.* **25**, 110–113.

Kramer, P. J. (1939). The forces concerned in the intake of water by transpiring plants. *Am. J. Bot.* **26**, 784–791.

Kramer, P. J. (1940a). Causes of decreased absorption of water by plants in poorly aerated media. *Am. J. Bot.* **27**, 216–220.

Kramer, P. J. (1940b). Root resistance as a cause of decreased water absorption by plants at low temperatures. *Plant Physiol.* **15**, 63–79.

Kramer, P. J. (1940c). Sap pressure and exudation. *Am. J. Bot.* **27**, 929–931.

Kramer, P. J. (1942). Species differences with respect to water absorption at low soil temperatures. *Am. J. Bot.* **29**, 828–832.

Kramer, P. J. (1945). Absorption of water by plants. *Bot. Rev.* **11**, 310–355.

Kramer, P. J. (1949). "Plant and Soil Water Relationships." McGraw-Hill, New York.

Kramer, P. J. (1950). Effects of wilting on the subsequent intake of water by plants. *Am. J. Bot.* **37**, 280–284.

Kramer, P. J. (1951). Causes of injury to plants resulting from flooding of the soil. *Plant Physiol.* **26**, 722–736.

Kramer, P. J. (1955). Bound water. *In* "Encyclopedia of Plant Physiology" (W. Ruhland, ed.), Vol. 1, pp. 223–242. Springer-Verlag, Berlin.

Kramer, P. J. (1963). Water stress and plant growth. *Agron. J.* **55**, 31–35.

Kramer, P. J. (1969). "Plant and Soil Water Relationships: A Modern Synthesis." McGraw-Hill, New York.

Kramer, P. J. (1983). "Water Relations of Plants." Academic Press, New York.

Kramer, P. J. (1987). The role of water stress in tree growth. *J. Arboric.* **13**, 33–38.

Kramer, P. J. (1988). Changing concepts regarding plant water relations. *Plant Cell Environ.* **11**, 565–568.

Kramer, P. J., and Bullock, H. C. (1966). Seasonal variations in the proportions of suberized and unsuberized roots of trees in relation to the absorption of water. *Am. J. Bot.* **53**, 200–204.

Kramer, P. J., and Clark, W. S. (1947). A comparison of photosynthesis in individual pine needles and entire seedlings at various light intensities. *Plant Physiol.* **22**, 51–57.

Kramer, P. J., and Coile, T. S. (1940). An estimation of the volume of water made available by root extension. *Plant Physiol.* **15**, 743–747.

Kramer, P. J., and Decker, J. P. (1944). Relation between light intensity and rate of photosynthesis of loblolly pine and certain hardwoods. *Plant Physiol.* **19**, 350–358.

Kramer, P. J., and Kozlowski, T. T. (1960). "Physiology of Trees." McGraw-Hill, New York.

Kramer, P. J., and Kozlowski, T. T. (1979). "Physiology of Woody Plants." Academic Press, New York.

Kramer, P. J., and Rose, R. W., Jr. (1986). Physiological characteristics of loblolly pine seedlings in relation to field performance. *In* "Proc. Int. Symp. on Nursery Management Practices for the Southern Pines" (D. B. South, ed.), pp. 416–440. Auburn University, Auburn, Alabama.

Kramer, P. J., and Wiebe, H. H. (1952). Longitudinal gradients of P^{32} absorption in roots. *Plant Physiol.* **27**, 661–674.

Kriedemann, P. E., and Downton, W. J. S. (1981). Photosynthesis. *In* "The Physiology and Biochemistry of Drought Resistance in Plants" (L. G. Paleg and D. Aspinall, eds.), pp. 283–314. Academic Press, Sydney.

Kriedemann, P. E., Loveys, B. R., Fuller, G. L., and Leopold, A. C. (1972). Abscisic acid and stomatal regulation. *Plant Physiol.* **49**, 842–847.

Krizek, D. T., and Dubik, S. P. (1987). Influence of water stress and restricted root volume on growth and development of urban trees. *J. Arboric.* **13**, 47–55.

Kuntz, I. D., and Kauzmann, W. (1974). Hydration of proteins and polypeptides. *Adv. Protein Chem.* **28**, 239–345.

Kuntz, J. E., and Riker, A. J. (1955). The use of radioactive isotopes to ascertain the role of root grafting in the translocation of water, nutrients, and disease-inducing organisms. *Proc. Int. Conf. Peaceful Uses At. Energy* **12**, 144–148.

Kurkova, E. B., and Motorina, M. V. (1974). Chloroplast ultrastructure and photosynthesis at different rates of dehydration. *Sov. Plant Physiol.* **21**, 28–31.

Kurtzman, R. H., Jr. (1966). Xylem sap flow as affected by metabolic inhibitors and girdling. *Plant Physiol.* **41**, 641–646.

Kutschera, L. (1960). "Wurzelatlas Mitteleuropäischer Ackerunkräuter und Kulturpflanzen." DLG Verlag, Frankfurt-am-Main.

Kutschera, U. (1989). Growth, *in vivo* extensibility and tissue tension in mung bean seedlings subjected to water stress. *Plant Sci.* **61**, 1–7.

Kutschera, U., and Briggs, W. R. (1987). Rapid auxin-induced stimulation of cell wall synthesis in pea internodes. *Proc. Natl. Acad. Sci. USA* **84**, 2747–2751.

Kutschera, U., and Briggs, W. R. (1988). Growth, *in vivo* extensibility, and tissue tension in developing pea internodes. *Plant Physiol.* **86**, 306–311.

Laan, P., and Blom, C. W. P. M. (1990). Growth and survival response of *Rumex* species to flooded and submerged conditions: The importance of shoot elongation, underwater photosynthesis and reserve carbohydrates. *J. Exp. Bot.* **41**, 775–783.

Laan, P., Tosserams, M., Blom, C. W. P. M., and Veen, B. W. (1990). Internal oxygen transport in *Rumex* spp. and its significance for respiration under hypoxic conditions. *Plant Soil* **122**, 39–46.

Labavitch, J. M., and Ray, P. M. (1974). Relationship between promotion of xyloglucan metabolism and induction of elongation by indoleacetic acid. *Plant Physiol.* **54**, 499–502.

Ladefoged, K. (1963). Transpiration of forest trees. *Physiol. Plant.* **16**, 378–414.

Laing, H. E. (1940). The composition of the internal atmosphere of *Nuphar advenum* and other water plants. *Am. J. Bot.* **27**, 861–868.

Laisk, A. (1983). Calculations of leaf photosynthetic parameters considering the statistical distribution of stomatal aperture. *J. Exp. Bot.* **34**, 1627–1635.

Laisk, A., Oja, V., and Kull, K. (1980). Statistical distribution of stomatal apertures of *Vicia faba* and *Hordeum vulgare* and the *Spannungsphase* of stomatal opening. *J. Exp. Bot.* **31**, 49–58.

Lamhamedi, M. S., Bernier, P. Y., and Fortin, J. A. (1992). Hydraulic conductance and soil water potential at the soil-root interface of *Pinus pinaster* seedlings inoculated with different dikaryons of *Pisolithus* sp. *Tree Physiol.* **10**, 231–244.

Landsberg, J. J., and Fowkes, N. D. (1978). Water movement through plant roots. *Ann. Bot.* **42**, 493–508.

Lang, A. (1990). Xylem, phloem and transpiration flows in developing apple fruits. *J. Exp. Bot.* **41**, 645–651.

Lange, O. L., Kappen, L., and Schulze, E.-D., eds. (1976). "Water and Plant Life," Vol. 19. Springer-Verlag, Berlin.

Lange, O. L., Lösch, R., Schulze, E.-D., and Kappen, L. (1971). Responses of stomata to changes in humidity. *Planta* **100**, 76–86.

Larson, P. R. (1980). Interrelations between phyllotaxis, leaf development and the primary-secondary vascular transition in *Populus deltoides*. *Ann. Bot.* **46**, 757–769.

LaRue, C. D. (1930). The water supply of the epidermis of leaves. *Papers Mich. Acad. Sci.* **13**, 131–139.

LaRue, C. D. (1952). Root grafting in tropical trees. *Science* **115**, 296.

Lassoie, J. P., Fetcher, N., and Solo, D. J. (1977a). Stomatal infiltration pressures versus potometer measurements of needle resistance in Douglas-fir and lodgepole pine. *Can. J. For. Res.* **7**, 192–196.

Lassoie, J. P., and Hinckley, T. M. (1991). "Techniques and Approaches in Forest Tree Ecophysiology." CRC Press, Boca Raton, FL.

Lassoie, J. P., Scott, D. R. M., and Fritschen, L. J. (1977b). Transpiration studies in Douglas-fir using the heat pulse technique. *For. Sci.* **23**, 377–380.

Lauer, M. J., and Boyer, J. S. (1992). Internal CO_2 measured directly in leaves: Abscisic acid and low leaf water potential cause opposing effects. *Plant Physiol.* **98**, 1310–1316.

Lawlor, D. W. (1970). Absorption of polyethylene glycols by plants and their effects on plant growth. *New Phytol.* **69**, 501–513.

Lawlor, D. W. (1993). "Photosynthesis: Molecular, Physiological and Environmental Processes," 2nd Ed. Longman Scientific & Technical, Essex.

Lawlor, D. W., and Milford, G. F. J. (1975). The control of water and carbon dioxide flux in water-stressed sugar beet. *J. Exp. Bot.* **26**, 657–665.

Layzell, D. B., Hunt, S., and Palmer, G. R. (1989). Mechanism of nitrogenase inhibition in soybean nodules. *Plant Physiol.* **92**, 1101–1107.

Lazof, D. B., Rufty, T. W., Jr., and Redinbaugh, M. G. (1992). Localization of nitrate absorption and translocation within morphological regions of the corn root. *Plant Physiol.* **100**, 1251–1258.

Lebedeff, A. F. (1928). The movement of ground and soil waters. *Proc. 1st Int. Cong. Soil Sci.* **1**, 459–494.

Lemon, E. R., ed. (1983). "CO_2 and Plants." Am. Assoc. Adv. Sci., Washington, D.C.

Leon, J. M., and Bukovac, M. J. (1978). Cuticle development and surface morphology of olive leaves with reference to penetration of foliar-applied chemicals. *J. Amer. Hort. Sci.* **103**, 465–472.

Leopold, A. C., Musgrave, M. E., and Williams, K. M. (1981). Solute leakage resulting from leaf desiccation. *Plant Physiol.* **68**, 1222–1225.

Leshem, B. (1965). The annual activity of intermediary roots of the Aleppo pine. *For. Sci.* **11**, 291–298.

Letey, J., Clark, P. R., and Amrhein, C. (1992). Water-sorbing polymers do not conserve water. *Calif. Agric.* **46**(3), 9–10.

Letey, J., Welch, N., Pelishek, R. E., and Osborn, J. (1962). Effects of wetting agents on irrigation of water repellent soils. *Calif. Agric.* **16**, 213.

Levitt, J. (1980). "Responses of Plants to Environmental Stresses," Vol. 2, pp. 25–92. Academic Press, New York.

Levitt, J., Scarth, G. W., and Gibbs, R. D. (1936). Water permeability of isolated protoplasts in relation to volume change. *Protoplasma* **26**, 237–248.

Levy, Y., and Kaufmann, M. R. (1976). Cycling of leaf conductance in citrus exposed to natural and controlled environments. *Can. J. Bot.* **54**, 2215–2218.

Lewis, F. J. (1945). Physical condition of the surface of the mesophyll cell walls of the leaf. *Nature* (London) **156**, 407–409.

Liebig, J. (1841). "Organic Chemistry in its Application to Agriculture and Physiology." English Translation by J. Owen, Cambridge, UK.

Lieffers, V. J., and Rothwell, R. L. (1987). Rooting of peatland black spruce and tamarack in relation to depth of water table. *Can. J. Bot.* **65**, 817–821.

Lin, C. H., and Lin, C. H. (1992). Physiological adaptation of waxapple to waterlogging. *Plant Cell Environ.* **15**, 321–328.

Lindblad, P., Rai, A. N., and Bergman, B. (1987). The Cycas revoluta-Nostoc symbiosis: Enzyme activities of nitrogen and carbon metabolism in the cyanobiont. *J. Gen. Microbiol.* **133**, 1695–1699.

Ling, G. N. (1969). A new model for the living cell: A summary of the theory and recent experimental evidence in its support. *Intern. Rev. Cytol.* **26**, 1–61.

Livingston, B. E. (1903). "The Role of Diffusion and Osmotic Pressure in Plants." University of Chicago Press, Chicago, IL.

Livingston, B. E. (1918). Porous clay cones for the auto-irrigation of potted plants. *Plant World* 2, 202–208.

Livingston, B. E. (1927). Plant water relations. *Quart. Rev. Biol.* 2, 494–515.

Livingston, B. E., and Beall, R. (1934). The soil as a direct source of carbon dioxide for ordinary plants. *Plant Physiol.* 9, 237–259.

Livingston, B. E., and Brown, W. H. (1912). Relation of the daily march of transpiration to variations in the water content of foliage leaves. *Bot. Gaz.* (Chicago) 53, 309–330.

Livingston, B. E., and Shreve, F. (1921). "The Distribution of Vegetation in the United States as Related to Climatic Conditions." Carnegie Inst. Wash. Publ. 284.

Lloyd, F. E. (1908). "The Behaviour of Stomata." Carnegie Inst. Washington Publ. 82.

Lloyd, J., Syvertsen, J. P., and Kriedemann, P. E. (1987). Salinity effects on leaf water relations and gas exchange of 'Valencia' orange on rootstock with different salt exclusion characteristics. *Aust. J. Plant Physiol.* 14, 605–617.

Lockard, R. G., and Schneider, G. W. (1981). Stock and scion growth relationships and the dwarfing mechanism in apples. *Hortic. Rev.* 3, 315–375.

Lockhart, J. A. (1965a). An analysis of irreversible plant cell elongation. *J. Theor. Biol.* 8, 264–275.

Lockhart, J. A. (1965b). Cell extension. *In* "Plant Biochemistry" (J. Bonner and J. E. Varner, eds.), pp. 826–849. Academic Press, New York.

Loehwing, W. F. (1934). Physiological aspects of the effect of continuous soil aeration on plant growth. *Plant Physiol.* 9, 567–583.

Loescher, W., and Nevins, D. L. (1972). Auxin-induced changes in *Avena* coleoptile cell wall composition. *Plant Physiol.* 50, 556–563.

Loescher, W. H., and Nevins, D. J. (1973). Turgor-dependent changes in *Avena* coleoptile cell wall composition. *Plant Physiol.* 52, 248–251.

Loftfield, J. V. G. (1921). "The Behavior of Stomata." Carnegie Inst. Washington Publ. 314.

Logsdon, S. D., Reneau, R. B., Jr., and Parker, J. C. (1987). Corn seedling root growth as influenced by soil physical properties. *Agron. J.* 79, 221–224.

Longstreth, D. J., and Kramer, P. J. (1980). Water relations during flower induction and anthesis. *Bot. Gaz.* 141, 69–72.

Lonnquist, J. H., and Jugenheimer, R. W. (1943). Factors affecting the success of pollination in corn. *J. Am. Soc. Agron.* 35, 923–933.

Loomis, R. S., Williams, W. A., and Hall, A. E. (1971). Agricultural productivity. *Annu. Rev. Plant Physiol.* 22, 431–468.

Loomis, W. E. (1935). The translocation of carbohydrates in maize. *Iowa State Coll. J. Sci.* 9, 509–520.

Lopez, F. B., and Nobel, P. S. (1991). Root hydraulic conductivity of two cactus species in relation to root age, temperature, and soil water status. *J. Exp. Bot.* 42, 143–149.

Lopushinsky, W. (1964). Effect of water movement on ion movement into the xylem of tomato roots. *Plant Physiol.* 39, 494–501.

Lopushinsky, W. (1980). Occurrence of root pressure exudation in Pacific Northwest conifer seedlings. *For. Sci.* 26, 275–279.

Lopushinsky, W. (1986). Seasonal and diurnal trends of heat pulse velocity in Douglas-fir and ponderosa pine. *Can. J. For. Res.* 16, 814–821.

Lopushinsky, W., and Klock, G. O. (1974). Transpiration of conifer seedlings in relation to soil water potential. *For. Sci.* 20, 181–186.

Lorio, P., Jr. (1993). Environmental stress and whole-tree physiology. *In* "Beetle-Pathogen Interactions in Conifer Forests" (T. D. Schowalter and G. M. Filip, eds.), pp. 81–101. Academic Press, London.

Lorio, P. L., Jr. (1994). The relationship of oleoresin exudation pressure (or lack thereof) to flow from wounds. *J. Sustainable For.* **1**, 81–93.

Loustalot, A. J. (1945). Influence of soil-moisture conditions on apparent photosynthesis and transpiration of pecan leaves. *J. Agr. Res.* **71**, 519–532.

Lovett, G. M., Reiners, W. A., and Olson, R. K. (1982). Cloud droplet deposition in subalpine Balsam fir forests: hydrological and chemical inputs. *Science* **218**, 1303–1304.

Lowry, M. W., Huggins, W. C., and Forrest, L. A. (1936). "The Effect of Soil Treatment on the Mineral Composition of Exuded Maize Sap at Different Stages of Development." Georgia Agr. Exp. Sta. Bull. 193.

Lucas, W. J., Ding, B., and van der Schoot, C. (1993). Plasmodesmata and the supracellular nature of plants. *New Phytol.* **125**, 435–476.

Ludevid, D., Höfte, H., Himelblau, E., and Chrispeels, M. J. (1992). The expression pattern of the tonoplast intrinsic protein τ-TIP in *Arabidopsis thaliana* is correlated with cell enlargement. *Plant Physiol.* **100**, 1633–1639.

Ludlow, M. M., and Björkman, O. (1984). Paraheliotropic leaf movement in Siratro as a protective mechanism against drought-induced damage to primary photosynthetic reactions: damage by excessive light and heat. *Planta* **161**, 505–518.

Lund, E. J. (1931). Electric correlation between living cells in cortex and wood in the Douglas fir. *Plant Physiol.* **6**, 631–652.

Lundegårdh, H. (1931). "Environment and Plant Development." English translation by Ashby Arnold, London.

Lüthen, H., Bigdon, M., and Böttger, M. (1990). Reexamination of the acid growth theory of auxin action. *Plant Physiol.* **93**, 931–939.

Luxmoore, R. J., King, A. W., and Tharp, M. L. (1991). Approaches to scaling up physiologically based soil-plant models in space and time. *Tree Physiol.* **9**, 281–292.

Luxmoore, R. J., Stolzy, L. H., and Letey, J. (1970). Oxygen diffusion in the soil-plant system. III. Oxygen concentration profiles, respiration rates, and the significance of plant aeration predicted for maize roots. *Agron. J.* **62**, 325–329.

Lyford, W. H., and Wilson, B. F. (1964). Development of the root system of *Acer rubrum* L. *Harvard For. Paper* 10.

Lynn, D. G., and Chang, M. (1990). Phenolic signals in cohabitation: implications for plant development. *Annu. Rev. Plant Physiol. Plant Mol. Biol.* **41**, 497–526.

Lyons, J. M. (1973). Chilling injury in plants. *Annu. Rev. Plant Physiol.* **24**, 445–466.

Lyons, J. M., Graham, D., and Raison, J. K., eds. (1979). "Low Temperature Stress in Crop Plants: The Role of the Membrane." Academic Press, New York.

Lyr, H., and Hoffman, G. (1967). Growth rates and growth periodicity of tree roots. *Int. Rev. For. Res.* **2**, 181–236.

Maas, E. V. (1985). Crop tolerance to saline sprinkling waters. *Plant Soil* **89**, 273–284.

Maas, E. V. (1993). Salinity and citriculture. *Tree Physiol.* **12**, 195–216.

Macallum, A. B. (1905). On the distribution of potassium in animal and vegetable cells. *J. Physiol.* (London) **32**, 95–128.

MacDougal, D. T. (1926). "The Hydrostatic System of Trees." Carnegie Inst. Washington Publ., 373.

MacFall, J. S. (1994). Effects of ectomycorrhizae on biogeochemistry and soil structure. *In* "Reappraisal of Mycorrhizae and Agriculture" (F. L. Pfleger and R. Linderman, eds.), pp. 213–238. APS Press, St. Paul, MN.

MacFall, J. S., and Johnson, G. A. (1994). Use of magnetic resonance imaging in the study of plants and soils. *In* "Tomography for Measurement of Soil Physical Properties and Processes" (J. W. Hopmans and S. Anderson, eds.), pp. 99–113. Soil Sci. Soc. Am., Madison, WI.

MacFall, J. S., Johnson, G. A., and Kramer, P. J. (1990). Observation of a water-depletion region surrounding loblolly pine roots by magnetic resonance imaging. *Proc. Natl. Acad. Sci. USA* **87**, 1203–1207.

MacFall, J. S., Johnson, G. A., and Kramer, P. J. (1991a). Comparative water uptake by roots of different ages in seedlings of loblolly pine (*Pinus taeda* L.). *New Phytol.* **119**, 551–560.

MacFall, J. S., Pfeffer, P. E., Rolin, D. B., MacFall, J. R., and Johnson, G. A. (1992). Observation of the oxygen diffusion barrier in soybean (*Glycine max*) nodules with magnetic resonance microscopy. *Plant Physiol.* **100**, 1691–1697.

MacFall, J. S., Slack, S. A., and Iyer, J. (1991c). Effects of *Hebeloma arenosa* and phosphorus fertility on root acid phosphatase activity of red pine (*Pinus resinosa*) seedlings. *Can. J. Bot.* **69**, 380–383.

MacFall, J. S., Slack, S. A., and Iyer, J. (1991b). Effects of *Hebeloma arenosa* and phosphorus fertility on growth of red pine (*Pinus resinosa*) seedlings. *Can. J. Bot.* **69**, 372–379.

MacFall, J. S., Slack, S. A., and Wehrli, S. (1992). Phosphorus distribution in red pine roots and the ectomycorrhizal fungus *Hebeloma arenosa*. *Plant Physiol.* **100**, 713–717.

Machlis L. (1944). The respiratory gradient in barley roots. *Am. J. Bot.* **31**, 281–282.

Madin, K. A. C., and Crowe, J. H. (1975). Anhydrobiosis in nematodes: Carbohydrate and lipid metabolism during dehydration. *J. Exp. Zool.* **193**, 335–342.

Maercker, U. (1965). Zur Kenntnis der Transpiration der Schliesszellen. *Protoplasma* **60**, 61–78.

Magistad, O. C., and Reitemeier, R. F. (1943). Soil solution concentrations at the wilting point and their correlation with plant growth. *Soil Sci.* **55**, 351–360.

Maier-Maercker, U. (1979a). Peristomatal transpiration and stomatal movement: A controversial view. I. Additional proof of peristomatal transpiration by hygrophotography and a comprehensive discussion in the light of recent results. *Z. Pflanzenphysiol.* **91**, 25–43.

Maier-Maercker, U. (1979b). Peristomatal transpiration and stomatal movement: A controversial view. III. Visible effects of peristomatal transpiration on the epidermis. *Z. Pflanzenphysiol.* **91**, 225–238.

Maier-Maercker, U. (1983). The role of peristomatal transpiration in the mechanism of stomatal movement. *Plant Cell Environ.* **6**, 369–380.

Malik, R. S., Dhankar, J. S., and Turner, N. C. (1979). Influence of soil water deficits on root growth of cotton seedlings. *Plant & Soil* **53**, 109–115.

Malone, M. (1993). Hydraulic signals. *Phil. Trans. Roy. Soc. London Ser. B* **341**, 33–39.

Mancino, C. F., and Pepper, I. L. (1992). Irrigation of turfgrass with secondary sewage effluent: soil quality. *Agron. J.* **84**, 650–654.

Maness, N. O., and McBee, G. G. (1986). Role of placental sac in endosperm carbohydrate import in sorghum caryopses. *Crop Sci.* **26**, 1201–1207.

Mansfield, T. A. (1986). The physiology of stomata: New insights into old problems. *In* "Plant Physiology" (F. C. Steward, J. F. Sutcliffe, and J. E. Dale, eds.), Vol. 9, pp. 155–224. Academic Press, New York.

Mansfield, T. A., Hetherington, A. M., and Atkinson, C. J. (1990). Some current aspects of stomatal physiology. *Annu. Rev. Plant Physiol. Plant Mol. Biol.* **41**, 55–75.

Mansfield, T. A., and Jones, R. J. (1971). Effects of abscisic acid on potassium uptake and starch content of stomatal guard cells. *Planta* **101**, 147–158.

Markhart, A. H., III, Fiscus, E. L., Naylor, A. W., and Kramer, P. J. (1979). Effect of temperature on water and ion transport in soybean and broccoli systems. *Plant Physiol.* **64**, 83–87.

Markhart, A. H., III, Peet, M. M., Sionit, N., and Kramer, P. J. (1980). Low temperature acclimation of root fatty acid composition, leaf water potential, gas exchange and growth of soybean seedlings. *Plant Cell Environ.* **3**, 435–441.

Marks, G. C., and Kozlowski, T. T., eds. (1973). "Ectomycorrhizae: Their Ecology and Physiology." Academic Press, New York.

Marschner, H. (1986). "Mineral Nutrition of Higher Plants." Academic Press, New York.

Marshall, D. C. (1958). Measurement of sap flow in conifers by heat transport. *Plant Physiol.* **21,** 95–101.

Martin, B., Bytnerowicz, A., and Thorstenson, Y. R. (1988). Effects of air pollutants on the composition of stable carbon isotopes, $\delta^{13}C$, of leaves and wood, and on leaf injury. *Plant Physiol.* **88,** 218–223.

Martin, B., Nienhuis, J., King, G., and Schaefer, A. (1989). Restriction fragment length polymorphisms associated with water use efficiency in tomato. *Science* **243,** 1725–1728.

Martin, B., and Thorstenson, Y. R. (1988). Stable carbon isotope composition ($\delta^{13}C$), water use efficiency, and biomass productivity of *Lycopersicon esculentum, Lycopersicon pennellii,* and the F_1 hybrid. *Plant Physiol.* **88,** 213–217.

Martin, C. E., and Schmitt, A. K. (1989). Unusual water relations in the CAM atmospheric epiphyte *Tillandsia usneoides* L. (Bromeliaceae). *Bot. Gaz.* **150,** 1–8.

Martin, C. K., Cassel, D. K., and Kamprath, E. J. (1979). Irrigation and tillage effects on soybean yields in a Coastal Plains soil. *Agron. J.* **71,** 592–594.

Martin, E. V., and Clements, F. E. (1935). Studies of the effect of artificial wind on growth and transpiration in *Helianthus annuus. Plant Physiol.* **10,** 613–636.

Maruyama, S., and Boyer, J. S. (1994). Auxin action on growth in intact plants: Threshold turgor is regulated. *Planta* **193,** 44–50.

Marvin, J. W. (1958). The physiology of maple sap flow. *In* "The Physiology of Forest Trees" (K. V. Thimann, ed.), pp. 95–124. The Ronald Press Company, New York.

Marx, D. H. (1980). Growth of loblolly and shortleaf pine seedlings after years on a strip-mined coal spoil in Kentucky is stimulated by *Pisolithus* ectomycorrhizae and "starter" fertilizer pellets. Abstr. North Am. Conf. Mycorrhizae, 3rd, 1977.

Marx, D. H., Cordell, C. E., Kenney, D. S., Mexal, J. G., Artman, J. D., Riffle, J. W., and Molina, R. J. (1984). Commercial vegetative inoculum of *Pisolithus tinctorius* and inoculation techniques for development of ectomycorrhizae on bare-root tree seedlings. *For. Sci. Monogr.* **25.**

Marx, D. H., Hedin, A., and Toe, P. S. (1985). Field performance of *Pinus caribaes,* var. *hondurensis* seedlings with specific ectomycorrhizae and fertilizer after 3 years on a savannah site in Liberia. *For. Ecol. Manag.* **13,** 1–25.

Masle, J., and Farquhar, G. D. (1988). Effects of soil strength on the relation of water-use efficiency and growth to carbon isotope discrimination in wheat seedlings. *Plant Physiol.* **86,** 32–38.

Masle, J., and Passioura, J. B. (1987). Effects of soil strength on the growth of wheat seedlings. *Aust. J. Plant Physiol.* **14,** 643–656.

Mason, A. C. (1958). The effect of soil moisture on the mineral composition of apple plants grown in pots. *J. Hort. Sci.* **33,** 202–211.

Mason, H. S., Guerrero, F. D., Boyer, J. S., and Mullet, J. E. (1988b). Proteins homologous to leaf glycoproteins are abundant in stems of dark-grown soybean seedlings: Analysis of proteins and cDNAs. *Plant Mol. Biol.* **11,** 845–856.

Mason, H. S., and Matsuda, K. (1985). Polyribosome metabolism, growth and water status in the growing tissues of osmotically stressed plant seedlings. *Physiol. Plant.* **64,** 95–104.

Mason, H. S., and Mullet, J. E. (1990). Expression of two soybean vegetative storage protein genes during development and in response to water deficit, wounding and jasmonic acid. *Plant Cell* **2,** 569–579.

Mason, H. S., Mullet, J. E., and Boyer, J. S. (1988a). Polysomes, messenger RNA and growth in soybean stems during development and water deficit. *Plant Physiol.* **86,** 725–733.

Masse, W. B. (1981). Prehistoric irrigation systems in the Salt River Valley, Arizona. *Science* **214,** 408–415.

Masuda, Y. (1985). Cell wall modifications during auxin-induced cell extension in monocotyledonous and dicotyledonous plants. *Biol. Plant. (Praha)* **27,** 119–124.

Materechera, S. A., Dexter, A. R., Alston, A. M., and Kirby, J. M. (1992). Growth of seedling roots in response to external osmotic stress by polyethylene glycol 20,000. *Plant Soil* **143**, 85–91.

Matthews, M. A., and Boyer, J. S. (1984). Acclimation of photosynthesis to low water potentials. *Plant Physiol.* **74**, 161–166.

Matthews, M. A., Van Volkenburgh, E., and Boyer, J. S. (1984). Acclimation of leaf growth to low water potentials in sunflower. *Plant Cell Environ.* **7**, 199–206.

Matyssek, R., Maruyama, S., and Boyer, J. S. (1991b). Growth-induced water potentials may mobilize internal water for growth. *Plant Cell Environ.* **14**, 917–923.

Matyssek, R., Maruyama, S., and Boyer, J. S. (1988). Rapid wall relaxation in elongating tissues. *Plant Physiol.* **86**, 1163–1167.

Matyssek, R., Tang, A.-C., and Boyer, J. S. (1991a). Plants can grow on internal water. *Plant Cell Environ.* **14**, 925–930.

Maugh, T. H., II (1978). Soviet science: A wonder water from Kazakhstan. *Science* **202**, 414.

Maurel, C., Reizer, J., Schroeder, J. I., and Chrispeels, M. J. (1993). The vacuolar membrane protein γ-TIP creates water specific channels in *Xenopus* oocytes. *EMBO J.* **12**, 2241–2247.

Mauro, A. (1965). Osmotic flow in a rigid porous membrane. *Science* **149**, 867–869.

Maximov, N. A. (1929). "The Plant in Relation to Water." English translation by R. H. Yapp. Allen and Unwin, London.

Mayoral, M. L., Atsmon, D., Shimshi, D., and Gromet-Elhanan, Z. (1981). Effect of water stress on enzyme activities in wheat and related wild species: Carboxylase activity, electron transport, and photophosphorylation in isolated chloroplasts. *Aust. J. Plant Physiol.* **8**, 385–393.

McAinsh, M. R., Brownlee, C., and Hetherington, A. M. (1990). Abscisic acid-induced elevation of guard cell crytosolic Ca²⁺ precedes stomatal closure. *Nature* (London) **343**, 186–188.

McArthur, D. A. J., and Knowles, N. R. (1992). Resistance responses of potato to vesicular-arbuscular mycorrhizal fungi under varying abiotic phosphorus levels. *Plant Physiol.* **100**, 341–351.

McCain, D. C., Croxdale, J., and Markley, J. L. (1988). Water is allocated differently to chloroplasts in sun and shade leaves. *Plant Physiol.* **86**, 16–18.

McCauley, G. N. (1993). Nonionic surfactant and supplemental irrigation of soybean on crusting soils. *Agron. J.* **85**, 17–21.

McComb, A. L., and Loomis, W. E. (1944). Subclimax prairie. *Torrey Bot. Club Bull.* **71**, 46–76.

McCree, K. J. (1974). Changes in stomatal response characteristics of grain sorghum produced by water stress during growth. *Crop Sci.* **14**, 273–278.

McCully, M. E., and Canny, M. J. (1988). Pathways and processes of water and nutrient movement in roots. *Plant Soil* **111**, 159–170.

McDermott, J. J. (1941). The effect of the method of cutting on the moisture content of samples from tree branches. *Am. J. Bot.* **28**, 506–508.

McDermott, J. J. (1945). The effect of the moisture content of the soil upon the rate of exudation. *Am. J. Bot.* **32**, 570–574.

McIntyre, G. I. (1987). The role of water in the regulation of plant development. *Can. J. Bot.* **65**, 1287–1298.

McIntyre, G. I., and Boyer, J. S. (1984). The effect of humidity, root excision, and potassium supply on hypocotyl elongation in dark-grown seedlings of *Helianthus annuus*. *Can. J. Bot.* **62**, 420–428.

McKenney, M. S., and Rosenberg, N. J. (1993). Sensitivity of some potential evapotranspiration estimation methods to climate change. *Agric. For. Meteorol.* **64**, 81–110.

McMichael, B. L., and Persson, H. (eds.) (1991). "Plant Roots and Their Environment." Elsevier, Amsterdam.

McNaughton, K. G., and Jarvis, P. G. (1983). Predicting effects of vegetation changes on transpiration and evaporation. *In* "Water Deficits and Plant Growth" (T. T. Kozlowski, ed.), Vol. 7, pp. 1–47. Academic Press, New York.

McNeil, D. L. (1976). The basis of osmotic pressure maintenance during expansion growth in *Helianthus annuus* hypocotyls. *Aust. J. Plant Physiol.* 3, 311–324.

McPherson, H. G., and Boyer, J. S. (1977). Regulation of grain yield by photosynthesis in maize subjected to a water deficiency. *Agron. J.* 69, 714–718.

McWilliam, J. R., and Kramer, P. J. (1968). The nature of the perennial response in Mediterranean grasses. I. Water relations and summer survival in *Phalaris. Aust. J. Agric. Res.* 19, 381–395.

McWilliam, J. R., Kramer, P. J., and Musser, R. L. (1982). Temperature-induced water stress in chilling-sensitive plants. *Aust. J. Plant Physiol.* 9, 343–352.

Mederski, H. J., and Jeffers, D. L. (1973). Yield response of soybean varieties grown at two soil moisture stress levels. *Agron. J.* 65, 410–412.

Meidner, H. (1975). Water supply, evaporation, and vapour diffusion in leaves. *J. Exp. Bot.* 26, 666–673.

Meidner, H. (1976a). Vapour loss through stomatal pores with the mesophyll tissue excluded. *J. Exp. Bot.* 27, 172–174.

Meidner, H. (1976b). Water vapour loss from a physical model of a substomatal cavity. *J. Exp. Bot.* 27, 691–694.

Meidner, H. (1986). Historical sketches 13. *J. Exp. Bot.* 37, 135–137.

Meidner, H. (1990). The absorption lag, epidermal turgor and stomata. *J. Exp. Bot.* 41, 1115–1118.

Meidner, H. (1992). Developments in mass flow porometry. *J. Exp. Bot.* 43, 1309–1314.

Meidner, H., and Mansfield, T. A. (1968). "Physiology of Stomata." McGraw-Hill, London.

Meinzer, F. C., and Grantz, D. A. (1991). Coordination of stomatal, hydraulic, and canopy boundary layer properties: Do stomata balance conductances by measuring transpiration? *Physiol. Plant.* 83, 324–329.

Meinzer, F. C., Grantz, D. A., and Smit, B. (1991). Root signals mediate coordination of stomatal and hydraulic conductance in growing sugarcane. *Aust. J. Plant Physiol.* 19, 329–338.

Meisner, C. A., and Karnok, K. J. (1991). Root hair occurrence and variation with environment. *Agron. J.* 83, 814–818.

Melchior, W., and Steudle, E. (1993). Water transport in onion (*Allium cepa* L.) roots. *Plant Physiol.* 101, 1305–1315.

Mendelssohn, I. A., McKee, K. L., and Patrick, W. H., Jr. (1981). Oxygen deficiency in *Spartina alterniflora* roots: Metabolic adaptation to anoxia. *Science* 214, 439–441.

Mendelssohn, I. A., and Postek, M. T. (1982). Elemental analysis of deposits on the roots of *Spartina alterniflora* Loieal. *Am. J. Bot.* 69, 904–912.

Mengel, D. B., and Barber, S. A. (1974). Development and distribution of the corn root system under field conditions. *Agron. J.* 66, 341–344.

Merkle, F. G., and Dunkle, E. C. (1944). Soluble salt content of greenhouse soils as a diagnostic aid. *Am. Soc. Agron. J.* 36, 10–19.

Merwin, H. E., and Lyon, H. (1909). Sap pressure in the birch stem. *Bot. Gaz.* 48, 442–458.

Meshcheryakov, A., Steudle, E., and Komor, E. (1992). Gradients of turgor, osmotic pressure, and water potential in the cortex of the hypocotyl of growing *Ricinus* seedlings. *Plant Physiol.* 98, 840–852.

Métraux, J.-P. (1982). Changes in cell-wall polysaccharide composition of developing *Nitella* internodes. *Planta* 155, 459–466.

Meyer, B. S. (1932). The daily periodicity of transpiration in the tulip poplar, *Liriodendron tulipifera* L. *Ohio J. Sci.* 32, 104–114.

Meyer, B. S. (1938). The water relations of plant cells. *Bot. Rev.* 4, 531–547.

Meyer, B S. (1945). A critical evaluation of the terminology of diffusion phenomena. *Plant Physiol.* 20, 142–164.

Meyer, B. S., and Anderson, D. B. (1952). "Plant Physiology." 2nd Ed. D. Van Nostrand & Co., New York.

Meyer, B. S., Anderson, D. B., Bohning, R. H., and Fratianne, D. G. (1973). "Introduction to Plant Physiology." D. Van Nostrand & Co., New York.

Meyer, R. F., and Boyer, J.S. (1972). Sensitivity of cell division and cell elongation to low water potentials in soybean hypocotyls. *Planta* 108, 77–87.

Meyer, R. F., and Boyer, J. S. (1981). Osmoregulation, solute distribution, and growth in soybean seedlings having low water potentials. *Planta* 151, 482–489.

Meyer, W. S., and Alston, A. M. (1978). Wheat responses to seminal geometry and subsoil water. *Agron. J.* 70, 981–986.

Meyer, W. S., and Ritchie, J. T. (1980). Water status of cotton as related to taproot length. *Agron. J.* 72, 577–580.

Michailides, T. J., Morgan, D. P., Grant, J. A., and Olson, W. H. (1992). Shorter sprinkler irrigations reduce *Botryosphaeria* blight of pistachio. *Calif. Agric.* 48, 28–32.

Michelena, V. A., and Boyer, J. S. (1982). Complete turgor maintenance at low water potentials in the elongating region of maize leaves. *Plant Physiol.* 69, 1145–1149.

Micke, W. C., Yeager, J. T., Vosson, P. M., Bethell, R. S., Foott, J. H., and Tyler, R. H. (1992). Apple rootstocks evaluated in California. *Calif. Agr.* 46, 23–25.

Milburn, J. A., and Johnson, R. P. C. (1966). The conduction of sap. II. Detection of vibrations produced by sap cavitation in *Ricinus* xylem. *Planta* 69, 43–52.

Milburn, J. A., Kallarackal, J., and Baker, D. A. (1990). Water relations of the banana. I. Predicting the water relations of the field-grown banana using the exuding latex. *Aust. J. Plant Physiol.* 17, 57–68.

Milburn, J. A., and Zimmermann, M. H. (1977). Preliminary studies on sapflow in *Cocos nucifera* L. II. Phloem transport. *New Phytol.* 79, 543–558.

Millard, P., and Chudek, J. A. (1993). Imaging the vascular continuity of *Prunus avium* during leaf senescence using nuclear magnetic resonance spectroscopy. *J. Exp. Bot.* 44, 599–603.

Miller, D. E., and Hang, A. N. (1980). Deficit, high-frequency irrigation of sugar beets with the line source technique. *Soil Sci. Soc. Am. J.* 44, 1295–1298.

Miller, D. M. (1985). Studies of root function in *Zea mays*. III. Xylem sap composition at maximum root pressure provides evidence of active transport into the xylem and a measurement of the reflection coefficient of the root. *Plant Physiol.* 77, 162–167.

Miller, E. C. (1938). "Plant Physiology," 2nd Ed. McGraw-Hill, New York.

Miller, E. C., and Saunders, A. R. (1923). Some observations on the temperature of the leaves of crop plants. *J. Agric. Res.* 26, 15–43.

Miller, E. E., and Salehzadeh, A. (1993). Stripper for bubble-free tensiometry. *J. Soil Sci. Soc. Am.* 57, 1470–1473.

Milligan, S. P., and Dale, J. E. (1988a). The effects of root treatments on growth of the primary leaves of *Phaseolus vulgaris* L.: General features. *New Phytol.* 108, 27–35.

Milligan, S. P., and Dale, J. E. (1988b). The effects of root treatments on growth of the primary leaves of *Phaseolus vulgaris* L.: Biophysical analysis. *New Phytol.* 109, 35–40.

Minshall, W. H. (1964). Effect of nitrogen-containing nutrients on exudation from detopped tomato plants. *Nature (London)* 202, 925–926.

Minshall, W. H. (1968). Effects of nitrogenous materials on translocation and stump exudation in root systems of tomato. *Can. J. Bot.* 46, 363–376.

Mittelheuser, C. J., and Van Steveninck, R. F. M. (1969) Stomatal closure and inhibition of transpiration induced by (RS)-abscisic acid. *Nature* (London) 221, 281–282.

Mohanty, P., and Boyer, J. S. (1976). Chloroplast response to low leaf water potentials. IV. Quantum yield is reduced. *Plant Physiol.* 57, 704–709.

Moinat, A. D. (1943). An auto-irrigator for growing plants in the laboratory. *Plant Physiol.* 18, 280–287.

Molisch, H. (1912). Das Offen- und Geschlossensein der Spaltöffnungen, veranschaulicht durch eine neue Methode (Infiltrations-methode). *Z. Bot.* 4, 106–122.

Molz, F. J., and Boyer, J. S. (1978). Growth-induced water potentials in plant cells and tissues. *Plant Physiol.* **62**, 423–429.

Molz, F. J., and Ferrier, J. M. (1982). Mathematical treatment of water movement in plant cells and tissues: a review. *Plant Cell Environ.* **5**, 191–206.

Molz, F. J., and Hornberger, G. M. (1973). Water transport through plant tissues in the presence of a diffusable solute. *Soil Sci. Soc. Am. Proc.* **37**, 833–837.

Molz, F. J., and Ikenberry, E. (1974). Water transport through plant cells and cell walls: Theoretical development. *Soil Sci. Soc. Am. Proc.* **38**, 699–704.

Molz, F. J., Kerns, D. V., Jr., Peterson, C. M., and Dane, J. H. (1979). A circuit analog model for studying quantitative water relations of plant tissue. *Plant Physiol.* **64**, 712–716.

Monteith, J. L. (1963). Dew: Facts and fallacies. *In* "The Water Relations of Plants" (A. J. Rutter and F. H. Whitehead, eds.), pp. 37–56. Wiley, New York.

Monteith, J. L. (1965). Evaporation and environment. *Symp. Soc. Exp. Biol.* **19**, 205–234.

Monteith, J. L., and Owen, P. C. (1958). A thermocouple method for measuring relative humidity in the range 95–100%. *J. Sci. Instrum.* **35**, 443–446.

Montfort, C. (1922). Die Wasserbilanz in Nährlösung, Salzlösung, und Hochmoorwasser. *Zeitschrf. Bot.* **14**, 98–172.

Mooney, H. A., Gulmon, S. L., Rundel, P. W., and Ehleringer, J. (1980). Further observations on the water relations of *Prosopis tamarugo* of the northern Atacama desert. *Oecologia* (Berlin) **44**, 177–180.

Moreland, D. E. (1950). A study of translocation of radioactive phosphorous in loblolly pine (*Pinus taeda* L.). *Elisha Mitchell Sci. Soc. J.* **66**, 175–181.

Morgan, J. M. (1980). Possible role of abscisic acid in reducing seed set in water-stressed wheat plants. *Nature* **285**, 655–657.

Morgan, J. M. (1983). Osmoregulation as a selection criterion for drought tolerance in wheat. *Aust. J. Agric. Res.* **34**, 607–614.

Morgan, J. M. (1984). Osmoregulation and water stress in higher plants. *Annu. Rev. Plant Physiol.* **35**, 299–319.

Morikawa, Y., Hattori, S., and Kiyono, Y. (1986). Transpiration of a 31-year-old *Chamaecyparis obtusa* Endl. stand before and after thinning. *Tree Physiol.* **2**, 105–114.

Morilla, C. A., Boyer, J. S., and Hageman, R. H. (1973). Nitrate reductase activity and polyribosomal content of corn (*Zea mays* L.) seedlings having low leaf water potentials. *Plant Physiol.* **51**, 817–824.

Morrison, J. C., Greve, L. C., and Richmond, P. A. (1993). Cell wall synthesis during growth and maturation of *Nitella* internodal cells. *Planta* **189**, 321–328.

Moss, D. N., and Rawlins, S. L. (1963). Concentration of carbon dioxide inside leaves. *Nature* (London) **197**, 1320–1321.

Moss, G. I., and Downey, L. A. (1971). Influence of drought stress on female gametophyte development in corn (*Zea mays* L.) and subsequent grain yield. *Crop Sci.* **11**, 368–372.

Mott, K. A. (1988). Do stomata respond to CO_2 concentrations other than intercellular? *Plant Physiol.* **86**, 200–203.

Mott, K. A., and Parkhurst, D. F. (1991). Stomatal responses to humidity in air and helox. *Plant Cell Environ.* **14**, 509–515.

Mozhaeva, L. V., and Pilschshikova, N. V. (1979). The motive force behind bleeding of plants. *Soviet Plant Physiol.* **26**, 802–807.

Muchow, R. C., and Sinclair, T. R. (1989). Epidermal conductance, stomatal density and stomatal size among genotypes of *Sorghum bicolor* (L.) Moench. *Plant Cell Environ.* **12**, 425–431.

Mudd, J. B., and Kozlowski, T. T., eds. (1975). "Responses of Plants to Air Pollution." Academic Press, New York.

Mujer, C. V., Rumpho, M. E., Lin, J.-J., and Kennedy, R. A. (1993). Constitutive and inducible aerobic and anaerobic stress proteins in the *Echinochloa* complex and rice. *Plant Physiol.* **101**, 217–226.

Muller, C. H. (1969). Allelopathy as a factor in ecological process. *Vegetation* **18**, 348–357.

Mundy, J., and Chua, N.-H. (1988). Abscisic acid and water-stress induce the expression of a novel rice gene. *EMBO J.* **7**, 2279–2286.

Munns, R. (1988). Why measure osmotic adjustment? *Aust. J. Plant Physiol.* **15**, 717–726.

Munns, R. (1990). Chemical signals moving from roots to shoots: the case against ABA. *In* "Importance of Root to Shoot Communication in the Response to Environmental Stress" (W. J. Davies and B. Jeffcoat, eds.), pp. 175–183. Monogr. 21. British Soc. Plant Growth Regulation, Bristol, U.K.

Munns, R. (1993). Physiological processes limiting plant growth in saline soils: Some dogmas and hypotheses. *Plant Cell Environ.* **16**, 15–24.

Munns, R., Brady, C. J., and Barlow, E. W. R. (1979). Solute accumulation in the apex and leaves of wheat during water stress. *Aust. J. Plant Physiol.* **6**, 379–389.

Munns, R., and King, R. W. (1988). Abscisic acid is not the only stomatal inhibitor in the transpiration stream of wheat plants. *Plant Physiol.* **88**, 703–708.

Munns, R., and Pearson, C. J. (1974). Effect of water deficit on translocation of carbohydrate in *Solanum tuberosum*. *Aust. J. Plant Physiol.* **1**, 529–537.

Munns, R., and Termaat, A. (1986). Whole plant responses to salinity. *Aust. J. Plant Physiol.* **13**, 143–160.

Munns, R., and Weir, R. (1981). Contribution of sugars to osmotic adjustment in elongating and expanded zones of wheat leaves during moderate water deficits at two light levels. *Aust. J. Plant Physiol.* **8**, 93–105.

Murphy, R., and Smith, J. A. C. (1994). A critical comparison of the pressure-probe and pressure-chamber techniques for estimating leaf cell turgor pressure in *Kalanchoe daigremontiana*. *Plant Cell Environ.* **17**, 15–29.

Musser, R. L., Thomas, S. A., and Kramer, P. J. (1983). Short and long term effects of root and shoot chilling of Ransom soybean. *Plant Physiol.* **73**, 778–783.

Mustafa, M. A., and Letey, J. (1970). Factors affecting effectiveness of two surfactants on water-repellent soils. *Calif. Agr.* **24**, 12–13.

Myers, B. A., Kuppers, M., and Neales, T. F. (1987). Effect of stem excision under water on bulk leaf water potential, leaf conductance, CO_2 assimilation and stemwood water storage in *Eucolyptus behriana* F. Muell. *Aust. J. Plant Physiol.* **14**, 135–145.

Myers, P. N., Setter, T. L., Madison, J. T., and Thompson, J. F. (1990). Abscisic acid inhibition of endosperm cell division in cultured maize kernels. *Plant Physiol.* **94**, 1330–1336.

Neales, T. F., Patterson, A. A., and Hartney, V. J. (1968). Physiological adaptation to drought in the carbon assimilation and water loss of xerophytes. *Nature* (London) **219**, 469–472.

Neher, H. V. (1993). Effects of pressures inside Monterey pine trees. *Trees* **8**, 9–17.

Neill, S. J., and Horgan, R. (1985). Abscisic acid production and water relations in wilty tomato mutants subjected to water deficiency. *J. Exp. Bot.* **36**, 1222–1231.

Némethy, G., and Scheraga, H. A. (1962). Structure of water and hydrophobic bonding in proteins. I. A model for the thermodynamic properties of liquid water. *J. Chem. Physics* **36**, 3382–3400.

Newman, E. I. (1966). Relationship between root growth of flax (*Linum usitatissimum*) and soil water potential. *New Phytol.* **65**, 273–283.

Newman, E. I. (1969). Resistance to water flow in soil and plant. I. Soil resistance in relation to amounts of root: Theoretical estimates. *J. appl. Ecol.* **6**, 1–12.

Newman, E. I. (1973). Permeability to water of the roots of five herbaceous species. *New Phytol.* **72**, 547–555.

Newman, E. I. (1974). Root-soil water relations. *In* "The Plant Root and its Environment" (E. W. Carson, ed.), pp. 363–440. University Press of Virginia, Charlottesville, VA.

Newman, E. I. (1976). Water movement through root systems. *Phil. Trans. R. Soc. London Ser. B* **273**, 463–478.

Newman, E. I., and Andrews, R. T. (1973). Uptake of phosphorus and potassium in relation to root growth and root density. *Plant Soil* **38**, 49–69.

Nir, I., Klein, S., and Poljakoff-Mayber, A. (1969). Effect of moisture stress on submicroscopic structure of maize roots. *Aust. J. Biol. Sci.* **22**, 17–33.

Nishiyama, I. (1975). A break on the Arrhenius plot of germination activity in rice seeds. *Plant Cell Physiol.* **16**, 535–536.

Nnyamah, J. U., and Black, T. A. (1977). Rates and patterns of water uptake in a Douglas fir forest. *Soil Sci. Soc. Am. J.* **41**, 972–979.

Nnyamah, J. U., Black, T. A., and Tan, C. S. (1978). Resistance to water uptake in a Douglas fir forest. *Soil Sci.* **126**, 63–76.

Nobel, P. S. (1974). "Biophysical Plant Physiology." W. H. Freeman, San Francisco, CA.

Nobel, P. S. (1983). "Biophysical Plant Physiology and Ecology." W. H. Freeman and Company, New York.

Nobel, P. S. (1991). "Physicochemical and Environmental Plant Physiology." Academic Press, London.

Nobel, P. S., and Cui, M. (1992). Hydraulic conductances of the soil, the root-soil air gap, and the root: changes for desert succulents in drying soil. *J. Exp. Bot.* **43**, 319–326.

Nobel, P. S., Miller, P. M., and Graham, E. A. (1992). Influence of rocks on soil temperature, soil water potential, and rooting patterns of desert succulents. *Oecologia* **92**, 90–96.

Nonami, H., and Boyer, J. S. (1987). Origin of growth-induced water potential: Solute concentration is low in apoplast of enlarging tissues. *Plant Physiol.* **83**, 596–601.

Nonami, H., and Boyer, J. S. (1989). Turgor and growth at low water potentials. *Plant Physiol.* **89**, 798–804.

Nonami, H., and Boyer, J. S. (1990a). Primary events regulating stem growth at low water potentials. *Plant Physiol.* **93**, 1601–1609.

Nonami, H., and Boyer, J. S. (1990b). Wall extensibility and cell hydraulic conductivity decrease in enlarging stem tissues at low water potentials. *Plant Physiol.* **93**, 1610–1619.

Nonami, H., and Boyer, J. S. (1993). Direct demonstration of a growth-induced water potential gradient. *Plant Physiol.* **102**, 13–19.

Nonami, H., and Schulze, E.-D. (1989). Cell water potential, osmotic potential, and turgor in the epidermis and mesophyll of transpiring leaves: Combined measurements with the cell pressure probe and nanoliter osmometer. *Planta* **177**, 35–46.

Nonami, H., Boyer, J. S., and Steudle, E. S. (1987). Pressure probe and isopiestic psychrometer measure similar turgor. *Plant Physiol.* **83**, 592–595.

Nonami, H., Schulze, E.-D., and Ziegler, H. (1991). Mechanisms of stomatal movement in response to air humidity, irradiance and xylem water potential. *Planta* **183**, 57–64.

Norberg, P., and Liljenberg, C. (1991). Lipids of plasma membranes prepared from oat root cells. *Plant Physiol.* **96**, 1136–1141.

Norby, R. J. (1987). Nodulation and nitrogenase activity in nitrogen-fixing woody plants stimulated by CO_2 enrichment of the atmosphere. *Physiol. Plant.* **41**, 77–82.

Norby, R. J., and Kozlowski, T. T. (1980). Allelopathic potential of ground cover species on *Pinus resinosa* seedlings. *Plant Soil* **57**, 363–374.

Norris, J. R., Read, D., and Varma, A. K. (1994). "Techniques for Mycorrhizal Research." Academic Press, San Diego.

Norris, R. F., and Bukovac, M. J. (1968). Structure of the pear leaf cuticle with special reference to cuticular penetration. *Am. J. Bot.* **55**, 975–983.

North, G. B., and Nobel, P. S. (1991). Changes in hydraulic conductivity and anatomy caused by drying and rewetting roots of *Agave deserti* (Agavaceae). *Am. J. Bot.* **78**, 906–915.

Nutman, F. J. (1933). The root-system of *Coffea arabica*. II. The effect of some soil conditions in modifying the "normal" root-system. *Emp. J. Exp. Agric.* **1**, 285–296.

Nutman, F. J. (1934). The root system of *Coffea arabica*. III. The spatial distribution of the absorbing area of the root. *Emp. J. Exp. Agric.* **2**, 293–302.

Nutman, F. J. (1941). Studies of the physiology of *Coffea arabica*. III. Transpiration rates of whole trees in relation to natural environmental conditions. *Ann. Bot.* **5**, 59–82.

Nye, P. H., and Tinker, P. B. (1977). "Solute Movement in the Soil-Root system." University of California Press, Berkeley, CA.

Oades, J. M. (1978). Mucilages at the root surface. *J. Soil Sci.* **29**, 1–16.

Oaks, A. (1992). A re-evaluation of nitrogen assimilation in roots. *Bioscience* **42**, 103–111.

Oertli, J. J. (1966). Active water transport in plants. *Physiol. Plant.* **19**, 809–817.

Ogawa, T., Grantz, D., Boyer, J. S., and Govindjee. (1982). Effects of cations and abscisic acid on chlorophyll *a* fluorescence in guard cells of *Vicia faba*. *Plant Physiol.* **69**, 1140–1144.

Ogawa, T., Ishikawa, H., Shimada, K., and Shibata, K. (1978). Synergistic action of red and blue light and action spectra for malate formation in guard cells of *Vicia faba* L. *Planta* **142**, 61–65.

Ögren, E., and Öquist, G. (1985). Effects of drought on photosynthesis, chlorophyll fluorescence and photoinhibition susceptibility in intact willow leaves. *Planta* **166**, 380–388.

Oke, T. R. (1987). "Boundary Layer Climates," 2nd Ed. Methuen, London.

O'Leary, J. W. (1969). The effect of salinity on permeability of roots to water. *Israel J. Bot.* **18**, 1–9.

O'Leary, J. W., and Knecht, G. W. (1971). The effect of relative humidity on growth, yield and water consumption of bean plants. *J. Am. Soc. Hort. Sci.* **96**, 263–265.

O'Leary, J. W., and Kramer, P. J. (1964). Root pressure in conifers. *Science* **145**, 284–285.

O'Leary, J. W., and Prisco, J. T. (1970). Response of osmotically stressed plants to growth regulators. *Adv. Front. Plant Sci.* **25**, 129–139.

Olesen, P., and Robards, A. W. (1990). The neck region of plasmodesmata: General architecture and some functional aspects. *In* "Parallels in Cell to Cell Junctions in Plants and Animals" (A. W. Robards, W. J. Lucas, J. D. Pitts, H. J. Jongsma, and D. C. Spray, eds.), pp. 145–170. Cell to Cell Signals in Plants and Animals, NATO Advanced Research Workshop, Springer-Verlag, Heidelberg, Germany.

Oliver, M. J. (1991). Influence of protoplasmic water loss on the control of protein synthesis in the desiccation-tolerant moss *Tortula ruralis*. *Plant Physiol.* **97**, 1501–1511.

Olszyk, D. M., and Tingey, D. T. (1986). Joint action of O_3 and SO_2 in modifying plant gas exchange. *Plant Physiol.* **82**, 401–405.

Omasa, K., Hashimoto, Y., Kramer, P. J., Strain, B. R., Aiga, I., and Kondo, J. (1985a). Direct observation of reversible and irreversible stomatal responses of attached sunflower leaves to SO_2. *Plant Physiol.* **79**, 153–158.

Omasa, K., Onoe, M., and Yamada, H. (1985b). NMR imaging for measuring root systems and soil water content. *Environ. Control Biol.* **23**, 99–102.

O'Neill, E. G., Luxmoore, R. J., and Norby, R. J. (1987). Increases in mycorrhizal colonization and seedling growth in *Pinus echinata* and *Quercus alba* in an enriched CO_2 atmosphere. *Can. J. For. Res.* **17**, 878–883.

Oosterhuis, D. M., Walker, S., and Eastham, J. (1985). Soybean leaflet movements as an indicator of crop water stress. *Crop Sci.* **25**, 1101–1106.

Oosting, H. J. (1956). "The Study of Plant Communities," 2nd Ed. W. H. Freeman, San Francisco, CA.

Oppenheimer, H. R. (1941). Root cushions, root stalagmites and similar structures. *Palestine J. Bot. Ser.* **4**, 11–19.

Orchard, P. W., and Jessop, R. S. (1984). The response of sorghum and sunflower to short-term waterlogging. I. Effects of stage of development and duration of waterlogging on growth and yield. *Plant Soil* **81**, 119–132.

Ordin, L., and Kramer, P. J. (1956). Permeability of *Vicia faba* root segments to water as measured by diffusion of deuterium hydroxide. *Plant Physiol.* **31**, 468–471.

Orians, G. H., and Solbrig, O. T. (1977). A cost-income model of leaves and roots with special reference to arid and semi-arid areas. *Am. Nat.* **111**, 677–690.

Ort, D. R., and Boyer, J. S. (1985). Plant productivity, photosynthesis and environmental stress. *In* "Changes in Eukaryotic Gene Expression in Response to Environmental Stress" (B. G. Atkinson and D. B. Walden, eds.), pp. 279–313. Academic Press, New York.

Ortega, J. K. E., Zehr, E. G., and Keanini, R. G. (1989). *In vivo* creep and stress relaxation experiments to determine the wall extensibility and yield threshold for the sporangiophores of *Phycomyces*. *Biophys. J.* **56**, 465–475.

Ortiz-Lopez, A., Ort, D. R., and Boyer, J. S. (1991). Photophosphorylation in attached leaves of *Helianthus annuus* at low water potentials. *Plant Physiol.* **96**, 1018–1025.

Oskamp, J., and Batjer, L. P. (1932). Soils in relation to fruit growing in New York. II. Size, production, and rooting habit of apple trees on different soil types in the Hilton and Morton Areas, Monroe County. Cornell Univ. Agr. Exp. Sta. Bull. 550.

Osmond, C. B. (1978). Crassulacean acid metabolism: A curiosity in context. *Annu. Rev. Plant Physiol.* **29**, 379–414.

Osmond, D. L., Wilson, R. F., and Raper, C. D., Jr. (1982). Fatty acid composition and nitrate uptake of soybean roots during acclimation to low temperature. *Plant Physiol.* **70**, 1689–1693.

Osonubi, O., Oren, R., Werk, K. S., Schulze, E.-D., and Heilmeier, H. (1988). Performance of two *Picea abies* (L.) Karst. stands at different stages of decline. IV. Xylem sap concentrations of magnesium, calcium, potassium, and nitrogen. *Oecologia* **77**, 1–6.

O'Toole, J. C. (1982). Adaptation of rice to drought-prone environments. *In* "Drought Resistance in Crops with Emphasis on Rice," pp. 195–213. International Rice Research Institute, Manila, Philippines.

O'Toole, J. C., and Baldia, E. P. (1982). Water deficits and mineral uptake in rice. *Crop Sci.* **22**, 1144–1150.

O'Toole, J. C., and Bland, W. L. (1987). Genotypic variation in crop plant root systems. *Adv. Agron.* **41**, 91–145.

O'Toole, J. C., Crookston, R. K., Treharne, K. J., and Ozbun, J. L. (1976). Mesophyll resistance and carboxylase activity. *Plant Physiol.* **57**, 465–468.

O'Toole, J. C., and Cruz, R. T. (1980). Response of leaf water potential, stomatal resistance and leaf rolling to water stress. *Plant Physiol.* **63**, 428–437.

O'Toole, J. C., Hsiao, T. C., and Namuco, O. S. (1984). Panicle water relations during water stress. *Plant Sci. Lett.* **33**, 137–143.

Oussible, M., Crookston, R. K., and Larson, W. E. (1992). Subsurface compaction reduces the root and shoot growth and grain yield of wheat. *Agron. J.* **84**, 34–38.

Outlaw, W. H., Jr. (1989). Critical examination of the quantitative evidence for and against photosynthetic CO_2 fixation by guard cells. *Physiol. Plant.* **77**, 275–281.

Outlaw, W. H., Jr., and Manchester, J. (1979). Guard cell starch concentration quantitatively related to stomatal aperture. *Plant Physiol.* **64**, 79–82.

Outlaw, W. H., Jr., Mayne, B. C., Zenger, V. E., and Manchester, J. (1981). Presence of both photosystems in guard cells of *Vicia faba* L. Implications for environmental signal processing. *Plant Physiol.* **67**, 12–16.

Owen, P. C. (1952). The relation of germination of wheat to water potential. *J. Exp. Bot.* **3**, 188–203.

Owston, P. W., Smith, J. L., and Halverson, H. F. (1972). Seasonal water movement in tree stems. *For. Sci.* **18**, 266–272.

Pallardy, S. G., and Kozlowski, T. T. (1979). Stomatal response of *Populus* clones to leaf water potential and environment. *Oecologia* **40**, 371–380.

Pallardy, S. G., and Kozlowski, T. T. (1980). Cuticle development in the stomatal region of *Populus* clones. *New Phytol.* **85**, 363–365.

Palzkill, D. A., and Tibbits, T. W. (1977). Evidence that root pressure flow is required for calcium transport to head leaves of cabbage. *Plant Physiol.* **60**, 854–856.

Pandey, R. K., Herrera, W. A. T., and Pendleton, J. W. (1984a). Drought response of grain legumes under irrigation gradient. I. Yield and yield components. *Agron. J.* **76**, 549–553.

Pandey, R. K., Herrera, W. A. T., Villegas, A. N., and Pendleton, J. W. (1984b). Drought response of grain legumes under irrigation gradient. III. Plant growth. *Agron. J.* **76**, 557–560.

Pankhurst, C. E., and Sprent, J. I. (1975). Effects of water stress on the respiratory and nitrogen-fixing activity of soybean root nodules. *J. Exp. Bot.* **26**, 287–304.

Pankhurst, C. E., and Sprent, J. I. (1976). Effects of temperature and oxygen tension on the nitrogenase and respiratory activities of turgid and water-stressed soybean and French bean root nodules. *J. Exp. Bot.* **27**, 1–9.

Paolillo, D. J., Jr. (1989). Cell and axis elongation in etiolated soybean seedlings are altered by moisture stress. *Bot. Gaz.* **150**, 101–107.

Pararajasingham, S., and Knievel, D. P. (1990). Nitrogenase activity, photosynthesis and total non-structural carbohydrates in cowpea during and after drought stress. *Can. J. Plant Sci.* **70**, 1005–1012.

Parker, J. (1949). Effects of variation in the root-leaf ratio on transpiration rate. *Plant Physiol.* **24**, 739–743.

Parker, J. (1950). The effects of flooding on the transpiration and survival of some southeastern forest tree species. *Plant Physiol.* **25**, 453–460.

Parkhurst, D. F. (1978). The adaptive significance of stomatal occurrence on one or both surfaces of leaves. *J. Ecol.* **66**, 367–383.

Parkhurst, D. F., and Loucks, O. L. (1972). Optimal leaf size in relation to environment. *J. Ecol.* **60**, 505–537.

Parrish, J. T. (1983). Paleozoic atmospheric circulation and oceanic upwelling. *In* "Paleoclimate and Mineral Deposits" (T. M. Cronin, W. F. Cannon, and R. Z. Poore, eds.), pp. 37–39. U.S. Geological Survey, Alexandria, VA.

Parsons, L. R., Combs, B. S., and Tucker, D. P. H. (1985). Citrus freeze protection with micro-sprinkler irrigation during an advective freeze. *Hort. Sci.* **20**, 1078–1080.

Parsons, L. R., and Kramer, P. J. (1974). Diurnal cycling in root resistance to water movement. *Physiol. Plant.* **30**, 19–23.

Parsons, L. R., Wheaton, T. A., Faryna, N. D., and Jackson, J. L. (1991). Elevated microsprinklers improve protection of citrus trees in an advective freeze. *Hort. Sci.* **26**, 1149–1151.

Passioura, J. B. (1972). The effect of root geometry on the yield of wheat growing on stored water. *Aust. J. Agric. Res.* **23**, 745–752.

Passioura, J. B. (1980a). The transport of water from soil to shoot in wheat seedlings. *J. Exp. Bot.* **31**, 333–345.

Passioura, J. B. (1980b). The meaning of matric potential. *J. Exp. Bot.* **31**, 1161–1169.

Passioura, J. B. (1982). Water in the soil-plant-atmosphere continuum. *In* "Encyclopedia of Plant Physiology" (O. L. Lange, P. S. Nobel, C. B. Osmond, and H. Ziegler, eds.), Vol. 12B, pp. 5–33. Springer-Verlag, New York.

Passioura, J. B. (1988a). Root signals control leaf expansion in wheat seedlings growing in drying soil. *Aust. J. Plant Physiol.* **15**, 687–693.

Passioura, J. B. (1988b). Water transport in and to roots. *Annu. Rev. Plant Physiol. Plant Mol. Biol.* **39**, 245–265.

Passioura, J. B. (1991). An impasse in plant water relations? *Bot. Acta* **104**, 405–411.

Passioura, J. B., and Tanner, C. B. (1985). Oscillations in apparent hydraulic conductance of cotton plants. *Aust. J. Plant Physiol.* **12**, 445–461.

Pate, J. S., Layzell, D. B., and Atkins, C. A. (1979). Economy of carbon and nitrogen in a nodulated and nonnodulated (NO_3-grown) legume. *Plant Physiol.* **64**, 1083–1088.

Patric, J. H., Douglass, J. E., and Hewlett, J. D. (1965). Soil water absorption by mountain and Piedmont forests. *Soil Sci. Soc. Am. Proc.* **29**, 303–308.

Patterson, D. T., Bunce, J. A., Alberte, R. S., and Van Volkenburgh, E. (1977). Photosynthesis in relation to leaf characteristics of cotton from controlled and field environments. *Plant Physiol.* **59**, 384–387.

Pavlychenko, T. K. (1937). Quantitative study of the entire root system of weed and crop plants under field conditions. *Ecology* **18**, 62–79.

Pearcy, R. W. (1990). Sunflecks and photosynthesis in plant canopies. *Annu. Rev. Plant Physiol. Plant Mol. Biol.* **41**, 421–453.

Pearcy, R. W., Ehleringer, J., Mooney, H. A., and Rundel, P. W., eds. (1989). "Plant Physiological Ecology." Chapman and Hall, New York.

Pearson, M., and Mansfield, T. A. (1993). Interacting effects of ozone and water stress on the stomatal resistance of beech (*Fagus sylvatica* L.). *New Phytol.* **123**, 351–358.

Pedersen, O. (1993). Long-distance water transport in aquatic plants. *Plant Physiol.* **103**, 1369–1375.

Peñuelas, J., and Matamala, R. (1990). Changes in N and S leaf content, stomatal density and specific leaf area of 14 plant species during the last three centuries of CO_2 increase. *J. Exp. Bot.* **41**, 1119–1124.

Pereira, J. S., and Kozlowski, T. T. (1977). Influence of light intensity, temperature, and leaf area on stomatal aperture and water potential of woody plants. *Can. J. For. Res.* **7**, 145–153.

Perry, D. A., Molina, R., and Amaranthus, M. P. (1987). Mycorrhizae, mycorrhizospheres, and reforestation: Current knowledge and research needs. *Can. J. For. Res.* **17**, 929–940.

Peters, D. B., and Russell, M. B. (1959). Relative water losses by evaporation and transpiration in field corn. *Soil Sci. Soc. Am. Proc.* **23**, 170–173.

Peterson, C. A. (1988). Exodermal Casparian bands: Their significance for ion uptake by roots. *Physiol. Plant.* **72**, 204–208.

Peterson, C. A., Emanuel, M. E., and Humphreys, G. B. (1981). Pathway of movement of apoplastic fluorescent dye tracers through the endodermis at the site of secondary root formation in corn (*Zea mays*) and broad bean (*Vicia faba*). *Can. J. Bot.* **59**, 618–625.

Peterson, C. A., and Perumalla, C. J. (1984). Development of the hypodermal Casparian band in corn and onion roots. *J. Exp. Bot.* **35**, 51–57.

Pfeffer, W. (1877). "Osmotische Untersuchunger." W. Englemann, Leipzig. English translation by G. R. Kepner and E. J. Stadelmann. 1985. "Osmotic Investigations: Studies on Cell Mechanics." Van Nostrand Reinhold, New York.

Pfeffer, W. (1900). "The Physiology of Plants." 2nd Ed. English translation by A. J. Ewart. Oxford at the Clarendon Press.

Philip, J. R. (1957). The physical principles of soil water movement during the irrigation cycle. *Proc. Int. Congr. Irrig. Drain.* **8**, 124–154.

Philip, J. R. (1958a). The osmotic cell, solute diffusibility, and the plant water economy. *Plant Physiol.* **33**, 264–271.

Philip, J. R. (1958b). Osmosis and diffusion in tissues: Half-times and internal gradients. *Plant Physiol.* **33**, 275–278.

Philip, J. R. (1958c). Correction to the paper entitled "Osmosis and diffusion in tissue: Half times and internal gradients." *Plant Physiol.* **33**, 443.

Philip, J. R. (1958d). Propagation of turgor and other properties through cell aggregations. *Plant Physiol.* **33**, 271–274.

Phillips, I. D. (1964). The importance of an aerated root system in the regulation of growth levels in the shoot of *Helianthus annuus*. *Ann. Bot.* (London) [N.S.] **28**, 17–36.

Pick, U., and Bassilian, S. (1982). Activation of magnesium ion specific adenosinetriphosphatase in chloroplast coupling factor 1 by octylglucoside. *Biochemistry* **24**, 6144–6152.

Pickard, W. F. (1973). A heat pulse method of measuring water flux in woody plant stems. *Math. Biosci.* **16**, 247–262.

Pier, P. A., and Berkowitz, G. A. (1987). Modulation of water stress effects on photosynthesis by altered leaf K^{+1}. *Plant Physiol.* **85**, 655–661.

Pierre, W. H., and Pohlman, G. G. (1934). The phosphorus concentration of the exuded sap of corn as a measure of the available phosphorus in the soil. *J. Am. Soc. Agron.* **25**, 160–171.

Pinter, P. J., Jr., Zipoli, G., Reginato, R. J., Jackson, R. D., Idso, S. B., and Hohman, J. P. (1990). Canopy temperature as an indicator of differential water use and yield performance among wheat cultivars. *Agr. Water Manag.* **18**, 35–48.

Piñero, D., Sarukhan, J., and Alberdi, P. (1982). The costs of reproduction in a tropical palm, *Astrocaryum mexicanum*. *J. Ecol.* **70**, 473–481.

Pisek, A., and Cartellieri, E. (1932). Zur Kenntnis des Wasserhaushaltes der Pflanzen. I. Sonnenpflanzen. *Jahrb. Wiss. Bot.* **75**, 195–251.

Pleasants, A. L. (1930). The effect of nitrate fertilizer on stomatal behavior. *J. Elisha Mitchell Sci. Soc.* **46**, 95–116.

Plymale, E. L, and Wylie, R. B. (1944). The major veins of mesomorphic leaves. *Am. J. Bot.* **31**, 99–106.

Poljakoff-Mayber, A., and Gale, J. (1975). "Plants in Saline Environments." Springer-Verlag, New York.

Pollard, J. K., and Sproston, T. (1954). Nitrogenous constituents of sap exuded from the sapwood of *Acer saccharum*. *Plant Physiol.* **29**, 360–364.

Portas, C. A. M., and Taylor, H. M. (1976). Growth and survival of young plant roots in dry soil. *Soil Sci.* **121**, 170–175.

Postlethwait, S. N., and Rogers, B. (1958). Tracing the path of the transpiration stream in trees by the use of radioactive isotopes. *Am. J. Bot.* **35**, 753–757.

Potrykus, I. (1991). Gene transfer to plants: Assessment of published approaches and results. *Annu. Rev. Plant Physiol. Plant Mol. Biol.* **42**, 205–225.

Potter, J. R., and Boyer, J. S. (1973). Chloroplast response to low leaf water potentials. II. Role of osmotic potential. *Plant Physiol.* **51**, 993–997.

Press, M. C., and Whittaker, J. B. (1993). Exploitation of the xylem stream by parasitic organisms. *Phil. Trans. Roy. Soc. London Ser. B* **341**, 101–111.

Preston, G. M., Carroll, T. P., Guggino, W. B., and Agre, P. (1992). Appearance of water channels in *Xenopus* oocytes expressing red cell CHIP28 protein. *Science* **256**, 385–387.

Preston, R. D. (1961). Theoretical and practical implications of the stresses in the water-conducting system. *In* "Recent Advances in Botany," pp. 1144–1149. University of Toronto Press, Toronto.

Preston, R. D. (1974). "The Physical Biology of Plant Cell Walls." Chapman and Hall, London.

Probine, M. C., and Preston, R. D. (1962). Cell growth and the structure and mechanical properties of the wall in internodal cells of *Nitella opaca*. II. Mechanical properties of the walls. *J. Exp. Bot.* **13**, 111–127.

Proebsting, E. L. (1943). Root distribution of some deciduous fruit trees in a California orchard. *Proc. Am. Soc. Hort. Sci.* **43**, 1–4.

Proebsting, E. L., and Gilmore, A. E. (1941). The relation of peach root toxicity to the reestablishing of peach orchards. *Am. Soc. Hort. Sci. Proc.* **38**, 21–26.

Putnam, A. R. (1983). Allelopathic chemicals. *Chem. Eng. News* **61**, 34–45.

Putnam, A. R., and Duke, W. B. (1978). Allelopathy in agroecosystems. *Annu. Rev. Phytopathol.* **16**, 431–451.

Putnam, A. R., and Tang, C. S., eds. (1986). "The Science of Allelopathy." Wiley New York.

Quadir, A., Harrison, P. J., and DeWreede, R. E. (1979). The effects of emergence and submergence on the photosynthesis and respiration of marine macrophytes. *Phycologia* 18, 83–88.

Queen, W. H. (1967). "Radial Movement of Water and ^{32}P through Suberized and Unsuberized Roots of Grape." Ph.D. Dissert., Duke University, Durham, NC.

Quick, W. P., Chaves, M. M., Wendler, R., David, M., Rodrigues, M. L., Passaharinho, J. A., Pereira, J. S., Adcock, M. D., Leegood, R. C., and Stitt, M. (1992). The effect of water stress on photosynthetic carbon metabolism in four species grown under field conditions. *Plant Cell Environ.* 15, 25–35.

Quisenberry, J. E., Cartwright, G. B., and McMichael, B. L. (1984). Genetic relationship between turgor maintenance and growth in cotton germplasm. *Crop Sci.* 24, 479–482.

Quispel, A. (1983). Dinitrogen-fixing symbioses with legumes, non-legume angiosperms and associative symbioses. *In* "Encyclopedia of Plant Physiology" (A. Läuchli and R. L. Bieleski, eds.), Vol. 15A, pp. 286–329. Springer-Verlag, Berlin/Heidelberg.

Raber, O. (1937). Water utilization by trees, with special reference to the economic forest species of the north temperate zone. U.S. Dept. Agr. Misc. Publ. 257.

Racker, E. (1977). Mechanisms of energy transformations. *Annu. Rev. Biochem.* 46, 1006–1014.

Radin, J. W. (1983). Control of plant growth by nitrogen: Differences between cereals and broadleaf species. *Plant Cell Environ.* 6, 65–68.

Radin, J. W. (1984). Stomatal responses to water stress and to abscisic acid in phosphorus-deficient cotton plants. *Plant Physiol.* 76, 392–394.

Radin, J. W., and Ackerson, R. C. (1981). Water relations of cotton plants under nitrogen deficiency. III. Stomatal conductance, photosynthesis, and abscisic acid accumulation during drought. *Plant Physiol.* 67, 115–119.

Radin, J. W., and Boyer, J. S. (1982). Control of leaf expansion by nitrogen nutrition in sunflower plants: Role of hydraulic conductivity and turgor. *Plant Physiol.* 69, 771–775.

Radin, J. W., and Matthews, M. A. (1989). Water transport properties of cortical cells in roots of nitrogen- and phosphorus-deficient cotton seedlings. *Plant Physiol.* 89, 264–268.

Radin, J. W., and Parker, L. L. (1979a). Water relations of cotton plants under nitrogen deficiency. I. Dependence upon leaf structure. *Plant Physiol.* 64, 495–498.

Radin, J. W., and Parker, L. L. (1979b). Water relations of cotton plants under nitrogen deficiency. II. Environmental interactions on stomata. *Plant Physiol.* 64, 499–501.

Radin, J. W., Reaves, L. L., Mauney, J. R., and French, O. F. (1992). Yield enhancement in cotton by frequent irrigation during fruiting. *Agron. J.* 84, 551–557.

Radler, F. (1965). Reduction of loss of moisture by the cuticle wax components of grapes. *Nature* (London) 207, 1002–1003.

Raison, J. K., Pike, C. S., and Berry, J. A. (1982). Growth temperature-induced alterations in the thermotropic properties of *Nerium oleander* membrane lipids. *Plant Physiol.* 70, 215–218.

Rand, R. (1983). Fluid mechanics of green plants. *Annu. Rev. Fluid. Mech.* 15, 29–45.

Rand, R. P. (1992). Raising water to new heights. *Science* 256, 618.

Raney, F., and Vaadia, Y. (1965a). Movement of tritiated water in the root system of *Helianthus annuus* in the presence and absence of transpiration. *Plant Physiol.* 40, 378–382.

Raney, R., and Vaadia, Y. (1965b). Movement and distribution of THO in tissue water and vapor transpired by shoots of *Helianthus* and *Nicotiana*. *Plant Physiol.* 40, 383–388.

Rao, A. S. (1990). Root flavonoids. *Bot. Rev.* 56, 1–84.

Rao, I. M., Sharp, R. E., and Boyer, J. S. (1987). Leaf magnesium alters photosynthetic response to low water potentials in sunflower. *Plant Physiol.* 84, 1214–1219.

Raper, C. D., Jr., and Barber, S. A. (1970a). Rooting systems of soybeans. I. Differences in root morphology among varieties. *Agron. J.* 62, 581–584.

Raper, C. D., Jr., and Barber, S. A. (1970b). Rooting systems of soybeans. II. Physiological effectiveness as nutrient absorption surfaces. *Agron. J.* 62, 686–588.

Raper, C. D., Jr., and Kramer, P. J., eds. (1983). "Crop Reactions to Water and Temperature Stresses in Humid, Temperate Climates." Westview Press, Boulder, CO.

Raper, C. D., Jr., and Kramer, P. J. (1987). Stress physiology. In "Soybeans" (J. R. Wilcox, ed.), 2nd Ed., pp. 589–641. American Society of Agronomy, Madison, WI.

Raper, C. D., Jr., Patterson, D. T., Parsons, L. R., and Kramer, P. J. (1977). Relative growth and nutrient accumulation rates for tobacco. *Plant Soil* **46**, 473–486.

Raschke, K. (1970). Leaf hydraulic system: Rapid epidermal and stomatal responses to changes in water supply. *Science* **167**, 189–191.

Raschke, K. (1972). Saturation kinetics of the velocity of stomatal closing in response to CO_2. *Plant Physiol.* **49**, 229–234.

Raschke, K. (1975). Stomatal action. *Annu. Rev. Plant Physiol.* **26**, 309–340.

Raschke, K. (1976). How stomata resolve the dilemma of opposing priorities. *Phil. Trans. Roy. Soc. London Ser. B* **273**, 551–560.

Raschke, K. (1986). The influence of the CO_2 content of the ambient air on stomatal conductance and the CO_2 concentration in leaves. In "Carbon Dioxide Enrichment of Greenhouse Crops" (H. Z. Enoch and B. A. Kimball, eds.), Vol. 2, pp. 87–102. CRC Press, Boca Raton, FL.

Raschke, K., and Hedrich, R. (1985). Simultaneous and independent effects of abscisic acid on stomata and the photosynthetic apparatus in whole leaves. *Planta* **163**, 105–118.

Rascio, A., Platani, C., Di Fonzo, N., and Wittmer, G. (1992). Bound water in durum wheat under drought stress. *Plant Physiol.* **98**, 908–912.

Raskin, I., and Kende, H. (1985). Mechanism of aeration in rice. *Science* **228**, 327–329.

Raven, J. A. (1983). Phytophages of xylem and phloem: a comparison of animal and plant sap-feeders. *Adv. Ecol. Res.* **13**, 135–234.

Ravina, I., and Magier, J. (1984). Hydraulic conductivity and water retention of clay soils containing coarse fragments. *Soil Sci. Soc. Am. J.* **48**, 736–740.

Rawlins, S. L. (1963). "Resistance to Water Flow in the Transpiration Stream," pp. 69–85. Bull. 664, Conn. Agric. Exp. Stn., New Haven.

Rawlins, S. L., and Dalton, F. N. (1967). Psychrometric measurement of soil water potential without precise temperature control. *Proc. Soil Sci. Soc. Am.* **31**, 297–301.

Ray, P. M. (1960). On the theory of osmotic water movement. *Plant Physiol.* **35**, 783–795.

Ray, P. M. (1967). Autoradiographic study of cell wall deposition in growing plant cells. *J. Cell Biol.* **35**, 659–674.

Rayan, A., and Matsuda, K. (1988). The relation of anatomy to water movement and cellular response in young barley leaves. *Plant Physiol.* **87**, 853–858.

Read, D. W. L., Vleck, S. V., and Pelton, W. L. (1962). Self-irrigating greenhouse pots. *Agron. J.* **54**, 457–4770.

Reckmann, U., Scheibe, R., and Raschke, K. (1990). Rubisco activity in guard cells compared with the solute requirement for stomatal opening. *Plant Physiol.* **92**, 246–253.

Reed, H. S., and MacDougal, T. (1937). Periodicity in the growth of the orange tree. *Growth* **1**, 371–373.

Reed, J. F. (1939). "Root and Shoot Growth of Shortleaf and Loblolly Pines in Relation to Certain Environmental Conditions." Duke Univ. Sch. For. Bull. 4.

Reich, P. B., Teskey, R. O., Johnson, P. S., and Hinckley, T. M. (1980). Periodic root and shoot growth in oak. *For. Sci.* **26**, 590–598.

Reicosky, D. C., and Ritchie, J. T. (1976). Relative importance of soil resistance and plant resistance in root water absorption. *Soil Sci. Soc. Am. J.* **40**, 293–297.

Reid, D. M., and Crozier, A. (1971). Effects of waterlogging on the gibberellin content and growth of tomato plants. *J. Exp. Bot.* **22**, 39–48.

Reid, E. W. (1890). Osmosis experiments with living and dead membranes. *J. Physiol.* (London) **11**, 312–351.

Reid, J. B., Sorensen, I., and Petrie, R. A. (1993). Root demography in kiwifruit (*Actinidia deliciosa*). *Plant Cell Environ.* **16**, 949–957.

Reimann, E. G., Van Doren, C. A., and Stauffer, R. S. (1946). Soil moisture relationships during crop production. *Soil Sci. Soc. Am. Proc.* **10**, 41–46.

Reisner, M. (1986). "Cadillac Desert." Viking Penguin, New York.

Renger, M., Strebel, O., Grimme, H., and Fleige, H. (1981). Nährstoffanlieferung an die Pflanzenwurzel durch Massenfluss. *Mitt. Dtsch. Bodenkd. Ges.* **30**, 63–70.

Renner, O. (1912). Versuche zur Mechanik der Wasserversorgung. 2. Über wurzeltätigkeit. *Ber. Dtsch. Bot. Ges.* **30**, 642–648.

Renner, O. (1915). Die Wasserversorgung der Pflanzen. *Handworterbuch Naturwiss.* **10**, 538–557.

Repp, G. I., McAlister, D. R., and Wiebe, H. H. (1959). Salt resistance of protoplasm as a test for salt tolerance of agricultural plants. *Agron. J.* **51**, 311–314.

Rhine, J. B. (1924). Clogging of stomata of conifers in relation to smoke injury and distribution. *Bot. Gaz.* **78**, 226–232.

Rhodes, L. H., and Gerdemann, J. W. (1975). Phosphate uptake zones of mycorrhizal and nonmycorrhizal onions. *New Phytol.* **75**, 555–561.

Rice, E. L. (1984). "Allelopathy," 2nd Ed. Academic Press, Orlando, FL.

Richards, J. H., and Caldwell, M. M. (1987). Hydraulic lift: Substantial nocturnal transport of water between soil layers by *Artemisia tridentata* roots. *Oecologia* **73**, 486–489.

Richards, L. A. (1949). Methods of measuring soil moisture tension. *Soil Sci.* **68**, 95–112.

Richards, L. A. (1954). "Diagnosis and Improvement of Saline and Alkaline Soils." U.S. Dept. Agr. Handbook 60.

Richards, L. A., and Ogata, G. (1958). Thermocouple for vapor pressure measurements in biological and soil systems at high humidity. *Science* **128**, 1089–1090.

Richards, L. A., and Wadleigh, C. H. (1952). Soil water and plant growth. *In* "Soil Physical Conditions and Plant Growth" (B. T. Shaw, ed.) pp. 73–251. Academic Press, New York.

Richards, L. A., and Weaver, L. B. (1944). Moisture retention by some irrigated soils as related to soil moisture tension. *J. Agric. Res.* **69**, 215–235.

Richards, S. J., and Marsh, A. W. (1961). Irrigation based on soil suction measurements. *Soil Sci. Soc. Am. Proc.* **25**, 65–69.

Richardson, M. D., Meisner, C. A., Hoveland, C. S., and Karnok, K. J. (1992). Time domain reflectometry in closed container studies. *Agron. J.* **84**, 1061–1063.

Richardson, S. D. (1953). Root growth of *Acer pseudoplatanus* L. in relation to grass cover and nitrogen deficiency. Meded. Landbouwhogesch. Wageningen.

Richmond, P. A., Métraux, J.-P., and Taiz, L. (1980). Cell expansion patterns and directionality of wall mechanical properties in *Nitella*. *Plant Physiol.* **65**, 211–217.

Richter, C., and Marschner, H. (1973). Umtausch von Kalium in verschiedener Wurzelzanen von Maiskeimpflanzen. *Z. Pflanzenphysiol.* **70**, 211–221.

Richter, D. D., King, K. S., and Witter, J. A. (1989). Moisture and nutrient status of extremely acid umbrechts in the Black Mountains of North Carolina. *Soil Sci. Soc. Am. J.* **53**, 1222–1228.

Rider, N. E. (1957). Water losses from various land surfaces. *Quart. J. Roy. Meteorol. Soc.* **83**, 181–183.

Ristaino, J. B., and Duniway, J. M. (1991). The impact of *Phytophthora* root rot on water extraction from the soil by roots of field-grown processing tomatoes. *J. Am. Soc. Hort. Sci.* **116**, 603–608.

Ristic, Z., and Ashworth, E. N. (1993). Ultrastructural evidence that intracellular ice formation and possibly cavitation are the sources of freezing injury in supercooling wood tissue of *Cornus florida* L. *Plant Physiol.* **103**, 753–761.

Ritman, K. T., and Milburn, J. A. (1991). Monitoring of ultrasonic and audible emissions from plants with or without vessels. *J. Exp. Bot.* **42**, 123–130.

Roach, W. A. (1934). Injection for the diagnosis and cure of physiological disorders of fruit trees. *Ann. Appl. Biol.* **21**, 333–343.

Robards, A. W., and Lucas, W. J. (1990). Plasmodesmata. *Annu. Rev. Plant Physiol. Plant Mol. Biol.* **41**, 369–419.

Robb, J., Busch, L., and Rauser, W. E. (1980). Zinc toxicity and xylem vessel wall alterations in white beans. *Ann. Bot.* **46**, 43–50.

Robb, J., Lee, S.-W., Mohan, R., and Kolattukudy, P. E. (1991). Chemical characterization of stress-induced vascular coating in tomato. *Plant Physiol.* **97**, 528–536.

Robb, J., Smith, A., Brisson, J. D., and Busch, L. (1979). Ultrastructure of wilt syndrome caused by *Verticillium dahliae*. VI. Interpretive problems in the study of vessel coatings and tyloses. *Can. J. Bot.* **57**, 795–821.

Robbins, E., and Mauro, A. (1960). Experimental study of the independence of diffusion and hydrodynamic permeability coefficients in collodion membranes. *J. Gen. Physiol.* **43**, 523–532.

Roberts, F. L. (1948). "A Study of the Absorbing Surfaces of the Roots of Loblolly Pine." M.A. Thesis, Duke University, Durham, NC.

Roberts, J. (1977). The use of tree-cutting techniques in the study of water relations of mature *Pinus sylvestris* L. *J. Exp. Bot.* **28**, 751–767.

Roberts, J. (1983). Forest transpiration: a conservative hydrological process? *J. Hydrol.* **66**, 133–144.

Roberts, J. K. M. (1984). Study of plant metabolism *in vivo* using NMR spectroscopy. *Annu. Rev. Plant Physiol.* **35**, 375–386.

Roberts, J. K. M., Andrade, F. H., and Anderson, I. C. (1985). Further evidence that cytoplasmic acidosis is a determinant of flooding intolerance in plants. *Plant Physiol.* **77**, 492–494.

Roberts, R. H., and Struckmeyer, B. E. (1946). The effect of top environment and flowering upon top-root ratios. *Plant Physiol.* **21**, 332–344.

Robeson, D. J., Bretschneider, K. E., and Gonella, M. P. (1989). A hydathode inoculation technique for the simulation of natural black rot infection of cabbage by *Xanthomonas campestris* pv. *campestris*. *Ann. Appl. Biol.* **115**, 455–459.

Robichaux, R. H., and Pearcy, R. W. (1984). Evolution of C_3 and C_4 plants along an environmental moisture gradient: Patterns of photosynthetic differentiation in Hawaiian *Scaevola* and *Euphorbia* species. *Am. J. Bot.* **71**, 121–129.

Robinson, D. G., and Cummins, W. R. (1976). Golgi secretion in plasmolyzed *Pisum sativum* L. *Protoplasma* **90**, 369–379.

Robinson, D. G., and Ray, P. M. (1977). The reversible cyanide inhibition of Golgi secretion in pea cells. *Eur. J. Cell Biol.* **15**, 65–77.

Robinson, S. P., Grant, W. J. R., and Loveys, B. R. (1988). Stomatal limitation of photosynthesis in abscisic acid-treated and in water-stressed leaves measured at elevated CO_2. *Aust. J. Plant Physiol.* **15**, 495–503.

Rogers, H. H., Bingham, G. E., Cure, J. D., Smih, J. M., and Surano, K. A. (1983). Responses of selected plant species to elevated CO_2 in the field. *J. Environ. Qual.* **12**, 569–574.

Rogers, H. H., and Bottomley, P. A. (1987). *In situ* nuclear magnetic resonance imaging of roots: Influence of soil type, ferromagnetic particle content and soil water. *Agron. J.* **79**, 957–965.

Rogers, H. H., Peterson, C. M., McCrimmon, J. N., and Cure, J. D. (1992). Response of plant roots to elevated atmospheric carbon dioxide. *Plant Cell Environ.* **15**, 749–752.

Rogers, W. S. (1929). Winter activity of the roots of perennial plants. *Science* **69**, 299–300.

Romberger, J. A. (1963). "Meristems, Growth, and Development in Woody Plants." U.S. Dep. Agr. Tech. Bull. 1293.

Rose, C. W., Stern, W. R., and Drummond, J. E. (1965). Determination of hydraulic conductivity as a function of depth and water content for soil *in situ*. *Aust. J. Soil Res.* **3**, 1–9.

Rosene, H. F. (1935). Proof of the principle of summation of cell E.M.F.'s. *Plant Physiol.* **10**, 209–224.

Rosenzweig, M. L. (1968). Net primary productivity of terrestrial communities: Prediction from climatological data. *Am. Nat.* **102**, 67–74.

Ruben S., Randall, M., and Hyde, J. L. (1941). Heavy oxygen (O^{18}) as a tracer in the study of photosynthesis. *J. Am. Chem. Soc.* **63**, 877–879.

Rudinsky, J. A., and Vitè, J. P. (1959). Certain ecological and phylogenetic aspects of the pattern of water conduction in conifers. *For. Sci.* **5**, 259–266.

Rufelt, H. (1956). Influence of the root pressure on the transpiration of wheat plants. *Physiol. Plant.* **9**, 154–164.

Rundel, P. W. (1973). The relationship between basal fire scars and crown damage in giant sequoia. *Ecology* **54**, 210–213.

Rundel, P. W. (1982). Water uptake by organs other than roots. *In* "Encyclopedia of Plant Physiology" (O. L. Lange, P. S. Nobel, C. B. Osmond and H. Ziegler, eds.), Vol. 12B, pp. 111–134. Springer-Verlag, New York.

Rundel, P. W., Ehleringer, J. R., and Nagy, K. A., eds. (1988). "Stable Isotopes in Ecological Research," Ecol. Studies 68. Springer-Verlag, New York.

Running, S. W. (1980). Field estimates of root and xylem resistance in *Pinus contorta* using root excision. *J. Exp. Bot.* **31**, 555–569.

Rupley, J. A., Gratton, E., and Careri, G. (1983). Water and globular proteins. *Trends Biochem. Sci.* **8**, 18–22.

Rushin, J. W., and Anderson, J. E. (1981). An examination of the leaf quaking adaptation and stomatal distribution in *Populus tremuloides* Michx. *Plant Physiol.* **67**, 1264–1266.

Russell, R., and Barber, D. A. (1960). The relationship between salt uptake and the absorption of water by intact plants. *Annu. Rev. Plant Physiol.* **11**, 127–140.

Rygol, J., and Lüttge, U. (1983). Water-relation parameters of giant and normal cells of *Capsicum annuum* pericarp. *Plant Cell Environ.* **6**, 545–553.

Rygol, J., Pritchard, J., Zhu, J. J., Tomos, A. D., and Zimmermann, U. (1993). Transpiration induces radial turgor pressure gradients in wheat and maize roots. *Plant Physiol.* **103**, 493–500.

Saab, I. N., and Sharp, E. R. (1989). Non-hydraulic signals from maize roots in drying soil: Inhibition of leaf elongation but not stomatal conductance. *Planta* **179**, 466–474.

Saab, I. N., Sharp, R. E., Pritchard, J., and Voetberg, G. S. (1990). Increased endogenous abscisic acid maintains primary root growth and inhibits shoot growth of maize seedlings at low water potentials. *Plant Physiol.* **93**, 1329–1336.

Sachs, J. (1873). Über das Wachstum der Haupt und Nebenwurzeln. *Arb. Bot. Inst. Wurzburg* **3**, 395–477, 584–634.

Sachs, J. von (1875). "History of Botany, 1530–1860." English Translation by Garnsay, Clarendon Press, Oxford.

Sachs, J. von (1882a). "Lectures On the Physiology of Plants," 2nd Ed., English Translation by Ward, Oxford University Press, 1887.

Sachs, J. von (1882b). "Textbook of Botany," 2nd Ed., English Translation by Virres. Clarendon Press, Oxford.

Sachs, T., and Novoplansky, N. (1993). The development and patterning of stomata and glands in the epidermis of *Peperomia*. *New Phytol.* **123**, 567–574.

Sachs, T., Novoplansky, A., and Cohen, D. (1993). Plants as competing populations of redundant organs. *Plant Cell Environ.* **16**, 765–770.

Safir, G. R. (ed.) (1987). "Ecophysiology of VA Mycorrhizal Plants." CRC Press, Boca Raton, FL.

Safir, G. R., Boyer, J. S., and Gerdemann, J. W. (1971). Mycorrhizal enhancement of water transport in soybean. *Science* **172**, 581–583.

Safir, G. R., Boyer, J. S., and Gerdemann, J. W. (1972). Nutrient status and mycorrhizal enhancement of water transport in soybean. *Plant Physiol.* **49**, 700–703.

Saini, H. S., and Aspinall, D. (1981). Effect of water deficit on sporogenesis in wheat (*Triticum aestivum* L.). *Ann. Bot.* **48**, 623–633.

Saini, H. S., and Aspinall, D. (1982). Sterility in wheat (*Triticum aestivum* L.) induced by water stress or high temperature: Possible mediation by abscisic acid. *Aust. J. Plant Physiol.* **9**, 529–537.

Saini, H. S., Sedgley, M., and Aspinall, D. (1984). Developmental anatomy in wheat of male sterility induced by heat stress, water deficit or abscisic acid. *Aust. J. Plant Physiol.* **11**, 243–254.

Sakuratani, T. (1981). A heat balance method for measuring water flux in the stem of intact plants. *J. Agric. Met.* **37**, 9–17.

Salé, G. (1983). Germination and establishment of *Viscum album* L. In "The Biology of Misteltoes" (M. Calder and P. Bernhardt, eds.), pp. 145–159. Academic Press, New York.

Salter, P. J., and Goode, J. E. (1967). "Crop Responses to Water at Different Stages of Growth." Commonwealth Agricultural Bureau, Franham Royal, Bucks, England.

Sammons, D. J., Peters, D. B., and Hymowitz, T. (1978). Screening soybeans for drought resistance. I. Growth chamber procedure. *Crop Sci.* **18**, 1050–1055.

Sammons, D. J., Peters, D. B., and Hymowitz, T. (1979). Screening soybeans for drought resistance. II. Drought box procedure. *Crop Sci.* **19**, 719–722.

Sampson, D. A., and Smith, F. W. (1993). Influence of canopy architecture on light penetration in lodgepole pine (*Pinus contorta* var. *latifolia*) forests. *Agric. For. Meteorol.* **64**, 63–79.

Sánchez-Díaz, M. F., and Kramer, P. J. (1971). Behavior of corn and sorghum under water stress and during recovery. *Plant Physiol.* **48**, 613–616.

Sánchez-Díaz, M., Pardo, M., Antolín, M., Peña, J., and Aguirreolea, J. (1990). Effect of water stress on photosynthetic activity in the *Medicago-Rhizobium-Glomus* symbiosis. *Plant Sci.* **71**, 215–221.

Sanders, F. E., and Tinker, P. B. (1971). Mechanism of absorption of phosphate from soil by *Endogone* mycorrhizas. *Nature* (London) **232**, 278–279.

Sanderson, J. (1983). Water uptake by different regions of the barley root: Pathways of radial flow in relation to development of the endodemis. *J. Exp. Bot.* **34**, 240–253.

Sandford, A. P., and Jarvis, P. G. (1986). Stomatal response to humidity in selected conifers. *Tree Physiol.* **2**, 89–103.

Sands, R., Fiscus, E. L., and Reid, C. P. P. (1982). Hydraulic properties of pine and bean roots with varying degrees of suberization, vascular differentiation and mycorrhizal infection. *Aust. J. Plant Physiol.* **9**, 559–569.

Sands, R., and Reid, C. P. P. (1980). The osmotic potential of soil water in plant/soil systems. *Aust. J. Soil Res.* **18**, 13–25.

Santarius, K. A. (1969). The effect of freezing and desiccation of chloroplasts in the presence of electrolytes. *Planta* **89**, 23–46.

Santarius, K. A., and Giersch, C. (1984). Factors contributing to inactivation of isolated thylakoid membranes during freezing in the presence of variable amounts of glucose and NaCl. *Biophys. J.* **46**, 129–139.

Sarquis, J. I., Jordan, W. R., and Morgan, P. W. (1992). Effect of atmospheric pressure on maize root growth and ethylene production. *Plant Physiol.* **100**, 2106–2108.

Sauter, J. J. (1971). "Physiology of Sugar Maple," pp. 10–11. Harvard For. Annu. Rept., 1970–1971.

Sayre, J. D. (1926). Physiology of stomata of *Rumex patientia*. *Ohio J. Sci.* **26**, 233–266.

Scandalios, J. G. (1993). Oxygen stress and superoxide dismutases. *Plant Physiol.* **101**, 7–12.

Scheibe, R., Reckmann, U., Hedrick, R., and Raschke, K. (1990). Malate dehydrogenases in guard cells of *Pisum sativum*. *Plant Physiol.* **93**, 1358–1364.

Schiefelbein, J. W., and Somerville, C. (1990). Genetic control of root hair development in *Arabidopsis thaliana*. *Plant Cell* **2**, 235–243.

Schildwacht, P. M. (1989). Is a decrease in water potential after withholding oxygen to roots the cause of the decline of leaf-elongation rates in *Zea mays* L. and *Phaseolus vulgaris* L.? *Planta* **177**, 178–184.

Schmalstig, J. G., and Cosgrove, D. J. (1990). Coupling of solute transport and cell expansion in pea stems. *Plant Physiol.* **94**, 1625–1633.

Schnabl, H., and Ziegler, H. (1977). The mechanism of stomatal movement in *Allium cepa* L. *Planta* **136**, 37–43.

Schneider, G. W., and Childers, N. F. (1941). Influence of soil moisture on photosynthesis, respiration, and transpiration of apple leaves. *Plant Physiol.* **16**, 565–583.

Scholander, P. F., Bradstreet, E. D., Hammel, H. T., and Hemmingsen, E. A. (1966). Sap concentrations in halophytes and some other plants. *Plant Physiol.* **41**, 529–532.

Scholander, P. F., Hammel, H. T., Bradstreet, E. D., and Hemmingsen, E. A. (1965). Sap pressure in vascular plants. *Science* **148**, 339–346.

Scholander, P. F., Hammel, H. T., Hemmingsen, E. A., and Bradstreet, E. D. (1964). Hydrostatic pressure and osmotic potential in leaves of mangroves and some other plants. *Proc. Natl. Acad. Sci. USA* **52**, 119–125.

Schönherr, J. (1976). Water permeability of isolated cuticular membranes: the effect of cuticular waxes on diffusion of water. *Planta* **131**, 159–164.

Schönherr, J., and Ziegler, H. (1980). Water permeability of *Betula* periderm. *Planta* **147**, 345–354.

Schopfer, P. (1989). pH dependence of extension growth in *Avena* coleoptiles and its implications for the mechanisms of auxin action. *Plant Physiol.* **90**, 202–207.

Schroeder, P. (1989). Characterization of a thermo-osmotic gas transport mechanism in *Alnus glutinosa* (L.) Gaert. *Trees* **3**, 38–44.

Schroeder, R. A. (1939). The effect of root temperature upon the absorption of water by the cucumber. *Univ. Missouri Res. Bull.* **309**, 1–27.

Schulte, P. J., and Castle, A. L. (1993). Water flow through vessel perforation plates: A fluid mechanical approach. *J. Exp. Bot.* **44**, 1135–1142.

Schultz, J. C., and Raskin, I., eds. (1994). "Plant Signals in Interactions with Other Organisms." Current Topics in Plant Physiology, Vol. 11. Amer. Soc. Plant Physiologists, Rockville, MD.

Schultz, R. P. (1972). Intraspecific root grafting in slash pine. *Bot. Gaz.* **133**, 26–29.

Schulz, K. E., Smith, M., and Wu, Y. (1993). Gas exchange of *Impatiens pallida* Nutt. (Balsaminaceae) in relation to wilting under high light. *Am. J. Bot.* **80**, 361–368.

Schulze, E.-D. (1986a). Whole-plant responses to drought. *Aust. J. Plant Physiol.* **13**, 127–141.

Schulze, E.-D. (1986b). Carbon dioxide and water vapor exchange in response to drought in the atmosphere and in the soil. *Annu. Rev. Plant Physiol.* **37**, 247–274.

Schulze, E.-D., and Bloom, A. J. (1984). Relationship between mineral nitrogen influx and transpiration in radish and tomato. *Plant Physiol.* **76**, 827–828.

Schulze, E.-D., Cermák, J., Matyssek, R., Penka, M., Zimmermann, R., Vasícek, F., Gries, W., and Kučera, J. (1985). Canopy transpiration and water fluxes in the xylem of the trunk of *Larix* and *Picea* trees: A comparison of xylem flow, porometer and cuvette measurements. *Oecologia* **66**, 475–483.

Schulze, E.-D., and Hall, A. E. (1982). Stomatal responses, water loss and CO_2 assimilation rates of plants in contrasting environments. *In* "Encyclopedia of Plant Physiology" (O. L. Lange, P. S. Nobel, C. B. Osmond, and H. Ziegler, eds.), Vol. 12B, pp. 181–230. Springer-Verlag.

Schulze, E.-D., Lange, O. L., Buschbom, U., Kappen, L., and Evenari, M. (1972). Stomatal responses to changes in humidity in plants growing in the desert. *Planta* **108**, 259–270.

Schussler, J. R., and Westgate, M. E. (1991a). Maize Kernel set at low water potential. I. Sensitivity to reduced assimilates during early kernel growth. *Crop Sci.* **31**, 1189–1195.

Schussler, J. R., and Westgate, M. E. (1991b). Maize kernel set at low water potential. II. Sensitivity to reduced assimilates at pollination. *Crop Sci.* **31**, 1196–1203.

Schwenke, H., and Wagner, E. (1992). A new concept of root exudation. *Plant Cell Environ.* **15**, 289–299.

Scott, F. M. (1963). Root hair zone of soil-grown plants. *Nature (London)* **199**, 1009–1010.

Scott, F. M. (1964). Lipid deposition in the intercellular space. *Nature (London)* **203**, 164–165.

Scott, F. M. (1966). Cell wall surface of the higher plants. *Nature (London)* **210**, 1015–1017.

Scott, L. I., and Priestley, J. H. (1928). The root as an absorbing organ. I. A reconsideration of the entry of water and salts in the absorbing region. *New Phytol.* **27**, 125–140.

Scott, N. S., Munns, R, and Barlow, E. W. R. (1979). Polyribosome content in young and aged wheat leaves subjected to drought. *J. Exp. Bot.* **30**, 905–911.

Sellin, A. (1993). Resistance to water flow in xylem of *Picea abies* (L.) Karst. trees grown under contrasting light conditions. *Trees* **7**, 220–226.

Sendak, P. E. (1978). Birch sap utilization in the Ukraine. *J. For.* **76**, 120–121.

Serpe, M. D., and Matthews, M. A. (1992). Rapid changes in cell wall yielding of elongating *Begonia argenteo-guttata* L. leaves in response to changes in plant water status. *Plant Physiol.* **100**, 1852–1857.

Serrano, E., Zeiger, E., and Hagiwara, S. (1988). Red light stimulates an electrogenic proton pump in *Vicia* guard cell protoplasts. *Proc. Natl. Acad. Sci. USA* **85**, 436–440.

Seyfried, M. S. (1993). Field calibration and monitoring of soil-water content with fiberglass electrical sensors. *J. Soil Sci. Soc. Am.* **57**, 1432–1436.

Shackel, K. A., Matthews, M. A., and Morrison, J. C. (1987). Dynamic relation between expansion and cellular turgor in growing grape (*Vitis vinifera* L.) leaves. *Plant Physiol.* **84**, 1166–1171.

Shahak, Y. (1986). Regulation of the chloroplast H^+-ATPase by light. *Eur. J. Biochem.* **154**, 179–185.

Shaner, D. L., and Boyer, J. S. (1976a). Nitrate reductase activity in maize (*Zea mays* L.) leaves. I. Regulation by nitrate flux. *Plant Physiol.* **58**, 449–504.

Shaner, D. L., and Boyer, J. S. (1976b). Nitrate reductase activity in maize (*Zea mays* L.) leaves. II. Regulation by nitrate flux at low leaf water potentials. *Plant Physiol.* **58**, 505–509.

Shantz, H. L. (1925). Soil moisture in relation to the growth of plants. *J. Am. Soc. Agron.* **17**, 705–711.

Sharkey, T. D., and Seemann, J. R. (1989). Mild water stress effects on carbon-reduction-cycle intermediates, ribulose bisphosphate carboxylase activity, and spatial homogeneity of photosynthesis in intact leaves. *Plant Physiol.* **89**, 1060–1065.

Sharp, R. E., and Boyer, J. S. (1986). Photosynthesis at low water potentials in sunflower: Lack of photoinhibitory effects. *Plant Physiol.* **82**, 90–95.

Sharp, R. E., and Davies, W. J.. (1979). Solute regulation and growth by roots and shoots of water-stressed maize plants. *Planta* **147**, 43–49.

Sharp, R. E., and Davies, W. J. (1985). Root growth and water uptake by maize plants in drying soil. *J. Exp. Bot.* **36**, 1441–1456.

Sharp, R. E., Hsiao, T. C., and Silk, W. K. (1990). Growth of maize primary root at low water potentials. II. Role of growth and deposition of hexose and potassium in osmotic adjustment. *Plant Physiol.* **93**, 1337–1346.

Sharp, R. E., Silk, W. K., and Hsiao, T. C. (1988). Growth of the maize primary root at low water potentials. I. Spatial distribution of expansive growth. *Plant Physiol.* **87**, 50–57.

Sheehy, J. E., Minchin, R. F., and Witty, J. F. (1985). Control of nitrogen fixation in a legume nodule: An analysis of the role of oxygen diffusion in relation to nodule structure. *Ann. Bot.* **55**, 549–562.

Sheriff, D. W. (1977a). Evaporation sites and distillation in leaves. *Ann. Bot.* **41**, 1081–1082.

Sheriff, D. W. (1977b). Where is humidity sensed when stomata respond to it directly? *Ann. Bot.* N.S. **41**, 1083–1084.

Sheriff, D. W. (1984). Epidermal transpiration and stomatal response to humidity: Some hypotheses explored. *Plant Cell Environ.* **7**, 669–677.

Shimazaki, K.-I., and Zeiger, E. (1985). Cyclic and noncyclic photophosphorylation in isolated guard cell chloroplasts from *Vicia faba* L. *Plant Physiol.* **78**, 211–214.

Shimshi, D. (1963a). Effect of chemical closure of stomata on transpiration in varied soil and atmospheric environments. *Plant Physiol.* **38**, 709–712.

Shimshi, D. (1963b). Effect of soil moisture and phenylmercuric acetate upon stomatal aperture, transpiration, photosynthesis. *Plant Physiol.* **38**, 713–721.

Shinozaki, K., Yoda, K., Hozumi, K., and Kira, T. (1964a). A quantitative analysis of plant form: The pipe model theory. I. Basic analyses. *Jap. J. Ecol.* **14**, 97–105.

Shinozaki, K., Yoda, K., Hozumi, K., and Kira T. (1964b). A quantative analysis of plant form: The pipe model theory. II. Further evidence of the theory and its application in forest ecology. *Jap. J. Ecol.* **14**, 133–139.

Shiraishi, M., Hashimoto, Y., and Kuraishi, S. (1978). Cyclic variations of stomatal aperture observed under the scanning electron microscope. *Plant Cell Physiol.* **19**, 637–645.

Shirazi, G. A., Stone, J. F., and Todd, G. W. (1976). Oscillatory transpiration in a cotton plant. *J. Exp. Bot.* **27**, 608–618.

Shirk, H. G. (1942). Freezable water content and the oxygen respiration in wheat and rye grain at different stages of ripening. *Am. J. Bot.* **29**, 105–109.

Shive, J. B., Jr., and Brown, K. W. (1978). Quaking and gas exchange in leaves of cottonwood (*Populus deltoides*, Marsh). *Plant Physiol.* **61**, 331–333.

Shmueli, E. (1971). The contribution of research to the efficient use of water in Israel agriculture. *Z. Bewässerungswirtsch.* **6**, 38–58.

Shone, M. G. T., and Clarkson, D. T. (1988). Rectification of radial water flow in the hypodemis of nodal roots of *Zea mays*. *Plant Soil* **111**, 223–229.

Shoup, S., and Whitcomb, C. (1981). Interaction between trees and ground covers. *J. Arboricult.* **7**, 186–187.

Shull, C. A. (1916). Measurement of the surface forces in soils. *Bot. Gaz.* (Chicago) **62**, 1–31.

Shull, C. A. (1930). Absorption of water and the forces involved. *J. Am. Soc. Agron.* **22**, 459–471.

Shumway, L. K., Weier, T. E., and Stocking, C. R. (1967). Crystalline structures in *Vicia faba* chloroplasts. *Planta* **76**, 182–189.

Silk, W. K., and Wagner, K. K. (1980). Growth-sustaining water potential distributions in the primary corn root. *Plant Physiol.* **66**, 859–863.

Simonneau, T., Habib, R., Goutouly, J.-P., and Huguet, J.-G. (1993). Diurnal changes in stem diameter depend upon variations in water content: Direct evidence in peach trees. *J. Exp. Bot.* **44**, 615–621.

Sinclair, T. R., and Ludlow, M. M. (1985). Who taught plants thermodynamics? The unfilled potential of plant water potential. *Aust. J. Plant Physiol.* **12**, 213–217.

Sinclair, W. B., and Bartholomew, E. T. (1944). Effects of rootstock and environment on the composition of oranges and grapefruit. *Hilgardia* **16**, 125–176.

Sionit, N., and Kramer, P. J. (1977). Effect of water stress during different stages of growth of soybean. *Agron. J.* **69**, 274–278.

Sionit, N., and Kramer, P. J. (1986). Woody plant reactions to CO_2 enrichment. *In* "Carbon Dioxide Enrichment of Greenhouse Crops" (H. Z. Enoch and B. A. Kimball, eds.), pp. 69–85. CRC Press, Boca Raton, FL.

Sionit, N., Teare, I. D., and Kramer, P. J. (1980). Effects of repeated application of water stress on water status and growth of wheat. *Physiol. Plant.* **50**, 11–15.

Skene, K. G. M. (1967). Gibberellin-like substances in root exudate of *Vitis vinifera*. *Planta* **74**, 250–262.

Skidmore, E. L., and Stone, J. F. (1964). Physiological role in regulating transpiration rate of the cotton plant. *Agron. J.* **56**, 405–410.

Skujins, J. J., and McLaren, A. D. (1967). Enzyme reaction rates at limited water activities. *Science* **158**, 1569–1570.

Slatyer, R. O. (1957). The significance of the permanent wilting percentage in studies of plant and soil water relations. *Bot. Rev.* **23**, 585–636.

Slatyer, R. O. (1967). "Plant–Water Relationships." Academic Press, New York.

Slatyer, R. O., and Bierhuizen, J. F. (1964a). A differential psychrometer for continuous measurements of transpiration. *Plant Physiol.* **39**, 1051–1056.

Slatyer, R. O., and Bierhuizen, J. F. (1964b). The influence of several transpiration suppressants on transpiration, photosynthesis, and water-use efficiency of cotton leaves. *Aust. J. Biol. Sci.* 17, 131–146.

Slatyer, R. O., and Taylor, S. A. (1960). Terminology in plant- and soil-water relations. *Nature* (London) 187, 922–924.

Slavik, B. (1974). "Methods of Studying Plant Water Relations." Springer-Verlag, Berlin.

Smit, B., and Stachowiak, M. (1988). Effects of hypoxia and elevated carbon dioxide concentration on water flux through *Populus* roots. *Tree Physiol.* 4, 153–165.

Smit, B., Stachowiak, M., and Van Volkenburgh, E. (1989). Cellular processes limiting leaf growth in plants under hypoxic root stress. *J. Exp. Bot.* 40, 89–94.

Smith, P. G., and Dale, J. E. (1988). The effects of root cooling and excision treatments on the growth of primary leaves of *Phaseolus vulgaris* L. *New Phytol.* 110, 293–300.

Smith, S., Weyers, J. D. B., and Berry, W. G. (1989). Variation in stomatal characteristics over the lower surface of *Commelina communis* leaves. *Plant Cell Environ.* 12, 653–659.

Snyder, R. L. (1992). When water is limited how many acres do you plant? *Calif. Agric.* 46, 7–9.

Sojka, R. E., and Stolzy, L. H. (1980). Soil-oxygen effects on stomatal response. *Soil Sci.* 130, 350–358.

Sorrell, B. K. (1991). Transient pressure gradients in the lacunar system of the submerged macrophyte *Egeria densa* Planch. *Aquatic Bot.* 39, 99–108.

Southwick, S. M., Shackel, K. A., Yeager, J. T., Asai, W. K., and Katacich, M., Jr. (1991). Over-tree sprinkling reduces abnormal shape in "Bing" sweet cheeries. *Calif. Agr.* 45, 24–26.

Southwick, S. W., and Childers, N. S. (1941). Influence of Bordeaux mixture and its component parts on transpiration and apparent photosynthesis of apple leaves. *Plant Physiol.* 16, 721–754.

Spalding, M. H., Spreitzer, R. J., and Ogren, W. L. (1983). Carbonic anhydrase-deficient mutant of *Chlamydomonas reinhardii* requires elevated carbon dioxide concentration for photoautotrophic growth. *Plant Physiol.* 73, 268–272.

Spanner, D. C. (1951). The Peltier effect and its use in the measurement of suction pressure. *J. Exp. Bot.* 2, 145–168.

Spanner, D. C. (1964). "Introduction to Thermodynamics." Academic Press, New York.

Sperry, J. S., Donnelly, J. R., and Tyree, M. T. (1988a). Seasonal occurrence of xylem embolism in sugar maple. *Am. J. Bot.* 75, 1212–1218.

Sperry, J. S., Holbrook, N. M., Zimmermann, M. H., and Tyree, M. T. (1987). Spring filling of xylem vessels in wild grapevine. *Plant Physiol.* 83, 414–417.

Sperry, J. S, and Sullivan, J. E. M. (1992). Xylem embolism in response to freeze-thaw cycles and water stress in ring-porous, diffuse porous, and conifer species. *Plant Physiol.* 100, 605–613.

Sperry, J. S., Tyree, M. T., and Donnelly, J. R. (1988b). Vulnerability of xylem to embolism in a mangrove vs. an inland species of Rhizophoraceae. *Physiol. Plant.* 74, 276–283.

Stahle, D. W., Cleaveland, M. K., and Hehr, J. G. (1988). North Carolina climate changes reconstructed from tree rings: A.D. 372 to 1985. *Science* 240, 1517–1519.

Stålfelt, M. G. (1932). Der stomatäre Regulator in der pflanzlichen Transpiration. *Planta* 17, 22–85.

Stålfelt, M. (1956a). Die stomatäre Transpiration und die Physiologie der Spaltöffnungen. *Encyl. Plant Physiol.* 3, 351–426. Springer-Verlag, Berlin.

Stålfelt, M. G. (1956b). Morphologie und Anatomie des Blatter als Transpirationsorganen. *Encycl. Plant Physiol.* 3, 324–341. Springer-Verlag, Berlin.

Stark, N., Spitzner, C., and Essig, D. (1985). Xylem sap analysis for determining nutritional status of trees: *Pseudotsuga menziesii. Can. J. For. Res.* 15, 429–437.

Staswick, P. E. (1988). Soybean vegetative storage protein structure and gene expression. *Plant Physiol.* 87, 250–254.

Staswick, P. E. (1989a). Correction. *Plant Physiol.* 89, 717.

Staswick, P. E. (1989b). Preferential loss of an abundant storage protein from soybean pods during seed development. *Plant Physiol.* 90, 1252–1255.

Staswick, P. E. (1989c). Developmental regulation and the influence of plant sinks on vegetative storage protein gene expression in soybean leaves. *Plant Physiol.* **89**, 309–315.

Steinberg, S. L., McFarland, M. J., and Worthington, J. W. (1990a). Comparison of trunk and branch sap flow with canopy transpiration in pecan. *J. Exp. Bot.* **41**, 653–659.

Steinberg, S. L., Miller, J. C., Jr., and McFarland, M. J. (1990b). Dry matter partitioning and vegetative growth of young peach trees under water stress. *Aust. J. Plant Physiol.* **17**, 23–36.

Steinberg, S. L., Van Bavel, C. H. M., and McFarland, M. J. (1989). A gauge to measure mass flow of sap in stems and trunks of woody plants. *J. Am. Soc. Hort. Sci.* **114**, 466–472.

Steudle, E. (1989). Water flow in plants and its coupling to other processes: An overview. *In* "Methods in Enzymology" (S. and B. Fleischer, eds.), Vol. 174, pp. 183–225. Academic Press, New York.

Steudle, E., and Boyer, J. S. (1985). Hydraulic resistance to water flow in growing hypocotyl of soybean measured by a new pressure-perfusion technique. *Planta* **164**, 189–200.

Steudle, E., and Brinckmann, E. (1989). The osmometer model of the root: Water and solute relations of roots of *Phaseolus coccineus*. *Bot. Acta* **102**, 85–95.

Steudle, E., and Frensch, J. (1989). Osmotic responses of maize roots: Water and solute relations. *Planta* **177**, 281–295.

Steudle, E., and Jeschke, W. D. (1983). Water transport in barley roots. *Planta* **158**, 237–248.

Steudle, E., Lüttge, U., and Zimmermann, U. (1975). Water relations of the epidermal bladder cells of the halophytic species *Mesembryanthemum crystallinum*: Direct measurements of hydrostatic pressure and hydraulic conductivity. *Planta* **126**, 229–246.

Steudle, E., Murrman, M., and Peterson, C. A. (1993). Transport of water and solutes across maize roots modified by puncturing the endodermis. *Plant Physiol.* **103**, 335–349.

Steudle, E., Oren, R., and Schulze, E.-D. (1987). Water transport in maize roots: Measurement of hydraulic conductivity, solute permeability, and of reflection coefficients of excised roots using the root pressure probe. *Plant Physiol.* **84**, 1220–1232.

Steudle, E., Smith, J., and Lüttge, U. (1980). Water-relation parameters of individual mesophyll cells of the Crassulacean acid metabolism plant *Kalanchoe daigremontiana*. *Plant Physiol.* **66**, 1155–1163.

Steudle, E., and Tyerman, S. D. (1983). Determination of permeability coefficients, reflection coefficients, and hydraulic conductivity of *Chara corallina* using the pressure probe: Effects of solute concentrations. *J. Membr. Biol.* **75**, 85–96.

Steudle, E., Ziegler, H., and Zimmermann, U. (1983). Water relations of the epidermal bladder cells of *Oxalis carnosa* Molina. *Planta* **159**, 38–45.

Steudle, E., and Zimmermann, U. (1974). Determination of the hydraulic conductivity and of refection coefficients in *Nitella flexilis* by means of direct cell-turgor pressure measurements. *Biochim. Biophys. Acta* **332**, 399–412.

Stevens, C. L., and Eggert, R. L. (1945). Observations on the causes of flow of sap in red maple. *Plant Physiol.* **20**, 636–648.

Steward, F. C., and Sutcliffe, J. F. (1959). Plants in relation to inorganic salts. *In* "Plant Physiology" (F. C. Steward, ed.), Vol. 2, pp. 253–478. Academic Press, New York.

Stewart, D. A., and Nielsen, D. R., eds. (1990). "Irrigation of Agricultural Crops." Agron. Mon. 30, American Society of Agronomy, Madison, WI.

Stillinger, F. H. (1980). Water revisited. *Science* **209**, 451–457.

Stolzy, L. H., Focht, D. D., and Fluehler, H. (1981). Indications of soil aeration. *Flora* **171**, 136–265.

Stone, E. C., and Fowells, H. A. (1955). The survival value of dew as determined under laboratory conditions. I. *Pinus ponderosa*. *For. Sci.* **1**, 183–188.

Stone, E. C., and Norberg, E. A. (1979). Root growth capacity: One key to bare root survival. *Calif. Agric.* **33**, 14–15.

Stone, J. E., and Stone, E. L. (1975a). Water conduction in lateral roots of red pine. *For. Sci.* **21**, 53–60.

Stone, J. E., and Stone, E. L. (1975b). The communal root system of red pine: Water conduction through root grafts. *For. Sci.* **21**, 255–262.

Strasburger, E. (1891). Ueber den Bau und die Verrichtungen der Leitungsbahnen in der Pflanzen. *Histol. Beitr.* **3**, 849–877.

Strogonov, B. P. (1964). "Physiological Basis of Salt Tolerance of Plants." English translation by A. Poljakoff-Mayber and A. M. Mayer. Israel Program for Sci. Translations, Jerusalem, Oldbourne Press, London.

Suhayda, C. G., and Goodman, R. N. (1981). Early proliferation and migration and subsequent xylem occlusion by *Erwinia amylovora* and the fate of its extracellular polysaccharide (EPS) in apple shoots. *Phytopathology* **71**, 697–707.

Sumner, D. R. (1982). Crop rotation and plant productivity. *In* "Handbook of Agricultural Productivity" (M. Rechcigl, Jr. ed.), Vol. 1, pp. 273–313. CRC Press, Boca Raton, FL.

Sung, F. J. M., and Krieg, D. R. (1979). Relative sensitivity of photosynthetic assimilation and translocation of [14]Carbon to water stress. *Plant Physiol.* **64**, 852–856.

Sureshi, K. K., and Rai, R. V. S. (1988). Allelopathic exclusion of understorey by a few multipurpose trees. *Int. Tree Crops J.* **5**, 143–151.

Surowy, T. K., and Boyer, J. S. (1991). Low water potentials affect expression of genes encoding vegetative storage proteins and plasma membrane proton ATPase in soybean. *Plant Mol. Biol.* **16**, 251–262.

Sutton, R. F. (1969). "Form and Development of Conifer Root systems." Tech. Commun. No. 7, Commonwealth Forestry Bureau, Oxford, UK.

Svenningsson, H., and Liljenberg, C. (1986). Membrane lipid changes in root cells of rape (*Brassica napus*) as a function of water deficit stress. *Physiol. Plant.* **68**, 53–58.

Svenson, S. E., and Davies, F. T., Jr. (1992). Comparison of methods for estimating surface area of water-stressed and fully hydrated pine needle segments for gas exchange analysis. *Tree Physiol.* **10**, 417–421.

Swank, W. T., and Douglas, J. E. (1974). Streamflow greatly reduced by converting hardwood stands to pine. *Science* **185**, 857–859.

Swanson, C. A. (1943). Transpiration in American holly in relation to leaf structure. *Ohio J. Sci.* **43**, 43–46.

Sylvia, D. M., Hammond, L. C., Bennett, J. M., Haas, J. H., and Linda, S. B. (1993). Field response of maize to a VAM fungus and water management. *Agron. J.* **85**, 193–198.

Syvertsen, J. P., and Graham, J. H. (1990). Influence of vesicular arbuscular mycorrhizae and leaf age on net gas exchange of *Citrus* leaves. *Plant Physiol.* **94**, 1424–1428.

Taiz, L. (1984). Plant cell expansion: Regulation of cell wall mechanical properties. *Annu. Rev. Plant Physiol.* **35**, 585–657.

Taiz, L., Métraux, J.-P., and Richmond, P. A. (1981). Control of cell expansion in the *Nitella* internode. *In* "Cell Biology Monographs, Vol. 8: Cytomorphogenesis in Plants" (O. Kiermayer, ed.), pp. 231–264. Springer, Wien.

Takahashi, H., and Scott, T. (1991). Hydrotropism and its interaction with gravitropism in maize roots. *Plant Physiol.* **96**, 558–564.

Tal, M. (1966). Abnormal stomatal behavior in wilty mutants of tomato. *Plant Physiol.* **41**, 1387–1391.

Talboys, P. W. (1978). Dysfunction of the water system. *In* "Plant Disease" (J. G. Horsfall and E. B. Cowling, eds.), Vol. 3, pp. 141–162. Academic Press, New York.

Tanford, C. (1963). The structure of water and aqueous solutions. *In* "Temperature: Its Measurement and Control in Science and Industry" (C. M. Herzfeld, ed.), Vol. 3, pp. 123–129. Reinhold Publishing Corp., New York.

Tanford, C. (1980). "The Hydrophobic Effect." Wiley, New York.

Tang, P. S., and Wang, J. S. (1941). A thermodynamic formulation of the water relations in an isolated living cell. *J. Physical Chem.* **45**, 443–453.

Tanner, W., and Beevers, H. (1990). Does transpiration have an essential function in long-distance ion transport in plants? *Plant Cell Environ.* 13, 745–750.

Tarczynski, M. C., Jensen, R. G., and Bohnert, H. J. (1993). Stress protection of transgenic tobacco by production of the osmolyte mannitol. *Science* 259, 508–510.

Tardieu, F. (1988). Analysis of spatial variability in maize root density. III. Effect of a wheel compaction on water extraction. *Plant Soil* 109, 257–262.

Tardieu, F., Katerji, N., Bethenod, O., Zhang, J., and Davies, W. J. (1991). Maize stomatal conductance in the field: Its relationship with soil and plant water potentials, mechanical constraints and ABA concentration in the xylem sap. *Plant Cell Environ.* 14, 121–126.

Taylor, H. M., Burnett, E., and Booth, G. D. (1978). Taproot elongation rates of soybeans. *Z. Acker. Pflanzenbau.* 146, 33–39.

Taylor, H. M., Jordan, W. R., and Sinclair, T. R. (1983). "Limitations to Efficient Water Use in Crop Production." American Society of Agronomy, Madison, WI.

Taylor, H. M., and Klepper, B. (1975). Water uptake by cotton root systems: An examination of assumptions in the single root model. *Soil Sci.* 120, 57–67.

Taylor, H. M., and Ratliff, L. F. (1969). Root elongation rates of cotton and peanuts as a function of soil strength and soil water content. *Soil Sci.* 108, 113–119.

Taylor, H. M., and Terrell, E. E. (1982). Rooting pattern and plant productivity. *In* "Handbook of Agricultural Productivity" (M. Rechcigl, Jr., ed.), Vol. 1, pp. 185–200. CRC Press, Boca Raton, FL.

Taylor, H. M., and Willatt, S. T. (1983). Shrinkage of soybean roots. *Agron. J.* 75, 818–820.

Teare, I. D., and Peet, M. M. (eds.) (1983). "Crop Water Relations." Wiley, New York.

Tepfer, M., and Taylor, I. E. P. (1981). The permeability of plant cell walls as measured by gel filtration chromatography. *Science* 213, 761–763.

Terashima, I. (1992). Anatomy of non-uniform leaf photosynthesis. *Photosynthesis Res.* 31, 195–212.

Terashima, I., Wong, S.-C., Osmond, C. B., and Farquhar, G. D. (1988). Characterisation of non-uniform photosynthesis induced by abscisic acid in leaves having different mesophyll anatomies. *Plant Cell Physiol.* 29, 385–394.

Terry, N., Waldron, L. J., and Ulrich, A. (1971). Effects of moisture stress on the multiplication and expansion of cells in leaves of sugar beet. *Planta* 97, 281–289.

Tesar, M. B. (1993). Delayed seeding of alfalfa avoids autotoxicity after plowing or glyphosate treatment of established stands. *Agron. J.* 85, 256–263.

Teskey, R. O., Grier, C. C., and Hinckley, T. M. (1985). Relations between root system size and water inflow capacity of *Abies amabilis* growing in a subalpine forest. *Can. J. For. Res.* 15, 669–672.

Teskey, R. O., and Hinckley, T. M. (1981). Influence of temperature and water potential on root growth of white oak. *Physiol. Plant.* 52, 363–369.

Teviotdale, B. L., Davis, R. M., Guerard, J. P., and Harper, D. H. (1990). Method of irrigation affects sour skin rot of onion. *Calif. Agric.* 44, 27–28.

Thielmann, J., Tolbert, N. E., Goyal, A., and Senger, H. (1990). Two systems for concentrating CO_2 and bicarbonate during photosynthesis by *Scenedesmus. Plant Physiol.* 92, 622–629.

Thoday, D. (1918). On turgescence and the absorption of water by the cells of plants. *New Phytol.* 17, 108–113.

Thomas, J. C., and Bohnert, H. J. (1993). Salt stress perception and plant growth regulators in the halophyte *Mesembryanthemum crystallinum. Plant Physiol.* 103, 1299–1304.

Thomas, R. B., and Strain, B. R. (1991). Root restriction as a factor in photosynthetic acclimation of cotton seedlings grown in elevated carbon dioxide. *Plant Physiol.* 96, 627–634.

Thomas, W. A. (1967). Dye and calcium ascent in dogwood trees. *Plant Physiol.* 42, 1800–1802.

Thompson, A. C., ed. (1985). "The Chemistry of Allelopathy," ACS Symp. 268. Am. Chem. Soc. Washington, D.C.

Thomson, C. J., and Greenway, H. (1991). Metabolic evidence for stelar anoxia in maize roots exposed to low O₂ concentrations. *Plant Physiol.* **96**, 1294–1301.

Thornley, J. H. M. (1972). A balanced quantitative model for root:shoot ratios in vegetative plants. *Ann. Bot.* **36**, 431–441.

Thornthwaite, C. W., and Mather, J. R. (1957). Instructions and tables for computing potential evapotranspiration and the water balance. *Drexel Inst. Technol. Lab. Climatology, Publ. Climatol.* **10**, 181–311.

Thut, H. F. (1932). The movement of water through some submerged plants. *Am. J. Bot.* **19**, 693–709.

Tiefer, M. A., Roy, H., and Moudrianakis, E. N. (1977). Binding of adenine nucleotides and pyrophosphate by the purified coupling factor of photophosphorylation. *Biochemistry* **16**, 2396–2404.

Ting, I. P. (1985). Crassulacean acid metabolism. *Annu. Rev. Plant Physiol.* **36**, 595–622.

Ting, I. P., and Loomis, W. E. (1965). Further studies concerning stomatal diffusion. *Plant Physiol.* **40**, 220–228.

Tinker, P. B. (1976). Roots and water: Transport of water to plant roots in soil. *Phil. Trans. Roy. Soc. London Ser. B* **273**, 445–461.

Tjepkema, J. D., Schwintzer, C. R., and Benson, D. R. (1986). Physiology of actinorhizal nodules. *Annu. Rev. Plant Physiol.* **37**, 209–232.

Todd, G. W., and Webster, D. L. (1965). Effects of repeated drought periods on photosynthesis and survival of cereal seedlings. *Agron. J.* **57**, 399–404.

Tomos, A. D., Steudle, E., Zimmermann, U., and Schulze, E.-D. (1981). Water relations of leaf epidermal cells of *Tradescantia virginiana. Plant Physiol.* **68**, 1135–1143.

Topp, G. C., and Davis, J. L. (1985). Measurement of soil water content using time domain reflectometry (TDR): A field evaluation. *Soil Sci. Soc. Am. J.* **49**, 19–24.

Torrey, J. G., and Clarkson, D. T., eds. (1975). "The Development and Function of Roots." Academic Press, New York.

Tranquillini, W. (1969). Photosynthese und Transpiration einiger Holzarten ber verschieden starkem Wind. *Centralbl. Gesamte Forsteve.* **85**, 43–49.

Transeau, E. N. (1905). Forest centers of eastern North America. *Am. Nat.* **39**, 875–889.

Traube, M. (1867). Experimente zur Theorie der Zellbildung und Endosmose. *Archiv. Anat. Physiol. wiss Medecin* **87**, 165.

Trejo, C. L., Davies, W. J., and Ruiz, L. M. P. (1993a). Sensitivity of stomata to abscisic acid: An effect of the mesophyll. *Plant Physiol.* **102**, 497–502.

Trejo, C. L., Gowing, D. J. G., and Davies, W. J. (1993b). Control of leaf growth and physiology: A link between climatic and edaphic effects. *In* "Plant Responses to Cellular Dehydration during Environmental Stress" (T. J. Close and E. A. Bray, eds.), pp. 48–56. American Society of Plant Physiologists, Rockville, MD.

Trewavas, A. (1981). How do plant growth substances work? *Plant Cell Environ.* **4**, 203–228.

Tripp, K. E., Peet, M. M., Pharr, D. M., Willits, D. H., and Nelson, P. V. (1991). CO₂-enhanced yield and foliar deformation among tomato genotypes in elevated CO₂ environments. *Plant Physiol.* **96**, 713–719.

Troughton, J., and Donaldson, L. A. (1981). "Probing Plant Structure." McGraw-Hill, New York.

Tsukahara, H., and Kozlowski, T. T. (1985). Importance of adventitious roots to growth of flooded *Platanus occidentalis* seedlings. *Plant Soil* **88**, 123–132.

Tubbs, F. R. (1973). Research field in the interaction of rootstocks and scions in woody perennials. Parts I and II. *Hort. Abstr.* **43**, 247–253, 325–335.

Tukey, H. B., Jr., Mecklenburg, R. A., and Morgan, J. V. (1965). A mechanism for the leaching of metabolites from foliage. *In* "Isotopes and Radiation in Soil-Plant Nutrition Studies," pp. 371–385. IAEA, Vienna.

Turner, L. M. (1936). Root growth of seedlings of *Pinus echinata* and *Pinus taeda. J. Agric. Res.* (Washington, D.C.) 53, 145–149.

Turner, N. C. (1970). Speeding the drying of alfalfa hay with fusicoccin. *Agron. J.* 62, 538–541.

Turner, N. C. (1986). Crop water deficits: A decade of progress. *Adv. Agron.* 39, 1–51.

Turner, N. C., Begg, J. E., Rawson, H. M., English, S. D., and Hearn, A. B. (1978). Agronomic and physiological responses of soybean and sorghum crops to water deficits. III. Components of water potential, leaf conductance, $^{14}CO_2$ photosynthesis, and adaptation to water deficits. *Aust. J. Plant Physiol.* 5, 179–194.

Turner, N. C., and Kramer, P. J., eds. (1980). "Adaptation of Plants to Water and High Temperature Stress." Wiley, New York.

Turner, N. C., Schulze, E.-D., and Gollan, T. (1985). The responses of stomata and leaf gas exchange to vapour pressure deficits and soil water contents. II. In the mesophytic herbaceous species *Helianthus annuus. Oecologia* 65, 348–355.

Turrell, F. M. (1936). The area of the internal exposed surface of dicotyledon leaves. *Am. J. Bot.* 23, 255–264.

Tyerman, S. D., and Steudle, E. (1982). Comparisons between osmotic and hydrostatic water flows in a higher plant cell: Determination of hydraulic conductivities and reflection coefficients in isolated epidermis of *Tradescantia virginiana. Aust. J. Plant Physiol.* 9, 461–480.

Tyree, M. T. (1973). An alternative explanation for the apparently active water exudation in excised roots. *J. Exp. Bot.* 24, 33–37.

Tyree, M. T., Dixon, M. A., Tyree, E. L., and Johnson, R. (1984). Ultrasonic acoustic emissions from the sapwood of cedar and hemlock: An examination of three hypotheses concerning cavitation. *Plant Physiol.* 75, 988–992.

Tyree, M. T., and Ewers, F. W. (1991). The hydraulic architecture of trees and other woody plants. *New Phytol.* 119, 345–360.

Tyree, M. T., Fiscus, E. L., Wullschlegel, S. D., and Dixon, M. A. (1986). Detection of xylem cavitation in corn under field conditions. *Plant Physiol.* 82, 597–599.

Tyree, M. T., Snyderman, D. A., Wilmot, T. R., and Machado, J.-L. (1991). Water relations and hydraulic architecture of a tropical tree (*Schefflera morototoni*). *Plant Physiol.* 96, 1105–1113.

Tyree, M. T., and Sperry, J. S. (1988). Do woody plants operate near the point of catastrophic xylem dysfunction caused by dynamic water stress? Answers from a model. *Plant Physiol.* 88, 574–580.

Tyree, M. T., and Sperry, J. S. (1989). Vulnerability of xylem to cavitation and embolism. *Annu. Rev. Plant Physiol. Plant Mol. Biol.* 40, 19–38.

Tyree, M. T., and Yang, S. (1992). Hydraulic conductivity recovery versus water pressure in xylem of *Acer saccharum. Plant Physiol.* 100, 669–676.

Tyree, M. T., and Yianoulis, P. (1980). The site of water evaporation from sub-stomatal cavities, liquid path resistances and hydroactive stomatal closure. *Ann. Bot.* (London) 46, 175–193.

U.S. Department of Agriculture (1965). Losses in Agriculture. Agriculture Handbook No. 291. Government Printing Office, Washington, DC.

U.S. Department of Agriculture (1979). Agricultural Statistics. Government Printing Office, Washington, DC.

U.S. Department of Agriculture, Soil Conservation Service (1975). Soil Taxonomy. Government Printing Office, Washington DC.

U.S. Department of the Interior (1977). Estimated Use of Water in the United States in 1975. U.S. Geol. Surv. Circ. No. 765.

Unrath, C. R. (1972). The quality of "Red Delicious" apples as affected by overtree sprinkler irrigation. *J. Am. Soc. Hort. Sci.* 97, 58–61.

Upchurch, D. R., and Ritchie, J. T. (1988). Root observations using a video recording system in mini-rhizotrons. *Agron. J.* 75, 1009–1015.

Ursprung, A. (1929). The osmotic quantities of the plant cell. Int. Bot. Congr. Proc. 4th, Vol. 2, 1081–1094.

Ursprung, A., and Blum, G. (1916). Zur Methode der Saugkraftmessung. *Ber. Deut. Bot. Ges.* **34**, 525–539.

Ussing, H. H. (1953). Transport through biological membranes. *Annu. Rev. Physiol.* **15**, 1–20.

Vaadia, Y. (1960). Autonomic diurnal fluctuations in rate of exudation and root pressure of decapitated sunflower plants. *Physiol. Plant.* **13**, 701–717.

Vaclavik, J. (1966). The maintaining of constant soil moisture levels lower than maximum capillary capacity in pot experiments. *Biol. Plant.* (Prague) **8**, 80–85.

Van Alfen, N. K., and Turner, N. C. (1975). Changes in alfalfa stem conductance induced by *Corynebacterium insidiosum* toxin. *Plant Physiol.* **55**, 559–561.

Van As, H., and Schaafsma, T. J. (1984). Noninvasive measurement of plant water flow by nuclear magnetic resonance. *Biophys. J.* **45**, 469–472.

Van Bavel, C. H. M. (1966). Potential evaporation: the combination concept and its experimental verification. *Water Resour. Res.* **2**, 455–468.

Van Bavel, C. H. M. (1967). Changes in canopy resistance to water loss from alfalfa induced by soil water depletion. *Agric. Meterol.* **4**, 165–176.

Van Bavel, C. H. M., and Ehrler, W. L. (1968). Water loss from a sorghum field and stomatal control. *Agron. J.* **60**, 84–86.

Van Bavel, C. H. M., Fritschen, L. J., and Lewis, W. E. (1963). Transpiration by Sudangrass as an externally controlled process. *Science* **141**, 269–270.

Van Bavel, C. H. M., and Verlinden, F. J. (1956). "Agricultural Drought in North Carolina." North Carolina Agric. Exp. Stn. Tech. Bull. 122.

Van der Post, C. J. (1968). Simultaneous observations on root and top growth. *Acta Hort.* **7**, 138–144.

van Eijk, M. (1939). Analyze der Wirkung des NaCl auf dis Entwicklung, Sukkulenz und Transpiration bei *Salicornia herbacea*, sowie Untersuchungen über den Einluss der Salzaufnahme auf die Wurzelatmung bei *Aster tripolium. Red. Trav. Bot. Neerl.* **36**, 559–657.

Van Gardingen, P. R., Grace, J., and Jeffree, C. E. (1991). Abrasive damage by wind to the needle surfaces of *Picea sitchensis* (Bong.) Carr. and *Pinus sylvestris* L. *Plant Cell Environ.* **14**, 185–193.

Van Gardingen, P. R., Jeffree, C. E., and Grace, J. (1989). Variation in stomatal aperture in leaves of *Avena fatua* L. observed by low-temperature scanning electron microscopy. *Plant Cell Environ.* **12**, 887–898.

Van Noordwijk, M., and Brouwer, G. (1988). Quantification of air-filled root porosity: A comparison of two methods. *Plant Soil* **111**, 255–258.

Van Overbeek, J. (1942). Water uptake by excised root systems of the tomato due to nonosmotic forces. *Am. J. Bot.* **29**, 677–683.

Van Rees, J. C. J., and Comerford, N. B. (1990). The role of woody roots of slash pine seedlings in water and potassium absorption. *Can. J. For. Res.* **20**, 1183–1191.

Van Volkenburgh, E., and Boyer, J. S. (1985). Inhibitory effects of water deficit on maize leaf elongation. *Plant Physiol.* **77**, 190–194.

Van Volkenburgh, E., and Davies, W. J. (1977). Leaf anatomy and water relations of plants grown in controlled environments and in the field. *Crop Sci.* **17**, 353–358.

Vance, C. P., Heichel, G. H., Barnes, D. K., Bryan, J. W., and Johnson, L. E. (1979). Nitrogen fixation, nodule development, and vegetative regrowth of alfalfa (*Medicago sativa* L.) following harvest. *Plant Physiol.* **64**, 1–8.

Vanderhoef, L. N., Findley, J. S., Burke, J. J., and Blizzard, W. E. (1977a). Auxin has no effect on modification of external pH by soybean hypocotyl cells. *Plant Physiol.* **59**, 1000–1003.

Vanderhoef, L. N., Shen Lu, T.-Y., and Williams, C.. (1977b). Comparison of auxin-induced and acid-induced elongation in soybean hypocotyl. *Plant Physiol.* **59**, 1004–1007.

Vanderhoef, L. N., Stahl, C. A., Williams, C. A., and Brinkmann, K. A. (1976). Additional evidence for separable responses to auxin in soybean hypocotyl. *Plant Physiol.* 57, 817–819.

Vartanian, N. (1981). Some aspects of structural and functional modifications induced by drought in root systems. *Plant Soil* 63, 83–92.

Vartanian, N., Marcotte, L., and Giraudat, J. (1994). Drought rhizogenesis in *Arabidopsis thaliana*. *Plant Physiol.* 104, 761–767.

Vegelin, S. J., White, I., and Jenkins, D. R. (1990). Improved field probes for soil water content and electrical conductivity measurements using time domain reflectometry. *Water Resour. Res.* 25, 2367–2376.

Veihmeyer, F. J. (1927). Some factors affecting the irrigation requirements of deciduous orchards. *Hilgardia* 2, 125–284.

Veihmeyer, F. J., and Hendrickson, A. H. (1928). Soil moisture at permanent wilting of plants. *Plant Physiol.* 3, 355–357.

Veihmeyer, F. J., and Hendrickson, A. H. (1938). Soil moisture as an indication of root distribution in deciduous orchards. *Plant Physiol.* 13, 169–177.

Veihmeyer, F. J., and Hendrickson, A. H. (1950). Soil moisture in relation to plant growth. *Annu. Rev. Plant Physiol.* 1, 285–304.

Veres, J. S., Johnson, G. A., and Kramer, P. J. (1991). *In vivo* magnetic resonance imaging of *Blechnum* ferns: Changes in T1 and N(H) during dehydration and rehydration. *Am. J. Bot.* 78, 80–88.

Vertucci, C. W. (1989). The effects of low water contents on physiological activities of seeds. *Physiol. Plant.* 77, 172–176.

Vertucci, C. W., and Leopold, A. C. (1987a). Water binding in legume seeds. *Plant Physiol.* 85, 224–231.

Vertucci, C. W., and Leopold, A. C. (1987b). The relationship between water binding and desiccation tolerance in tissues. *Plant Physiol.* 85, 232–238.

Vertucci, C. W., and Roos, E. E. (1990). Theoretical basis of protocols for seed storage. *Plant Physiol.* 94, 1019–1023.

Vesque, M. J. (1884). Mouvement de la sève ascendante. *Ann. Sci. Naturelles,* VI Series, Tome XIX, 159–199.

Vessey, J. K., Raper, C. D., Jr., and Henry, L. T. (1990). Cyclic variations in nitrogen uptake rate in soybean plants: Uptake during reproductive growth. *J. Exp. Bot.* 41, 1579–1584.

Vieira da Silva, J. V., Naylor, A. W., and Kramer, P. J. (1974). Some ultrastructural and enzymatic effects of water stress in cotton (*Gossypium hirsutum* L.) leaves. *Proc. Natl. Acad. Sci. USA* 71, 3243–3247.

Viets, F. G., Jr. (1972). Water deficits and nutrient availability. *In* "Water Deficits and Plant Growth" (T. T. Kozlowski, eds.), Vol. 4, pp. 217–239. Academic Press, New York.

Voesenek, L. A. C. J., Banga, M., Thier, R. H., Mudde, C. M., Harren, F. J. M., Barendse, G. W. M., and Blom, C. W. P. M. (1993). Submergence-induced ethylene synthesis, entrapment, and growth in two plant species with contrasting flooding resistances. *Plant Physiol.* 103, 783–791.

Voetberg, G. S., and Sharp, R. E. (1991). Growth of the maize primary root at low water potentials. III. Role of increased proline deposition in osmotic adjustment. *Plant Physiol.* 96, 1125–1130.

Vogel, S. (1981). "Life in Moving Fluids." Princeton University Press, Princeton, New Jersey.

Vogel, S. (1989). Drag and reconfiguration of broad leaves. *J. Exp. Bot.* 40, 941–948.

Volpin, H., Elkind, Y., Okon, Y., and Kapulnik, Y. (1994). A vesicular arbuscular mycorrhizal fungus (*Glomus intraradix*) induces a defense response in alfalfa roots. *Plant Physiol.* 104, 683–689.

Wadleigh, C. H. (1946). The integrated soil moisture stress upon a root system in a large container of saline soil. *Soil Sci.* 61, 225–238.

Wadleigh, C. H., and Ayers, A. D. (1945). Growth and biochemical composition of bean plants as conditioned by soil moisture tension and salt concentration. *Plant Physiol.* 20, 106–132.

Wadleigh, C. H., Gauch, H. G., and Magistad, O. C. (1946). "Growth and Rubber Accumulation in Guayule as Conditioned by Soil Salinity and Irrigation Regime." U.S. Dept. Agric. Tech. Bull. 925.

Waisel, V., Eshel, A., and Kafkafi, U., eds. (1991). "Plant Roots." Dekker, New York.

Walker, D. A., and Zelitch, I. (1963). Some effects of metabolic inhibitors, temperature, and anaerobic conditions on stomatal movement. *Plant Physiol.* 38, 390–396.

Walker, R. F., West, D. C., McLaughlin, S. B., and Amundsen, C. C. (1989). Growth, xylem pressure potential, and nutrient absorption of loblolly pine on a reclaimed surface mine as affected by an induced *Pisolithus tinctorios* infection. *For. Sci.* 35, 569–581.

Walker, R. R. (1986). Sodium exclusion and potassium-sodium selectivity in salt-treated trifoliate orange (*Poncirus trifoliata*) and Cleopatra mandarin (*Citrus reticulata*) plants. *Aust. J. Plant Physiol.* 13, 293–303.

Walsh, K. B., Vessey, J. K., and Layzell, D. B. (1987). Carbohydrate supply and N_2 fixation in soybean: The effect of varied daylength and stem girdling. *Plant Physiol.* 85, 137–144.

Walter, H. D. (1931). "Die Hydratur der Pflanze und ihre physiologische-ökologische Bedeutung," pp. 118–121. G. Fischer, Jena.

Walter, H. D. (1965). Klarung des spasifischerr Wasserzustandes in Plasma. *Ber. Dtsch. Bot. Ges.* 78, 104–114.

Walton, D. C. (1980). Biochemistry and physiology of abscisic acid. *Annu. Rev. Plant Physiol.* 31, 453–489.

Wang. J., Hesketh, J. D., and Woolley, J. T. (1986). Preexisting channels and soybean rooting patterns. *Soil Sci.* 141, 432–437.

Ward, D. A., and Bunce, J. A. (1986). Novel evidence for a lack of water vapour saturation within the intracellular airspace of turgid leaves of mesophytic species. *J. Exp. Bot.* 37, 504–516.

Wardlaw, I. F. (1967). The effect of water stress on translocation in relation to photosynthesis and growth. I. Effects during grain development in wheat. *Aust. J. Biol. Sci.* 20, 25–39.

Wareing, P. F., and Phillips, I. D. J. (1981). "Growth and Differentiation in Plants," 3rd Ed. Pergamon Press, New York.

Waring, R. H., and Running, S. W. (1978). Sapwood and water storage: Its contribution to transpiration and effect on water conductance through the stems of old-growth Douglas-fir. *Plant Cell Environ.* 1, 131–140.

Waring, R. H., and Schlesinger, W. H. (1985). "Forest Ecosystems: Concepts and Management." Academic Press, Orlando, FL.

Waring, R. H., Whitehead, D., and Jarvis, P. G. (1979). The contribution of stored water to transpiration in Scot Pine. *Can. J. For. Res.* 10, 555–558.

Waters, I., Armstrong, W., Thompson, C. J., Setter, T. L., Adkins, S., Gibbs, J., and Greenway, H. (1989). Diurnal changes in radial oxygen loss and ethanol metabolism in roots of submerged and non-submerged rice seedlings. *New Phytol.* 113, 439–451.

Waters, I., Kuiper, P. J. C., Watkins, E., and Greenway, H. (1991). Effects of anoxia on wheat seedlings. I. Interaction between anoxia and other environmental factors. *J. Exp. Bot.* 42, 1427–1435.

Watson, B. T., and Wardlaw, I. F. (1981). Metabolism and export of 14C-labelled photosynthate from water-stressed leaves. *Aust. J. Plant. Physiol.* 8, 143–153.

Weatherley, P. E. (1982). Water uptake and flow in roots. *In* "Encyclopedia of Plant Physiology" (O. L. Lange, P. S. Nobel, C. B. Osmond, H. Ziegler, eds.), Vol. 12B, pp. 70–109. Springer, Berlin/Heidelberg/New York.

Weatherspoon, C. P. (1968). "The Significance of the Mesophyll Resistance in Transpiration." Ph.D. Dissert., Duke University, Durham, NC.

Weaver, J. E. (1919). The ecological relations of roots. *Carnegie Inst. Washington Publ.* 286.

Weaver, J. E. (1920). Root development in the grassland formation. *Canegie Inst. Washington Publ.* 292.

Weaver, J. E. (1925). Investigations on the root habits of plants. *Am. J. Bot.* **12**, 502–509.

Weaver, J. E. (1926). "Root Development of Field Crops." McGraw-Hill, New York.

Weaver, J. E., and Bruner, W. E. (1927). "Root Development of Vegetable Crops." McGraw-Hill, New York.

Weaver, J. E., and Clements, F. E. (1938). "Plant Ecology," 2nd Ed., McGraw-Hill, New York.

Weaver, J. E., and Zink, E. (1946). Length of life of roots of ten species of perennial range and pasture grasses. *Plant Physiol.* **21**, 201–217.

Webb, L. J., Tracey, J. G., and Haydock, K. P. (1967). A factor toxic to seedlings of the same species associated with living roots of the non-gregarious rain forest tree *Guevillea robusta*. *J. Appl. Ecol.* **4**, 13–25.

Weisz, P. R., Denison, R. F., and Sinclair, T. R. (1985). Response to drought stress of nitrogen fixation (acetylene reduction) rates by field-grown soybeans. *Plant Physiol.* **78**, 525–530.

Weisz, P. R., and Sinclair, T. R. (1987). Regulation of soybean nitrogen fixation in response to rhizosphere oxygen. II. Quantification of nodule gas permeability. *Plant Physiol.* **84**, 906–910.

Welbaum, G. E., and Bradford, K. J. (1988). Water relations of seed development and germination in muskmelon (*Cucumis melo* L.). *Plant Physiol.* **86**, 406–411.

Wenger, K. F. (1955). Light and mycorrhiza development. *Ecology* **36**, 518–520.

Wenkert, W. (1980). Measurement of tissue osmotic pressure. *Plant Physiol.* **65**, 614–617.

Went, F. W. (1943). Effect of the root system on tomato stem growth. *Plant Physiol.* **18**, 51–65.

Went, F. W. (1975). Water vapor absorption in *Prosopis*. *In* "Physiological Adaptation to the Environment" (F. J. Vernberg, ed.), pp. 67–75. Intext Educational Publications, New York.

Westgate, M. E., and Boyer, J. S. (1984). Transpiration- and growth-induced water potentials in maize. *Plant Physiol.* **74**, 882–889.

Westgate, M. E., and Boyer, J. S. (1985a). Carbohydrate reserves and reproductive development at low water potentials in maize. *Crop Sci.* **25**, 762–769.

Westgate, M. E., and Boyer, J. S. (1985b). Osmotic adjustment and the inhibition of leaf, root, stem and silk growth at low water potentials in maize. *Planta* **164**, 540–549.

Westgate, M. E., and Boyer, J. S. (1986a). Silk and pollen water potentials in maize. *Crop Sci.* **26**, 947–51.

Westgate, M. E., and Boyer, J. S. (1986b). Reproduction at low silk and pollen water potentials in maize. *Crop Sci.* **26**, 951–956.

Westgate, M. E., and Boyer, J. S. (1986c). Water status of the developing grain of maize. *Agron. J.* **78**, 714–719.

Westgate, M. E., and Steudle, E. (1985). Water transport in the midrib tissue of maize leaves. *Plant Physiol.* **78**, 183–191.

Westgate, M. E., and Thomson Grant, D. L. (1989). Water deficits and reproduction in maize: Response of the reproductive tissue to water deficits at anthesis and mid-grain fill. *Plant Physiol.* **91**, 862–867.

Weyers, J. D. B., and Johansen, L. G. (1985). Accurate estimation of stomatal aperture from silicone rubber impressions. *New Phytol.* **101**, 109–115.

Weyers, J. D. B., and Meidner, H. (1990). "Methods of Stomatal Research." Longman Scientific and Technical, New York.

White, J. W., and Castillo, J. A. (1989). Relative effect of roots and shoot genotypes on yield of common bean under drought stress. *Crop Sci.* **29**, 360–362.

White, J. W., and Mastalerz, J. W. (1966). Soil moisture as related to container capacity. *Proc. Am. Soc. Hort. Sci.* **89**, 758–768.

White, J. W. C. (1989). Stable hydrogen isotopes in plants: A review of current theory and some potential applications. *Ecol. Studies* **68**, 142–162.

White, J. W. C., Cook, E. R., Lawrence, J. R., and Broecker, W. S. (1985). The D/H ratio of sap in trees: Implications for water sources and tree ring D/H ratios. *Geochim. Cosmochim. Acta* **49**, 237–246.

White, L. M., and Ross, W. H. (1939). Effect of various grades of fertilizers on the salt content of the soil solution. *J. Agr. Res.* **59**, 81–100.

White, P. R. (1938). "Root-pressure": An unappreciated force in sap movement. *Am. J. Bot.* **25**, 223–227.

White P. R., Schuker, E., Kern, J. R., and Fuller, F. H. (1958). "Root pressure" in gymnosperms. *Science* **128**, 308–309.

Whitehead, D., Edwards, W. R. N., and Jarvis, P. G. (1984). Conducting sapwood area, foliage area, and permeability in mature trees of *Picea sitchensis* and *Pinus contorta. Can. J. For. Res.* **14**, 940–947.

Whitehead, D., and Hinckley, T. M. (1991). Models of water flux through forest stands: Critical leaf and stand parameters. *Tree Physiol.* **9**, 35–57.

Whitehead, D., and Kelliher, F. M. (1991a). A canopy water balance model for a *Pinus radiata* stand before and after thinning. *Agric. For. Meteorol.* **55**, 109–126.

Whitehead, D., and Kelliher, F. M. (1991b). Modeling the water balance of a small *Pinus radiata* catchment. *Tree Physiol.* **9**, 17–33.

Whiteman, P. C., and Koller, D. (1964). Saturation deficit of the mesophyll evaporating surfaces in a desert halophyte. *Science* **146**, 1320–1321.

Whittaker, R. H., and Likens, G. E. (1975). The biosphere and man. *In* "Primary Productivity of the Biosphere" (H. Lieth and R. H. Whittaker, eds.), pp. 305–328. Springer-Verlag, New York.

Wiebe, H. H., and Kramer, P. J. (1954). Translocation of radioactive isotopes from various regions of roots of barley seedlings. *Plant Physiol.* **29**, 342–348.

Wiegand, K. M. (1906). Pressure and flow of sap in the maple. *Am. Nat.* **40**, 409–453.

Wiersma, J. V., and Bailey, T. B. (1975). Estimation of leaflet, trifoliate, and total leaf areas of soybeans. *Agron. J.* **67**, 26–30.

Wiggans, C. C. (1936). The effect of orchard plants on subsoil moisture. *Proc. Am. Soc. Hort. Sci.* **33**, 103–107.

Wiggans, C. C. (1938). Some results from orchard irrigation in eastern Nebraska. *Proc. Am. Soc. Hort. Sci.* **36**, 74–76.

Wilcox, H. (1962). Growth studies of the root of incense cedar *Libocedrus decurrens*. II. Morphological features of the root system and growth behavior. *Am. J. Bot.* **49**, 237–245.

Wild, A., ed. (1988). "Russell's Soil Conditions and Plant Growth," 11th Ed. Longman Group, Harlow, Essex, England. John Wiley & Sons, New York.

Will, G. M. (1966). Root growth and dry-matter production in a high-producing stand of *Pinus radiata. New Zealand For. Serv. Res. Note* **44**.

Willey, C. R. (1970). Effects of short periods of anaerobic and near-anaerobic conditions on water uptake by tobacco roots. *Agron. J.* **62**, 224–229.

Williams, H. F. (1933). Absorption of water by the leaves of common mesophytes. *J. Elisha Mitchell Sci. Soc.* **48**, 83–100.

Williams, R. J., and Leopold, A. C. (1989). The glassy state in corn embryos. *Plant Physiol.* **89**, 977–981.

Williams, S. (1809). "The Natural and Civil History of Vermont." 2nd Ed., Vol. 1, pp. 87–97. Samuel Mills, Burlington, VT.

Williams, W. T. (1950). Studies in stomatal behavior. IV. The water relations of the epidermis. *J. Exp. Bot.* **1**, 114–131.

Williamson, V. M., and Colwell, G. (1991). Acid phosphatase-1 from nematode resistant tomato. *Plant Physiol.* **97**, 139–146.

Wilson, B. F. (1967). Root growth around barriers. *Bot. Gaz.* **128**, 79–82.

Wilson, C. C. (1947). The porometer method for the continuous estimation of dimensions of stomates. *Plant Physiol.* **22**, 582–589.

Wilson, C. C. (1948). The effect of some environmental factors on the movements of guard cells. *Plant Physiol.* **23**, 5–37.

Wilson, C. C., Boggess, W. R., and Kramer, P. J. (1953). Diurnal fluctuations in the moisture content of some herbaceous plants. *Plant Physiol.* **36**, 762–765.

Wilson, J. B. (1988). A review of evidence on the control of shoot:root ratio, in relation to models. *Ann. Bot.* **61**, 433–449.

Wilson, J. R., and Ludlow, M. M. (1984). Time trends of solute accumulation and the influence of potassium fertilizer on osmotic adjustment of water-stressed leaves of three tropical grasses. *Aust. J. Plant Physiol.* **10**, 523–537.

Wilson, R. H., and Evans, H. J. (1968). The effect of potassium and other univalent cations on the conformation of enzymes. *In* "The Role of Potassium in Agriculture" (V. J. Kilmer, S. E. Younts, and N. C. Brady, eds.), pp. 189–202. American Society of Agronomy, Madison, WI.

Wilson, T. P., Canny, M. J., and McCully, M. E. (1991). Leaf teeth, transpiration and the retrieval of apoplastic solutes in balsam poplars. *Physiol. Plant.* **83**, 225–232.

Wind, G. P. (1955). Flow of water through plant roots. *Netherlands J. Agr. Sci.* **3**, 259–264.

Winneberger, J. H. (1958). Transpiration as a requirement for growth of land plants. *Physiol. Plant.* **11**, 56–61.

Wise, R. R., Ortiz-Lopez, A., and Ort, D. R. (1992). Spatial distribution of photosynthesis during drought in field-grown and acclimated and nonacclimated growth chamber-grown cotton. *Plant Physiol.* **100**, 26–32.

Wittwer, S. H. (1975). Food production: Technology and the resource base. *Science* **188**, 579–584.

Wolf, F. A. (1962). "Aromatic or Oriental Tobaccos." Duke University Press, Durham, NC.

Wong, S. C., Cowan, I. R., and Farquhar, G. D. (1979). Stomatal conductance correlates with photosynthetic capacity. *Nature* (London) **282**, 424–426.

Wood, C. (1988). Urban waste water irrigates Florida citrus. *Citrus Ind.* **69**, 14–16.

Woods, F. W. (1957). Factors limiting root penetration in deep sands of the southeastern Coastal Plain. *Ecology* **38**, 357–359.

Woods, F. W., and Brock, K. (1970). Interspecific transfer of inorganic materials by root systems of woody plants. *Ecology* **45**, 886–889.

Woods, T. E. (1980). Biological and Chemical Control of Phosphorus Cycling in a Northern Hardwood Forest." Ph.D. Dissert. Yale University, New Haven, CT.

Wooley, J. T. (1967). Relative permeabilities of plastic films to water and carbon dioxide. *Plant Physiol.* **42**, 641–643.

Wray, F. J. (1971). "Changes in the Ionic Environment around Plant Roots." D. Phil. Thesis, Oxford.

Wright, J. P., and Fisher, D. B. (1983). Estimation of the volumetric elastic modulus and membrane hydraulic conductivity of willow sieve tubes. *Plant Physiol.* **73**, 1042–1047.

Wright, L. N., and Jordan, G. L. (1970). Artificial selection for seedling drought tolerance in boer lovegrass (*Eragrostis curvula* Nees). *Crop Sci.* **10**, 99–102.

Wright, S. T. C. (1969). An increase in the "inhibitor-β" content of detached wheat leaves following a period of wilting." *Planta* **86**, 10–20.

Wright, S. T. C., and Hiron, R. W. P. (1969). (+)-Abscisic acid, the growth inhibitor induced in detached wheat leaves by a period of wilting. *Nature* (London) **224**, 719–720.

Wuenscher, J. E., and Kozlowski, T. T. (1971). The response of transpiration resistance to leaf temperature as a desiccation resistance mechanism in tree seedlings. *Physiol. Plant.* **24**, 254–259.

Wylie, R. B. (1943). The role of the epidermis in foliar organization and its relations to the minor venation. *Am. J. Bot.* **30**, 273–280.

Wyn Jones, R. G. (1980). An assessment of quaternary ammonium and related compounds as osmotic effectors in crop plants. *In* "Genetic Engineering of Osmoregulation" (D. W. Rains, R. C. Valentine, and A. Hollaender, eds.), pp. 155–170. Plenum Press, New York.

Xu, X., and Bland, W. L. (1993). The short life and replant problems of deciduous fruit trees. *Agron. J.* **85**, 384–388.

Yadava, V. L., and Doud, S. L. (1980). The short life and replant problems of deciduous fruit trees. *Hort. Rev.* **2**, 1–116.

Yakir, D. (1992). Water compartmentation in plant tissue: Isotopic evidence. *In* "Water and Life" (G. N. Somero, C. B. Osmond, and C. L. Boli, eds.), pp. 205–223. Springer-Verlag, Berlin.

Yakir, D., DeNiro, M. J., and Gat, J. R. (1990). Natural deuterium and oxygen-18 enrichment in leaf water of cotton plants grown under wet and dry conditions: Evidence for water compartmentation and its dynamics. *Plant Cell Environ.* **13**, 49–56.

Yakir, D., DeNiro, M. J., and Rundel, P. W. (1989). Isotopic inhomogeneity of leaf water: Evidence and implications for the use of isotopic signals transduced by plants. *Geochem. Cosmochim. Acta* **53**, 2769–2773.

Yancey, P. H., Clark, M. E., Hand, S. C., Bowlus, R. D., and Somero, G. N. (1982). Living with water stress: Evolution of osmolyte systems. *Science* **217**, 1214–1222.

Yang, S., and Tyree, M. T. (1993). Hydraulic resistance in *Acer saccharum* shoots and its influence on leaf water potential and transpiration. *Tree Physiol.* **12**, 231–242.

Yelenosky, G. (1964). Tolerance of trees to deficiencies of soil aeration. *Proc. Inst. Shade Tree Conf.* **40**, 127–147.

Younis, H. M., Boyer, J. S., and Govindjee. (1979). Conformation and activity of chloroplast coupling factor exposed to low chemical potential of water in cells. *Biochim. Biophys. Acta* **548**, 328–340.

Younis, H. M., Weber, G., and Boyer, J. S. (1983). Activity and conformational changes in chloroplast coupling factor induced by ion binding: Formation of a magnesium-enzyme-phosphate complex. *Biochemistry* **22**, 2505–2512.

Yu, G. H. (1966). "A Study of Radial Movement of Salt and Water in Roots." Ph.D. Dissert., Duke University, Durham, NC.

Yu, P. T., Stolzy, L. H., and Letey, J. (1969). Survival of plants under prolonged flooded conditions. *Agron. J.* **61**, 844–846.

Zabadal, T. J. (1974). A water potential threshold for the increase of abscisic acid in leaves. *Plant Physiol.* **53**, 125–127.

Zeevaart, J. A. D., and Creelman, R. A. (1988). Metabolism and physiology of abscisic acid. *Annu. Rev. Plant Physiol. Plant Mol. Biol.* **39**, 439–473.

Zeiger, E. (1983). The biology of stomatal guard cells. *Annu. Rev. Plant Physiol.* **34**, 441–475.

Zeiger, E., Armond, P., and Melis, A. (1980). Fluorescence properties of guard cell chloroplasts: Evidence for linear electron transport and light-harvesting pigments of photosystems I and II. *Plant. Physiol.* **67**, 17–20.

Zeiger, E., Farquhar, G. D., and Cowan, I. R. (1987). "Stomatal Function." Stanford University Press, Stanford, CA.

Zeikus, J. G., and Ward, J. C. (1974). Methane formation in living trees: A microbial origin. *Science* **184**, 1181–1183.

Zekri, M., and Parsons, L. R. (1989). Grapefruit leaf and fruit growth in response to dry, microsprinkler, and overhead sprinkler irrigation. *J. Am. Soc. Hort. Sci.* **114**, 25–29.

Zelitch, I. (1961). Biochemical control of stomatal opening in leaves. *Proc. Natl. Acad. Sci. USA* **47**, 1423–1433.

Zelitch, I., and Waggoner, P. E. (1962a). Effect of chemical control of stomata on transpiration and photosynthesis. *Proc. Natl. Acad. Sci. USA* **48**, 1101–1108.

Zelitch, I., and Waggoner, P. E. (1962b). Effect of chemical control of stomata on transpiration and photosynthesis of intact plants. *Proc. Natl. Acad. Sci. USA* **48**, 1297–1299.

Zhang, J., and Davies, W. J. (1989a). Sequential response of whole plant water relations to prolonged soil drying and the involvement of xylem sap ABA in the regulation of stomatal behaviour of sunflower plants. *New Phytol.* **113**, 167–174.

Zhang, J., and Davies, W. J. (1989b). Abscisic acid produced in dehydrating roots may enable the plant to measure the water status of the soil. *Plant Cell Environ.* **12**, 73–81.

Zhang, J., and Davies, W. J. (1990). Changes in the concentration of ABA in xylem sap as a function of changing soil water status can account for changes in leaf conductance and growth. *Plant Cell Environ.* **13**, 277–285.

Zhu, G.-L., and Boyer, J. S. (1992). Enlargement in *Chara* studied with a turgor clamp: Growth rate is not determined by turgor. *Plant Physiol.* **100**, 2071–2080.

Zhu, G. L., and Steudle, E. (1991). Water transport across maize roots. *Plant Physiol.* **95**, 305–315.

Ziegler, A. M., Bambach, R. K., Parrish, J. T., Barrett, S. F., Gierlowski, E. H., Parker, W. C., Raymond, A., and Sepkoski, J. J., Jr. (1981). Paleozoic biogeography and climatology. *In* "Paleobotany, Paleoecology and Evolution" (K. J. Niklas, ed.), pp. 231–266. Praeger Press, New York.

Zimmermann, M. H. (1964). Effect of low temperature on ascent of sap in trees. *Plant Physol.* **39**, 568–572.

Zimmermann, M. H. (1978). Structural requirements for optimal water conduction in tree stems. *In* "Tropical Trees as Living Systems" (P. B. Tomlinson and M. H. Zimmermann, eds.), pp. 517–532. Cambridge University Press, Cambridge.

Zimmermann, M. H. (1983). "Xylem Structure and the Ascent of Sap." Springer-Verlag, Berlin.

Zimmermann, M. H., and McDonough, J. (1978). Dysfunction in the flow of food. *In* "Plant Disease" (J. G. Horsfall and E. B. Cowling, eds.), Vol. 3, pp. 117–140. Academic Press, New York.

Zimmermann, M. H., and Milburn, J. A., eds. (1975). "Transport in Plants. I. Phloem Transport." Encyclopedia of Plant Physiology, N.S. Vol. 1, Springer-Verlag, Berlin.

Zimmermann, M. H., and Tomlinson, P. B. (1974). Vascular patterns in palm stems: Variation of the *Raphis* principle. *J. Arnold Arbor.* **55**, 402–424.

Zimmermann, U., Balling, A., Rygol, J., Link, A., and Haase, A. (1991). Comments on the article of J. B. Passioura, "An impasse in plant water relations." *Bot. Acta* **104**, 412–415.

Zimmermann, U., Haase, A., Langbein, D., and Meinzer, F. (1993). Mechanisms of long distance water transport in plants: A reexamination of some paradigms in the light of new evidence. *Phil. Trans. Roy. Soc. London Ser. B* **344**, 19–31.

Zimmermann, U., Hüsken, D., and Schulze, E.-D. (1980). Direct turgor pressure measurement in individual leaf cells of *Tradescantia virginiana. Planta* **149**, 445–453.

Zimmermann, U., Rygol, J., Balling, A., Klöck, G., Metzler, A., and Haase, A. (1992). Radial turgor and osmotic pressure profiles in intact and excised roots of *Aster tripolium. Plant Physiol.* **99**, 186–196.

Zimmermann, U., and Steudle, E. (1978). Physical aspects of water relations of plant cells. *Adv. Bot. Res.* **6**, 45–117.

Zur, B. (1967). Osmotic control of the matric soil-water potential. II. Soil-plant systems. *Soil Sci.* **103**, 30–38.

Index